Biodiversity, Ecosystem Functioning, and Human Wellbeing

Biodiversity, Ecosystem Functioning, and Human Wellbeing

An Ecological and Economic Perspective

EDITED BY

Shahid Naeem,
Daniel E. Bunker,
Andy Hector,
Michel Loreau,
and
Charles Perrings

OXFORD
UNIVERSITY PRESS

Great Clarendon Street, Oxford OX2 6DP

Oxford University Press is a department of the University of Oxford.
It furthers the University's objective of excellence in research, scholarship,
and education by publishing worldwide in

Oxford New York

Auckland Cape Town Dar es Salaam Hong Kong Karachi
Kuala Lumpur Madrid Melbourne Mexico City Nairobi
New Delhi Shanghai Taipei Toronto

With offices in

Argentina Austria Brazil Chile Czech Republic France Greece
Guatemala Hungary Italy Japan Poland Portugal Singapore
South Korea Switzerland Thailand Turkey Ukraine Vietnam

Oxford is a registered trade mark of Oxford University Press
in the UK and in certain other countries

Published in the United States
by Oxford University Press Inc., New York

© Oxford University Press 2009

The moral rights of the authors have been asserted
Database right Oxford University Press (maker)

First published line 2009
Reprinted 2010, 2011, 2012, 2013

All rights reserved. No part of this publication may be reproduced,
stored in a retrieval system, or transmitted, in any form or by any means,
without the prior permission in writing of Oxford University Press,
or as expressly permitted by law, or under terms agreed with the appropriate
reprographics rights organization. Enquiries concerning reproduction
outside the scope of the above should be sent to the Rights Department,
Oxford University Press, at the address above

You must not circulate this book in any other binding or cover
and you must impose the same condition on any acquirer

British Library Cataloguing in Publication Data
Data available

Library of Congress Cataloging in Publication Data
Data available

Typeset by Newgen Imaging Systems (P) Ltd., Chennai, India
Printed in Great Britain
on acid-free paper by
CPI Group (UK) Ltd, Croydon, CR0 4YY

ISBN 978-0-19-954795-1 (Hbk.)
ISBN 978-0-19-954796-8 (Pbk.)

10 9 8 7 6 5

Contents

List of contributors	viii
Preface	xi
Shahid Naeem, Daniel E. Bunker, Andy Hector, Michel Loreau, and Charles Perrings	
Acknowledgments	xiv
Part 1: Introduction, background, and meta-analyses	1
1 Introduction: the ecological and social implications of changing biodiversity. An overview of a decade of biodiversity and ecosystem functioning research	3
Shahid Naeem, Daniel E. Bunker, Andy Hector, Michel Loreau, and Charles Perrings	
2 Consequences of species loss for ecosystem functioning: meta-analyses of data from biodiversity experiments	14
Bernhard Schmid, Patricia Balvanera, Bradley J. Cardinale, Jasmin Godbold, Andrea B. Pfisterer, David Raffaelli, Martin Solan, and Diane S. Srivastava	
3 Biodiversity-ecosystem function research and biodiversity futures: early bird catches the worm or a day late and a dollar short?	30
Martin Solan, Jasmin A. Godbold, Amy Symstad, Dan F. B. Flynn, and Daniel E. Bunker	
Part 2: Natural science foundations	47
4 A functional guide to functional diversity measures	49
Owen L. Petchey, Eoin J. O'Gorman, and Dan F. B. Flynn	
5 Forecasting decline in ecosystem services under realistic scenarios of extinction	60
J. Emmett Duffy, Diane S. Srivastava, Jennie McLaren, Mahesh Sankaran, Martin Solan, John Griffin, Mark Emmerson, and Kate E. Jones	
6 Biodiversity and the stability of ecosystem functioning	78
John N. Griffin, Eoin J. O'Gorman, Mark C. Emmerson, Stuart R. Jenkins, Alexandra-Maria Klein, Michel Loreau, and Amy Symstad	

7 The analysis of biodiversity experiments: from pattern toward mechanism — 94
Andy Hector, Thomas Bell, John Connolly, John Finn, Jeremy Fox, Laura Kirwan, Michel Loreau, Jennie McLaren, Bernhard Schmid, and Alexandra Weigelt

8 Towards a food web perspective on biodiversity and ecosystem functioning — 105
Bradley Cardinale, Emmett Duffy, Diane Srivastava, Michel Loreau, Matt Thomas, and Mark Emmerson

9 Microbial biodiversity and ecosystem functioning under controlled conditions and in the wild — 121
Thomas Bell, Mark O. Gessner, Robert I. Griffiths, Jennie McLaren, Peter J. Morin, Marcel van der Heijden, and Wim van der Putten

10 Biodiversity as spatial insurance: the effects of habitat fragmentation and dispersal on ecosystem functioning — 134
Andrew Gonzalez, Nicolas Mouquet, and Michel Loreau

Part 3: Ecosystem services and human wellbeing — 147

11 Incorporating biodiversity in climate change mitigation initiatives — 149
Sandra Díaz, David A. Wardle, and Andy Hector

12 Restoring biodiversity and ecosystem function: will an integrated approach improve results? — 167
Justin Wright, Amy Symstad, James M. Bullock, Katharina Engelhardt, Louise Jackson, and Emily Bernhardt

13 Managed ecosystems: biodiversity and ecosystem functions in landscapes modified by human use — 178
Louise Jackson, Todd Rosenstock, Matthew Thomas, Justin Wright, and Amy Symstad

14 Understanding the role of species richness for crop pollination services — 195
Alexandra-Maria Klein, Christine Müller, Patrick Hoehn, and Claire Kremen

15 Biodiversity and ecosystem function: perspectives on disease — 209
Richard S. Ostfeld, Matthew Thomas, and Felicia Keesing

16 Opening communities to colonization – the impacts of invaders on biodiversity and ecosystem functioning — 217
Katharina Engelhardt, Amy Symstad, Anne-Helene Prieur-Richard, Matthew Thomas, and Daniel E. Bunker

17 The economics of biodiversity and ecosystem services   **230**
Charles Perrings, Stefan Baumgärtner, William A. Brock, Kanchan Chopra, Marc Conte, Christopher Costello, Anantha Duraiappah, Ann P. Kinzig, Unai Pascual, Stephen Polasky, John Tschirhart, and Anastasios Xepapadeas

18 The valuation of ecosystem services   **248**
Edward B. Barbier, Stefan Baumgärtner, Kanchan Chopra, Christopher Costello, Anantha Duraiappah, Rashid Hassan, Ann P. Kinzig, Mark Lehmann, Unai Pascual, Stephen Polasky, and Charles Perrings

19 Modelling biodiversity and ecosystem services in coupled ecological–economic systems   **263**
William A. Brock, David Finnoff, Ann P. Kinzig, Unai Pascual, Charles Perrings, John Tschirhart, and Anastasios Xepapadeas

Part 4: Summary and synthesis   **279**

20 TraitNet: furthering biodiversity research through the curation, discovery, and sharing of species trait data   **281**
Shahid Naeem and Daniel E. Bunker

21 Can we predict the effects of global change on biodiversity loss and ecosystem functioning?   **290**
Shahid Naeem, Daniel E. Bunker, Andy Hector, Michel Loreau, and Charles Perrings

References   **299**
Index   **357**

Contributors

Patricia Balvanera, Centro de Investigaciones en Ecosistemas, Universidad Nacional Autónoma de México, Apdo. Postal 27-3, Sta. Ma. de Guido, Morelia, Michoacán, México 58090; pbalvane@oikos.unam.mx

Edward B. Barbier, University of Wyoming, Department of Economics and Finance, Department 3985, Ross Hall 123, Laramie, WY 82071, USA; ebarbier@uwyo.edu

Stefan Baumgärtner, Leuphana Universität Lüneburg Centre for Sustainability ManagementPostfach 2440 D-21314 Lüneburg, Germany; baumgaertner@leuphana.de

Thomas Bell, Department of Zoology, University of Oxford, South Parks Road, Oxford OX1 3PS, UK; thomas.bell@zoo.ox.ac.uk

Emily Bernhardt, Department of Biology, Box 90388, Duke University, Durham, NC 27708, USA; ebernhar@duke.edu

William A. Brock, University of Wisconsin, Madison – Department of Economics, 1180 Observatory Drive, Madison, WI 53706, USA; wbrock@ssc.wisc.edu

James Bullock, Centre for Ecology & Hydrology, Benson Lane, Wallingford, OX10 8BB, UK; jmbul@ceh.ac.uk

Daniel E. Bunker, Department of Biological Sciences, New Jersey Institute of Technology, 433 Colton Hall, University Heights, Newark, NJ 07102-1982, USA; dbunker@njit.edu

Bradley J. Cardinale, Department of Ecology, Evolution and Marine Biology, University of California at Santa Barbara, Santa Barbara, California 93106, USA; cardinale@lifesci.ucsb.edu

Kanchan Chopra, Institute of Economic Growth, Delhi University Enclave, Delhi – 110 007, India; kanchan@iegindia.org

John Connolly, Environmental and Ecological Modelling Group, UCD School of Mathematical Sciences, Dublin, Ireland; john.connolly@ucd.ie

Marc Conte, Environmental Science & Management, University of California Santa Barbara, 4410 Bren Hall,Santa Barbara, CA 93106-5131, USA; conte@bren.ucsb.edu

Christopher Costello, Donald Bren School of Environmental Science & Management, University of California Santa Barbara, 4410 Bren Hall,Santa Barbara, CA 93106-5131, USA; costello@bren.ucsb.edu

Sandra Díaz, Instituto Multidisciplinario de Biología Vegetal (CONICET-UNC) and FCEFyN, Universidad Nacional de Córdoba, Casilla de Correo 495, 5000 Córdoba, Argentina; sdiaz@com.uncor.edu

J. Emmett Duffy, Virginia Institute of Marine Science, The College of William and Mary, Gloucester Point, VA 23062-1346, USA; jeduffy@vims.edu

Anantha Duraiappah, Ecosystem Services Economics Unit, Division of Environmental Policy Implementation, United Nations Environment Programme (UNEP),United Nations Avenue, Gigiri, PO Box 30552, 00100Nairobi, Kenya;anantha.duraiappah@unep.org

Mark C. Emmerson, Environmental Research Institute, University College Cork, Lee Road, Cork, Ireland, and Department of Zoology, Ecology and Plant Science Distillery Fields, North Mall, University College Cork, Ireland; emerson@ucc.ie

Katharina Engelhardt, University of Maryland Center for Environmental Science, Appalachian Laboratory, 301 Braddock Road, Frostburg, MD 21532, USA; engelhardt@al.umces.edu

John Finn, Teagasc, Environment Research Centre, Johnstown Castle, Wexford Ireland; john.finn@teagasc.ie

David Finnoff, University of Wyoming, Department of Economics and Finance, Department 3985, Ross Hall 123, Laramie, WY 82071, USA; Finnoff@uwyo.edu

Dan F. B. Flynn, Department of Ecology, Evolution and Environmental Biology (E3B), Columbia University, Schermerhorn Extension, 10th Floor, Mail Code 5557, 1200 Amsterdam Avenue, New York, NY 10027, USA; dff2101@columbia.edu

Jeremy Fox, Department of Biological Sciences, University of Calgary, 2500 University Drive NW, Calgary, Alberta T2N 1N4 Canada; jefox@ucalgary.ca

Mark O. Gessner, Department of Aquatic Ecology, Eawag: Swiss Federal Institute of Aquatic Science & Technology, 8600 Dübendorf, Switzerland and Institute of Integrative Biology (IBZ), ETH Zurich, 8600 Dübendorf, Switzerland

Jasmin A. Godbold, Oceanlab, University of Aberdeen, Main Street, Newburgh, Aberdeenshire, AB41 6AA, UK; j.a.godbold@abdn.ac.uk

CONTRIBUTORS

Andrew Gonzalez, Department of Biology, McGill University, 1205 Dr., Penfield Avenue, Montreal, H3A 1B1, Canada; andrew.gonzalez@mcgill.ca

John N. Griffin, Marine Biological Association of the United Kingdom, The Laboratory, Citadel Hill, Plymouth PL1 2PB, UK and Marine Biology and Ecology Research Centre, School of Biological Sciences, University of Plymouth, Plymouth PL4 8AA, UK

Robert I. Griffiths, Molecular Microbial Ecology Section, Centre for Ecology and Hydrology (Oxford), Mansfield Road, Oxford OX1 3SR, UK

Rashid Hassan, Dept of Agricultural Economics Extension and Rural Development, University of Pretoria, PRETORIA 0002, South Africa;rashid.hassan@up.ac.za

Andy Hector, Institute of Environmental Sciences, University of Zurich, CH-8057, Zurich, Switzerland; ahector@uwinst.uni.ch

Patrick Hoehn, Department of Crop Science, Agroecology, University of Göttingen, Waldweg 26, 37073 Göttingen, Germany

Louise Jackson, Department of Land, Air and Water Resources, University of California, Davis, CA 95616, USA; lejackson@ucdavis.edu

Stuart R. Jenkins, School of Ocean Sciences, University Bangor Menai Bridge, Anglesey LL59 5AB, UK

Kate E. Jones, Institute of Zoology, Zoological Society of London and Cambridge University, Regent's Park, London NW1 4RY, UK

Felicia Keesing, Biology Program, Bard College, Annandale-on-Hudson, NY 12504, USA

Ann P. Kinzig, ecoSERVICES Group, School of Life Sciences, Arizona State University, Box 874501, Tempe, AZ 85287-4501, USA; kinzig@asu.edu

Laura Kirwan, Teagasc, Environment Research Centre, Johnstown Castle, Wexford Ireland; Laura.Kirwan@teagasc.ie

Alexandra-Maria Klein, Department of Environmental Science, Policy and Management, 137 Mulford Hall, University of California at Berkeley, California 94720-3114, USA and Department of Crop Science, Agroecology, University of Göttingen, Waldweg 26, 37073 Göttingen, Germany

Claire Kremen, Department of Environmental Science, Policy and Management, 137 Mulford Hall, University of California at Berkeley, California 94720-3114, USA

Markus Lehmann, Secretariat of the Convention on Biological Diversity, 413, Saint Jacques Street, suite 800 Montreal QC, H2Y 1N9, Canada; markus.lehmann@cbd.int

Michel Loreau, Department of Biology, McGill University, 1205 ave Docteur Penfield, Montreal, Québec H3A 1B1, Canada; michel.loreau@mcgill.ca

Jennie R. McLaren, Department of Botany, University of British Columbia, #3529-6270 University Boulevard, Vancouver, BC, V6T 1Z4, Canada; jmclaren@interchange.ubc.ca.

Peter J. Morin, Department of Ecology, Evolution, and Natural Resources, Rutgers Cook College, 148 ENRS Building, Cook Campus, 14 College Farm Road, New Brunswick, New Jersey, USA

Nicolas Mouquet, ISEM-UMR 5554, University of Montpellier II, Place Eugene Bataillon, CC065, 34095 Montpellier Cedex 05, France

Christine Müller, Institute of Environmental Sciences, University of Zürich, Winterthurerstrasse 190, CH-8057 Zürich, Switzerland

Shahid Naeem, Department of Ecology, Evolution, and Environmental Biology, Columbia University, 1200 Amsterdam Ave, MC 5557, New York, NY 10025, USA; sn2121@columbia.edu

Eoin J. O'Gorman, Environmental Research Institute, Lee Road Cork, Ireland, and Department of Zoology, Ecology and Plant Science, Distillery Fields, North Mall, University College Cork, Ireland; e.ogorman@mars.ucc.ie

Richard S. Ostfeld, Cary Institute of Ecosystem Studies, PO Box AB, Millbrook, NY 12545, USA

Unai Pascual, Department of Land Economy, University of Cambridge, 19 Silver Street, Cambridge, CB3 9EP, UK; up211@cam.ac.uk

Charles Perrings, ecoSERVICES Group, School of Life Sciences, Arizona State University, Box 874501, Tempe, AZ 85287-4501, USA; Charles.Perrings@asu.edu

Owen L. Petchey, Department of Animal and Plant Sciences, Alfred Denny Building, University of Sheffield, Western Bank, Sheffield S10 2TN, UK; o.petchey@sheffield.ac.uk

Andrea B. Pfisterer, Institute of Environmental Sciences, Universität Zürich, Winterthurerstrasse 190, CH-8057 Zürich, Switzerland; pfisterer@uwinst.unizh.ch

Stephen Polasky, Department of Applied Economics, University of Minnesota, 1994 Buford Avenue, St Paul, MN 55108, USA; polasky@umn.edu

Anne-Helene Prieur-Richard, DIVERSITAS, Muséum National d'Histoire Naturelle (MNHN), 57 Rue Cuvier – CP 41, 75231 Paris Cedex 05, France; anne-helene@diversitas-international.org

David Raffaelli, Environment Department, University of York, York, UK; dr3@york.ac.uk

Todd Rosenstock, Department of Plant Sciences, University of California, Davis, CA 95616, USA; trosenstock@ucdavis.edu

Mahesh Sankaran, Institute of Integrative and Comparative Biology, Faculty of Biological Sciences, University of Leeds, Leeds LS2 9JT, UK

Bernhard Schmid, Institute of Environmental Sciences, Universität Zürich, Winterthurerstrasse 190, CH-8057 Zürich, Switzerland; bernhard.schmid@uwinst.uzh.ch

Martin Solan, Oceanlab, University of Aberdeen, Main Street, Newburgh, Aberdeenshire, Scotland AB41 6AA, UK; m.solan@abdn.ac.uk

Diane S. Srivastava, Department of Zoology, University of British Columbia, Vancouver, British Columbia V6T 1Z4, Canada; srivast@zoology.ubc.ca

Amy Symstad, U.S. Geological Survey, Northern Prairie Wildlife Research Center, 26611 U.S. Highway 385, Hot Springs, SD 57747, USA; asymstad@usgs.gov

Matthew Thomas, Center for Infectious Disease Dynamics and Department of Entomology, 1 Chemical Ecology Lab, Penn State, University Park 16802, PA, USA; mbt13@psu.edu and Matthew Thomas, CSIRO Entomology, GPO Box 1700, Canberra, ACT 2601, Australia; matthew.thomas@csiro.au

John Tschirhart, University of Wyoming, Department of Economics and Finance, Department 3985, Ross Hall 123, Laramie, WY 82071, USA; tsch@uwyo.edu

Marcel van der Heijden, Ecological Farming systems Research Station ART, Agroscope Reckenholz Tanikon, Reckenholzstrasse 191, 8046 Zurich, Switzerland and Vrije Universiteit Amsterdam, Faculty of Earth and Life Sciences, Institute of Ecological Science, Department of Animal Ecology, De Boelelaan 1085, 1081 HV Amsterdam, The Netherlands

Wim H. van der Putten, Netherlands Institute for Ecology (NIOO-KNAW), Centre for Terrestrial Ecology, P.O. Box 40, 6666 ZG Heteren, The Netherlands and Laboratory of Nematology, Wageningen University and Research Centre, PO Box 8123, 6700 ES Wageningen, The Netherlands

David A. Wardle, Department of Forest Ecology and Management, Swedish University of Agricultural Sciences, SE901-83 Umeå, Sweden

Alexandra Weigelt, Institute of Ecology, University of Jena, Dornburgerstr. 159, 07743 Jena, Germany; alexandra.weigelt@uni-jena.de.

Justin Wright, Department of Biology, Box 90338, Duke University, Durham, NC 27708, USA; jw67@duke.edu

Anastasios Xepapadeas, University of Economics and Business Department of International and European Economic Studies 76 Patission Street, 104 34 Athens, Greece; xepapad@aueb.gr

Preface

This volume serves as an introduction, reference, and survey both of the profound transformation experienced in the last decade by ecology's fast-growing field of biodiversity and ecosystem functioning and of the economics of ecosystem services. Motivated in the early 1990s by environmental concerns over worldwide declines in biodiversity, the biodiversity and ecosystem functioning research area originated as a synthesis of the relatively disparate fields of community and ecosystem ecology. Neither discipline by itself could adequately describe the wide array of possible ecological consequences of biodiversity loss (Loreau et al. 2001, Naeem et al. 2002, Hooper et al. 2005). The first generation of research on biodiversity and ecosystem functioning rapidly grew into a discipline that can be characterized by several features (Loreau et al. 2002). First, species or functional group richness was the primary way of operationally defining and manipulating biodiversity. Second, many studies often worked within a single trophic level (usually plants), though microcosm and mesocosm studies using microbes and invertebrates proved exceptions. Third, research efforts considered only biogeochemical processes, especially primary productivity, as ecosystem functions. Fourth, the prevailing mechanisms were limited to niche complementarity (i.e. niche differences lead to greater exploitation of available resources that lead to greater levels of ecosystem functioning) and selection effects (i.e. higher diversity communities invariably contain one or a few dominant species with disproportionate influences over ecosystem function) that were often viewed as opposing hypotheses vying for supremacy. Fifth, local extinction or biodiversity loss was largely considered a random process and experiments focused on producing as many randomly constructed species combinations as possible to explore how biodiversity loss influenced ecosystem functioning. Sixth, the research was largely experimental, complex, abstract, and confirmatory in nature (i.e. simply confirming that changes in biodiversity did indeed change ecosystem functioning). Finally, work on biodiversity and ecosystem functioning was colored by a tremendous debate over interpretation of its findings.

Over the last few years, however, biodiversity and ecosystem functioning research has evolved dramatically. This volume provides a thorough review of the new face presented by the second generation of biodiversity and ecosystem functioning research. Its 21 chapters are written by more than 60 authors who have been at the forefront of this transition. Virtually everything that characterized the first generation of biodiversity and ecosystem functioning research has changed. First, rather than species or functional group richness, the new focus is on trait-based, functional biodiversity, as well as on community composition. Second, biodiversity and ecosystem functioning studies are increasingly multi-trophic and span both terrestrial and marine ecosystems in comparison to the dominance of terrestrial plant studies that typified earlier biodiversity and ecosystem functioning work. Third, trait-based mechanisms of ecosystem functioning have become a major thrust for contemporary biodiversity and ecosystem functioning research, while niche complementarity and selection effects are considered to be co-occurring (not conflicting) mechanisms. Fourth, rather than assuming random local extinctions, much new work on biodiversity and ecosystem functioning employs trait-based extinction probabilities or increasingly uses empirical extinction scenarios to establish its biodiversity gradients. Fifth, compared to the more abstract deliberations of the first generation of biodiversity and ecosystem functioning research, there is now much more attention to the role of biodiversity and ecosystem functioning in restoration ecology, agriculture, invasions, disease, pollination, climate change, and other ecosystem-service-related environmental issues. Finally, consensus has been achieved (Loreau *et al.* 2001, Hooper *et al.* 2005) and the debate that once clouded the interpretation of biodiversity and ecosystem functioning findings has largely abated.

There are also entirely new features of the second generation of biodiversity and ecosystem functioning research as well. Enough studies have now accumulated to allow meta-analyses, which obviate the sometimes subjective interpretation of trends in biodiversity and ecosystem functioning experiments expressed during the earlier contentious period. Second, *in silico*, trait-based simulation modeling of biodiversity and ecosystem functioning relationships at larger scales has augmented the complex and costly combinatorial experimental approach and represents an entirely new and promising method for large-scale biodiversity and ecosystem functioning research. Third, metacommunity theory applied to biodiversity and ecosystem functioning provides additional understanding of ecosystem complexity and stability.

Beyond the basic science of biodiversity and ecosystem functioning, this volume also explores the current state of the economics of biodiversity and ecosystem services. With antecedents in both natural resource and ecological economics, this field of economics incorporates insights from ecology to build an understanding of the ways in which biodiversity and ecosystem functioning contribute to human wellbeing. The field received a major stimulus from the Millennium Ecosystem Assessment's (2005b) focus on ecosystem services – the benefits that people derive from the processes and functioning of both 'natural' and 'managed' ecosystems. By conceptualizing ecosystem processes and functioning as factors in the production of ecosystem services that directly or indirectly benefit people, the Millennium Ecosystem Assessment has brought many ecological questions within the realm of economics. For example, it has made it natural to analyze the trade-offs (in terms of ecosystem services) of alternative ecological configurations. At the same time it has compelled economists to pay serious attention to the ecological stocks and flows that underpin the production of many ecosystem services. This volume explains and expands upon the ways in which the new face of biodiversity and ecosystem functioning research is interfacing with research into the decisions that people make about how to use the resources of the environment.

The contents of this volume

In 2000, the National Science Foundation (NSF) funded a Research Coordinating Network (RCN) entitled 'Biotic Mechanisms of Ecosystem Regulation in the Global Environment' (BioMERGE) to foster collaboration and usher biodiversity and ecosystem functioning research through its maturation phase (Naeem *et al.* 2007). The relationship between biodiversity and ecosystem functioning is also the central theme of the ecoSERVICES core project of DIVERSITAS (http://www.diversitas-international.org/), an international programme that promotes biodiversity science and aims to bridge the science and policy interface. This volume is the final product of a five-year collaboration between BioMERGE and DIVERSITAS.

The volume is divided into four sections. The first section, *Introduction, Background, and Meta-Analyses*, provides the background for the volume. The editors provide the background, historical context, and an overview of the volume's content in Chapter 1, followed by a meta-analysis by Schmid *et al.* (Chapter 2) that quantitatively tests several biodiversity and ecosystem functioning hypotheses using the enormous body of published experimental studies. The last chapter in this section is an historical and quantitative analysis of the impact of biodiversity and ecosystem functioning research by Solan *et al.* (Chapter 3) that quantitatively tests several biodiversity and ecosystem functioning hypotheses using the enormous body of published experimental studies.

The second section, *Natural Science Foundations*, consists of seven chapters. In Chapter 4, Petchey *et al.* describe one of the major contributions of biodiversity and ecosystem functioning research to ecology: an increasing emphasis on functional diversity. Petchey *et al.* illustrate both the advantages and challenges of focusing on functional diversity by reviewing how authors have attempted to quantify functional diversity. Duffy *et al.* (Chapter 5), consider how functional diversity has transformed biodiversity and ecosystem functioning research from a largely confirmatory science to one that is increasingly predictive.

The remaining chapters of the second section address universal challenges for all of ecology, in the context of biodiversity and ecosystem functioning. These are stability and complexity (Chapter 6 by Griffin *et al.*), identifying the mechanisms generating ecological relationships (Chapter 7 by Hector *et al.*), the importance of trophic structure (Chapter 8 by Cardinale *et al.*), microbial ecology (Chapter 9 by Bell *et al.*), and the importance of the spatial dimension and metacommunities in determining the effects of diversity on ecosystem functioning (Chapter 10 by Gonzalez *et al.*).

The third section takes research on biodiversity and ecosystem functioning further than it has ever gone into the human dimension. The first six chapters cover the most pressing environmental challenges humanity faces. Notably, these chapters also highlight a new emphasis on ecosystem services that go beyond the historic focus on primary productivity. Díaz *et al.* consider the effects of biodiversity on the carbon cycle (Chapter 11) as a way to shed light on anthropogenic climate change that has been largely devoid of considerations of biodiversity. Wright *et al.* consider the role that diversity may play in fostering the restoration of degraded or abandoned habitats (Chapter 12). Jackson *et al.* (Chapter 13) consider the importance of biodiversity in the agricultural ecosystems that now cover one third of Earth's terrestrial surfaces, and focus on biological control as a case study. Klein *et al.* (Chapter 14) discuss the critical ecosystem service of pollination, which is equally important for many crops as well as unmanaged or restored systems. The mitigation of disease (Chapter 15 by Ostfeld *et al.*) and biological invasions (Chapter 16 by Engelhardt *et al.*) are two other biotic ecosystem services that are strongly influenced by biodiversity.

What truly makes this volume unique are the chapters of Section 3, which consider the economic perspective. Perrings *et al.* (Chapter 17) provide a synthesis of the economics of ecosystem services and biodiversity, and the options open to policy-makers to address the failure of markets to account for the loss of ecosystem services. Barbier *et al.* (Chapter 18) examine the challenges of valuing ecosystem services and, hence, to understanding the human consequences of decisions that neglect these services. Brock *et al.* (Chapter 19) examine the ways in which economists are currently incorporating biodiversity and ecosystem functioning research into decision models for the conservation and management of biodiversity.

The fourth and final section consists of two chapters, one describing the new, ambitious direction of biodiversity and ecosystem functioning research to become a global science (Chapter 20) and a synthesis of this volume (Chapter 21) by the editors that describes the nature of the progress made thus far and the future directions and challenges that have been covered by the many authors of this volume.

Acknowledgments

This volume is the summation of five years of cooperation among biodiversity and ecosystem functioning researchers and environmental economists fostered through joint meetings between BioMERGE and BESTNet (NSF-funded Research Coordination Networks) and the ecoSERVICES project of DIVERSITAS. This collaboration was founded on the principles of inclusiveness (i.e. including participants irrespective of their position on the issues), attention to balance across the various stages in scientific careers (i.e. include graduate students, postdoctoral researchers, junior and senior faculty), and gender balance.

Justin Wright, the first associate director of BioMERGE, coordinated meetings in Seattle (2002) and the Missouri Botanical Garden (2003). Daniel Bunker, the second associate director of BioMERGE, and Andy Hector, the then co-chair of DIVERSITAS' ecoSERVICES core project, coordinated meetings in Borneo (2005) and Switzerland (2006) with the help of Chris Philipson, Glen Reynolds, Philipppe Saner, and Maja Weilenmann. Two ecoSERVICES workshops in Paris, coordinated by John Tschirhardt (2005) and Charles Perrings (2007), laid the groundwork for the economic chapters included in the volume.

The bulk of the funding for BioMERGE came from NSF grants # 0130289 and 0435178 with additional support from the University of Washington, Seattle, and Columbia University. Funding for BESTNet came from NSF grant # 0639252. DIVERSITAS contributed both financial and logistical support to a number of the preparatory workshops, and in particular supported the participation of non-US participants. This volume and its contents serves as a testament to the value of supporting international cooperation, integration, and synthesis among social and natural scientists in basic and applied research.

PART 1
Introduction, background, and meta-analyses

CHAPTER 1

Introduction: the ecological and social implications of changing biodiversity. An overview of a decade of biodiversity and ecosystem functioning research

Shahid Naeem, Daniel E. Bunker, Andy Hector, Michel Loreau, and Charles Perrings

1.1 Biodiversity, ecosystem functioning, and human wellbeing: An unconventional perspective

Conventional approaches to ecology often lack the necessary integration to make a compelling case for the critical importance of biodiversity to ecosystem functioning and human wellbeing. Traditional ecology textbooks (e.g. Ricklefs and Miller 1999, Krebs 2001, Smith and Smith 2005, Begon et al. 2006), for example, often begin with species adaptations to local environmental conditions and then proceed through topics such as the population biology of single species, the dynamics of interacting populations (e.g. competitors, predator–prey, host–parasite, mutualisms, food webs), the relationship between stability and complexity, biogeography, and biomes, with little mention of ecosystem ecology. Ecosystem ecology is included, but treated separately. Topics such as C, N, P, and S biogeochemistry, primary and secondary production, decomposition, trophic pyramids, and energy flow make sparse reference to population or community ecology. Today, most ecology texts also include treatment of environmental issues such as pollution, the ozone hole, climate change, collapsing fisheries, disappearing forests, the adverse consequences of unbounded human population growth, emerging diseases, and conservation biology; this last topic being where the value of biodiversity dominates. These topics, however, are often tacked on as final chapters that are poorly integrated with the earlier 'pure' ecology. This approach obscures the inextricable links between biodiversity, ecosystems, and human wellbeing.

This linear march through the biological hierarchy, loosely coupled with its significance to human wellbeing, while of some pedagogical merit, does not prepare one for understanding and applying ecology in the context of the modern world. In today's world, almost everything, especially biodiversity, has been impacted by human activities (Millennium Ecosystem Assessment 2005c, Kareiva et al. 2007). A different, rather unconventional approach is needed for understanding ecology and environmental biology, one that asks the question that is rarely asked by ecology texts – *What is the significance of biodiversity to human wellbeing?*

Rather than the conventional perspective, which sees biodiversity as a culmination of population and community ecological processes with ecosystem processes being separate, ecologists at a conference in 1992 in Bayreuth, Germany, considered an alternative perspective, one that added biotic feedback from biodiversity to ecosystem processes (Schulze and Mooney 1993). Although this concept of biotic feedback was unconventional and controversial, it actually dates back to Darwin. In the *Origin*, Darwin (1859) hypothesizes, based on his *principle of divergence*, that as diversity evolves and fills niche space it will lead to an increase in productivity and other ecosystem processes due to the ecological 'division of labour' (Hector and Hooper 2002). There are also a few later echoes of this idea before it was fully reborn at the 1992 Bayreuth conference. Carlander (1952) found a positive relationship between the diversity of freshwater fishes and their overall secondary productivity, which he interpreted as coming about due to more complete filling of niche space. Similarly, in two papers that are under-appreciated in the biodiversity and ecosystem functioning literature, Bell (1990, 1991) found that complementary differences amongst species of *Chlamydomonas* led to increased productivity and greater temporal stability of production. Nevertheless, it was only following the 1992 Bayreuth conference that investigation of the effects of biodiversity on ecosystem functioning coalesced as a focused research area.

The first generation of research on the relationship between biodiversity and ecosystem functioning consisted largely of experimental confirmations that the two were indeed linked with one another – changes in biodiversity had predictable effects on ecosystem functioning (Loreau *et al.* 2002). Both the design of the experiments and interpretation of the results, however, were surrounded by much debate (e.g. Guterman 2000, Kaiser 2000, Naeem 2000, Tilman 2000, Wardle *et al.* 2000b). Solan *et al.* (Chapter 3) review the history and impacts of biodiversity and ecosystem functioning research, providing an in depth analysis of what began as a rather unconventional approach to ecological research, but which has since grown into a major paradigm in ecology.

The simple heuristic of plotting a trajectory of ecosystem functioning against a gradient in biodiversity in a bivariate plot (Fig. 1.1), an approach begun by Vitousek and Hooper (1993), and asking what the shape of the trajectory might be, provoked much research and discussion, but such simple plots belie the underlying complexity of the problem. *Biodiversity* and *ecosystem function* are both difficult to define and quantify; thus trajectories in such a poorly defined bivariate space are difficult

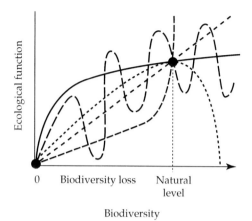

Figure 1.1 A plethora of trajectories of ecosystem function in response to changes in biodiversity. Historically, initial discussions of the relationship between biodiversity and ecosystem functioning were facilitated by the heuristic device of plotting hypothetical responses of ecosystem functions, such as primary production or nutrient cycling rates, against a gradient in biodiversity that was considered to embody taxonomic, functional, phylogenetic, and even the spatial and temporal dimensions of biological diversity. 'Natural' indicates the maximum level of diversity typical for an ecosystem, whereas positions to the right of 'natural' indicate the addition of exotic or domestic species that lead to higher levels of diversity. The solid line reflects the hypothesis that many species were redundant, which would lead to a ceiling in gains per species added. The dotted, hump-shaped line reflects a linear increase in which species contributed similarly to functioning, followed by a decline when exotic species were added. The small-dash, linear line indicates improvements in ecosystem functioning with every species added, suggesting no redundancy. The dashed, concave up curve indicates a dramatic decline in function as soon as diversity drops below natural levels due to the loss of keystone species. The long-dashed line indicates an idiosyncratic or unpredictable response of functioning to biodiversity loss. Over fifty different hypothetical trajectories have been described.

to interpret. Furthermore, biodiversity does not exist in nature outside of ecosystems; nor does an ecosystem exist without biodiversity – plotting one orthogonal to the other as dependent and independent variables is a strange thing to do. The bivariate plot, various biodiversity and ecosystem functioning trajectories, and their interpretation are reviewed elsewhere (e.g. Schläpfer and Schmid 1999, Naeem 2001b, Naeem *et al.* 2002), thus we will not dwell on them here. Research on the relationship between biodiversity and ecosystem functioning has moved well beyond this early framework. Today, the complexities underlying the relationships between biodiversity and ecosystem functioning consume contemporary research. The thought experiment that follows will serve to illustrate these complexities.

1.2 Sterilizing Earth: a thought experiment in three parts

As an introduction to this topic and to provide the context for this volume, let us perform the following thought experiment in three parts. First, consider a space anywhere on Earth's surface at any scale – a park, city, farm, lake, river, wetland, sea, biome, or the biosphere – and then sterilize it. Every plant, animal, and microorganism that occupies this space is destroyed, leaving nothing behind but rocks, sand, water, dead organic matter, and a variety of atmospheric gasses. Second, humans are spared but now find themselves in the barren space, their wellbeing entirely dependent on how one restores the ecosystem. Third, we erect a barrier to all living organisms and we use this barrier to control the functional, phylogenetic, and biogeographic identity of the species we allow to enter. We also allow for the direct importation of species, be they native, exotic, domestic, or genetically engineered. We also use the barrier to control the timing and order of entry, the abundance, and spatial distribution of the species that enter. In other words, we, the human occupants of the space, fully control every aspect of the biodiversity of organisms that will re-populate the sterile space. The importance of biodiversity, ecosystem functioning, and the wellbeing of the humans occupying the space would begin to be revealed as biodiversity is reestablished. If the experiment sounds a little bizarre, recall that the ill-fated Biosphere II had much the same aims.

Once we have fixed this image in our mind, we can immediately see that there is a near infinite number of ways to go about reestablishing biodiversity in a sterile space and if human wellbeing is at stake, the decisions we make take on enormous importance. Figure 1.2 illustrates this thought experiment and summarizes the key elements of biodiversity that we have to consider. First, each species we introduce possesses functional traits which reflect their tolerances and responses to (e.g. drought or salt tolerance) and impacts on (e.g. nitrogen-fixing or sulfur-reducing) environmental factors such as soil moisture, salinity, and nutrient availability (e.g. Lavorel and Garnier 2002). The species we introduce will be related to one another by their functional traits, ranging from being nearly redundant (having the full set of traits in common) or nearly singular (possessing largely unique traits) (e.g. Naeem 1998). Third, species will also possess homologous characters that reflect their shared evolutionary history or phylogeny and will be either closely or distantly related (e.g. Ackerly 2004, Edwards et al. 2007). Fourth, species will either consume, be consumed by, compete with, parasitize, or facilitate other species in a web of interactions that vary in strength (inset in Fig. 1.2) (e.g. McCann et al. 1998, Thébault and Loreau 2006). Fifth, the abundance of species, in terms of either density or mass, will vary depending on each species' growth rates, body size, metabolism and life history (e.g. Brown et al. 2004), resource availability (e.g. Tilman 1982), stoichiometry (Elser and Sterner 2002), interactions with other species, top down and bottom up controls within the food web (Pimm 1982, De Ruiter et al. 2005, McCann et al. 2005), and spatial factors (e.g. Tilman and Kareiva 1997, Loreau et al. 2003). Sixth, species are assembled by biogeographic processes (e.g. MacArthur and Wilson 1967, MacArthur 1972, Hubbell 2001, Lomolino and Heaney 2004). Seventh, the timing, order of entry, and other factors affecting assembly also influence biodiversity (e.g. Weiher and Keddy 1999, Fukami and Morin 2003, Larsen et al. 2005). Collectively, these many factors determine the biodiversity one finds in a community, all of them influencing flows of nutrients into and out of the inorganic pool, the use and return of water, and the flow of energy sequestered by primary producers and lost through respiration (De Angelis 1992, Loreau 1994, Loreau 1995, Grover and Loreau 1996, De Mazancourt et al. 1998, Hulot et al. 2000, Norberg et al. 2001).

The preceding long list of factors is meant to emphasize the overwhelming complexity of what is embodied in the structure and function of biodiversity in ecosystems and the dilemma we face if we have to construct an ecosystem from the ground up. There are three approaches we could take in

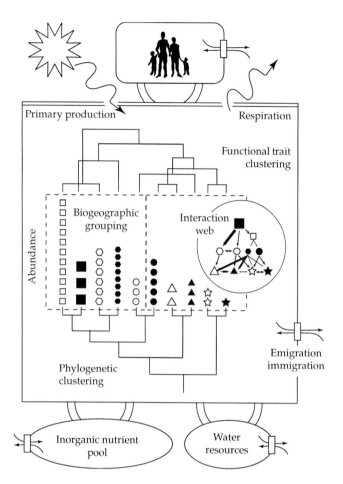

Figure 1.2 Graphic model of the relationship between biodiversity, ecosystem function, and human wellbeing. This figure reflects the thought experiment described in the text in which humans are dependent on the services derived from the functioning ecosystem within which they reside. In the central box, species are portrayed as different shapes and colors and their abundance is indicated by the number of each shape. The upper dendrogram reflects species clustering by similarity in functional traits. The lower dendrogram reflects species relations by phylogeny. The ovals below the box represent the inorganic nutrient pool and water resources. All compartments can potentially have inputs and outputs (immigration and emigration in the case of biodiversity and humans). Curved, parallel lines indicate conduits between compartments. The inset represents a web of biotic interactions (arrows linking species) of different strengths (width of arrows). The figure motivates the thought exercise of imagining how biodiversity is related to ecosystem functioning and human wellbeing by considering how such relationships are affected by the species selected to reside in the ecosystem, and the resulting mix of traits, phylogenetic relationships, biotic interactions, and other factors. The main purpose of the figure is to illustrate the complexity of what underlies otherwise simple relationships typically plotted in biodiversity and ecosystem function bivariate space (Fig. 1.1).

repopulating a sterile space, all of which reflect the rapidly evolving field of biodiversity and ecosystem functioning. We could simply restore biodiversity to what it was at the time of sterilization, under the assumption that that was the best biodiversity for the people who lived there. We could, alternatively, avoid making the assumption that resident biodiversity is the best for humanity and employ ecological principles to engineer the construction of a biota that would maximize the wellbeing of its human residents. Finally, we could explore numerous combinations of species at different relative densities and monitor ecosystem function and human wellbeing until we obtained the biodiversity that optimizes human wellbeing. We will refer to these three approaches as *restorative*, *eco-engineered*, and *explorative*, respectively.

This thought experiment demonstrates that every species contributes to ecosystem functioning and human wellbeing in complex ways, though obviously at different levels (from negligible to enormous) and with different impacts on humans (from beneficial to harmful). We know that each species we admit into the space will change its ecosystem functions (e.g. primary productivity, nutrient cycling, decomposition), its biotic functions (e.g. its susceptibility to invasion, the likelihood that an emerging disease might spread, or the dynamic stability of its populations), and the wellbeing of its human occupants (e.g. the prevalence of poverty, hunger, and economic prosperity). (Note that we distinguish between ecosystem and biotic functions, the former referring to biogeochemical functions and the latter referring to the influences of species on one another. Some researchers, however, use *ecosystem function* to refer to both biogeochemical and biotic functions.) Obviously some species additions will have little effect while others will have dramatic impacts. Further, because all species interact with one another directly or indirectly, with strengths of interaction that vary in sign and magnitude (inset in Fig. 1.2), the impact of each species will be within the context of the community present.

Of the three approaches to restore our sterilized space, restorative, eco-engineered, and explorative, the most neutral approach, or one that requires the least commitment to any particular ecological framework, is the explorative approach that conducts hundreds (if not thousands or millions) of experiments in which each experiment randomly creates a different community, sometimes rich in biodiversity, other times poor or intermediate. With each species selection, one imports a set of traits, modifies the functional and phylogenetic clustering of the community, modifies the community web, and, depending on the density and mass of individuals initially imported, alters the relative abundance of species. In short, every biotic function is modified in some way. Immediately upon introduction, every entering species will also affect the flow of nutrients, energy, and water through the system in some way, again in ways ranging from small to large.

If during an explorative biodiversity and ecosystem functioning approach to repopulating a sterile habitat we simultaneously plot some metric of the diversity of the species we include (e.g. functional diversity, see Chapter 4) against some metric of ecosystem functioning (e.g. net primary production) or against some metric of a biotic function (e.g. resistance to the spread of an emerging disease) or against some measure of human wellbeing, what would that plot look like? Would the values for biodiversity, ecosystem function, biotic function, or human wellbeing show any correlations? Would the maximum for human wellbeing coincide with maximum biodiversity? Biodiversity and ecosystem functioning research attempts to shed light on these and many other questions.

Note that the end product of such an exercise, repeatedly plotting ecosystem function measured for different communities differing in biodiversity, is a plot like Fig. 1.1. The exercise, however, illustrates how difficult it is to obtain such plots, how complex the underlying processes are, and how difficult interpretation of such plots might be.

1.3 The evolution of biodiversity and ecosystem functioning research

Our thought experiment makes transparent how any attempt to experimentally explore the relationship between biodiversity and ecosystem functioning requires the researcher to make several decisions that ultimately determine what can and cannot be inferred from the results. What should be manipulated, for example? Should species richness, functional diversity, genetic diversity, relative abundance, or some combination of these or other factors be manipulated? Should species be selected based on biogeography (e.g. only species known to coexist in nature), or could one use any species that is likely to survive in the experiment (e.g. a series of exotic species or domestic species)? How extensive should the biodiversity gradient of the experiment be? Should the lowest level of diversity in the experimental biodiversity gradient be a sterile plot, a single species, or a complete food web with only one species per trophic level? Should the highest level of biodiversity gradient be all species that co-occur in nature, a subset, or more than what is typically found in nature, and should several trophic levels be used with as many species per trophic level as possible?

Most decisions in biodiversity and ecosystem functioning experiments concern two things; the rationale for species selection and the extent of the biodiversity gradient to be used in re-populating the replicates. One of the first biodiversity and ecosystem functioning experiments (Naeem *et al.* 1994, Naeem *et al.* 1995), for example, was a complex, multi-trophic terrestrial mesocosm built in a system of growth chambers known as the Ecotron (Lawton *et al.* 1993, Lawton 1996). Researchers selected species known to co-occur in the field, and were likely to survive in the growth chambers; they paid little attention to whether the species were exotic or not. The researchers also decided to use a biodiversity gradient of just three levels of species richness, but to retain multiple trophic levels throughout. Experiments that followed took similar approaches – they began with a pool of species that were known to co-occur in nature, were not overly concerned with whether they were exotic or not, and established replicate systems (microcosms, artificial ponds, flower pots, and grassland plots) cleansed of the species they wanted to manipulate. The experimenters then repopulated their cleansed replicates with species. Each experiment, however, took slightly different approaches. For example, the experiments of Cedar Creek, Minnesota, focused exclusively on plants (Tilman *et al.* 1996, Reich *et al.* 2001) as did the European BIODEPTH experiments (Hector *et al.* 1999), though one BIODEPTH site manipulated insect abundance (Mulder *et al.* 1999). Hooper manipulated only plants in Californian serpentine grasslands, but focused on functional groups rather than species (Hooper and Vitousek 1997). In fact, most grassland experiments did a combination of both functional group and species manipulations (Tilman *et al.* 1997b, Naeem *et al.* 1999, Wardle *et al.* 2000a, Reich *et al.* 2004).

Early theory and experiments were confirmatory; they primarily sought to examine the possibility that diversity could indeed affect ecosystem functioning. As biodiversity and ecosystem functioning research shifted out of its early confirmatory phase, however, it increased the scope of organisms and systems it investigated and became more exploratory in scope. Hundreds of experiments explored freshwater, stream, wetland, microbial, coral reefs, marine ecosystems, and grassland ecosystems, many of which are discussed directly in this volume or indirectly as parts of meta-analyses. Although each experiment is unique, common to all of them was the need to decide rationally which species to select for inclusion or exclusion and what would constitute the gradient in biodiversity.

The expansion in extent, scope, and complexity of biodiversity and ecosystem functioning experiments led the field to become more predictive rather than confirmatory and exploratory, with an increased emphasis, first, on developing theory with which to interpret these experimental results (Hector *et al.* Chapter 7, Loreau and Hector 2001, Fox and Harpole 2008), and later on applying these findings to real-world problems associated with biodiversity loss. Key features of this new research include greater precision in its metrics as well as increasing integration of the many processes and factors known to impact biodiversity. Greater precision in metrics, for example, involved the evolution of better, more appropriate and applicable measures of biodiversity. In this regard, functional diversity has emerged as the frontrunner for the most relevant component of biodiversity with respect to ecosystem functioning (Díaz and Cabido 2001, Naeem 2002a, Petchey and Gaston 2002a). How one defines, quantifies, and uses functional diversity to interpret how changes in biodiversity can impact ecosystem functioning has become a dominant part of biodiversity and ecosystem functioning research (see Petchey *et al.*, Chapter 4).

Biodiversity and ecosystem functioning research has also expanded to include important elements of ecological systems previously understudied. Among them, trophic complexity has become a major theme in contemporary biodiversity and ecosystem functioning research. Rather than working within a single trophic level like many early plant-only experiments, trophic complexity has become an important part of biodiversity and ecosystem functioning experiments and theory (see Cardinale *et al.*, Chapter 8). Additionally, rather than black-boxing the microbial world or using them for microcosm tests of theory, biodiversity and ecosystem functioning is expanding to uncover the role of microbes in more complex and natural systems (see Bell *et al.*, Chapter 9).

Research on the relationship between biodiversity and stability has also gone beyond the initial attempt to simply confirm if there was or was not a relationship between the two to recognizing multiple relationships (mostly positive, some neutral, and a few negative) and multiple mechanisms (see Griffin *et al.*, Chapter 6). Theory has also gone well beyond resource-based or Lotka–Volterra type models to explore multitrophic systems, metacommunities, and other theoretical advances (see e.g. Gonzalez *et al.*, Chapter 10).

With advances in experiments, observational studies, precision in metrics, tools, and theory, knowledge on biodiversity and ecosystem functioning is now sufficiently developed that the first projections of the ecosystem consequences of biodiversity loss have begun. Developing means for biodiversity and ecosystem functioning projections was the goal of the National Science Foundation research coordinating network known as Biotic Mechanisms of Ecosystem Regulation in the Global Environment (BioMERGE) (Naeem and Wright 2003, Naeem et al. 2007). This is an ambitious and important trend in biodiversity and ecosystem functioning research. It is ambitious because it incorporates many factors that impact biodiversity (see the long list we provided above) into simulations to project future states of ecosystem functioning. It is important because these projections are meant to be realistic and large-scale and useful to researchers, managers, and policymakers alike. Currently, studies that have used this approach are limited, but show promise. Duffy et al. (Chapter 5) review this emerging field in biodiversity and ecosystem functioning research.

1.4 Biodiversity and humanity: strains in a productive partnership

The minute humanity began to manipulate nature beyond what is common for ecosystem engineers like beavers or termites (Jones et al. 1994), or what might be expected from niche construction (Laland and Sterelny 2006), humans began a path that would lead to extraordinary success, but at an extraordinary price. Success can be seen in terms of humans becoming the dominant geomorphic (Wilkinson 2005), biogeochemical (Vitousek et al. 1997), and consumer species (Imhoff et al. 2004). In the last two centuries, aided by access to fossil fuel and fossil water, humanity has appropriated an ever-increasing proportion of terrestrial surfaces (Foley et al. 2005, Worm et al. 2005), net primary productivity (Haberl et al. 2007), fresh water (Vörösmarty et al. 2000, Oki and Kanae 2006), and marine resources (Worm et al. 2006). In so doing, levels of output and consumption have grown everywhere – albeit at very uneven rates. Success has meant a process of economic development that has built up stocks of 'produced' and 'human' capital (infrastructure, buildings, equipment, and financial assets on the one side, technology, skills, education, and learning on the other), whilst running down stocks of 'natural' capital (Dasgupta 2001). Running down natural capital does not only mean the depletion of non-renewable resources like oil, minerals, fossil water, it also means the loss of biodiversity (Sala et al. 2000, Millennium Ecosystem Assessment 2005b) and ecosystem services (Millennium Ecosystem Assessment 2005a, Kareiva et al. 2007). In some cases, this has left people little better off than they were before. Many of the two billion people in poverty, or the one billion in hunger, are dependent on common-pool environmental resources for their livelihoods. In other cases, the costs associated with the loss of natural capital are simply not taken into account by those whose actions have caused it. Indeed, there is a widespread view that our systematic neglect of the human costs of the erosion of natural capital stocks has to change (World Commission on Environment and Development 1987, United Nations Environmental Program 2007, Holdren 2008). For change to be well-informed, we need to better understand the consequences of eroding the natural capital base.

The replacement of naturally occurring animals and plants with domesticated species began in terrestrial systems over ten thousand years ago and is now increasing in aquatic and marine systems as well (Duarte et al. 2007). When human populations were small, such substitutions of naturally diverse systems with smaller numbers of more manageable, higher-yielding species, was not likely to have major impacts on ecosystem functioning at large scales. As the scale of human activity has increased, however, so has its ecological impact.

The initial conditions of the thought experiment, that of complete human domination of ecosystems, are admittedly extreme, but they represent an important endpoint in a continuum that structures the conceptual framework of biodiversity and ecosystem functioning research. At one end, all of humanity's needs are met by managing ecosystems using only the species necessary to maximize human wellbeing. At the other end of the continuum is a state in which every need of humanity is met by nature – a state that probably last existed six million years ago when our primate ancestors started down the evolutionary pathway that would lead to the origination of the human species.

Where in the gradient, from complete human domination of ecosystems to being no different from other species in their impacts, is human wellbeing optimized? The right balance remains unknown, but it is clear that humanity is shifting to the end where all of humanity's needs are met by heavily managed ecosystems. In the case of agro-ecosystems, Jackson et al. (Chapter 13) note such systems now comprise 65 per cent of terrestrial ecosystems, with 10 per cent in high-input agriculture, 15 per cent in low-input agriculture, and 40 per cent in mixed use. By 2050 an additional 10^9 hectares of wildlands are likely to be converted to managed lands to feed our growing population.

Managed ecosystems reflect a production-simplification tradeoff in which the production of utilitarian biomass (i.e. edible plants and animals, biofuel, lumber) is increased at the expense of native biodiversity that may appear at first glance to have less utility, though its full utility has yet to be understood or inventoried. The transformation of complex landscapes that typically housed hundreds of species of plants and vertebrates, thousands of species of invertebrates, and untold numbers of species of microorganisms, to managed systems always lowers local (e.g. species per square meter) taxonomic richness to a tenth or hundredth of its original value. More importantly, biodiversity loss associated with simplification often brings with it concomitant reductions in trait diversity and reductions in the number, type, strengths, and arrangement of biotic interactions among species in the community web. Simplification refers more specifically to such reductions in functional diversity and complexity than it does to taxonomic loss.

Initially, biodiversity and ecosystem functioning research focused on the single function of production which would prove to be neither a persuasive argument for conserving biodiversity nor an accurate reflection of the true costs of ecological simplification. Biodiversity and ecosystem functioning studies generally found a positive, asymptotic relationship between biodiversity and production, suggesting that biodiversity loss meant loss in production and implying that human wellbeing would decline in the face of such declining production. The demonstrated relationships, however, typically described strong gains in production with just a few species and vanishingly small gains in production with each species added (Schmid, Chapter 2). Furthermore, monocultures or combinations of just two or three species could often out-produce communities that were much more species-rich. In terms of production, it seemed one could get by with far fewer species in an ecosystem than was typically found in nature.

Lost production due to simplification in natural systems (which was counter to the production-simplification tradeoff in managed systems where production appears to improve under simplification) was only one part of the picture; stability could also be affected by biodiversity loss. Proof that stability and diversity were positively related, however, was much more difficult to demonstrate (McCann 2000, Cottingham et al. 2001). Biodiversity was seen as a means of enhancing system reliability (Naeem and Li 1997, Naeem 1998, Naeem 2003) and a means of improving and stabilizing long-term gains in ecosystem function (e.g. Doak et al. 1998, Tilman et al. 1998, Yachi and Loreau 1999).

An important point that many researchers have made, but which is seldom demonstrated, was that production was not the only function that was affected by biodiversity loss. There is increasing evidence that the maintenance of multiple ecosystem processes requires many more species than does the maintenance of a single process (Eviner and Chapin 2003, Hector and Bagchi 2007, Gamfeldt et al. 2008). An important example of a multiple function is the role of biomass production as both a provisioning ecosystem service (i.e. timber and non-timber forest products) as well as regulatory service in terms of carbon storage (Díaz et al., Chapter 11).

While understanding the true costs of the production-simplification tradeoff in terms of changes in the magnitude, reliability, and stability of multiple biogeochemical functions is a major thrust in biodiversity and ecosystem functioning research, of equal importance are the impacts of simplification on biotic functions. Jackson et al. (Chapter 13) note that biocontrol and pollination (see Klein et al., Chapter 14), both biotic functions, have received attention in the biodiversity and ecosystem functioning literature and are frequently cited as examples of ecosystem services relevant to agro-ecosystems (Balvanera et al. 2005, Tscharntke et al. 2005, Philpott and Armbrecht 2006, Kremen et al. 2007, Priess et al. 2007). Two other examples of biotic functions include the influences of biodiversity over invasive species (see Engelhardt *et al.*, Chapter 16) and diseases (see Ostfeld *et al.*, Chapter 15).

The question of restoration, especially restoration targets, takes on new meaning in light of the production-simplification tradeoff and the relationship between biodiversity and ecosystem functioning. Most habitats designated for restoration have lost biodiversity either due to simplification (e.g. for agriculture) or degradation by pollution or unsustainable extraction, such as clear cutting lumber or over harvesting fish. Restoring simplified or degraded habitats to some version of their former self requires thinking about restoring not only lost diversity, but lost functioning and services as well (Wright *et al.*, Chapter 12).

1.5 The emergence of a unified natural–social biodiversity and ecosystem functioning framework

Biodiversity and ecosystem functioning research can and should supply managers, conservation biologists, policy makers, and other interested parties, with the information they need to make the best decisions they can regarding their effects on biodiversity. Although it has so far done poorly at informing management and policy (Solan *et al.*, Chapter 3), it is founded on a central construct that clearly indicates that it can do so. This construct is written simply as,

Biodiversity → Ecosystem Functioning →
Ecosystem Services → Human Wellbeing,

where each arrow represents a causal relationship and *ecosystem services* are ecosystem functions that benefit humans. This framework, in fact, became the central framework for the Millennium Assessment (2003).

Typically, biodiversity and ecosystem functioning researchers assumed that if they demonstrated that biodiversity was important to the magnitude and stability of any ecosystem function, then it would automatically follow that biodiversity is important to the magnitude and stability of ecosystem services and, by extension, to the magnitude and stability of human wellbeing. That is, they took for granted that if the left-hand side of the construct was demonstrated, then the right-hand part of the construct, the link between ecosystem services and human wellbeing, would follow. And if it did not follow automatically, then it was up to economists to separately pursue the right-hand side of the construct.

In principle, the logic of working on individual parts of the construct was sound and in keeping with the tradition of ecologists and economists working separately. In practice, however, the result was that the natural science of biodiversity and ecosystem functioning, though published in high-profile scientific journals, failed to carry through to management and policy (Solan *et al.*, Chapter 3). The apparent simplification–biodiversity tradeoff, which was the hallmark of human development, was pitted against scientific cautions about hidden costs. The value of land, water, farms, lumber, fisheries, and other natural resources are, however, far greater than the potential gains suggested by biodiversity and ecosystem functioning research. Vanishingly small gains in production in abstract experimental systems or arguments about improved stability were not translated into ecosystem services (i.e. the right-hand side of the construct). Indeed, in spite of the adoption of the biodiversity → ecosystem functioning → ecosystem services → human wellbeing framework by the Millennium Assessment and over a decade of biodiversity and ecosystem functioning research, only a handful of case studies were available to support the Assessment's conclusions that greater biodiversity provides more ecosystem services. A case study approach was similarly used by Balmford *et al.* (2002). Early attempts to estimate the economic value of the ecosystem services supported by biodiversity received considerable attention (e.g. Costanza *et al.* 1997, Costanza and Folke 1997, Pimentel *et al.* 1997), but because they rested on questionable methodology were dismissed by most economists. Nevertheless, they did serve to emphasize that non-marketed ecosystem services were more important than previously believed. At the same time, there is little evidence from over a decade and a half of research – comprising hundreds of biodiversity and ecosystem functioning and economic analyses and the adoption of the principles by the Millennium Assessment – that biodiversity conservation as a route to improve human wellbeing has become a strong part of the private or public consciousness.

While each link in the biodiversity → ecosystem functioning → ecosystem services → human wellbeing framework is important in its own right, these links do need to be developed in unison, as neither the natural science underlying the influence of biodiversity on ecosystem functioning nor the social science underlying the link between ecosystem services and human wellbeing can carry the day on their own. Economists' perspectives on the importance of biodiversity have contributed significantly to understanding the social implications of biodiversity loss (Barbier *et al.* 1994, Perrings 1995, Perrings *et al.* 1995, Swanson 1995, Folke *et al.* 1996, Chichilisky and Heal 1998, Hollowell 2001); thus the foundation for a unified framework exists.

The economic literature on biodiversity and ecosystem services is rapidly growing (Heal 2005, Carson 2008). There are three major thrusts to this literature, which are reflected in the chapters included in this volume. One thrust addresses the reasons why markets fail to allocate biological resources efficiently, and identifies corrective measures. Perrings *et al.* (Chapter 17) identify the externality and public good problems that lie at the heart of biodiversity loss, and survey the range of corrective mechanisms discussed in the literature. These include the development of markets for services such as ecotourism or bioprospecting. But they also include a number of instruments designed to encourage resource users to take the biodiversity consequences of their actions into account, such as taxes, access charges, user fees, payment for ecosystem services, direct compensation payment, and transferable development rights.

A second thrust addresses the valuation of ecosystem services and, through this, of the biodiversity and ecosystem functioning that underpins the production of services. Barbier *et al.* (Chapter 18) review the economics of ecosystem service valuation, and illustrate the way in which the demand for basic ecosystem components may be derived from the demand for ecosystem services. They show how the approach can be used to value the biological resources that support not only provisioning services (e.g. the production of foods, fuels and fibres) and cultural services (e.g. the non-consumptive enjoyment of landscapes for recreational, educational, scientific, spiritual, or cultural reasons), but also regulating services. In the last case, the economic theory of portfolio choice provides a natural way to investigate the implications of biodiversity for risk management.

A third thrust addresses the incorporation of ecosystem components into economic decision models. Brock *et al.* (Chapter 19) review the ways in which economists model decision problems in coupled ecological–economic systems that are subject to varying levels of anthropogenic impact. They also discuss the consequences of the different objectives that motivate people, ranging from the preservation of naturalness to the management of food production systems.

Central to this unification, as both Perrings *et al.* (Chapter 17) and Brock *et al.* (Chapter 19) note, is an understanding of (a) the mechanisms that connect biodiversity and ecosystem functioning to the production of valued ecosystem services, and (b) the set of incentives that lead individuals to behave in ways that are more or less closely aligned with the social interest. All too frequently, decisions made by private resource users neglect costs that are displaced onto others.

There is a complex array of social and natural feedbacks that the simple biodiversity → ecosystem functioning → ecosystem services → human wellbeing construct does not capture. Incorporating these feedbacks will be necessary if effective economic instruments based on biodiversity and ecosystem services are to be designed to ensure that private decisions are compatible with the social interest. The chapters on economics make it clear that the emerging natural–social unified approach can occur if ecologists and economists work together.

Summary

In this introduction, we have reviewed the basic ideas that have structured the revolution in the natural and social sciences that inextricably links biodiversity with human wellbeing. Our emphasis is on the scientific basis for biodiversity's influence over ecosystem functioning and its concomitant effects on human wellbeing. Although the contemporary field of biodiversity and ecosystem functioning emerged only in 1992 (the year of the Earth Summit in Rio and the establishment of the United Nations Convention on Biological Diversity and the United Nations Framework Convention on Climate Change), the field of biodiversity and ecosystem functioning has evolved rapidly through three stages. It first survived the contentious confirmatory years of the late 1990s, moved through an exploratory phase at the beginning of this century, and is now in the throes of building a new, joint, natural–social model for humanity. The authors of these chapters are those who have spearheaded this change and are driving its leading edge. What follows are descriptions of those achievements, advances, and future directions. The reader will find what unfolds to be as scientifically fascinating as it is relevant to solving our most pressing environmental problems.

CHAPTER 2

Consequences of species loss for ecosystem functioning: meta-analyses of data from biodiversity experiments

Bernhard Schmid, Patricia Balvanera, Bradley J. Cardinale, Jasmin Godbold, Andrea B. Pfisterer, David Raffaelli, Martin Solan, and Diane S. Srivastava

2.1 Introduction

2.1.1 Two meta-analyses of biodiversity studies published in 2006

The study of patterns in the distribution and abundance of species in relation to environmental variables in nature (e.g. Whittaker 1975), and to species interactions (Krebs 1972), has had a long tradition in ecology. With increasing concern about the consequences of environmental change for species extinctions, researchers started to assess the potential of a reversed causation: does a change in species diversity affect environmental factors and species interactions, such as soil fertility or species invasion? Manipulative experiments that explicitly tested the new paradigm started in the early 1990s and since then the number of such studies has been increasing exponentially (Balvanera et al. 2006, Chapter 3).

In 2006, two meta-analysis papers were published which together provided the most comprehensive quantitative assessment of the overall trends observed in manipulative biodiversity experiments to date. Both studies showed that, on average, random reductions in diversity resulted in reductions of ecosystem functions, but differed in the covariates examined. First, Balvanera et al. (2006) analyzed studies published from 1974–2004. This meta-analysis showed that biodiversity effects, measured as correlation coefficients between some measure of biodiversity (usually species richness) and a representative response at the ecosystem, community, or population level, were significantly influenced by several factors; the specifics of experimental designs, the type of system studied, and the category of response measured. For example, biodiversity effects were particularly strong when the experimental designs included high-diversity mixtures (>20 species) and in well-controlled systems (i.e. laboratory mesocosm facilities).

A second meta-analysis was conducted by Cardinale et al. (2006a) which focused on experiments, published from 1985–2005, where species richness was manipulated at a focal trophic level and either standing stock (abundance or biomass) at that same trophic level, or resource depletion (nutrients or biomass) at the level 'below' the focal level was measured. Cardinale et al. (2006a) used log ratios of responses to characterize biodiversity effects. Their analyses showed that species-rich communities achieved higher stocks and depleted resources more fully than species-poor communities, but that diverse communities did not necessarily capture more resources or achieve more biomass than the most productive species in monoculture. Cardinale et al. (2006a) also fitted data from experiments to a variety of functional relationships, and found that experiments were usually best approximated by a saturating function. The results from both meta-analyses were remarkably consistent across different trophic levels and between terrestrial and aquatic ecosystems. In this chapter we present further analyses of the two meta-data sets, in parallel, and attempt a joint interpretation.

2.1.2 The two meta-data sets used in this chapter

The two meta-data sets assembled by Balvanera *et al.* (2006) and Cardinale *et al.* (2006a) are hereafter referred to as B and C, respectively. Together, the two databases contain more than 900 published effects of biodiversity on ecosystem functioning (Schmid *et al.* 2009, Cardinale *et al.* 2009). In B, these effects were extracted directly from the publications and therefore rely on the analysis (assumed to be correctly executed) carried out by the original authors. In more than half of the cases, the extracted biodiversity effects were correlation coefficients (Balvanera *et al.* 2006). For these, and for additional cases, significance, direction, and shape of the relationship between biodiversity and each response variable could be extracted. In C, the mean values of response variables were available for each level of species richness. This allowed the authors to decide whether a linear, log-linear, or saturating curve (Michaelis–Menten) was the best fitting relationship (see Cardinale *et al.* 2006a). For ease of comparison with B, the correlation coefficients obtained using the log-linear fit in C are used for this chapter. These were very closely correlated with the correlation coefficients on the Michaelis–Menten scale ($r = 0.99$, $n = 105$). The significance was not assessed in C because the relationships were calculated from means.

If the same response variable was measured repeatedly in an experiment, it was only entered once in each of the two meta-databases: B focused on the first date on which measurements were taken in a study (excluding establishment phases of experiments) while C selected the last date of published measurements. Although about half of the measurements contained in C are also in B, the two data sets were kept separate for our new analyses because of the different ways in which biodiversity effects were initially extracted or calculated.

We speak of a 'biodiversity effect' if a function varies among different levels of biodiversity. Because different levels of biodiversity can be ordered from low to high, in most cases a biodiversity effect can be more specifically defined as a positive or negative relationship between variations in biodiversity as the explanatory variable and a function as response variable. Thus, a positive diversity effect occurs when a relationship is positive and a negative biodiversity effect occurs when a relationship is negative.

2.1.3 Hypotheses

The goal of meta-analyses of biodiversity–ecosystem functioning experiments is to assess to what extent biodiversity effects reported in single studies can be generalized across different design variables, system types, and response categories. Ideally, hypotheses about variation between studies should be derived, *a priori*, from underlying mathematical theory about mechanisms responsible for biodiversity effects. In practice, however, it is often only possible to look for patterns in variation of biodiversity effects and then develop explanatory hypotheses in retrospect. This is primarily due to the fact that the majority of biodiversity experiments included in our meta-databases focused on demonstrating biodiversity effects rather than attempting to test specific mechanistic hypotheses (for an exception, see e.g. Dimitrakopoulos and Schmid 2004). The hypotheses presented in this chapter are derived from patterns found in the previous meta-analyses of B and C. To avoid repetition of results reported in Balvanera *et al.* (2006), we omit hypotheses relating to the influence of specific experimental designs. Instead, we consider several new hypotheses (see below). We also consider the shape of the relationship between biodiversity and specific response variables.

Our **first hypothesis** is that biodiversity effects differ among ecosystem types (Hooper *et al.* 2005). Differences in biodiversity effects among ecosystems could arise, for example, from variation in the ratios of producer/consumer stocks, or the size, generation times, or growth rates of dominant organisms. For example, Giller *et al.* (2004) suggested that biodiversity–ecosystem functioning relationships differ between aquatic and terrestrial ecosystems because of more rapid turnover of material and individuals in aquatic systems. However, despite the often expressed concern that extrapolation from one ecosystem type to another is unwarranted (Hooper et al. 2005, Balvanera et al. 2006), we were unable to find specific predictions about the direction of differences in biodiversity effects between ecosystem types.

We distinguish between population-level functions, recorded for individual target species, such as density, cover or biomass; community-level functions, recorded for multi-species assemblages, such as density, biomass, consumption, diversity; and ecosystem-level functions, which could not be assigned to population- or community-level and included abiotic components such as nutrients, water or CO_2/O_2. Our **second hypothesis** then is that species richness enhances community (and ecosystem) responses but affects population responses negatively (Balvanera et al. 2006). This follows from basic Lotka–Volterra dynamics (see e.g. Kokkoris et al. 1999, Loreau 2004), and the assumption of a maximum community response given by the total availability of resources in the environment. Consider for example a system with s species, where the population growth rate (r_i) of species i, with carrying capacity K_i in monoculture, will be reduced by its own population size (N_i) as well as the populations of $s-1$ competing species ($N_1 \ldots N_s$):

$$r_i = r_{i,max} \bullet (K_i - N_i - \alpha_{1,i} \bullet N_1 - \alpha_{2,i} \bullet N_2 - \ldots - \alpha_{s,i} \bullet N_s)/K_i$$

Every addition to a community with species i of a species j with an inter-specific competition coefficient $\alpha_{j,i} > 0$ will reduce the growth rate r_i and thus negatively affect the population size of species i. However, if $\alpha_{j,i} < 1 > \alpha_{i,j}$, the sum of the two species i and j can produce a larger community size $N_i + N_j$ than each species by itself. That is, if inter-specific competition coefficients are generally smaller than 1, the community size can increase with increasing species richness according to Lotka–Volterra dynamics. Hypothetically, with increasing species richness, total community responses can be summed over more populations, but individual populations will each be under increasing pressure (McGrady-Steed and Morin 2000, Brown et al. 2001, Bunker et al. 2005).

Our **third hypothesis** predicts that standing stocks should respond differently to species richness manipulations than rates (or depletion of resources). However, as with differences between ecosystem types, it is difficult to predict the direction of the differences. Using the argument made above that, for example, community size (as a measure of standing stock) may have upper limits due to the total availability of resources in the environment, whereas rates of change in community size should not be restricted in this way, it follows that rates should be affected more strongly than stocks. This argument is used by researchers who claim that plant species richness may well increase plant productivity but not carbon storage (see e.g. Körner 2004). On the other hand, the theory developed by Michel Loreau (personal communication) predicts that stocks should be more responsive than rates.

Whereas the above hypotheses can already be applied to biodiversity studies focusing on a single trophic level, our **fourth hypothesis** specifically concerns biodiversity effects observed in multi-trophic studies. We consider the effect of changing biodiversity at one trophic level on functions carried out by a different (mostly adjacent) trophic level. If the latter is above the manipulated level, we speak of bottom-up biodiversity effects; if it is below the manipulated level, we speak of top-down biodiversity effects. Despite some similarities between systems with one versus two trophic levels (Ives et al. 2005), biodiversity effects may be more difficult to generalize and predict in multi-trophic systems because of the many possibilities for positive and negative feedback (see e.g. Petermann et al. 2008), as well as differences between generalist and specialist interactions (Petchey et al. 2004a, Thébault and Loreau 2006, Petchey et al. 2008). Theory and some empirical results suggest that bottom-up effects of biodiversity should usually be negative because higher diversity increases resistance to disease and predation (Koricheva et al. 2000, Loreau 2001, Fox 2004a, Petchey et al. 2004a, Keesing et al. 2006, Duffy et al. 2007). However, some empirical results suggest opposite trends (e.g. Koricheva et al. 2000, Pfisterer et al. 2003, Gamfeldt et al. 2005). Top-down biodiversity effects should also be negative because a more diverse community at trophic level t should be able to deplete the community at trophic level $t-1$ more completely, thus reducing functions such as standing stock at the this lower trophic level (Fox 2004b, Petchey et al. 2004a, Fox 2005a, Duffy et al. 2007). This leaves positive effects of biodiversity for within-trophic level (a large number of studies surveyed in the two meta-analyses) and for symbiont relationships. However, these patterns may differ between green (living plant-based) and brown (detrital-based) food webs. A recent meta-analysis of top-down and bottom-up effects in detrital food webs (Srivastava et al., 2009) showed that detrital processing (top-down effects) was increased by high detritivore diversity, but showed variable responses to detrital diversity (bottom-up effects).

Finally, positive effects may also be expected if the trophic distance between the level at which biodiversity is varied and the level at which the response is measured is two or a multiple of two (e.g. top-down from secondary consumers at level t to primary producers at level $t-2$, or, conversely, from trophic levels t to $t+2$), because two negative interactions can together lead to a positive one. Such effects are implicit in the Hairston–Smith–Slobodkin (1960) hypothesis and can be seen in some of the output from simulation models (Petchey et al. 2004a). In the previous meta-analysis of Balvanera et al. (2006), however, we observed that biodiversity effects tend to get weaker the greater the trophic distance is between the level at which diversity is manipulated and the level at which a function is measured.

Our **fifth and final hypothesis** considers the premise that if increasing species richness of a community increases total resource and space use, then less of the resource or space should be available to potential invaders unless they are competitive dominants that displace the existing native species. That is, if the number of species that can fit in the community depends on the 'niche dimensionality' of the environment (Harpole and Tilman 2007), then the more species that are already there in a community, the more difficult it will be for further species to successfully colonize (Fargione et al. 2003, Mwangi et al. 2007).

2.1.4 Shape of the biodiversity–ecosystem functioning relationship

In the second part of the analyses, we focus on the expected shape of the relationship between biodiversity and response functions. Using a survey, Schläpfer et al. (1999) canvassed expert opinions as to whether the relationship was either constant (i.e. no relationship), idiosyncratic, linear, non-linear (logistic, optimum), log-linear, or asymptotic.

The simplest hypothesis about the shape of biodiversity–ecosystem functioning relationships is that of a constant response for all species richness levels, either including or excluding a species richness level of zero. However, including zero species provides a stricter hypothesis that has almost never been tested empirically (in most experiments the zero-richness level was not included) and will therefore not be discussed further. A problem with the hypothesis of constant response is that it cannot be tested for statistical significance.

The experts in Schläpfer et al. (1999) predicted log-linear or saturating shapes for relationships between biodiversity and primary production, nutrient cycling, or water cycling. Such shapes are also predicted by niche theory, which assumes complementarity in resource use among species, but increasing niche overlap with increasing species richness (Tilman 1997, Loreau 1998a, Schmid et al. 2002b). Linear, logistic, or even exponential relationships may be expected between biodiversity and bioregulation (e.g. biocontrol or resistance to the spread of disease), if interactions among species are highly specialized (Stephan et al. 2000, cf. gene-for-gene interactions in host–parasite systems). Indeed, about half of the experts in the survey of Schläpfer et al. (1999) predicted an exponential or logistic shape for relationships between biodiversity and bioregulation.

In the last part of our analyses, we ask, if a function asymptotes at high diversity, how many species are required for a 50 per cent of the maximum function. If the biodiversity–function relationship is log-linear, we ask how much a 50 per cent or 75 per cent reduction in species richness changes the function.

2.1.5 Methods of analysis

Our new analyses of the influence of explanatory terms on both the variation and shape of biodiversity effects were based on the data descriptions and methods presented in B and C. For B, we added data on the significance, direction, and shape of biodiversity effects to the correlation coefficients used in the original analysis. We distinguished the following shapes of biodiversity effects in B: negative, negative linear, negative log-linear, no relationship, positive, positive linear, positive log-linear, positive but not linear, and none of these conditions. For C, we calculated correlation coefficients after fitting log-linear relationships, excluding studies with only two species richness levels (where correlation coefficients can only be 1 or –1). Furthermore, we used log-linear fits because these were often also used in the original papers. To assess the shape of biodiversity effects in C, we fitted linear, log-linear, and saturating (Michaelis–Menten) curves.

In B, we used three different measures of the relationship between biodiversity and response to analyze differences in biodiversity effects: (1) correlation coefficients (r) standardized to Zr values, (2) significances (0 for relationships with $P \geq 0.05$, 1 for relationships with $P < 0.05$), and (3) signs (–/+, only significant relationships). In C, we used only correlation coefficients standardized to Zr values in the analysis. Correlation coefficients were converted into Zr values to improve normality (correlation coefficients are bound between –1 and 1 and thus not normally distributed). The formula for the conversion is (Rosenberg et al. 2000):

$$Zr = 0.5 \ln\left((1+r)/(1-r)\right).$$

The number of plots, N, used for the determination of each single biodiversity effect in the original publications, corrected by the degree of freedom, was used as weighting variable in B (note that n, as opposed to N, will be used later to refer to the number of effects rather than the number of plots used to calculate a single effect). Because the correlation coefficients in C were calculated from the means at each level of species richness, Zr values were weighted by the number of species richness levels used in fitting the relationship. In both B and C, analyses with unweighted Zr values yielded similar results and are therefore not presented.

We used linear mixed-model analyses to test the influence of explanatory terms on the Zr values. Study site and publication were used as random terms. Latitude and longitude were tested against site as an error term. Explanatory terms which varied within sites (but not within publications) were tested against publication as an error term (as in B and C). To avoid problems of confounding and correlated responses, all explanatory terms were fitted both individually and in a combined analysis. Only if a fixed term was significant in both cases (comparing the likelihood of a model with and without the term) was it retained for further analysis. With these stringent rules, we tried to ensure that hypothesis tests were robust across an entire data set and not due to influences of correlated variation in other factors. Interactions between explanatory terms were also tested, but were seldom retained in the model under the stringent rules mentioned above.

Logistic mixed models were used to analyze differences in significances (probability of observing significant biodiversity effects) and signs (probability of observing a significant positive biodiversity effect among the significant effects) in relationships between biodiversity and response in B. Significance corresponds to the finding that a standardized correlation coefficient is significantly different from zero. Even if information about the correlation coefficient was not available, the direction of the effect could be extracted from the original publications, and therefore positive versus negative significance could be distinguished. To avoid overrating studies with small sample sizes in the logistic models, the number of experimental units per study divided by the mean number of experimental units across all studies was used as a weighting variable for each data point. To test the five hypotheses, we used ratios of mean deviances as approximate F-values (McCullagh and Nelder 1989). This allowed us to use publication as the appropriate error term for corresponding fixed terms that did not vary within publications.

All presented means of Zr values, percentages of significant biodiversity effects or percentages of positive directions among significant biodiversity effects, are weighted means using the weighting variables mentioned above. Values and significance levels that are not presented in figures or tables are given in the text.

2.2 Hypotheses to explain variation in biodiversity effects

Before the detailed presentation of the results, an overview of the analyses, including all the explanatory terms discussed below, is provided (Table 2.1). This table first lists the fixed terms in the different models (generally in descending order of F-values) followed by the random terms. In the analysis of meta-data set C, which contained more homogeneous data and thus fewer candidate explanatory terms than the larger meta-data set B, only one fixed term was retained in the model. In B, terms for finer categories of responses were fitted to reduce the amount of unexplained variance (residual).

2.2.1 Distribution of studies

The reported biodiversity effects came from more than 100 independent experiments, mainly carried out in North America and Europe (Fig. 2.1).

Table 2.1 Multivariate mixed-model analyses for (a) data in B (cf. Balvanera et al. 2006) and (b) data in C (cf. Cardinale et al. 2006a).

Source of variation	Degree of freedom	% var. explained	F-ratio	P
(a)				
Proportion of significant effects				
Responses of communities ≠ ecosystems ≠ populations (see 2.2.3)	2	5.4	31.5	< 0.001
Responses vary among ecosystem types (see 2.2.2)	7	8.5	6.4	< 0.001
Responses of residents > invaders (see 2.2.6)	1	0.5	6.0	0.015
Responses vary among response groups	27	4.7	2.0	0.002
Responses vary among study sites (random term)	92	22.4	1.3	0.189
Responses vary among publications (random term)	41	7.4	2.2	< 0.001
Residual	595	51.1		
Proportion of positive within significant effects				
Responses of residents > invaders (see 2.2.6)	1	11.0	145.7	< 0.001
Responses of stocks ≠ rates (see 2.2.4)	1	5.3	70.6	< 0.001
Responses of communities ≠ ecosystems ≠ populations (see 2.2.3)	2	5.9	39.1	< 0.001
Top-down and bottom-up responses ≠ others (see 2.2.5)	5	4.7	12.3	< 0.001
Responses vary among ecosystem types (see 2.2.2)	7	12.1	5.3	< 0.001
Responses vary among response groups	24	8.5	4.7	< 0.001
Responses vary among study sites (random term)	81	17.7	0.7	0.936
Responses vary among publications (random term)	40	12.9	4.3	< 0.001
Residual	291	21.9		
Zr-values				
Responses of communities ≠ ecosystems ≠ populations (see 2.2.3)	2	18.6	88.9	< 0.001
Responses of residents > invaders (see 2.2.6)	1	4.2	40.4	< 0.001
Responses decrease with cos(latitude)	1	2.5	11.6	0.001
Responses of stocks ≠ rates (see 2.2.4)	2	3.3	15.8	< 0.001
Responses vary among ecosystem types (see 2.2.2)	7	12.9	6.6	< 0.001
Top-down and bottom-up responses ≠ others (see 2.2.5)	5	1.2	2.3	0.042
Responses vary among response groups	24	4.1	1.7	0.031
Responses vary among study sites (random term)	63	13.3	0.8	0.834
Responses vary among publications (random term)	30	8.4	2.7	< 0.001
Residual	302	31.5		
(b)				
Zr-values				
Responses of stocks ≠ resource depletion (see 2.2.4)	1	4.2	9.5	0.003
Responses vary among publications (random term)	43	67.8	3.6	< 0.001
Residual	63	28.0		

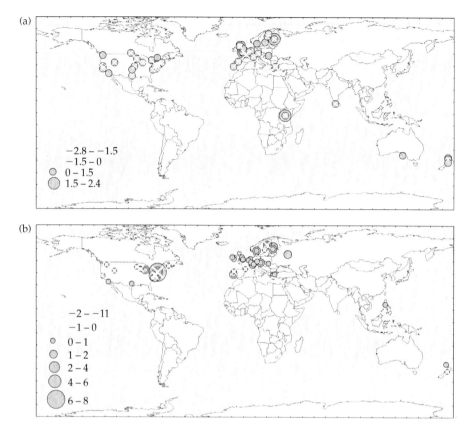

Figure 2.1 Geographical distribution of studies in B from which standardized correlation coefficients (Zr) between biodiversity and response variables could be extracted (see Balvanera et al. 2006). The size of the symbols is proportional to Zr. ● symbols indicate positive, × symbols negative biodiversity effects.
(b) Geographical distribution of studies in C from which log-ratios of responses at highest species richness and responses of average monocultures could be extracted (measure 'LLR1' in Cardinale et al. 2006). The size of the symbols is proportional to LLR1. ● symbols indicate positive, × symbols negative biodiversity effects.

Interestingly, Zr values declined significantly with increasing cosine of latitude in the larger data set B (Table 2.1(a)). However, this trend was largely explained by higher values in Europe ($n = 187$, mean $Zr = 0.26$) relative to those from North America ($n = 234$, mean $Zr = -0.6$), with the latter representing lower latitudes (higher cosines). Biodiversity effects did not vary across altitudes of study sites, but most of these were located < 500 m above sea level. The geographical distribution of studies indicates a strong bias towards locations where the major funding is, rather than where the science is most warranted (see Chapter 3). Experiments undertaken in the tropics (e.g. Potvin and Gotelli 2008) and at high latitudes are urgently needed to further test biodiversity effects under more contrasting conditions. To what extent the difference in Zr values between European and North American studies may be confounded by researcher preferences remains unknown.

A larger number of biodiversity effects are documented for terrestrial studies than are documented for aquatic studies (Table 2.2(a)). Of the terrestrial fraction, a large number of the biodiversity effects reported are from studies carried out in grassland ecosystems (60 per cent in B and 45 per cent in C). It is therefore not surprising that a large number of biodiversity effects are reported from studies manipulating plant diversity, with detritivores plus mycorrhizae coming second, herbivores third, and carnivores fourth (Table 2.2(a)). In B, 154 of all biodiversity effects could be classified as an ecosystem

Table 2.2 (a) Number of studies in which different trophic groups have been manipulated in aquatic and terrestrial ecosystems and (b) number of studies in which different trophic groups have been manipulated and stocks or rates/depletion of resources have been measured.

	Consumers	Detritivores or mycorrhizae	Herbivores	Plants	Multitrophic manipulations	Total
(a)						
Database B (cf. Balvanera et al. 2006)						
Aquatic	3	12	41	54	40	150
Terrestrial	4	87	5	510	15	621
Total	7	99	46	564	55	771
Database C (cf. Cardinale et al. 2006)						
Aquatic	17	22	21	13		73
Terrestrial	8	12	3	51		74
Total	25	34	24	64		147
(b)						
Database B (cf. Balvanera et al. 2006)						
Stocks	4	67	32	415	33	551
Rates	3	32	14	147	22	218
Unclassified	0	0	0	2	0	2
Total	7	99	46	564	55	771
Database C (cf. Cardinale et al. 2006)						
Stocks	19	28	10	14		76
Depletion of resources	6	6	14	50		71
Total	25	34	24	64		147

response, 434 as a community response and 183 as a population response. Ecosystem service groups covered in B included 251 biodiversity effects on primary production (including abundance and standing biomass), 254 on bioregulation, 195 on soil fertility, 46 on nutrient and water cycling, and 10 on climate regulation.

The majority of biodiversity effects concerned responses of standing stock, but a considerable number was also related to rates (in B) or resource depletion (in C; Table 2.2(b)). Resource depletion in C was measured as (1) instantaneous rate ($n = 5$) or as a reduction in resource compared with (2) control without species ($n = 44$, or $n = 28$ for experiments that lasted less than one generation time) or (3) the beginning of the experiment ($n = 21$, or $n = 13$ for experiments that lasted less than one generation time); one value was not classified. In B, relationships between the trophic level at which species richness was manipulated and the one at which a response was measured (above, below, ecosystem, same, symbiont, within [multitrophic diversity manipulations]) were distributed more or less regularly across both stocks ($n = 99, 11, 66, 321, 40, 14$, respectively) and rates ($n = 86, 6, 30, 83, 5, 8$, respectively). However, there were only a few relations that could be classified as top-down biodiversity effects (category 'below', $n = 17$). Among the bottom-up biodiversity effects (category 'above', $n = 185$), about a third were cases where detrital diversity of primary producers was manipulated and decomposer functions were measured ($n = 65$).

2.2.2 Biodiversity effects vary among ecosystem types (hypothesis 1, Table 2.3)

In both B and C biodiversity effects came mainly from four broadly defined ecosystem types: grassland, fresh-water, marine, and forest. Note that the breadth of definition is narrower for those ecosystem types in which a larger number of studies have been carried out (grassland) than in those with fewer studies (aquatic marine). The remaining biodiversity effects were represented by approximately the same number of other ecosystem types (bacterial microcosm, crop/

Table 2.3 Tests of hypotheses about variation in biodiversity effects (see Sections 2.1.3 and 2.1.5). 'Cardinale' and 'Balvanera' refer to data in C (cf. Cardinale et al. 2006a) and in B (cf. Balvanera et al. 2006), respectively.

Number	Hypothesis	Cardinale Zr ($n = 108$)	Balvanera Zr ($n \leq 449$)	Balvanera P (sign.)($n \leq 766$)	Balvanera P (pos. sign.)($n \leq 766$)
1	Biodiversity effects vary among ecosystem types	no ($P > 0.1$)	yes ($P < 0.001$)	yes ($P < 0.001$)	yes ($P < 0.001$)
2	Biodiversity effects differ between ecosystem, community and population level	–	yes ($P < 0.001$)	yes ($P < 0.001$)	yes ($P < 0.001$)
3	Biodiversity effects differ between stocks and rates or depletion of resources	yes ($P = 0.003$)	yes ($P < 0.001$)	yes ($P < 0.001$)	yes ($P < 0.001$)
4	Biodiversity effects depend on trophic relationships	– (confounded with above)	yes ($P < 0.001$)	yes ($P < 0.001$)	yes ($P = 0.035$)
5	Biodiversity affects residents and invaders differently	–	yes ($P < 0.001$)	yes ($P = 0.004$)	yes ($P < 0.001$)

successional, ruderal/salt marsh, soil community). Zr values varied significantly between ecosystem types in B, but not in C (Table 2.3). When tested, the significance remained when ecosystem service groups (refer to previous paragraph) or finer categories of responses (see Table 2.1(a)) were fit before ecosystem type in the analyses. However, because the significant variation in B was, at least partly, due to stronger biodiversity effects in the ecosystem types with lower values of n (see B), it is too early to draw any general conclusions. More importantly, there was no overall tendency in any of the analyses for biodiversity effects to be more (or less) frequently positive (or more or less often significant) in terrestrial systems than it was for aquatic ecosystems. This supports the view that similar mechanistic processes underpin the biodiversity–ecosystem functioning relationship under terrestrial and aquatic conditions.

2.2.3 Biodiversity effects differ among ecosystem, community, and population levels (hypothesis 2, Table 2.3)

Our results strongly suggest that while increasing species richness often enhances the performance of entire communities, it also often reduces the average contributions of individual species. Biodiversity effects on ecosystem-level (abiotic) responses also tend to be positive, but not as much and not as often as the biodiversity effects on community-level responses (Fig. 2.2(a)). This suggests a more direct mechanistic link in the latter case. In the meta-analysis of B, the difference among ecosystem-, community-, and population-level responses was identified as the strongest explanatory factor for variation in biodiversity effects. In the new analysis presented here, this is true for both Zr values and significances (Table 2.3, Fig. 2.2(a)). The result remained highly significant ($F_{2,209} = 82.5$, $P < 0.001$; mixed model with site and publication as random terms) even if only significant Zr values were analyzed ($n = 307$).

2.2.4 Biodiversity effects differ between stocks and rates (or depletion of resources) (hypothesis 3, Table 2.3)

In B, biodiversity effects on stocks and rates were distinguished: stocks referred to levels of an ecosystem property (e.g. standing biomass) whilst rates referred to changes in such levels over time. In C, biodiversity effects on stocks and depletion of resources were distinguished: stocks referred to levels of an ecosystem property only at a focal trophic group (see Table 2.2(b)) whilst depletion referred to direct rates of resource depletion, or to differences between the consumed and unconsumed resource levels at a trophic group below the focal one (see Section 2.2.1). In the following test of our third hypothesis, we treat resource depletion as equivalent to rates.

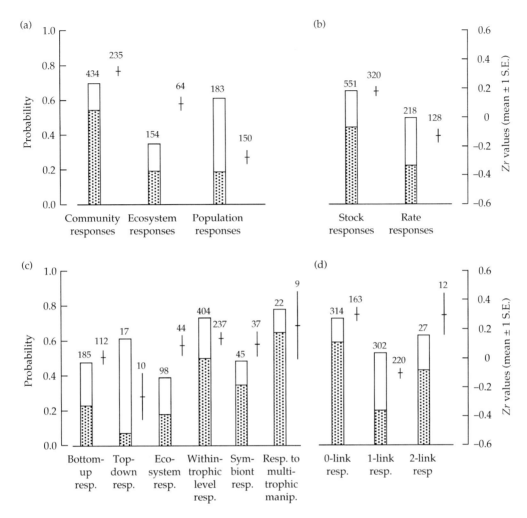

Figure 2.2 Probability (left y-axis) of observing significant responses to biodiversity (height of shaded plus unshaded column) and proportion of positive responses among them (height of shaded column relative to height of shaded plus unshaded column). If the shaded bar is shorter than the unshaded bar, then significant biodiversity effects have a greater probability to be negative than positive. The average Zr value ($\pm$ 1 standard error) is also shown for each category (axis on the right). The number of measured responses in each group is listed. Response categories for test of (a) hypothesis 2 (Section 2.2.3), (b) hypothesis 3 (Section 2.2.4), (c) hypothesis 4 (Section 2.2.5) and (d) a corollary to hypothesis 4. Data from meta-data set B (Balvanera et al. 2006).

In B, we found more significant and more positive effects of species richness on stocks than we did on rates (Table 2.3, Fig. 2.2(b)). Overall, 55 per cent of the significant biodiversity effects on rates were negative, as were the average Zr values for those rates. For example, resource extraction from primary producers (disease severity, consumed biomass) usually declined with increasing plant species richness. These results were robust to changes in fitting sequence of other hypotheses.

A similar result was found in C, where Zr values for stocks (mean $Zr = 1.53$) were significantly ($F_{1,63} = 9.52$, $P < 0.01$; mixed model with publication as random term) larger than Zr values for depletion of resources (mean $Zr = 1.01$). If those depletion measures which were taken after a time interval of a generation or more were excluded (see Section 2.2.1), then the difference between Zr values for stocks and depletion of resources (mean $Zr = 0.82$) became even larger ($F_{1,49} = 13.48$, $P < 0.001$; mixed model with publication as random term). The Zr values were higher in C than in B. This was probably due to two differences between the analyses: in B, Zr values were averaged over a more heterogeneous set of responses, whilst in C, Zr values were calculated from means at each species richness level, thereby excluding variation around means. When compared to stocks, the significantly lower Zr values for depletion of resources in C may in part be due to a difference in trophic distance: stocks in C were measured at the trophic level where diversity was manipulated and there was depletion of resources at the trophic level below (see hypothesis 4).

2.2.5 Biodiversity effects depend on trophic relationships (hypothesis 4, Table 2.3)

Our findings support the predictions that top-down biodiversity effects should be negative (i.e. increased diversity at one trophic level reduces the function performed by the trophic level below) and biodiversity effects within trophic levels, and for symbiont relationships, should be positive (Table 2.3; Fig. 2.2(c)). Bottom-up biodiversity effects (i.e. increased diversity at one trophic level changes the function performed by the trophic level above) were on average close to zero, although we had expected them to be negative. This was due to the fact that bottom-up biodiversity effects on functions measured for detritivores were generally positive (28 out of 36 significant effects), whereas other bottom-up biodiversity effects were more often negative (37 out of 58 significant effects) than positive. Analysis of a larger dataset found equal numbers of positive and negative bottom-up biodiversity effects of detrital diversity on functions measured for detritivores (Srivastava et al. 2009).

When we analyzed differences among responses at increasing trophic distance ($t \pm 0 < t \pm 1 < t \pm 2$) to the level whose biodiversity was manipulated, we found that effects within trophic levels and at a trophic distance of two were positive (higher function values at higher biodiversity levels), in contrast to those at a distance of one (Fig. 2.2(d); $F_{2,504} = 13.4/P < 0.001$ for differences in frequency of significant effects, $F_{2,271} = 63.7/P < 0.001$ for differences in frequency of positive effects, and $F_{2,293} = 44.9/P < 0.001$ for differences in Zr values). This result is consistent with our hypothesis that two negative effects between adjacent trophic groups should, on balance, multiply to a positive effect between groups at a trophic distance of two.

2.2.6 Biodiversity affects residents and invaders differently (hypothesis 5, Table 2.3)

In B, 93 of the 771 biodiversity effects concerned the response of invaders to the species richness of the residents in a community. Of these, 76 per cent were significant biodiversity effects. The vast majority (89 per cent) of the significant biodiversity effects were negative, as were the average Zr values. That is, invader functions were reduced at higher biodiversity of residents, corresponding to an increased invasion resistance of more diverse resident communities. This contrasts with 59 per cent significant biodiversity effects on responses of residents, of which only 31 per cent were negative (Fig. 2d). These results demonstrate that one of the most general effects of high biodiversity is increased invasion resistance (Knops et al. 1999, Hector et al. 2001, van Ruijven et al. 2003, Fargione and Tilman 2005, Spehn et al. 2005, Mwangi et al. 2007), thereby corroborating our fifth hypothesis.

2.3 Hypotheses about the shape of the relationship between biodiversity and responses (Table 2.4)

The first five hypotheses in Table 2.4 specify alternative shapes of biodiversity effects found in the previous meta-analyses and expected by experts (Schläpfer et al. 1999).

2.3.1 Alternative shapes of biodiversity effects

Four out of 23 experts predicted that the general shape of biodiversity effects would be a horizontal line in Schläpfer et al. (1999). Out of the 771 biodiversity effects assembled in B, 286 (37 per cent) could not be assigned a specific shape and were therefore considered as horizontal. Although none of the experts predicted the linear relationship as a general

Table 2.4 Tests of hypotheses and questions about the shape of biodiversity effects. 'Schläpfer', 'Cardinale', and 'Balvanera' refer to data in Schläpfer et al. (1999), in C (cf. Cardinale et al. 2006a), and in B (cf. Balvanera et al. 2006), respectively.

Number	Hypothesis or question	Schläpfer experts ($n < 38$)	Cardinale ($n \leq 108$)	Balvanera ($n \leq 771$)
1	Biodiversity effects have no discernible shape	4 of 23 experts	–	286 of 771 cases
2	Biodiversity effects are linear	0 of 23 experts	34 of 108 cases	164 of 771 cases
3	Biodiversity effects have a discernible shape but are not linear	18 of 23 experts	74 of 108 cases	321 of 771 cases
4	Biodiversity effects are log-linear	15 of 22 experts	25 of 108 cases	101 of 771 cases
5	Biodiversity effects saturate		49 of 108 cases	–
6	Shape of biodiversity effects on primary production (incl. biomass, abundance)	17 log-lin./sat., 8 no rel., 3 lin., 3 logist., 1 optimum	17 saturating, 15 log-linear, 11 linear	90 no/idios., 81 other, 46 log-linear, 34 linear
7	Shape of biodiversity effects on nutrient and water cycling	16 log-lin./sat., 8 no rel., 4 logist., 2 lin., 1 optimum	–	18 other, 15 no/idios., 11 linear, 2 log-linear
8	Shape of biodiversity effects on responses related to bioregulation	15 logist./expon., 10 log-lin./sat., 4 lin.	–	83 linear, 72 no/idios., 60 other, 39 log-linear
9	Number of species required for 50% response (positive, saturating relationship)	5–6	1.2	–
10	Reduction in response with reduction from 16 → 8 species (positive, log-linear relationship)	ca. 30%	10.90%	–

shape for biodiversity effects, the linear function actually did fit best in 34 out of 108 cases (32 per cent) in C; and it was observed in 164 out of 771 cases (21 per cent) in B (Table 2.4). Where the highest species richness was < 10, linear relationships were observed in 27 per cent and 40 per cent of the studies in B and C, respectively. Among studies in which the highest species richness was ≥ 10, 19 per cent in B and 28 per cent in C were linear.

The majority of biodiversity effects that were assigned a shape, however, were not linear, reflecting the predictions of the majority (78 per cent) of experts and consistent with theoretical expectations (Section 2.1.4; see also Chapter 8). Experts and authors of the publications used in B did not distinguish between non-linear curves that do (Michaelis–Menten) or do not saturate (log-linear). In C, however, this distinction could be made and showed that the average R^2 value was 0.690 and 0.682 for the saturating Michaelis–Menten and log-linear relationship, respectively. Furthermore, if the linear relationship was included in the comparison, the log-linear was the best fitting in only 25 studies, whereas the saturating curve was the best fitting in 49 studies (Table 2.4). However, the log-linear was the worst-fitting relationship in only seven cases, compared with 66 for the linear and 35 for the saturating relationship. Thus the log-linear relationship has an intermediate position: it crudely fits a large number of biodiversity effects. This may reflect a mixture of operating mechanisms, including complementarity and selection effects (Schmid et al. 2002b). Nevertheless, most theoretical models (see e.g. Tilman et al. 1997c, Loreau 1998a, Cardinale et al. 2004) show that biodiversity effects should saturate, at least at high levels of species richness, which are seldom ascertained in experimental studies. It should be noted that R^2 values and vote counting are very crude measures for distinguishing between functions of different shape.

2.3.2 Shapes of biodiversity effects differ between major response categories

Whilst the previous section focused on the general shape of all analyzed biodiversity effects, the following section will investigate differences in the shapes of biodiversity effects between studies. In particular, we want to test the hypothesis that biodiversity effects on primary production and nutrient and water cycling are log-linear (or saturating), whereas those on bioregulation are more often linear or logistic. The data in B and C provide some support for these hypotheses (6–8 in Table 2.4). A large number of observed effects of plant diversity on primary production, or responses related to it, including all types of abundance measures, were log-linear or saturating, whereas the few examples for responses related to nutrient and water cycling did not reveal any clear pattern (Table 2.4). Responses related to bioregulation in B showed the lowest proportion of log-linear relationships and the highest proportion of linear ones. This is broadly consistent with the hypothesis and the expectation of the experts (Schläpfer et al. 1999), who predicted the smallest amount of log-linear biodiversity effects or, in other words, redundancy for these responses. Despite the large body of literature on biocontrol there is, to our knowledge, no general theory about how the diversity of hosts should be related to diversity of enemies. This is the case even though empirical work on quantitative interaction webs across varying diversity levels has been done (e.g. Albrecht et al. 2007) and the importance of distinguishing between interactions with generalists versus specialists has been demonstrated in models of multitrophic diversity manipulations (Petchey et al. 2004a, Thébault and Loreau 2005).

2.3.3 Consequences of observed shapes of biodiversity effects

Under the assumption of a positive, saturating relationship, the experts in Schläpfer et al. (1999) greatly overestimated the number of species required to reach 50 per cent of the maximum response. The average estimate was between 5–6 species (Schläpfer et al. 1999), whereas analysis of the data in C suggested that an average of only 1.2 species are needed (Table 2.4). This result suggests that the presence of a single species results in almost half of the response. However, one limitation of fitting Michaelis–Menten curves is that they assume a zero response for the species richness level of zero, which is not always appropriate, as for example in the case of evapotranspiration of an ecosystem. We think a more interesting question is how many species does one need relative to a one-species monoculture to obtain some percentage of maximal function.

When experts were asked to predict the consequences of a 50 per cent species loss from 16 to 8 species, they overestimated the reduction in the response by a factor of three under the assumption of a positive, log-linear relationship (Table 2.4). The empirical log-linear results from C suggest that, with each halving of species richness, the response would be reduced by about 11 per cent of the 16-species richness level. Reduction in the number of species from 16 to 1 species would involve four halving events, corresponding to a reduction by about 44 per cent. Similarly, if we assume a saturating Michaelis–Menten relationship the average reduction in C from 16 to 8 species would be 5.2 per cent and from 16 to 1 species would be 38.8 per cent. Comparing the expert predictions for the reduction from 16 to 8 (30 per cent) and from 16 to 4 (40 per cent; see Schläpfer et al. 1999) suggests that the experts assumed a linear relationship between species richness and response, even though they selected a weaker relationship more often (see Section 2.3.1).

2.4 What have we learned from biodiversity manipulation experiments?

The joint interpretation of results from of the new analyses of the two meta-data sets of B and C demonstrates that, despite the large heterogeneity of data, biodiversity effects are a general feature of most biological systems. Recent discussions have focused on the details of experimental design and analyses, as well as the mechanisms underpinning biodiversity effects (see e.g. Cardinale et al. 2007). It is gratifying to see that these issues could not mask the influence and importance of major biological factors in explaining the variation in biodiversity effects.

2.4.1 Hypotheses to explain variation in biodiversity effects

The major biological factors used in the analyses presented in Table 2.1 involve multilevel factors such as different ecosystem types or types of response variables (i.e. response groups), where we could only state the existence of significant variation, and factors, with few well-interpretable levels and contrasts between these levels. We tested five hypotheses regarding the influence of these biological factors on the strength and direction of biodiversity effects (see Table 2.3). The first hypothesis that biodiversity effects vary between ecosystem types and therefore restrict the potential for generalizations from one ecosystem to another, was confirmed (Section 2.2.2). However, the surprising (cf. Giller *et al.* 2004) similarity of responses between terrestrial and aquatic ecosystems (and among the ecosystem types studied most often) shows that there are very likely to be common processes and patterns operating among different ecosystems.

The second hypothesis, that an increasing diversity of species positively affects responses at the community (and ecosystem) level and negatively affects responses at the population level, was also confirmed (Section 2.2.3). This supports predictions from basic Lotka–Volterra theory and arguments about density compensation (McGrady-Steed and Morin 2000). If total resource or energy inputs from the environment fix the response at community or ecosystem level (see e.g. Bunker *et al.* 2005), and if these inputs are distributed among several species, the average response of species at population level must go down as diversity goes up. The evidence for this effect in the current analysis was very strong and robust across the large range of biodiversity effects in the meta-dataset of B. It would be interesting to explore whether a theoretical relationship can be found between positive/negative effects of biodiversity on community/population-level responses, as found here, and the better-known positive/ negative effects of biodiversity on temporal variation in community/population-level responses (May 1974, Tilman 1996, Flynn *et al.* 2008). The comparatively weak influence of biodiversity on ecosystem-level responses may reflect an indirect relationship between biotic components, whose biodiversity was manipulated, and abiotic components of which functional responses were measured.

Our third hypothesis was that biodiversity effects on stocks might differ from biodiversity effects on rates, but we could not predict the direction of the difference (Section 2.2.4). Nevertheless, we clearly showed that there were differences between stocks and rates and that, in fact, biodiversity influenced stocks more strongly and more positively than rates (or depletions of resources). This result was consistent in both analyses of the B and C meta-data sets, despite some differences between the two. In C, stocks were measured directly in the diversity-manipulated group, whereas depletions of resources were measured at the trophic level below. In B, stocks and rates were measured at the same or at different trophic levels above or below the one manipulated. Our results are still difficult to understand, but we can at least conclude that the assumption that rates or depletion of resources should be more responsive to biodiversity than stocks is wrong. This becomes relevant, for example, in the context of rates and stocks in ecosystem carbon cycling (Körner 2003, Körner 2004). According to our results, there is no longer a reason to believe that high biodiversity will simply increase turnover rates rather than storage.

The fourth hypothesis predicted that increased biodiversity at one trophic level reduces functions at other trophic levels (negative bottom-up and top-down biodiversity effects), whereas it increases functions at the same trophic level or for symbionts (Section 2.2.5). Indeed, these predictions were met, with the exception that bottom-up biodiversity effects (mainly detrital diversity of primary producers) on detritivore functions were mostly positive. In a new meta-analysis using a larger number of such studies, Srivastava *et al.* (2009) found equal numbers of positive and negative effects of detrital diversity on detritivore functions. Although other bottom-up biodiversity effects and top-down biodiversity effects were mostly negative in the present analysis, this was not the case when two trophic levels separated the manipulated and the measured groups. This indicates that two negative biodiversity effects between adjacent trophic levels can multiply to a positive effect between more distant levels. It is gratifying to see that even for multi-trophic biodiversity studies predictions made on theoretical grounds (e.g. Loreau 2001, Fox 2004a, Fox 2004b, Fox 2005a, Petchey et al. 2004a, Keesing et al. 2006, Duffy et al. 2007) are broadly supported by data, although it is still too early to derive further generalizations.

Our fifth hypothesis, that increasing biodiversity should affect the responses of residents positively and the responses of invaders negatively, was strongly supported by the data in B and was highly robust across the span of measurements and ecosystems (Section 2.2.6). Nevertheless, there has been some debate whether this is a general trend or a specific feature of experiments (Levine and D'Antonio 1999, Fridley et al. 2007). The problem here is that in non-experimental situations it is hard to distinguish invaders from residents unless the invasion process is directly observed; also, conditions that favour diversity in general cannot be dissected from those that promote invaders in particular (Espinosa-García et al. 2004). Nevertheless, niche theory predicts a lowered availability of free niche space with increased species richness (Fargione et al. 2003, Harpole and Tilman 2007, Mwangi et al. 2007), as well as the results from the experiments presented here, both of which are consistent with positive biodiversity effects on invasion resistance.

2.4.2 Alternative shapes of biodiversity effects

Although a large number of biodiversity effects have the shape of a log-linear or a saturating curve, these shapes are by no means the only ones (e.g. Chapter 1) – especially if responses related to bioregulation are considered (both beneficial and detrimental interactions between species whose diversity was manipulated and those species whose responses were measured). For the latter, biodiversity effects often did not diminish or saturate over the range of species richness levels tested. In contrast, responses related to primary production and nutrient or water cycling did show evidence of deceleration or reaching saturation (Section 2.3.2). This difference between biodiversity effects on bioregulation and biodiversity effects on water or element cycling was expected on theoretical grounds as well as being predicted by experts (Schläpfer et al. 1999). Saturating relationships for resource uptake and conversion are consistent with increasing overlap of resource niches that are expected with increasing diversity (Tilman 1997, Loreau 1998a). Such limitations may not affect relationships between biodiversity and bioregulation. However, it should be noted that studies of bioregulation tend to manipulate just a few species across minimal levels of species richness and this may provide an alternative explanation for the differences.

With the detailed metadata contained in C, it is possible to calculate how severe reductions of species richness might be in comparison to estimates provided by the experts 10 years ago. When doing so, we were surprised to find that experts assumed that a much larger number of species (5–6) would be needed to maintain responses at half-saturation level than the empirical investigations estimate. The empirical estimates suggest that the average monoculture should already reach the half-saturation level. Similarly, in comparison to empirical findings, experts overestimated the consequences of halving species richness by a factor of three. This suggests that experts often do not think about the difference between systems with no species and a system with a single species, perhaps because experts do not consider it meaningful to measure ecosystem properties at a species richness level of zero.

At this juncture, one fundamentally important caveat should be considered. As Hector and Bagchi (2007) have shown, it is likely that more than one species will be needed to maintain multiple responses at half-level. Thus it could well turn out that as the number of responses considered are increased, the number of species needed to maintain multi-response half-levels is also likely to increase to (or above) an expert-estimated saturation of around 5–6 species.

2.4.3 Recommendations for the next-generation biodiversity experiments

It was only possible to review biodiversity effects reported until summer 2005. In the meantime, the number of studies has increased further and new meta-analyses could be started. We hope that some of the new and future studies will look

Table 2.5 Some variables that should be included in publications and meta-data bases of biodiversity experiments.

Number	Variable
1	Reference (author and date)
2	Experiment/study identification
3	Locality (logitude, latitude)
4	Level of control (enclosed, field)
5	Ecosystem type
6	Cause of diversity gradient
7	The species diversity measure used
8	Type of experiment (substitutive vs. additive)
9	Trophic group for diversity gradient
10	Lowest species richness
11	Highest species richness
12	Number of species richness levels
13	Total number of species in pool
14	Total number of different species compositions
15	Total number of experimental units (M)
16	Response measured
17	Trophic level of response
18	Mean response
19	Mean response at lowest richness
20	Standard error of mean response at lowest richness
21	Mean response at highest richness
22	Standard error of mean response at highest richness
23	Significance level
24	Direction of effect
25	Correlation coefficient
26	Type of correlation coefficient (univariate or multivariate)
27	Shape of functional response to biodiversity
...	
...	Further variables indicating additional experimental treatments
...	etc.

more specifically at mechanisms generating biodiversity effects. If so, future meta-analyses can go beyond the testing of rather phenomenological hypotheses and begin to understand mechanistic processes.

We suggest that the old and new data should be combined in an open-access data table that would allow continuous monitoring of overall trends and further analysis. Comparing the variables used in the two existing databases showed that a similar reduced set was independently derived by the two groups of authors from a multitude of candidate variables (Table 2.5). Values for this set of variables should be reported, if possible, by every new study on biodiversity–ecosystem functioning relationships. The latter will require a change in ethos and a willingness to share data both nationally and internationally (a trend which is increasing across many disciplines, such as molecular biology), but would significantly bolster crosscutting analyses aimed at identifying the generalities of biodiversity effects.

Acknowledgements

We thank Michel Loreau, Peter Morin, Shahid Naeem, and an anonymous reviewer for very useful comments on the manuscript of this chapter. We thank the Swiss Agency for the Environment, Forests and Landscape (SAEFL) for financial support.

CHAPTER 3

Biodiversity-ecosystem function research and biodiversity futures: early bird catches the worm or a day late and a dollar short?

Martin Solan, Jasmin A. Godbold, Amy Symstad, Dan F. B. Flynn, and Daniel E. Bunker

3.1 Introduction

In the early 1990s, an increasing number of ecologists began to challenge the view that biodiversity was merely an expression of the abiotic conditions of the environment and, instead, started to recognize that the properties of ecosystems are also mediated by the biota (e.g. Lubchenco et al. 1991, Chapin et al. 1992). Consequently, research foci shifted from elucidating the effects of abiotic conditions on biodiversity to investigating biodiversity's effects on ecosystem processes (Naeem et al. 2002). The development of this central tenet was formalized at a conference held in Bayreuth, Germany in 1991 (Schulze and Mooney 1993) and led to an extended range of hypotheses (Schläpfer and Schmid 1999) that collectively formed a framework within which the relationship between biodiversity and ecosystem function (hereafter, BEF) could be experimentally tested (Schmid et al. 2002a, Raffaelli et al. 2003). The emerging BEF ideology received widespread attention and has become one of the central research agendas in contemporary ecology (Loreau et al. 2001, Loreau et al. 2002), spurred by the anticipation that anthropogenic global change will dramatically reduce biodiversity in most ecosystems and will have considerable ecological consequences and affect human wellbeing within the next century (Vitousek et al. 1997, Sala et al. 2000, Diaz et al. 2006, Fischlin et al. 2007).

Although initial experiments successfully articulated BEF hypotheses by manipulating biodiversity under controlled laboratory conditions (for reviews, see Loreau et al. 2001, Covich et al. 2004, Hooper et al. 2005), critics were quick to assert that laboratory studies lack realism because they tend to include a few 'non-representative' taxa, often from only one trophic level, and attributes of the system are measured infrequently and in the absence of the appropriate environmental context. Strong debate followed (for summary, see Mooney 2002) over the applicability of such data to the real world and its capacity to inform policy-relevant issues (Carpenter 1996, Srivastava and Vellend 2005). Further, discussion surrounding fundamental differences in opinion over experimental methodology (e.g. Huston 1997, Wardle 1998, Wardle 1999, Doak et al. 1998) and the emergence of empirical studies that were not consistent with theoretical predictions (e.g. Wardle et al. 1997a, Gastine et al. 2003) gave the impression to the wider ecological community that opinion was divided and advice contradictory (Kaiser 2000, Cameron 2002, Schmid 2002); arguably what was actually being witnessed was the ordinary, albeit rapid, evolution and maturation of a new paradigm (Naeem 2002b). Unlike other ecological debates, however, the BEF community became disproportionately distracted by a minority of detractors and went to extraordinary lengths to defend its position (e.g. Naeem et al. 1999,

Srivastava *et al.* 2004, Bulling *et al.* 2006, Benton *et al.* 2007), attract and encourage other ecologists (Austin 1999, Emmerson and Huxham 2002, Raffaelli *et al.* 2003, Prosser *et al.* 2007), and demonstrate reconciliation between opposing camps, most prominently at a conference entitled 'Biodiversity and Ecosystem Functioning: Synthesis and Perspectives' held in Paris, France, in 2000 (Hughes and Petchey 2001, Loreau *et al.* 2001, Loreau *et al.* 2002). This legacy of distrust has been slow to fade and continues to reinforce the view that the core BEF literature lacks credibility and is irrelevant to global ecological issues, a dialogue likely to hinder the dissemination of information to managers and policymakers (Benton *et al.* 2007).

Articulating the appropriate interpretation of BEF research is of fundamental importance if we are to provide a tenable solution to the biodiversity crisis. Achieving this goal is proving difficult despite overwhelming support for the notion that biological diversity regulates ecosystem processes (Schläpfer *et al.* 1999). Syntheses of the first 15 years of research provide ample evidence that increasing biodiversity can have positive effects on ecosystem processes and that these average effects are best explained by the loss of the most productive species from the community (Balvanera *et al.* 2006, Cardinale *et al.* 2006a). Part of the problem in communicating the significance of these results is undoubtedly historical (Raffaelli *et al.* 2005a), but there is also a real difficulty in expressing a consistent position on BEF matters because of the wide variation in how ecosystems respond to altered species diversity. As more realism is incorporated into experimental studies, the effects of biodiversity on ecosystem processes tend to be weaker because ecosystem processes are a product of multiple biological and environmental variables (e.g. Petchey *et al.* 1999, O'Connor and Crowe 2005, Balvanera *et al.* 2006, Dyson *et al.* 2007).

In order to capture such complexity a more holistic approach needs to be adopted (e.g. Naeem and Wright 2003) that collates a portfolio of evidence from several lines of enquiry (e.g. Keer and Zedler 2002, Muotka and Laasonen 2002). Looking across the literature, it is clear that BEF science has attempted to achieve this in a series of phases (Fig. 3.1). Although the timing of publications from each phase shows that different approaches have been used in the science base almost since its beginning, the general trend shows a growing appreciation of the power of these different approaches to incorporate more realism. Following Paris, it became clear that BEF experiments needed to integrate processes that are responsible for the generation, maintenance, and loss of biodiversity at local and regional scales. The first phase of BEF research had generated the necessary theory and developed the framework for laboratory experiments (Phase 1, Fig. 3.1), but it lacked sufficient environmental context to remove high levels of uncertainty when referencing the likely consequences of future global change. Consequently, the scope of BEF research broadened (Reality filter, Fig. 3.1). The next phase of BEF research added realism using a range of strategies including connection to the real world (e.g. Duffy *et al.* 2003) and the use of real assemblages in the laboratory (e.g. Widdecombe *et al.* 2000, Emmerson *et al.* 2001), field experiments using species addition or removal (e.g. Hector *et al.* 1999, Symstad and Tilman 2001), *in situ* manipulation (e.g. O'Connor and Crowe 2005), or by increasing system complexity through the incorporation of multiple trophic levels (Petchy *et al.* 1999, France and Duffy 2006a), community complexity (France and Duffy 2006b), or multiple ecosystem processes (Hector and Bagchi 2007). At the same time, theoretical models were extended to include environmental fluctuations (e.g. Yachi and Loreau 1999, Ives and Cardinale 2004) and further validated using empirical data (e.g. Emmerson and Raffaelli 2000, Swan and Palmer 2005, Fox 2006). A move away from the use of random extinction scenarios to the consideration of biodiversity futures based on available data (e.g. Solan *et al.* 2004, Zavaleta and Hulvey 2004, Bunker *et al.* 2005, McIntyre *et al.* 2007, Bracken *et al.* 2008) accompanied the innovative use of field observations (Phase 3, Fig. 3.1) (e.g. Wardle *et al.* 1997a, Troumbis and Memtsas 2000, Cardinale *et al.* 2005, Ruesink *et al.* 2006), including natural and anthropogenic gradients (Vitousek *et al.* 1994, Austin 2002, Fukami and Wardle 2005). This third phase is currently in its infancy (see Naeem and Wright 2003, Naeem 2006b), but its genesis heralds a point in BEF research history where the discipline has matured and a full suite of evidence (i.e. theory, methodology, laboratory and field experiments, field

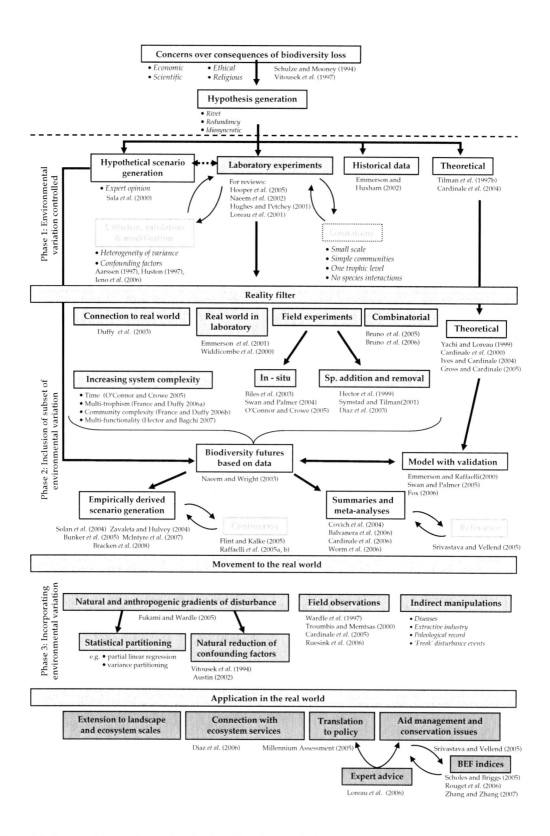

Figure 3.1 Summary of the research approaches adopted to address the relationship between biodiversity and ecosystem function in the peer-reviewed scientific literature. Modified from Godbold (2008). See Plate 1.

observations, and natural and anthropogenic experiments) has become available in a form that can be presented to managers and policymakers.

At a time when understanding the consequences of biodiversity loss is of critical value (Pimm et al. 1995, Chapin et al. 2000c), the large number and broad range of papers contributing to BEF research raise important questions about whether the BEF literature is sufficiently visible to effectively inform front-line managers and policymakers. Whilst there are several authoritative reviews on the present status (Loreau et al. 2001, Covich et al. 2004, Hooper et al. 2005, Balvanera et al. 2006, Cardinale et al. 2006a) and clear ideas have been articulated regarding future research needs (e.g. Bulling et al. 2006, Raffaelli 2006, Naeem 2006b), it remains unclear what the large arsenal of scientific papers in the BEF archive has collectively accomplished in terms of information supply. In this chapter, the core peer-reviewed BEF literature is identified and described. The sequential flow of information from this literature base is then tracked through the scientific literature to other scientific disciplines. Trends in funding dedicated to BEF research are examined to gauge government support for the topic, and trends in the appearance of biodiversity as a topic in popular literature and the legal and legislative arenas are quantified to estimate the incorporation of research findings into the public arena. Finally, the application of the science by end users tasked with managing the environment is explored. The notion that BEF research forms a primarily academic field that contributes little to environmental concerns is critically examined and recommendations are offered that may enhance dialogue among researchers, managers and policy makers.

3.2 Methodology

3.2.1 The biodiversity-ecosystem function literature base

To obtain the most representative sample of peer-reviewed publications examining the BEF relationship, citation data were retrieved from the *ISI Web of Knowledge* using the *Science Citation Index Expanded* and *Social Sciences Citation Index* databases. A 'general search' using the search term *"("biodiversity"*

OR "species diversity" OR "species richness") AND ("ecosystem function" OR "ecosystem proces*")" in the titles and keywords of all document types, in all languages, was performed. Only publications from 1990–2006 (inclusive) were used, as these span records published from immediately before the Bayreuth conference (1991, proceedings published in Schulze and Mooney 1993) to the most recent adaptive synthesis workshop (BioMERGE, Biotic Mechanisms of Ecosystem Regulation in the Global Environment; an NSF funded research coordination network) held in Switzerland in December, 2006. The results of this screening are hereafter referred to as the BEF database. In all subsequent analyses, search returns were refined by publication year to remove entries published in 2007.

As the focus is to determine how BEF research and ideology propagates through the peer-reviewed literature, publications in the BEF database were examined, using the *ISI Web of Science* results analysis tool, by publication year, source title, subject category, country of research origin, or document type. For tracking the relative influence of previously published sources of information, the annual citation rate was used as an estimate of scientific importance. To adjust for differences in publication year, the annual citation rate was calculated by dividing the total citation count by the number of years since publication. Self-citations (authors citing themselves) were not excluded (but were differentiated) because it is recognized that such contributions are important in the dissemination of information across the scientific community. To examine the relationship between year of publication, impact factor of the journal in which it is published, and number of authors with the total number of citations and the annual citation rate, simple correlations were calculated. In addition, variation in annual citation rate among publications categorized by their subject matter (discussion, theory, methodology, laboratory experiments, field experiments, field observations, or practical application in the real world), country affiliation of first author, and ecological habitat type (not habitat-specific, freshwater, terrestrial, or marine) were examined graphically. Correlations and graphical examinations are preferential to other methods (such as linear regression or analysis of variance) in this case, because other factors that may influence why a paper is cited cannot be controlled for or incorporated into the analysis. Thus, any observed relationships are indicative and not necessarily causative.

Not all publications will carry the same weight of influence in informing the wider scientific community. Publications that have been cited more frequently are likely to have a disproportionate influence on the dissemination of information and ideology through the literature. The six most cited publications from the BEF database were used to gain an impression of how heavily cited publications directly and indirectly influence the wider scientific community. Following Benton et al. (2007), publications that directly cited each of these six publications ($n = 2546$) were assigned to categories (discussion, theory, methodology, laboratory experiments, field experiments, field observations, or practical application in the real world) that reflect how the information contained within each paper was used in the subsequent generation of publications. As these six first-generation publications are themselves cited by other highly cited publications, it is also possible to gain an impression of the information flow as it is disseminated across multiple generations of publications. For each generation of publications, the five most highly cited publications citing the previous generation were determined and linked, either directly or indirectly (through subsequent generations), to the six most highly cited publications from the BEF database. The horizontal extension of such linkages was continued until either all of the top five citing publications were from publications not listed within the BEF database or until the publications were yet to be cited. The properties of the resulting network provide an indication of the rate, form and connectivity of information flow within the BEF community and to multiple disciplines.

3.2.2 Biodiversity and BEF science impact outside the scientific literature

The proportion of funding from national science budgets allocated to BEF-related research provides a measure of how high BEF sits on national scientific agendas and whether its status is gaining momentum or waning over time. In the UK, research proposals submitted to NERC (Natural Environment Research Council) are required to classify the research against an Environmental and Natural Resource Issues (ENRI) classification. The total expenditure on research within the ENRI category of 'biodiversity' for 2001–2006 was collated. For the USA, the National Science Foundation (NSF) website (http://www.nsf.org/) was searched for all awards granted between 1 January 1990 and 31 December 2006, using the search term: (biodiversity OR "species diversity" OR "species richness") AND ecosystem AND (function OR proces*).

To investigate how BEF research and ideology is used outside the peer-reviewed literature, the Lexis-Nexis Butterworth online service (http://www.lexisnexis.com/) was searched using the search term "biodiversity OR species diversity OR species richness OR ecosystem function OR ecosystem process" for the period 1990–2006. For both the UK and USA, all occurrences of the search terms in newspaper stories, legislation (including hearings and enacted laws), and court case documents were determined for each year.

Another broad category central to the application of BEF research are planning documents used and generated by managers and agencies. The Final Environmental Impact Statement (FEIS) for general management plans of three US national parks (USDI-National Park Service 2006a, b, 2007) and three US national forest/grassland complexes (USDA Forest Service 1997, 2001a, b) were manually searched for language indicating the acceptance of BEF ideology. The references cited within five of these documents (citations for the sixth were unavailable) were categorized by topic (e.g. taxon natural history and distribution, methodology, planning documents), source (journal, book, government agency research report), and age (the difference between their publication year and the year the FEIS was published). These documents were completed between 1997–2007 and cover approximately 6.9 million ha of public lands in the northern Great Plains, Colorado Rocky Mountains, and Sierra Nevada in the USA.

3.3 Results

3.3.1 Description of BEF database

Between 1990 and 2006, a total of 942 publications in the peer-reviewed literature have either directly or indirectly considered concepts surrounding the

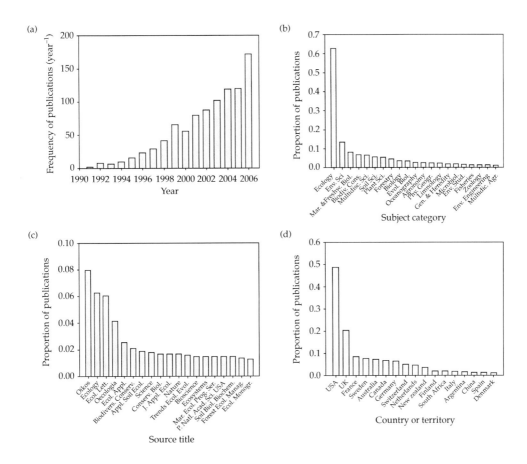

Figure 3.2 Summary of the BEF database (n = 942) in terms of (a) the frequency of publications per year and (b)–(d), the ranked proportion of publications per subject category, source title, or country/territory of research origin. For clarity, subject categories or countries/territories are only illustrated if,

effects of biodiversity on ecosystem function. The majority of these contributions are based on studies in terrestrial systems (60 per cent), but aquatic systems are also well represented (13 per cent freshwater, 11 per cent marine). The remaining 17 per cent (n = 158) of contributions in the database are not specific to a particular habitat and mainly consist of generic discussion (n = 94) or the development of theory (n = 24) and methodology (n = 34). The number of publications per annum increased from no publications in 1990 to 172 in 2006 (Fig. 3.2(a)). Collectively, the BEF database spans 50 ISI subject categories; the majority of publications fall within ecology (63 per cent), environmental sciences (14 per cent), marine and freshwater biology (8 per cent), or biodiver-

sity and conservation (7 per cent) (Fig. 3.2(b)). Approximately a quarter (n = 235) of publications in the BEF database are published in *Oikos* (8 per cent), *Ecology* (6 per cent), *Ecology Letters* (6 per cent), or *Oecologia* (4 per cent), with the remaining contributions distributed across an additional 210 source titles (Fig. 3.2(c)). The majority of publications include authors/co-authors from Europe (65 per cent) or the USA (49 per cent), although a total of 59 countries or territories are represented (Fig. 3.2(d)). Closer examination reveals that the majority of BEF research is led by authors in the USA (43 per cent) and the UK (13 per cent), with other significant lead authors emanating from France (6 per cent), Canada (5 per cent) and Sweden (4 per cent). The content of individual publications varies, from the discussion of BEF ideology (31 per cent) to reporting the results from manipulative experiments carried out in the laboratory (14 per cent) or in the field (18 per cent). Although discussion and reports of manipulative experiments dominate the literature, it is important to emphasize that these data are complemented by a substantial number (21 per cent) of correlative field observations. A smaller proportion of publications are concerned with the development of theory (5 per cent) and methodology (6 per cent), and fewer still (6 per cent) use BEF ideology to underpin management philosophy or applications in the real world.

3.3.2 What determines citation frequency?

Contributions in the BEF database have collectively been cited > 22,000 times (= 23.9 cites per item) and have an h-index – the number n of papers that have all received at least n citations (Hirsch 2005) – of 72, indicating high quality science. Of these, 49 per cent are incidences of self-citation (= 11.7 self-cites per item) suggesting a rather insular science base. However, although the propensity to self-cite has risen dramatically in recent years, the proportion of citations by researchers outside of the BEF database is increasing (inferred from the widening gap between self vs. non-self citations, Fig. 3.3(a)). Analyses of the BEF database demonstrate that several other variables may have an effect on whether a paper is cited more or less frequently. Of the variables examined in this chapter, the total number of times a paper is cited is most strongly related to journal impact factor ($r = 0.45$) and the year of publication ($r = -0.39$); total citation rate is only weakly, if at all, related to the publication's number of authors ($r = 0.15$). For annual citation rate, the relative strength of the three variables is the same, but journal impact factor ($r = 0.47$; Fig. 3.3(c)) is the only one that accounts for a reasonable amount of the variance (year of publication, $r = -0.12$, Fig. 3.3(b); number of authors, $r = 0.26$, Fig. 3.3(d)). There is also an indication that the subject matter ($r = -0.14$, Fig. 3.3(e)) and broad habitat type ($r = -0.13$, Fig. 3.3(f)) of the individual study are of lesser importance. The median number of citations to discursive papers is relatively low, although some of these papers are cited frequently. In contrast, laboratory experiments and theoretical papers have the highest median citation rates and papers that attempt to apply BEF ideology to the real world are rarely cited, possibly reflecting differences in motivation between those tasked with academic science and those tasked with managing or conserving natural resources. The median citation rate is not greatly affected by habitat type, most likely because the most cited publications are predominantly theoretical and a few discursive publications that are of generic value.

3.3.3 Who is taking notice?

The six most cited papers within the BEF database (hereafter referred to as the BEF-6) consist of a general review (Loreau *et al.* 2001), the development of methodology (Huston 1997) and key laboratory (Naeem *et al.* 1994) or field experiments (Hooper and Vitousek 1997, Tilman *et al.* 1997b, Hector *et al.* 1999) based on terrestrial plant communities. Collectively, these papers have been cited over 2500 times and, individually, their mean citation rate (424 ± 62.6, $\bar{x} \pm 1\,s.d.$) is greater than that of major developments in other fields published at approximately the same time (e.g. the discovery of the first planet outside our solar system, Wolszczan and Frail 1992 [cited 298 times]; or the discovery of the oldest known hominid, White *et al.* 1994 [cited 220 times]). In each case, the percolation of ideas through the literature has followed a similar path, from discursive contributions through to applications of ideology in the real world (Fig. 3.4). Interestingly, however, as the BEF discipline has matured (particularly post-Paris) there is some indication that the time taken for BEF ideology to influence applications in the real world has shortened and that the emphasis of papers has moved away from phase one and towards phase two and three (in Fig. 3.4, compare pink in panel d, a paper published in 2001, to the earlier papers in other panels). Building on previous contributions decreases the time required for gaining confidence in the relevance and applicability of research findings (i.e. closing the credibility gap, *sensu* Benton *et al.* 2007), circumventing

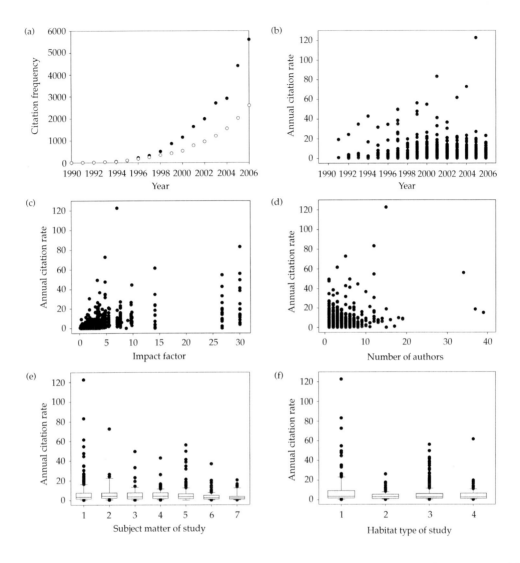

Figure 3.3 Citations to the BEF database between 1990 and 2006: (a) citation frequency by year including (solid circles) and excluding (open circles) self-citations, (b)–(f) annual citation rate in relation to (b) the year of publication, (c) impact factor, (d) number of authors, (e) publication subject matter and (f) habitat type. Subject matter refers to the content of each paper: 1 = discussion, 2 = theory, 3 = methodology, 4 = laboratory experiments, 5 = field experiments, 6 = field observation, and 7 = practical application in the real world. Habitat type refers to the studied ecosystem, such that 1 = not habitat specific, 2 = freshwater, 3 = terrestrial and 4 = marine.

the need for an extended period of critique before conclusions are generally accepted by the wider community.

The BEF-6 influence the next generation of publications either directly (solid lines, Fig. 3.5) or indirectly along the linkages between citation nodes (horizontal branches, Fig. 3.5) and via other citations (dotted lines, Fig. 3.5). The length of the pathway from the BEF-6 to the external scientific community, however, varies considerably. In the first generation of publications, the most influential papers directly citing the BEF-6 tend to be from within the BEF database, indicating a rapid integration and absorption of philosophy by the BEF community but a lack of penetration to other disciplines. This process is reversed in subsequent

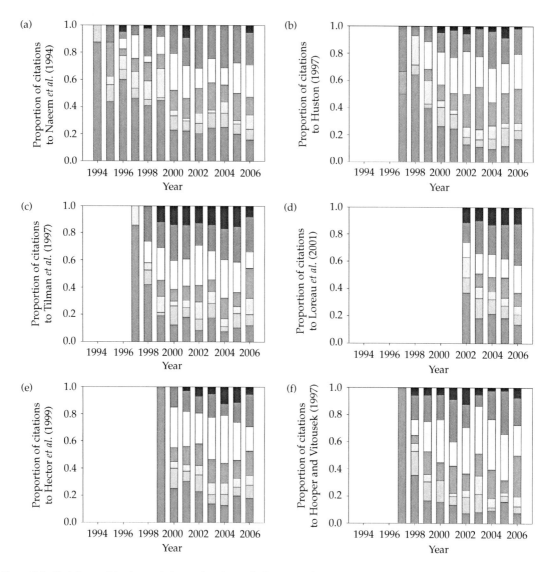

Figure 3.4 The influence of the six most cited papers from the BEF database since publication to December 2006 for (a) Naeem et al. (1994), (b) Huston (1997), (c) Tilman et al. (1997), (d) Loreau et al. (2001), (e) Hector et al. (1999) and (f) Hooper and Vitousek (1997). These publications influenced subsequent publications by contributing to discussion (red), development or reinforcement of theory (orange) or methodology (yellow), or by initiating or informing laboratory experiments (green), field manipulations (turquoise), or field observations (blue), or they were used to underpin practical applications in the real world (pink). Expanded from Benton et al. (2007). See Plate 2.

generations such that, by the fourth generation of publications, the most influential publications citing BEF ideology are all outside of the BEF database (or are yet to be cited). These tend to be contributions relating to invasive species, food web, or landscape ecology and a significant proportion of these tend to be review articles.

3.3.4 Is BEF research being supported?

In the USA, the mean annual sum (±1 s.d.) granted annually to BEF-related research over the years examined (1995–2006) was US$6,563,952 ± 8,727,303 or, from 2001–2006 (post-Paris), U$11,817,500 ± 9.965,911 (× 0.5 for approximate conversion to £GBP), but the range of award sizes increased considerably

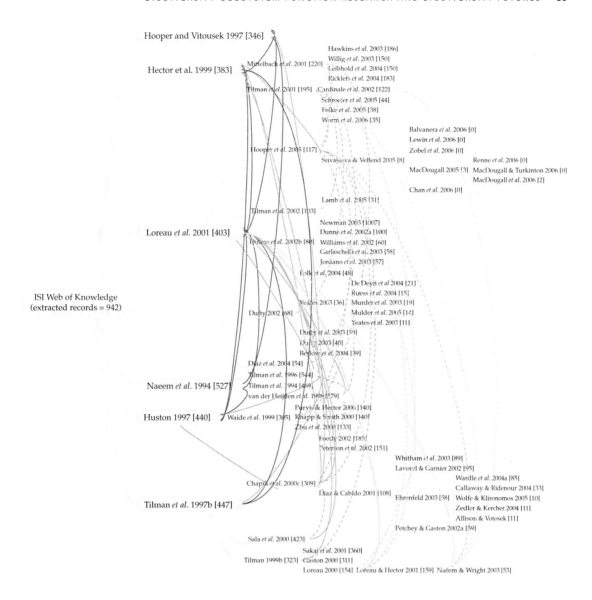

Figure 3.5 The rate, form, and connectivity of information flow within the BEF community and to multiple disciplines from the six most highly cited papers (=BEF-6) within the BEF database. For each generation of publications, the five most highly cited publications citing the previous generation were determined and linked, either directly (solid lines) or indirectly (dotted lines) to the BEF-6 via other highly cited publications. Line colour indicates the generation sequence (blue → red → green → orange). Publications not included in the BEF database are presented as a citation. The number of cites (from publication until December 2006) are indicated in square brackets. References listed are available in the electronic appendix. See Plate 3.

during this period (Fig. 3.6(a)). Interestingly, although the total NSF research budget plateaus after 2004, the percentage of grant money to BEF research projects increased sharply (Fig. 3.6(b)), indicating a shift in the perceived importance of BEF research. In the UK, between 2001 and 2006, the mean (± 1 s.d.) annual amount of funding for biodiversity-related research was £12,188,321 ± 1,505,590 (× 2 for approximate conversion to US$), or 23.3 ± 3.2 per cent of the total

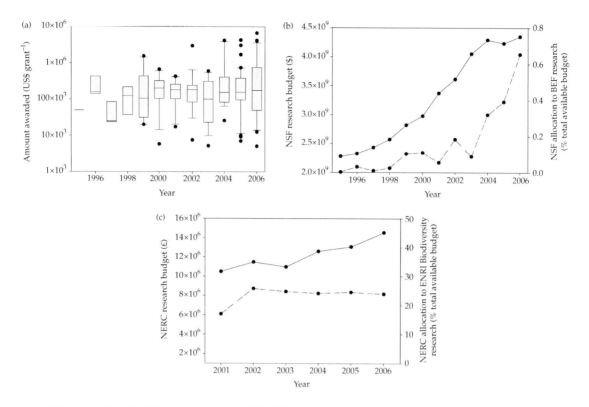

Figure 3.6 Funding allocated to BEF-related research by the National Science Foundation in the USA and to all biodiversity related research by the Natural Environment Research Council in the UK from 1995–2006 (USA) and 2000–2006 (UK). (a) Annual individual award amounts in the USA shown in boxplots by year, on a log axis; boxes range between the 25th and 75th percentiles, with lines at the median, and whiskers range to the 5th and 95th percentiles. Points are outliers. (b)–(c) Total funds awarded to all research (solid lines) and percentage of the total funding awarded to BEF-related research (dashed lines) within each calendar year for the USA (b) and UK (c).

NERC funding awarded. This allocation has increased from £10,940,384 in 2001 to £14,540,280 in 2006 (Fig. 3.6(c)), representing 17.0–23.1 per cent of total NERC funding. In contrast to the USA, however, the BEF budget in the UK has essentially remained constant despite the fact that the total research budget has been increasing. Clearly, the academic interest in BEF research has been translated effectively and has a major impact on these two funding agencies.

3.3.5 Export of BEF ideology

It is clear that news stories relating to biodiversity in both the UK and the USA were initiated by the signing of the Convention on Biological Diversity during the United Nations Conference on Environment and Development, held in Rio de Janeiro, Brazil in 1992 (Fig. 3.7). Although the signing of the treaty had an immediate impact, media coverage was quick to wane. However, biodiversity remained firmly in the news, albeit at low but slowly increasing frequency over the next 5–7 years. Following a series of major natural disasters around the globe in the late 1990s, the number of newspaper stories concerning the environment and biodiversity increased dramatically. This pattern of media interest paralleled the rise in legislative enactments and court cases related to environmental protection (Fig. 3.7), although the total number of cases is low. In the USA, the steady increase in biodiversity-related documents in court cases appears from a qualitative analysis to be driven by cases involving habitat protection,

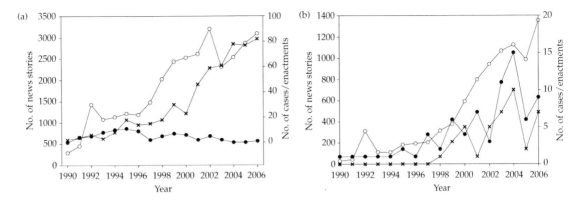

Figure 3.7 An indication of the export of BEF research and ideology from the peer-reviewed scientific literature to the public domain for (a) the USA and (b) the UK in media (open circles), legislation (crosses), and court cases (closed circles) for the period 1990–2006.

with several non-governmental organizations (e.g. the Center for Biological Diversity, the Natural Resources Defense Council, and the Sierra Club) appearing frequently. Similarly, habitat protection issues appear to compose the bulk of the US legislative documents concerning biodiversity, although these terms also appear in hearings on the Clean Air and Clean Water acts.

It seems that the timing of the emergence of BEF concepts in the public domain coincided with major events or announcements that raise public awareness of environmental issues, rather than any direct lobbying by the BEF community to increase political attention to the consequences of biodiversity loss (Loreau et al. 2006 provides one exception). Interestingly, environmental issues in general received less attention in the UK during the 2005 electoral campaigns, rebounding in the following year, suggesting that issues with direct impacts closer to home may override more indirect global concerns.

3.3.6 Transfer of concepts to end users and stakeholders

The evaluation of the transfer of BEF concepts to those making management decisions is decidedly less quantitative than the other portions of this chapter, but it reveals some interesting information. In the USA, the Department of the Interior's National Park Service oversees all national parks, whose purpose is to conserve, unimpaired, the natural and historic resources therein for the enjoyment of future generations. In contrast, the national forests and grasslands, administered by the Department of Agriculture's Forest Service, were established expressly for multiple uses, and until recently, their emphasis in management has been on extractive uses, especially timber harvesting and cattle grazing. In response to changing scientific and public views, the Chief of the Forest Service announced in 1992 that an ecological approach would be adopted so that national forests and grasslands would 'represent diverse, healthy, productive, and sustainable ecosystems' (USDA Forest Service 1997, p. 3). Although all the planning documents that were reviewed used this approach, terms such as 'process' and 'function' were generally not equivalent to the same terms used in the BEF database (e.g. fire, flooding, or insect or pathogen outbreaks in the planning documents; primary productivity, nutrient retention, or carbon storage in the BEF database). All of the documents also followed their respective agencies' policies in emphasizing biodiversity as something to preserve, protect, and, where necessary, restore, and they often stated that one way to do this is to maintain 'naturally functioning ecosystems.' Thus, although management seems to emphasize biodiversity as a function of the environment, rather than the reverse, there is scope for BEF science to influence management.

Interestingly, BEF concepts are more prevalent in the Forest Service documents than in the Park Service documents. In Park Service plans, the closest statement to the BEF assertion that biodiversity improves ecosystem function was an aspiration to 'restore native ecological processes by reintroducing native plants and animals and removing non-native species where practicable' (USDI National Park Service 2006a, p. 100). In Forest Service plans, the BEF ideology was relatively common. For example, the decisions made in developing the management plan for the northern Great Plains units emphasized the need to '...sustain ecosystems by ensuring their health, diversity, and productivity' (USDA Forest Service 2001a). More explicitly, the FEIS for the Sierra Nevada forest plan states that 'an ecosystem that is well supplied with native species, has a level of biomass that is not maximizing available nutrients and water, and is adapted to disturbance, is typically a resilient one' (USDA Forest Service 2001b, p. 104), consistent with the experimental findings of Pfisterer and Schmid (2002). The planning document was, however, completed before this publication (*loc. cit.*), indicating that either the management philosophy in Sierra Nevada was ahead of BEF research or that it was alluding to untested concepts.

Although BEF concepts do appear in these management plans, those tasked with preparing the documents are not directly citing the BEF literature. It is clear that, for most management units, individual species are still of major concern, since their natural history, distribution, and management comprise major portions of the references cited (Fig. 3.8(a–b)). References to ecological concepts (such as BEF) are relatively rare, comprising just 19 of the 227 published ecological references cited across the five plans. These findings are consistent with the survey undertaken by Pullin *et al.* (2004) which showed that only 23 per cent of managers 'always' or 'usually' used scientific publications when compiling management plans. The latest (< 5 years old) research is often not cited (Fig. 3.8c), most likely due to the long time taken to prepare and finalize these documents (> 2 years). Interestingly, periodical articles are the most common type of citation (56 per cent), but books or book chapters (24 per cent) are an important source of ecological information. Symposium proceedings (11 per cent), government research reports (5 per cent), and theses or dissertations (4 per cent) comprise the rest. Only 16 of the 227 ecological references cited are in journals in which BEF research is most often published (those in Fig. 3.1(c) in decreasing order of citations: *Conservation Biology, Ecological Applications, Ecology, Ecosystems, Bioscience*).

3.4 Conclusions and recommendations

The two great sins of innovation are to be either too early or too late with the solution to a problem, an adage of particular relevance when assembling the evidence base needed to understand the extent to which changes in biodiversity are causally linked to ecosystem processes. A naïve view is that ecologists are perfectly placed to answer BEF hypotheses, as human civilizations have long understood the importance of biotic interactions and the environment (McNeely 1994, Rozzi 2004) and much of the scientific information required to understand the BEF relationship, including experiments that directly manipulate diversity, have been in place since the mid-19th century (Hector and Hooper 2002, Statzner and Moss 2004). Contemporary BEF studies build on these older perspectives and the analyses in this chapter confirm that that these syntheses are readily absorbed by the wider scientific community. Since Paris, the discipline has matured and gained scientific credibility. Indeed, contrary to criticisms voiced in the late 1990s, a suite of evidence is now available based on theory, empirical investigations, field observations, and practical application, backed by large-scale (e.g. 19,485 km^2, Kennedy *et al.* 2003) and long-term observations in terrestrial plant (19 years, O'Connor *et al.* 2001), soil (10 years, Lindberg *et al.* 2002) and marine systems (7 years, Paine 2002). Yet in spite of this wealth of evidence, the scientific consensus that biodiversity is critical for ecosystem function has only partially been embedded in the policy arena. Indeed, if future generations are presented with the BEF science base as it now stands, it will be incomprehensible to them if policy makers have not fully adopted this consensus and utilized it to moderate and mitigate global scale problems caused by biodiversity loss.

Although the evidence base is well resourced and available, establishing generality by substantiating hypotheses with multiple lines of enquiry is limited in practice (good examples include Keer and Zedler

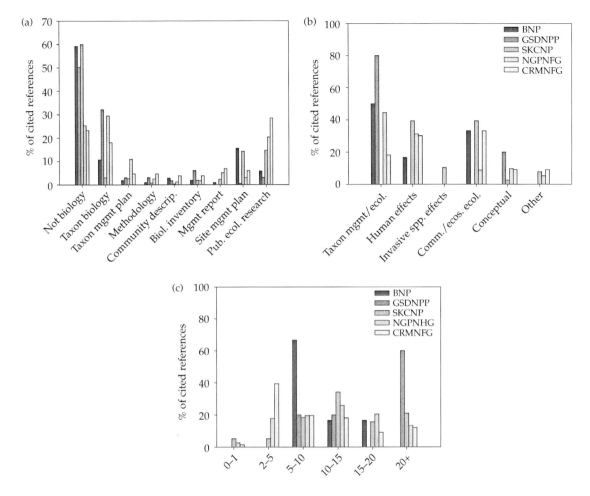

Figure 3.8 The types and age of references cited by five Final Environmental Impact Statements (FEIS) for national park and national forest/grassland general management plans. In (a), all references are categorized by subject matter: 'Not Biology' = physical science, paleontology, economics, and census information; 'Taxon Biology' = natural history, distribution, and conservation status of species or select taxa, and field guides/taxonomic keys; 'Taxon Mgmt Plan' = management plans/guidelines for specific taxa, and reports on anthropogenic impacts on specific taxa; 'Methodology' = guidelines on monitoring, habitat delineation, and range evaluation; 'Community Desc' = descriptions of plant communities and vegetation maps; 'Biol Inventory' = biological inventory and survey results, including disease incidences, for a given region; 'Mgmt Report' = unpublished site-specific biological evaluations and management or monitoring reports; 'Site Mgmt Plan' = management plans for parks, forests, grasslands, and refuges; and 'Pub Ecol Research' = published ecological research (not including natural history, distribution, etc. in the first category). In (b), references from the 'Pub Ecol Research' category are further categorized as taxon-specific management or ecology, effects of human-caused disturbances (i.e. unintentional management; e.g. vehicle or foot traffic, climate change, acid deposition) on taxa or communities, invasive species effects, community or ecosystem ecology, conceptual, or other. In panel (c), the percentages of cited references since publication year of the FEIS are presented.

2002, Muotka and Laasonen 2002). Even when achieved, an effective interface between policy-makers and scientists is, at best, underdeveloped – despite some lobbying for the installation of such a mechanism (e.g. Loreau *et al.* 2006). It seems that the primary problem for the BEF community is not the science base, but one of communication. The emergence of the BEF paradigm occurred when concerns relating to environmental change were beginning to gain momentum and at a time when there was heightened public awareness of the general impacts that human activities have on the biosphere (e.g. Halpern *et al.* 2008). Whilst BEF investigators undoubtedly contributed to and shaped opinion, it is evident that BEF science was catapulted into the public domain because it was absorbed into general discussions about global change, rather than because of any top down coordinated effort by the BEF community to directly emphasize BEF concepts. Where opportunities for high-profile communication have taken place (e.g. *Issues in Ecology* publication and the subsequent attempt to revoke; Naeem et al. 1999, Wardle et al. 2000b), the ambiguity caused by competing scientific views over the relevance and extrapolation potential of controlled experiments (which represent < 6 per cent of the BEF database) may have prevented effective communication of the science to the media and policymakers at a time that mattered most.

It is clear that there are also major deficiencies in communication between those carrying out the fundamental science and those tasked with generating and implementing management objectives, conservation strategy, or policy. Cultural differences between the way interdisciplinary fields go about their science (attendance of different conferences, publication in and readership of different journals, access to information sources, technical terminology) form a significant barrier (Raffaelli et al. 2005a), such that policymakers are often unaware of developments in BEF science and vice versa. Direct appeals from one community to another, although rare, have been made (Minns et al. 2001 provide a notable example), but tend to remain in a format and/or source that remains inaccessible to the target audience. A recent analysis of views on biodiversity research in Europe (Hildén et al. 2006) confirms that these problems are significant, concluding that 'policy makers do not understand biodiversity, its value or research results' and that 'relevant policy processes are not researched' because 'researchers do not understand management needs' (p. 16). This stalemate appears to be a product of subtle differences in opinion over what information, if any, should underpin effective ecosystem management strategies; some ecological concepts do not always translate well to conservation measures (Gascon and Lovejoy 1998). Indeed, BEF science has tended to concentrate on the consequences of biodiversity loss for specific ecosystem processes, whereas managers and policy makers have focussed on the values of ecosystem services whilst failing to incorporate the link between biodiversity and ecosystem service provision at a landscape level (Kremen 2005).

There is some reassurance that BEF science is being noticed outside of the BEF community, albeit that the primary routes of dissemination (discussion of wider issues in the public arena or scholarly reviews of associated fields in the scientific literature) are incidental rather than directed. Irrespective of the mechanism of dissemination, the effect (since the mid-1990s) has been a five-fold reduction in the time taken for the latest research findings to be implemented in real-world conservation efforts. In some cases the motivation for such rapid adoption of the science has been commercial gain (Bullock et al. 2001, Skelton and Barrett 2005), but in most cases three main drivers seem more likely to be responsible. First, longer exposure time of the general topic places the subject area on the agenda and increases capacity to build on prior knowledge, short-circuiting the inherent time lag associated with the credibility process that needs to be in place before action can be taken (Benton et al. 2007). Second, the maturation of the discipline towards experimental designs and approaches that incorporate more realism (Phases 2–3 and beyond, Fig. 3.1) makes the science base less nebular and more applicable to large-scale real-world environmental problems, and therefore more accessible and tractable to environmental managers. Increasingly, academic scientists are teaming up with environmental managers to perform large-scale experiments that otherwise would not be possible (e.g. Bison–prairie dog–plant impacts on grassland ecosystem processes in the Badlands National Park, USA; Fahnestock and Detling 2002). Third, the more widespread adoption of sustainable ecological concepts (for example, the ecological approach adopted by the US Forest Service) provides strong impetus to converge towards a standardized, practical approach to ecosystem-based management that fulfils, or at least incorporates, conservation principles and strategies required by law and desired by the public.

It is clear that decision making is being informed by BEF ideology, but the influence that the scientific community has on this process appears to be small relative to the force of social change in attitudes. If the BEF research community is to effectively inform managers and policymakers in time to mitigate impending global problems, the needs of these customers have to be considered and impediments to communication must be recognized and overcome. Only an extended science–policy consultation involving the interaction of BEF scientists and decision makers (i.e. evidence-based conservation, *sensu* Sutherland *et al.* 2004) is likely to bridge this divide, lead to the appropriate exchange of information, and hasten the development of policy based on a portfolio of evidence that incorporates levels of uncertainty (Benton *et al.* 2007). Whether future generations consider the BEF science as the early bird that caught the worm, or as a primarily academic field that was a day late and a dollar short, will depend on whether the next phase of BEF research enhances the science–policy interface.

PART 2
Natural science foundations

CHAPTER 4

A functional guide to functional diversity measures

Owen L. Petchey, Eoin J. O'Gorman, and Dan F. B. Flynn

4.1 Introduction

4.1.1 What is functional diversity and why is it important?

Functional diversity is a component of biodiversity that concerns what organisms do, rather than, for example, their taxonomic identity (Tilman 2001, Naeem and Wright 2003, Hooper et al. 2005, Petchey and Gaston 2006). What kinds of resources do organisms exploit, where do they exploit them, and when do they exploit them? Each of these characteristics, and many others, can be a component of functional diversity. Since ecosystem processes ultimately result from the actions of organisms, knowing about these actions and summarizing them in a measure of diversity should inform about ecosystem processes (Tilman 2001, Hooper et al. 2005). Therefore, functional diversity has the potential to link morphological, physiological, and phenological variation at the individual level to ecosystem processes and patterns.

One critical reason that functional diversity might link organisms and ecosystems is that it implicitly contains information about how species will compensate for the loss of another. For illustration, consider some species that are functionally rather similar, and some that are rather dissimilar. The similar species each access the same pool of resources. If one of these species is lost, no reduction in the use of that resource pool results, as the other species are present and will accordingly simply increase their use of that same resource. There will also be only a small loss of functional diversity, since the lost species was not very unique in its functional characteristics. If, however, the lost species was quite dissimilar to others in its functional traits, both functional diversity and use of the resource pool will decline in concert. Other changes associated with loss of access to that resource pool might include a reduction in ecosystem processes, such as net primary production.

Another illustration of the interactions that are implicitly represented in measures of functional diversity is that the effect on functional diversity of losing a particular species (or adding a particular species) is context-dependent. The context here is the other species present in the community. In one context, loss of a particular species can have little effect on functional diversity. This would be if the community contains, and continues to contain, functionally similar species to the lost species. If there are no similar species, however, the loss of the same species could have a large effect.

Thus functional diversity is a measure of diversity that implicitly incorporates some mechanisms of ecological interactions between species. In doing so, it provides a general (and yes, assumption-laden) approach for scaling from characteristics of individuals to properties of communities and ecosystems. This makes functional diversity a potentially powerful concept in ecology. At present, much attention is focused on how to measure it, and that is the broad subject of this chapter.

4.1.2 Why not functional group richness?

One method for quantifying functional diversity is to examine the characteristics of some species, and to somehow assign species to functional groups (or guilds) based on these characteristics (Root 1967). This is a popular method, because of the perceived ease of assigning species groups, and has been used repeatedly in biodiversity-ecosystem function experiments. However, this method has several major drawbacks, all linked to the central issue of how to assign species to groups. The details of the assignment is relatively unimportant here, and methods range from the very subjective to more objective (Holmes *et al.* 1979, Chapin *et al.* 1996, Díaz and Cabido 1997). What is critical here is the decision about how similar species need to be in order to belong to the same functional group, or rather where the boundaries of the groups should be placed. This decision defines the number of functional groups and the average number of species in a functional group. The number of functional groups represented by the species in a local assemblage (functional group richness) is the measure of functional diversity.

The first problem with functional group richness is that any interspecific differences within a group are ignored. This is less of a problem if species form very distinct groups; that is, if species' functional characteristics are distributed discontinuously. However, if variation is more continuous, the amounts of variation that are ignored by functional group richness could be large. Since many important functional traits are continuous, this is a large problem. Even traits which are often considered discontinuous, like N-fixing or not, are in fact continuous at some scale, with the efficiency of nitrogen fixation varying under different soil conditions (Vitousek *et al.* 2002).

The second problem is that the behaviour of functional group richness depends greatly on the number of functional groups that are defined (Petchey & Gaston 2002a, Fonseca and Ganade 2001). Defining many functional groups results in few species per functional group, and strong effects of species richness (and extinctions) on functional diversity, as the number of functional groups begins to approximate the number of species. Defining few functional groups results in many species per functional group, and weak effects of species richness on functional diversity.

Third, no objective method exists for deciding how different species should be in order to belong to different functional groups. Consequently, the number of functional groups, number of species per functional group, and, for example, the effects of species richness and extinction on functional group richness are arbitrary.

Finally, evidence exists that some real assignments of species to functional groups are little better than random assignment of species to groups. In particular, correlations between functional group richness and ecosystem processes in a well-studied grassland plant ecosystem were often higher with random assignment of species groups than the actual grouping used (Petchey 2004, Wright *et al.* 2006).

4.1.3 Non-grouping measures of functional diversity

The recognition that functional group richness is a problematic measure of functional diversity prompted the search for better alternatives. Given some information about what species do, how can one transform this into a measure of the diversity of what the species do? If we record what species do quantitatively, and call this information their traits, or functional traits, then the problem is how to measure the diversity of trait values. Thinking of the species as points in n-dimensional trait space, the question becomes how to measure the diversity of a cloud of points in n-dimensional space.

One of the first solutions was by Walker *et al.* (1999), followed by Petchey and Gaston's (2002a) use of a method of measuring phylogenetic diversity to solve the problem. Since then, a small industry has sprung up, with frequent critiques, suggestions of new measures, and supposed improvements to old measures (Mason *et al.* 2003, Heemsbergen *et al.* 2004, Botta-Dukát 2005, Mason *et al.* 2005, Mouillot *et al.* 2005a, Pavoine and Doledec 2005, Ricotta 2005a, b, Cornwell *et al.* 2006, Lepš *et al.* 2006, Podani and Schmera 2006, Petchey and Gaston 2007, Fox and Harpole 2008, Mouchet *et al.* 2008, Quintana *et al.* 2008, Walker *et al.* 2008). The impression one might come away with from so much activity in this field is lack of consensus about the most appropriate measure of functional diversity.

The aim of this chapter is to provide a functional guide to functional diversity measurements by two complementary approaches. First, through illustrations of the uses that functional diversity can be put to: what types of ecological questions can functional diversity help us answer? And second, by showing potential users how to choose among six different measures of functional diversity. Examples show how several of the metrics work (i.e. what they do), in particular how they respond to changes in species richness and species composition, highlighting crucial aspects of the metrics, and how they apply to two simulated datasets. We will also get a sense of where the field needs to go through this analytical review.

4.2 What metrics are out there?

We will focus on six measures of functional diversity.

- Convex hull volume (CHV) (Cornwell *et al.* 2006)
- FD_{var} (Mason *et al.* 2003)
- Functional attribute diversity (FAD) (Walker *et al.* 1999)/Mean dissimilarity (MD) (Heemsbergen *et al.* 2004)
- Rao's quadratic entropy (Q) (Botta-Dukát 2005)
- FD (Petchey and Gaston 2002a)
- Podani and Schmera's modification to FD (here termed FD_{LD}) (Podani and Schmera 2006)

These can be divided among three categories: measures that work on trait values directly (CHV and FD_{var}), measures that work on the distance matrix (FAD/MD and Q), and measures that work on the functional dendrogram (FD and FD_{LD}).

4.2.1 Measures directly using the trait values (CHV and FD_{var})

The convex hull volume (CHV) is the smallest convex shape (set) that can enclose a set of points in *n*-dimensional space, where each trait provides a new dimension. It was first employed as a measure of the dispersion of species in *n*-dimensional trait space by Cornwell *et al.* (2006). In one-dimensional space, convex hull volume is the range of the data (maximum value minus the minimum). This measure has only been used for continuous traits.

FD_{var} (Mason *et al.* 2003), when used with a single trait, can be thought of as the sum of squared deviations of species from the weighted mean of the species. It is therefore relatively similar to variance. The weighting of trait values by species can be by abundance, and so gives a measure of functional diversity that can account for differences in abundance between species.

This metric was designed solely to work with single traits, which considerably restricts its utility, but its authors recognized this limitation and suggested averaging for multiple traits. The solution used in the simulations below follows this suggesting and averages FD_{var} values across multiple traits. It is also possible to weight the FD_{var} values for each trait by the degree of statistical independence from other traits, a technique which has been proposed by researchers working on related questions in functional ecology, although not previously for FD_{var}. The notion is to discount traits which are highly correlated with other traits. One formulation of this weighting parameter is $w_t = \frac{1}{2} + \sum_{l=1}^{T}\left(1 - \frac{r_{tl}^2}{2}\right)$, for *T* traits, considering the relationship between trait *t* and all other traits *l* (Kark *et al.* 2002, Mouillot *et al.* 2005c).

4.2.2 Distance-based measures (MD/FAD and Q)

MD (mean dissimilarity) (Heemsbergen *et al.* 2004) and FAD (Functional Attribute Diversity) (Walker *et al.* 1999) are both calculated as the mean distance between species in multivariate space. Note that another distance-based measure is now available (Schmera *et al.* 2009). This requires a decision about the distance measure used: for example, Euclidean, Manhattan, or Gower. Hereafter we use MD to include both MD and FAD.

Q, representing Rao's quadratic entropy, was proposed as a measure of functional diversity by Botta-Dukát (2005). It is the sum, across species pairs, of the product of the distance between the two species in trait space and their two relative abundances. However, other researchers have implemented a modified version of Q, using arbitrary weighting values rather than using relative abundances, which are then modified until the maximum explanatory power for the ecosystem function of interest is achieved (Weigelt *et al.* 2008).

Table 4.1 Number of papers citing Walker et al. (1999)/Heemsbergen et al. (2004) for FAD/MD, Petchey and Gaston (2002a) for FD, Mason et al. (2003) for FD_{var}, Botta-Dukát (2005) for Q, Cornwell et al. (2006) for CHV and Podani and Schmera (2006) for FD_{LD}. The greater number of citations for FAD/MD and FD are a reflection of the age of the papers.

Theme of paper	Number of citations						
	FAD/MD	FD	FD_{var}	Q	CHV	FD_{LD}	Total*
Unrelated studies	96	21	4	1	7	1	124
Use of discrete groupings or standard diversity indices	46	8	1	1	1	0	55
Methods associated with calculating functional diversity	11	11	3	4	0	3	20
New measure of functional diversity	7	11	1	0	0	1	18
Application to empirical data	3	8	4	3	1	0	16
Benefits of looking at functional diversity	9	11	1	1	0	0	15
Pro's and con's of indices	7	7	4	2	0	0	8
Potential use as ecological indicator	1	5	4	0	0	0	6
Total number of times cited	180	82	22	12	9	5	262

* There is some overlap in papers cited, so 'Total' refers to the actual number of papers cited under each heading.

Researchers should thus take care when interpreting studies using Q to examine specifically what implementation was used.

4.2.3 Dendrogram-based measures (FD, FD_{LD})

FD (Petchey and Gaston 2002a) is the branch length across the regional functional dendrogram that is required to join the set of species present in an assemblage. The regional functional dendrogram describes the functional relationships among the species in a region, and is constructed from the traits of species. It is important to note that only one dendrogram is constructed: the regional dendrogram (regional because it contains all the species of interest in the study region). The FD of local assemblages is the length of branches required to connect across this regional dendrogram the species in a local assemblage.

In FD_{LD} (Podani and Schmera 2006), each local assemblage has its own dendrogram constructed. This is very different from FD (Petchey and Gaston 2002a), in which only one dendrogram, the regional dendrogram, is constructed. This has no effect on what the measures can be used for. For example, both FD and FD_{LD} can be used to measure the relative functional diversity of local assemblages; both can be used to assess the effect of new species (invaders) on functional diversity.

4.3 Applications of functional diversity

We compiled a list of papers that cite seven of the key functional diversity methods papers: FAD/MD (Walker et al. 1999, Heemsbergen et al. 2004), FD (Petchey and Gaston 2002a), FD_{var} (Mason et al. 2003), Q (Botta-Dukát 2005), CHV (Cornwell et al. 2006) and FD_{LD} (Podani and Schmera 2006) from the ISI database. The list of citing papers was examined to see how continuous functional diversity measures are being used (see Table 4.1). Each citing paper was classified into one of eight categories according to their main focus. In total, these seven methods papers were cited by 262 unique papers (FAD 137, MD 45, FD 82, FD_{var} 22, Q 12, CHV 9, FD_{LD} 5).

One striking pattern in papers citing the seven methods papers is the low application to empirical work. Just 16 (out of 262) studies have attempted to apply one of the six functional diversity indices to empirical data. Heemsbergen et al. (2004) manipulated the functional dissimilarity of detritivore communities in soil microcosms (measured using MD). They found that functional dissimilarity, and not species number, led to community compositional effects on key ecosystem processes (loss of leaf litter mass and soil respiration). MD was also used to assess the functional similarity of plants in an Australian rangeland community (Walker and Langridge 2002). This study followed on from the methods used in Walker et al. (1999); however, this time they found that the most dominant species in the community were no more dissimilar to each other than to all other species.

FD and MD have been compared to species richness (SR) and functional group richness (FGR) using data from the BIODEPTH experiments (Petchey et al. 2004b). Here, it was found that FD and MD had greater explanatory power because they use a much greater amount of trait information and can allow for small differences between species that functional groups ignore. SR, FGR, FD, and MD have also been analyzed for roadside data from Bibury in the UK (Thompson et al. 2005). FD has been used to test if functional diversity of exotic mammalian predators leads to extinction of island bird species (Blackburn et al. 2005) and if there is a loss in functional diversity of a tropical amphibian community after logging activities (Ernst et al. 2006). De Bello et al. (2006) used a variation of Q to calculate the functional diversity of plots exposed to various levels of sheep grazing.

Recently (2007–2008), the number of papers incorporating these continuous measures of functional diversity into empirical studies has notably risen. This is perhaps a sign that they are being increasingly recognized as a useful way to examine applied questions. Schamp et al. (2008) used CHV as one of six metrics to estimate the dispersion of traits (biomass, height, and seed mass) in their target plant community. Epps et al. (2007) used Q to calculate the chemical diversity of litter and foliar mixtures. They then examined the relationship between species richness and chemical diversity using published data from temperate and tropical forest systems. Jiang et al. (2007) compared the effects of plant functional diversity (measured using FD_{var}) on community productivity and soil water content in an experiment on artificial plant communities. Mason et al. (2007, 2008) examined the functional diversity of French lake communities using FD_{var} as a measure of functional divergence. Mouillot et al. (2007) used both FD and FD_{var} to determine whether assembly rules in lagoon fish communities are driven by functional traits and to seek relationships between functional diversity of fish and environmental gradients. FD has also been used to calculate the functional diversity of zooplankton in Canadian lakes (Barnett and Beisner 2007), applied to long-term avian distribution datasets (Petchey et al. 2007), and used to determine the functional significance of forest diversity in a long term biodiversity experiment (Scherer-Lorenzen et al. 2007b).

Another subset of citing papers consider the pros and cons of these functional diversity indices. Petchey and Gaston (2006) provide a review of the most popular functional diversity measures in BEF (Biodiversity-Ecosystem Functioning) research, some of which are evaluated by Ricotta (2005a). Podani and Schmera (2006) review dendrogram-based measures of functional diversity. Mason et al. (2003) highlight that FD does not account for species abundance, and claim incorrectly that it is a simple surrogate for SR. They acknowledge the shortcomings of FD_{var} in only accounting for one character at a time. Botta-Dukát (2005) suggests that Q is the best alternative as it uses both species abundance and pairwise functional differences between species, a view echoed by other authors (Lepš et al. 2006). It has been proposed that functional diversity can be split into functional richness, evenness, and divergence (Mason et al. 2005). Here, FD can be considered a measure of functional richness, while FD_{var} and Q can be considered measures of functional divergence. A functional diversity framework for BEF research has also been suggested, based around use of response and effect traits (Lavorel and Garnier 2002, Naeem and Wright 2003).

There still appears to be a sizeable portion of studies that make use of functional group richness, despite this method being discouraged in many of the continuous functional diversity papers. This is particularly true of older studies (2000–2003) that cite Walker et al. (1999). This is unsurprising given that, at this stage, continuous measures of functional diversity were not yet widely recognized. For example, Allen et al. (2003), Anderson et al. (2000), Davic (2003), Decocq and Hermy (2003), Hector et al. (2000), Lepš et al. (2001), Lloret and Vila (2003), Ni (2003), and Richardson et al. (2002) all used discrete functional groups or plant functional types as measures of functional diversity. More recently, trophic groupings have been empirically shown to be inappropriate for grassland species (Heisse et al. 2007), arable plants (Hawes et al. 2005), lake-dwelling shredders (Bjelke and Herrmann 2005), and larval anuran predators (Chalcraft and Resetarits 2003).

Discrete functional groupings continue to be employed in studies of functional diversity. For example, Berg et al. (2004), Bret-Harte et al. (2004), Chu et al. (2006), Dimitrakopoulos et al. (2006), Downing (2005), Fischer et al. (2007), Krab et al. (2008), Micheli and Halpern (2005), Moretti (2006), and Zavaleta and Hulvey (2007) all used trophic groups, functional groups or grass/forb/legume classifications. In many cases, the use of discrete functional groupings seems to be a result of the difficulty in applying trait-based functional diversity measures to real systems. The large amounts of data collection necessary for appropriate trait clustering may prove impractical for many studies. Additionally, there may be a perception that the subjectivity surrounding the choice and number of traits makes continuous measures as arbitrary as discrete measures of functional diversity. This point is further highlighted by the number of authors that have used variations of the main continuous functional diversity measures (Roscher et al. 2004, Fukami et al. 2005, Pavoine and Doledec 2005) or proposed additional functional diversity indices (Dumay et al. 2004, Ricotta 2004, Bady et al. 2005, Mouillot et al. 2005a, Thuiller et al. 2006a).

In spite of this apparent reluctance to embrace continuous measures of functional diversity in empirical BEF studies, there seems to be an increasing appreciation for the benefits of these measures over discrete functional groupings. A bootstrap analysis on empirical data from the BIODEPTH experiments shows that FGR has relatively poor power for explaining variation in an ecosystem process (Petchey 2004). Naeem and Wright (2003), Jax (2005), Ricotta (2005a) and Bulling et al. (2006) have all recommended the use of functional diversity for advancing the understanding of how biodiversity affects ecosystem functioning, rather than focusing just on species diversity. There has been a call for studies of parasite biodiversity to look to taxonomic and functional diversity for patterns and processes that have escaped notice to date (Poulin 2004). Attempts have also been made to encourage phytoplankton ecologists to make use of the new methods for studying functional diversity in lakes (Weithoff 2003). This could be important for BEF research given the short generation times associated with phytoplankton, allowing true succession to take place in one season. Finally, it has been demonstrated that even for species-rich systems like coral reefs, a single functionally important family, parrotfish, are the only creatures carrying out significant bioerosion, a major reef process (Bellwood et al. 2003). Without attempting to quantify the functional diversity of ecosystems and the consequences of loss of that diversity in an appropriate way, we are in danger of overlooking a crucial aspect of BEF relationships.

Other studies that cite these key functional diversity methods papers discuss the use of traits (Cornelissen et al. 2003, Norberg 2004, Poff et al. 2006, Barnett et al. 2007) and species abundances (de Bello et al. 2007, Lavorel et al. 2008) in quantifying functional diversity. There is also a significant body of work that recommends the use of functional diversity as an ecological indicator in environmental impact assessment (Bady et al. 2005, Heino 2005, Mouillot et al. 2005b, Dierssen 2006, Henle et al. 2006, Mouillot et al. 2006). Here, in the presence of increased environmental stress, the range of functional attributes is likely to be narrow as only the most adapted species survive, with the niche filtering concept suggesting that surviving species will share many biological characteristics (Franzén 2004, Statzner and Moss 2004).

4.4 Differences between measures. Effects of species richness on functional diversity

4.4.1 Background

Biodiversity can affect ecosystem processes by a variety of routes (Tilman 1999b, Chapin et al. 2000c). As biodiversity changes, so does composition and richness. For example, an invasion adds a novel species with a novel identity, thereby changing composition; it also increases species richness by one. During the late 1990s and into the first decade of the twenty-first century an acrimonious debate flared around these effects (Loreau *et al.* 2001). Which was more important: the effects of the identity of a novel species, or the effects of having one more species (Loreau *et al.* 2001)?

Experimental and theoretical results suggested that both can be important in different situations (Hooper and Vitousek 1997, Huston 1997, Tilman 1997). One investigation focused on the effects of species richness and composition on functional diversity (Petchey and Gaston 2002a). When is functional diversity affected strongly by species richness, and when strongly by species composition? Since functional diversity should relate with ecosystem processes, the answer to this question also informs about the relative importance of species richness and composition for ecosystem processes.

Since the then-dominant measure of functional diversity, namely functional group richness, was problematic for answering questions about the importance of species richness, Petchey and Gaston (2002a) developed a continuous measure (FD – see above). Results of this and a related study (Petchey and Gaston 2002b) indicated that the effective dimensionality of trait space determined the relative importance of species richness and composition. The effects of species richness were strongest in high dimensional space; the effects of composition dominated in low dimensional space.

The effective dimensionality of trait space results from the number of traits used to measure functional diversity and the correlation (or lack thereof) between them. Many uncorrelated traits means high effective dimensionality, while many correlated traits or few traits (regardless of correlation) means low effective dimensionality.

Thus the nature of the variation among species was critically important for understanding patterns of functional diversity. Furthermore, Petchey and Gaston (2002a) found that this conclusion was robust to details of how their functional dendrogram was constructed. Here, we extend Petchey and Gaston's simulations and ask how robust is their conclusion in the face of variation in how functional diversity is measured.

4.4.2 Methods

A collection of local assemblages varying in species composition and richness were simulated. The species richness gradient, from six to 20 species, was constructed by randomly selecting species from a regional species pool containing 20 species. At each species richness level there were up to five distinct communities, each containing a different random composition of species. Obviously, there was only one composition possible for the 20 species community, and some random draws within a richness level may have contained the same set of species. Five sets of assemblages that each represented an extinction trajectory from 20 to six species were also constructed. These trajectories are useful to know how loss (or gain) of a species can affect a functional diversity measure.

The functional diversity of each local community was calculated by the methods described in Section 2 (CHV, FD_{var}, MD, Q, FD, FD_{LD}). Simulations and measurements of all measures of functional diversity were made in R (R Development Core Team 2008). Where required, distances were Euclidean and the clustering algorithm was UPGMA. Contact OLP for R script to calculate the measures of functional diversity, including convex hull volume.

Calculating convex hull volume is complex (Cornwell *et al.* 2006). In the R library *geometry* (Grasman and Gramacy 2008) the function *convhulln* interfaces the Qhull library (Barber *et al.* 1996). For one dimensional data (one trait) we made CHV the range of the data (maximum value minus the minimum value). CHV requires that the number of species is greater than the number of traits, since each species is a vertex in the convex hull, hence the lowest richness in the simulations was six species.

We made two separate sets of simulations. In one, species differed in only one trait, and in the other they differed in five traits. Trait values were drawn from a normal distribution with a mean of zero and standard deviation of one. When there were five traits, there were no correlations between them.

4.4.3 Results

The effect of trait number (one versus five) was consistent across all six diversity measures. Increasing the number of traits caused a greater

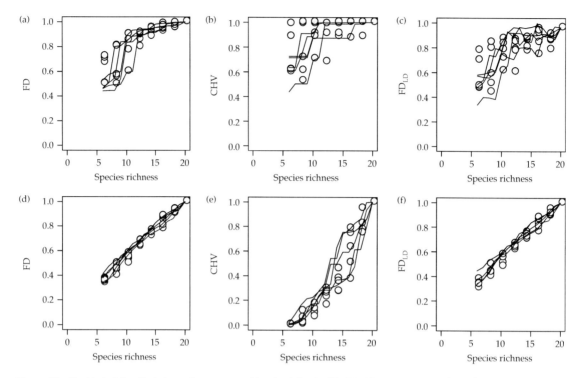

Figure 4.1 Simulated relationships between three measures of functional diversity (FD, CHV, FD_{LD}; see Section 4.4 for explanation) when species in the regional pool differ in one trait dimension (top row) or five trait dimensions (bottom row). At each richness level there are five different communities (filled circles), each a random draw of species from the regional pool. There are also five extinction trajectories, from 20 species to six species, shown by unbroken lines.

effect of species richness on functional diversity values, and a weaker effect of species composition. The effect was clearest for FD, CHV, and FD_{LD} (Fig. 4.1), and was also present for MD, Q, and FD_{var} (Fig. 4.2). With one trait, values of functional diversity varied greatly within a level of species richness (panels (a), (b), and (c) in Figs. 4.1 and 4.2); there was much reduced variation with five traits (panels (d), (e), and (f) in Figs. 4.1 and 4.2). Similarly, with one trait, different sequences of species loss had very different effects on the loss of functional diversity. With five traits, loss of functional diversity was more similar among different sequences of species loss.

The general relationship between species richness and functional diversity differed greatly among the six measures. Three (FD, CHV, and FD_{LD}) exhibited an overall positive relationship with species richness (Fig. 4.1). Two (MD and Q) exhibited little relationship with species richness, and one (FD_{var}) appeared to show a negative relationship (Fig. 4.2).

Effects of species loss differed qualitatively between the six measures. Two (FD and CHV) either decreased or remained the same when a species was lost (Fig. 4.1). The other four (FD_{LD}, MD, Q, FD_{var}) decreased, increased, or remained unchanged (Figs. 4.1 and 4.2).

4.4.4 Conclusions

The effect of number of traits on the relative importance of species richness and composition (Petchey and Gaston 2002a), appears robust to different measures of functional diversity. Therefore, to understand the effects of richness and composition, the choice of which traits to include in measures of functional diversity is obviously critical. The choice of which functional diversity measure to use with this question is less important. This

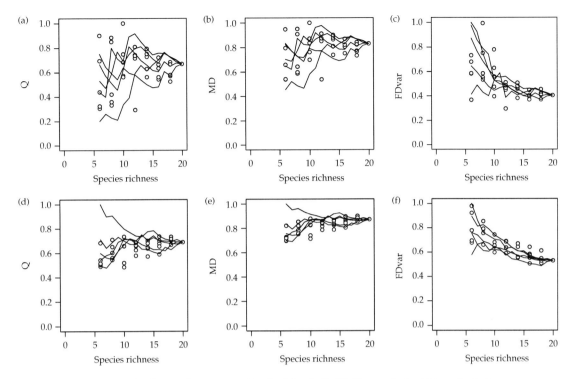

Figure 4.2 Simulated relationships between three measures of functional diversity (Q, MD, FD$_{var}$; see Section 4.4 for explanation) when species in the regional pool differ in one trait dimension (top row) or five trait dimensions (bottom row). All other details are the same as in Fig. 4.1.

reinforces previous findings (e.g. Petchey and Gaston 2002a) that details of the functional diversity measure (such as distance measure and clustering method) may have little importance relative to choosing appropriate sets of traits.

The qualitative effect of species richness on functional diversity does, however, differ between diversity measures. For example, FD, CHV, and FD$_{LD}$ generally increase with increases in species richness, while the others (MD, Q, FDvar) do not. This broad difference results from the fact that loss of a species can cause increases in some measures of functional diversity (MD, Q, FDvar), but cannot cause an increase in other measures (FD, CHV). FD$_{LD}$ can increase, but only rarely and by small amounts (Figs. 4.1(c) and (f)). Should a measure of functional diversity be able to increase when an extinction occurs? Should a measure be able to decrease when a species is added?

Suggestions that the answer to the last two questions is 'no' mean that a positive relationship is correct (Ricotta 2005a); only FD and CHV meet this criteria. However, others suggest that measures of functional diversity should not be closely correlated with species richness, or even that functional diversity should be independent of species richness (e.g. Mason et al. 2003). It remains unclear, however, how to reconcile these ideas with the suggestions that increases (or decreases) in species richness cannot cause a decrease or (increase) in functional diversity. Furthermore, there may be multiple 'facets' to functional diversity that are not included in our analyses. For example, how regularly are species distributed in trait space, and the regularity of the distribution of species abundances in trait space (Villéger et al. 2008). Indeed, adding species abundances to measures of functional diversity greatly increases the complexity of measurement.

Why do some measures behave very differently from others? Consider the idea of the species in an assemblage forming a cloud of data in

n-dimensional trait space. FD and CHV are intuitively similar, and both measure, in some broad sense, the volume of trait space that is represented by the data cloud. As they both measure volume (in some sense), adding a species cannot cause a decrease, and removing a species cannot cause an increase. They do, however, differ in the sense that volume is measured. CHV is the minimum (convex) volume that includes all the species (points in the data cloud). Whether any species, or how many species, occupy the space inside the convex hull has no effect on CHV. This is clear in Fig. 4.1(b), where loss (or addition) of species often has no effect on CHV. Furthermore, CHV is influenced perhaps more than other measures by the most extreme (i.e. minimum and maximum) trait values, since these, by definition, determine the volume. In contrast, FD measures, in a sense, a combination of the size of the data cloud, and how filled in is the data cloud. That is, FD is affected both by the range of trait values present, and by the different species that might occur within the range. (FD_{LD} is in these respects similar to FD and CHV.)

MD, Q, and FD_{var} are also intuitively similar to each other. These each represent, in some sense, the average distance between species in n-dimensional space. Here, adding a species that lies close to several other species can decrease the average distance, or increase it if the new species is distant from most others. Conversely, removing a species that lies close to other species will likely increase the average distance.

4.5 Discussion

4.5.1 Which measure?

First, this question is secondary to that of which traits to include (Naeem and Wright 2003, Lepš et al. 2006, Petchey and Gaston 2006). Indeed, the question of what qualifies as a 'functional trait' may be complex and hierarchical (Violle et al. 2007). The measure can be less important than the traits included (Figs. 4.1 and 4.2), and this phenomenon is probably quite general. A consequence is that we risk focusing too much effort on the rather 'easy' practice of inventing new measures and modifying old ones (as demonstrated by the high proportion of new measures of functional diversity in Table 4.1), and far too little effort thinking about which traits to include. What traits of plants are important determinants of biomass production (Díaz and Cabido 1997)? What traits of stream invertebrates are important determinants of leaf litter breakdown (Jonsson and Malmqvist 2000)? To stand any chance of measuring functional diversity correctly, these questions (and those about the traits of other organisms) must be answered. Which traits to include is a great challenge – 'great' in the sense that it gives a reason and opportunity to spend time learning more about what the organisms in ecosystems do.

If we are confident that we have the functionally important traits, only then can we ask 'which measure?' There certainly isn't one answer to this question, but here are some thoughts on how to decide. First, decide if you (i.e. your ecological question) need a measure of functional diversity that includes evenness in relative abundances of species (Q and FD_{var}) (Ricotta 2007). In doing so, note that a 'yes' answer means that addition of a species can decrease functional diversity and loss of a species can increase it (Figs. 4.1 and 4.2, panels (d) and (e)). If you need a measure that does not include species' relative abundances, and does not allow addition (or loss) of species to decrease (or increase) functional diversity, then FD or CHV are appropriate. If you need a measure that incorporates how much of a range of trait space is filled choose FD; otherwise CHV is likely sufficient.

If you can't make up your mind about any of these questions, or are curious, see how the answer to your particular question depends on the measure (as in Section 4.4). You may find, as here, that the choice of measure is less important than which traits are included in the measure.

4.5.2 Conclusions

What is the significance of this enormous activity in developing metrics for the quantification of functional diversity? Functional diversity is a vital component of research about the functional consequences of biodiversity (Biodiversity-Ecosystem Functioning, or BEF research). Indeed, the

Millennium Ecosystem Assessment concluded that species *per se* is less important than what the species do. Arguably this knowledge was implicit in the design of some of the earliest biodiversity experiments. For example, Naeem *et al.* (1994) chose for the Ecotron experiments a functionally balanced set of plant species. Consequently, the ecosystem services that humans rely upon are more likely related with the functional diversity of organisms, rather than taxonomic or phylogenetic diversity. The examples in Section 4.3 provide some empirical evidence for this. The explosion of different measures may represent the recognized importance of creating an accurate and predictive measure. It has always been clear that we first need to know which traits are important for each ecosystem service we might choose to focus on. Our analyses in Section 4.4 show that the choice of traits may hold more sway than the choice of a functional diversity index. The conclusion must therefore be that understanding species' natural histories (traits) and how these interact with their environment is a prerequisite for functional diversity to be a concept that can lead to management and conservation of ecosystem services.

Other important challenges remain. The data demands of including intraspecific variability in measures of functional diversity (Cianciaruso *et al.* 2009). It seems clear, however, that variation among the individuals in a species could have important effects on community and ecosystem processes (e.g. Pachepsky *et al.* 2007). Another challenge is how to estimate total functional diversity from a sample. Rarefaction methods can be applied to functional diversity measures and have potential (e.g. Walker *et al.* 2008). Perhaps most importantly, we need to carry out experiments that test the assumptions and predictions of functional diversity. The knowledge from these will help us understand the broad ecological questions to which functional diversity can make a significant contribution.

4.6 Acknowledgements

We thank participants at the BioMERGE-DIVERSITAS meeting in Ascona, Switzerland, in December 2006, including Justin Wright, Martin Solan, Katia Engelhardt, Mahesh Sankaran, and Dan Bunker.

CHAPTER 5

Forecasting decline in ecosystem services under realistic scenarios of extinction

J. Emmett Duffy, Diane S. Srivastava, Jennie McLaren, Mahesh Sankaran, Martin Solan, John Griffin, Mark Emmerson, and Kate E. Jones

5.1 Introduction

Estimates suggest that Earth is in the midst of the sixth mass extinction in its history, with rates of species loss 3–4 orders of magnitude above the average over geologic time (Pimm *et al.* 1995, Dirzo and Raven 2003). Unlike previous mass extinctions caused by asteroid impacts, intense volcanism, or geochemical changes (Vermeij 2004), the current crisis is driven by an entirely unprecedented phenomenon: human transformation of the global environment (Vitousek *et al.* 1997, Sala *et al.* 2000). The principal drivers of this extinction are habitat loss, over-exploitation, pollution, and impacts of invasive species (Purvis *et al.* 2000c). These drivers have not only dramatically increased the background rate of extinction, but they have also generally reduced biodiversity at local scales, and altered the species composition of communities. Superimposed on global and local declines in diversity, the distributions of many remaining species are being radically altered as human activities transport them, intentionally or otherwise, to new regions and habitats. Establishment of these non-native species can increase local species richness, but further alters the composition of communities (Sax and Gaines 2003) and ultimately leads to homogenization and decline of biodiversity across the globe.

These changes in biodiversity have raised serious concerns about their consequences for human welfare. Partly for this reason, in the early 1990s a concerted research effort arose to identify and understand the relationships between changing biodiversity and functioning of ecosystems (Schulze and Mooney 1994). Using synthesized communities of different species richness under controlled conditions in field plots and in the lab, researchers focused explicitly on the question of how the number of species (or functional groups) within a community influences aggregate ecosystem properties such as productivity and resource use (Kinzig *et al.* 2001, Loreau *et al.* 2001, Loreau *et al.* 2002, Hooper *et al.* 2005, and Chapters 1–3). Theoretical, experimental and observational validation studies of biodiversity-ecosystem functioning linkages have subsequently developed into a growth industry, as the present volume attests.

Motivated in part by conservation concerns, initial studies focused on the likely environmental consequences of biodiversity loss, but the complicated experiments and mixed interpretations soon refocused efforts on nuances of design and statistics (Huston 1997, Allison 1999, Loreau and Hector 2001). One early result was that experiments examining biodiversity-ecosystem function links converged on a 'random assembly' design, which synthesized simple – but often artificial – communities of varying species richness under controlled conditions and measured changes in ecosystem properties along the gradient in species richness (Schmid *et al.* 2002a). While this approach is capable of rigorously partitioning the effects of species number and interactions on ecosystem processes (Chapter 7), the factors producing diversity change were unspecified and, implicitly, regarded as unimportant. As a consequence, the focus of experiments became disconnected from the problems of applied conservation science (Srivastava 2002, Duffy 2003, Srivastava and Vellend 2005). Despite their statistical advantages, it has been recognized for some time that the random assembly designs common in biodiversity and ecosystem functioning experiments usually do not mimic real-world processes of diversity loss (for a counter-example, see the sensitivity to stress scenario in Solan *et al.* 2004), which arise as traits of organisms mediate responses to specific drivers of environmental change (Huston 1997, Grime 1998, Srivastava 2002, Giller *et al.* 2004, Lepš 2004). In essence, random assembly designs trade off realism for precision in interpretation. To assess the implications of this trade-off, theory (Gross and Cardinale 2005), simulations (Petchey and Gaston 2002b, Ostfeld and LoGiudice 2003, Solan *et al.* 2004, McIntyre *et al.* 2007), and experiments (Jonsson *et al.* 2002, Zavaleta and Hulvey 2004) have begun to explore effects on ecosystem properties of non-random extinction, simulating explicit loss scenarios based on vulnerability of species with certain traits or mimicking natural gradients of diversity loss in nature (see Fig. 3.1).

In this chapter we explore the prospects for a more rigorous linkage of biodiversity and ecosystem functioning research to empirically based processes of extinction. The key to developing a general predictive framework for relating environmental change, through biodiversity change, to effects on ecosystem services is a focus on organismal traits (Chapin *et al.* 1997, Hooper *et al.* 2002, Lavorel and Garnier 2002, Naeem and Wright 2003, McGill *et al.* 2006), and specifically the correspondence, or lack thereof, between (1) *response* traits that influence risk of decline or extinction in response to a driver, (2) traits – involving both response and effect – that mediate community reorganization after primary extinction(s), and (3) *effect* traits that influence ecosystem functioning. The term 'trait' has been used in a variety of ways and its history is fraught with some confusion (Violle *et al.* 2007). Darwin used the term to refer to a morphological, behavioural, or physiological characteristic of an organism that affects performance. Thus, in this original sense, it is inherently related to function. We use the term in essentially this sense here, emphasizing that a trait is a property that can, in principle, be measured on an individual organism.

Making the connection from environmental change to altered ecosystem services can be thought of as a process involving three overlapping stages. Given a particular driver of environmental change, we need first to know whether it will cause an *extinction* or decline of species: trait-based models of vulnerability can make testable predictions about what kinds of species will be lost (and gained). The second stage is *structural reorganization*: basic principles of trait-based community interactions and succession can be used to predict how a community reorganizes in response to species loss and gain. Third, and finally, is *functional reorganization*: trait-based principles of physiology and ecology and insights from biodiversity and ecosystem functioning research can, in principle, predict consequences of the changed community structure for biomass distributions, and rates of productivity and biogeochemical processes.

A simple example of application of this trait-based framework involves the trophic cascade: (1) the correlated traits of large body size, carnivory, and associated 'slow' life history (low fecundity and population growth rate, long generation time) render species vulnerable to overexploitation and harassment, (2) consequent loss of large predators causes relatively predictable structural changes in a community, typically increasing abundance of their herbivore prey and indirectly reducing plant abundance, and (3) these structural changes produce functional changes, typically including reduced primary productivity. This general phenomenon has been observed in a wide range of ecosystem types in response to habitat alteration and hunting (e.g. Pace *et al.* 1999), showing that some degree of generalization is possible in linking realistic impacts of environmental change, through changing biodiversity to changing ecosystem functioning.

Although we are a long way from generalizing such an integrated approach, important advances have been made. We can begin such an integration by summarizing what is known about how organismal traits affect risk of extinction under major drivers of environmental change, how traits influence community reorganization following extinctions, how traits affect major ecosystem processes, and the degree of covariance among these response and effect traits. To date, efforts to 'connect the dots' between these components have focused mainly on plant assemblages (e.g. Díaz and Cabido 1997, Lavorel and Garnier 2002, Díaz et al. 2007). Here we begin to extend trait-based approaches linking extinction and ecosystem functioning to multitrophic ecosystems and we review the few empirical studies that have explicitly addressed how realistic extinction scenarios affect ecosystem functioning.

5.2 Extinction filters: environmental change and response traits mediating decline

5.2.1 Background

Recent years have seen major growth in understanding how organismal traits predispose species to endangerment and extinction, across a range of environmental drivers and taxa including plants (Suding et al. 2005, Wiegmann and Waller 2006), amphibians (Lips et al. 2003, Cooper et al. 2008), reptiles (Foufopoulos and Ives 1999), birds (Gaston and Blackburn 1995, Davies et al. 2007), mammals (Purvis et al. 2000a, Fisher et al. 2003, Jones et al. 2003, Blackburn et al. 2005, Cardillo et al. 2005), and marine fishes (Jennings et al. 1998, Jennings et al. 1999, Dulvy et al. 2004), as well as some synthesis across taxa (Purvis et al. 2000c). The probability that a species will go extinct depends both on intrinsic traits, such as body size and degree of habitat specialization, and on extrinsic factors. There is abundant evidence that low population density, small geographic range, and proximity to human populations increase risk of decline and extinction across a broad spectrum of plant and animal taxa (e.g. Laurance 1991, Foufopoulos and Ives 1999, Purvis et al. 2000b, Harcourt et al. 2002, Jones et al. 2003, Blackburn et al. 2004, Cardillo et al. 2004, Henle et al. 2004, Hero et al. 2005, Kiessling and Aberhan 2007). Despite their clear importance, these characters are, at least in part, emergent properties of the organisms interacting with their environments. We focus here primarily on intrinsic traits of individual organisms.

5.2.2 Empirical results

As a first pass at finding general patterns linking response traits to extinction risk, we searched the literature for data relating intrinsic organismal traits to species decline as a basis for developing empirically based extinction trajectories and predicting their ecosystem impacts. We focused on several drivers of environmental change known, suspected, or anticipated to be important in affecting species abundance and distribution; these included loss or fragmentation of habitat, nitrogen deposition, atmospheric CO_2 concentration, climate warming, precipitation, and exploitation. Most of the empirical studies we found correlated intrinsic organismal traits with species decline or rarity in response to these factors (Tables 5.1 and 5.2).

A notable feature of the results is the contrast between patterns for plants and animals. The studies of plant population decline that we found considered nearly the full spectrum of drivers, including nitrogen deposition, habitat fragmentation and loss, atmospheric CO_2 concentration, warming, and precipitation (Table 5.1). The animal studies spanned a narrower range of drivers, mostly involving habitat loss (including fragmentation and conversion), which was also a major driver for plant decline, and overharvesting (Table 5.2). These different foci may reflect real differences in processes driving decline of animal versus plant species, but it is also possible that research effort or perception of which drivers are most important may differ among animal and plant ecologists, or among marine and terrestrial researchers (Raffaelli et al. 2005b). For example, despite the paucity of studies linking plant extinction to harvesting, it's clear that a number of medicinal and ornamental plant species have been endangered by harvesting, including the rosy periwinkle (*Catharanthus roseus*) of Madagascar and the Pacific yew (*Taxus brevifolia*) in the USA. The plant studies also tended to focus on functional groups (grasses, forbs, trees), life history syndromes (perennial, annual), and taxonomic groups (dicots, legumes) rather than individual species, such that the plant analyses often

Table 5.1 Plant traits associated with population change in response to environmental change in empirical studies. Response shows the sign of population change associated with the listed trait.

Driver	Trait	Response	Reference
N-deposition	Life history: perennial	−	1
	Taxonomy: legumes	−	1
	Pollination: self	−	2
	Growth form: forbs	−	3
	Growth form: trees	−	4
	Growth form: grasses	+	3
Habitat loss and fragmentation	Stature: short	−	5, 6, 7
	Seed longevity: short	−	2
	Seed size: large	−	8
	Dispersal: wind	−	5
	Dispersal: ant	−	5
	Seed size: small	−	6
	Taxonomy: legumes	−	6
	Taxonomy: dicots	−	7
	Pollination: animal-assisted	−	9, 10
Global change (increased CO_2)	Growth form: C4-grasses	−	11, 12
	Growth form: C4-grasses	0	13
	Growth form: C3-grasses	−	13, 14, 15
	Growth form: C3-grasses	+	11
	Growth form: forbs	+	11, 12, 13, 14, 15
	Growth form: C3-trees	+	16, 17
	Taxonomy: legumes	+	11, 14
Global change (warming)	Growth form: shallow-rooted forbs	−	18
	Growth form: forbs	−	19
	Growth form: forbs	+	3
	Growth form: grasses	0	3
	Growth form: tap-rooted forbs	0/+	18
	Growth form: shrubs	+	19
Global change (increased precipitation)	Growth form: grasses	0	3
	Growth form: trees	+	4, 20, 21
	Growth form: forbs	+	3

1 Suding et al. (2005); 2 Freville et al. (2007); 3 Zavaleta et al. (2003); 4 Davis et al. (1999); 5 Williams et al. (2005); 6 Leach and Givnish (1996); 7 Duncan and Young (2000); 8 Cramer et al. (2007); 9 Sodhi et al. (2008); 10 Aguilar et al. (2006); 11 Reich et al. (2001); 12 Polley et al. (2003); 13 Owensby et al. (1999); 14 Teyssonneyre et al. (2002); 15 Potvin and Vasseur (1997); 16 Polley et al. (1997); 17 Bond and Midgley (2000); 18 Cross and Harte (2007); 19 Harte and Shaw (1995); 20 Kraaij and Ward (2006); 21 Davis et al. (1998).

considered suites of co-occurring traits. Although animal studies more commonly addressed single traits such as body size and habitat specialization, many also identified one or more components of a multi-trait 'slow life history' syndrome as a risk factor. Finally, several animal studies found that extinction risk was greater at higher 'trophic level', a complex concept defined by the organism's interaction with its environment, rather than an organismal trait *per se*. Despite its ambiguous nature, we have retained trophic level as a 'trait' in our analyses, since it is at least partly dictated by individual-level traits (body morphology, dentition, digestive physiology, etc.), and is relatively conserved within clades. We recognize, however, that future research would benefit from more rigorous identification of the intrinsic traits that dictate an organism's trophic level.

Table 5.2 Animal traits associated with population change in response to environmental change in empirical studies. Response shows the sign of population change associated with the listed trait. We follow the original authors in treating trophic level as a 'trait', although this is a proxy for a syndrome of intrinsic traits and interactions with the external environment (see text).

Driver	Trait	Fish Response	Fish Reference	Reptiles Response	Reptiles Reference	Mammals Response	Mammals Reference	Birds Response	Birds Reference	Invertebrates Response	Invertebrates Reference
Habitat loss and fragmentation	Body size: large	–	1			– – – –	2, 3, 4, 5	– – 0	6, 7, 8	0 0 0	9, 10, 11
	Body size: small			– – – –	13, 14, 15	– 0	2, 16	0	8, 12		
	Specialization	–	1			– – – – 0	16, 22, 23	– – – –	7, 12, 17, 18, 19	– –	20, 21
	Life history: 'slow'	–	1			– – 0	22		18, 24, 12		
	Trophic level: high									– – – – – 0	9, 10, 25, 26, 27, 11
Climate warming	Specialization					–	28		29		
Exploitation	Body size: large	– – – –	30, 31, 32, 33			–			12		
	Body size: small					0			12		
	Life history: 'slow'	– –	30, 33			–	34		12		
	Trophic level: high	– –	35, 36								
Undifferentiated	Body size: large	0	37			0	38	– – –	6, 39, 40		
	Body size: small			–	41						
	Specialization	–	37		14			– – –	39, 40, 42	– –	43, 44
	Life history: 'slow'			–	41	0	38		42		
	Trophic level: high					0	38	– – –	39, 40, 42		

1 Olden et al. (2008); 2 Harcourt et al. (2002); 3 Cardillo et al. (2005); 4 Collen et al. (2006); 5 Johnson et al. (2002); 6 Gaston and Blackburn (1995); 7 Feeley et al. (2007); 8 Foufopoulos and Mayer (2007); 9 Didham et al. (1998a); 10 Davies et al. (2000); 11 Starzomski and Srivastava (2007); 12 Owens and Bennett (2000); 13 Foufopoulos and Ives (1999); 14 Segura et al. (2007); 15 Watling and Donnelly (2007); 16 Laurance (1991); 17 Shultz et al. (2005); 18 Jiguet et al. (2007); 19 Sekercioglu et al. (2002); 20 Kotiaho et al. (2005); 21 Charrette et al. (2006); 22 Purvis et al. (2000b); 23 Jones et al. (2003); 24 Amano and Yamaura (2007); 25 Didham et al. (1998b); 26 Gilbert et al. (1998); 27 Hoyle (2004); 28 Laidre et al. (2008); 29 Sekercioglu et al. (2008); 30 Jennings et al. (1998); 31 Jennings et al. (1999); 32 Dulvy et al. (2004); 33 Reynolds et al. (2005); 34 Bodmer et al. (1997); 35 Pauly et al. (1998); 36 Hutchings and Baum (2005); 37 Angermeier (1995); 38 Brashares (2005); 39 Gillespie (2001); 40 Boyer (2008); 41 Cooper et al. (2008); 42 Kruger and Radford (2008); 43 Biesmeijer et al. (2006); 44 Gonzalez-Megias et al. (2008)

Our survey of response traits suggests a few potential generalizations. For plants, these are more tentative than for animals: there is abundant evidence that commonly recognized plant growth forms respond differently to a variety of environmental change drivers. The most consistent such pattern we found is that rising atmospheric CO_2 tends to increase the abundance of forbs at the expense of grasses, and possibly favours C3 over C4 plants. There is also some evidence that warming and increased precipitation favour shrubs and trees over grasses. More generally, however, the responses of plants to environmental change appear idiosyncratic. This likely reflects the context-dependency of trait associations with response to environmental change (Lavorel and Garnier 2002, Eviner and Chapin 2003, Brook *et al.* 2008).

For animals, the patterns appear somewhat clearer. The major drivers of decline and extinction of animal species are generally recognized to be habitat loss (including fragmentation and conversion) and overexploitation (Purvis *et al.* 2000c). For most animal taxa studied, analyses considered how traits correlated with rarity in space or decline through time, assuming that these are functions of both habitat loss and exploitation/persecution. For marine fishes, there is some confidence that over-harvesting is indeed a major driver of the patterns observed (Dulvy *et al.* 2005, Olden *et al.* 2007). The studies we reviewed suggest that the response traits that render animal species vulnerable to these drivers, across a disparate range of taxa including mammals, birds, fishes, reptiles, and even some insects, are surprisingly consistent and include large body size, high trophic level, and specialization in diet and/or habitat (Table 5.2). A similar result also emerges from a very different source: in experimental microbial food webs exposed to environmental warming, extinctions occurred disproportionately among top predators and herbivores (Petchey *et al.* 1999).

5.2.3 Summary

The vulnerability of large animals and high trophic levels emphasizes that studies of predator loss and trophic cascades are directly relevant to biodiversity and ecosystem functioning research, though they have seldom been considered in this context (Duffy 2003, Thebault and Loreau 2003, Thebault and Loreau 2006, Duffy *et al.* 2007). The long and rich history of research linking plant traits to environmental gradients is also highly relevant to biodiversity and ecosystem functioning research. Although plant biodiversity and ecosystem functioning research has historically focused on species and functional groups, a more explicit focus on traits better approaches the mechanistic processes linking environmental change to ecosystem processes (e.g. McGill *et al.* 2006). Such trait-focused research is providing important generalizations linking environmental change through biodiversity change to ecosystem functioning (Lavorel and Garnier 2002, Diaz *et al.* 2004, Díaz *et al.* 2007, Suding *et al.* 2008).

Some caveats are in order. We, like most previous authors, have considered both drivers of environmental change and traits singly (with the exception of trait syndromes such as 'slow life history' and 'trophic level'). However, multiple drivers typically operate simultaneously and synergistically (Sala *et al.* 2000, Brook *et al.* 2008), and different traits can also interact to increase risk of extinction above that expected from individual trait values, as demonstrated for desert fishes (Davies *et al.* 2004, Olden *et al.* 2008). Moreover, the importance of intrinsic biological traits in predicting extinction can be sharpened in proximity to dense human populations, as demonstrated for carnivores (Cardillo *et al.* 2004). Thus, a pressing direction for future work is study of how synergisms between single and multiple drivers of extinction, and among traits, influence community reorganization and, thus, ecosystem functioning (Sala *et al.* 2000, Vinebrooke *et al.* 2004, Starzomski and Srivastava 2007).

5.3 Structural reorganization: traits and community organization

5.3.1 Background

Predicting how a community reorganizes in response to environmental change encompasses the whole of community ecology, so generalizing the role of organismal traits in this process is accordingly challenging. By analogy with ecosystem functioning, community reorganization can be considered to involve both response and effect traits, although these have received less attention in community ecology (McGill *et al.* 2006). In the context of community organization, response traits are those that mediate an organism's sensitivity to influences such as interspecific competition (e.g. plant biomass and height, Gaudet and Keddy 1988), grazing (e.g. plant height and life history, Díaz *et al.* 2007), and changing resources (e.g. specific leaf area, Ackerly and Cornwell 2007); effect traits are those that mediate an organism's influence on other species, such as predator hunting mode and habitat breadth (Schmitz *et al.* 2004, Schmitz 2008). In principle, community organization stems from interactions mediated by such response and effect traits, and the positive or negative feedbacks from these changes that influence community structure.

One indirect approach to linking traits with species interactions is to focus on how distributions of traits within communities are influenced by key community interactions. An example on a large scale is the global synthesis of data on how terrestrial plant assemblages respond to ungulate grazing. Meta-analysis of 197 studies from all over the world revealed that grazing favoured annuals over perennials, short plants over tall plants, prostrate over erect plants, and stoloniferous and rosette architecture over tussock architecture (Díaz *et al.* 2007). Such comparative studies provide insight on how traits are linked to interactions but do not reveal the dynamic responses of interacting species.

In the context of biodiversity and ecosystem functioning research, two aspects of community structural reorganization have received most attention to date. The first is biomass or numerical *compensation*, i.e. growth or expansion of remaining species to fill the void left by a species that goes extinct. Most studies of community response to extinction, like studies of biodiversity and ecosystem functioning generally, have focused on competitive assemblages within a trophic level, so the process of primary interest has been compensation by surviving species after loss of competitors. In general, theory shows that compensation tends to buffer community- and ecosystem-level properties from species loss, and that the order of extinctions is important in determining whether or not this compensation effect decays with further extinctions (Ives and Cardinale 2004). But competition and other interactions also complicate predictions. In a dynamic theoretical simulation, Ives and Cardinale (2004) confirmed that, in the absence of interactions among species, extinction order could be perfectly predicted from species' direct tolerances to the theoretical stressor. However, extinction order was less predictable in the presence of strong interspecific interactions because compensation among remaining species after an initial extinction changed the interaction pathways in the web. Moreover, the mechanism by which species diversity affects function is also important here. Specifically, compensation amplifies differences between extinction scenarios when diversity–function effects are driven by resource partitioning, but has less effect when they are driven by facilitation or sampling mechanisms (Gross and Cardinale 2005).

5.3.2 Empirical results: compensation

Several simulation studies have examined the influence of biomass compensation in mediating the impacts of primary extinctions on ecosystem processes. Solan *et al.* (2004) showed that the degree of numerical compensation among marine benthic invertebrates can have a large impact on predicted changes in biogenic mixing depth under different extinction scenarios (Fig. 5.1). Specifically, when extinction order was random, compensation by remaining species maintained community-level function (biogenic mixing depth) down to very low levels of species richness; interestingly, this compensation was much less effective when extinction proceeded in a more realistic order by body size (Fig. 5.1). Simulations of freshwater fish assemblages similarly found that the presence or absence of compensation among tropical freshwater fishes had a strong effect on the rates of nutrient recycling (McIntyre *et al.* 2007). Thus the degree of compensation can have major consequences for how and whether species loss affects ecosystem functioning.

Two important questions are: (1) How common is compensation in nature? and (2), Is there any consistent relationship between organismal traits and the strength of compensation? Regarding the

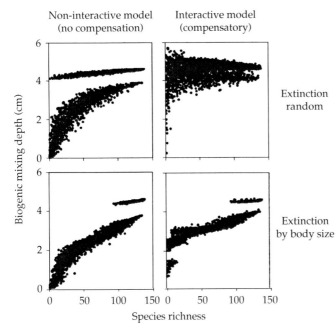

Figure 5.1 Evidence from simulation studies that extinction impacts on ecosystem functioning (biogenic mixing depth in marine sediments) depend strongly on the order of marine invertebrate species loss and the interactions among species in the reorganizing community. Simulations based on empirical data show that declining diversity has very different effects on mixing depth depending on whether extinction order is random (top) versus ordered by loss of largest-bodied species first (bottom), and whether remaining species show biomass compensation (right) or not (left). The two clouds of points in each panel represent sequences with and without the very large, mobile deposit-feeding brittle star in this system. After Solan et al. (2004), used by permission.

first question, Larsen et al. (2005) found no evidence of compensation in bee and beetle assemblages as species were lost along gradients of human impact in the field; together with a positive correlation between response (vulnerability) and effect (ecosystem process) traits, this led to decline of function as species were lost from the assemblages (Fig. 5.2). It is not clear how common such situations are. Moreover, species are not equal in their ability to compensate (Lyons and Schwartz 2001, Smith and Knapp 2003), meaning that different extinction scenarios, which leave behind a community with different species, also result in differing potential for compensation. For example, among grassland plants, dominant species compensated for removal of the rarest species, whereas rare species could not compensate for reductions in the density of the dominant species (Smith and Knapp 2003). In answer to the second question, competitive ability has been linked to particular traits, for example, to individual biomass, height, canopy architecture, and leaf shape among wetland plants (Gaudet and Keddy 1988). And compensation following lake acidification is reported to diminish at higher trophic levels (Vinebrooke et al. 2003). But we are unaware of studies explicitly addressing relationships of traits to compensation in competitive communities.

5.3.3 Background: cascading extinction

The second aspect of structural reorganization that has received attention in biodiversity and ecosystem functioning research involves the effects of species loss on subsequent loss of additional species, i.e. 'cascading (or secondary) extinction'. In natural multitrophic ecosystems the general problem of community reorganization involves not only compensation among competitors but also interactions among trophic levels (e.g. Ives and Cardinale 2004). These interactions clearly depend on traits of the species involved, as well as of the abiotic environment, implying that they are context-dependent. Several studies have addressed the effects on multitrophic communities of losing species by focusing on summary descriptors of these interactions such as trophic level, connectance, and interaction strength. Disruption of feeding links or positive interactions, such as pollination, can precipitate secondary extinctions (Pimm 1980), with further implications for

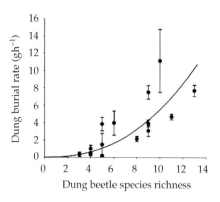 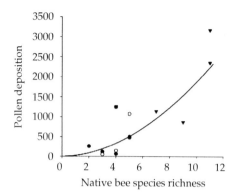

Figure 5.2 Rates of ecosystem services increase with species richness in surveys across gradients in human disturbance for two different systems, taxa, and ecosystem processes. In both cases insect abundance was positively related to species richness, suggesting weak density compensation among species. After Larsen *et al.* (2005), used by permission.

ecosystem functioning. Loss of a consumer species can release certain of its prey species from control, drive others to extinction as released competitors proliferate, and/or deprive specialized higher predators of food, fostering their extinction.

5.3.4 Results: cascading extinction

Secondary or cascading extinctions that result from loss of a focal species have been the subject of considerable attention in theory and simulation studies (Table 5.3). Recent studies have identified trophic position (Borrvall *et al.* 2000, Christianou and Ebenman 2005, Eklöf and Ebenman 2006), connectivity (Solé and Montoya 2001, Dunne *et al.* 2002b, Memmott *et al.* 2004, Eklöf and Ebenman 2006), and the configuration of trophic interactions (Christianou and Ebenman 2005) as important properties in determining which primary species deletions are most likely to cause secondary extinctions, and which species are most likely to go extinct secondarily (reviewed by Ebenman and Jonsson 2005). The theoretical results are difficult to generalize, in part because of variation among studies in assumptions, including strength and nature of competition, redundancy among species within groups, and skew in interaction strengths. Recent studies confirm that such assumptions are important to how loss of a focal species affects secondary extinction and its ecosystem consequences (Thebault *et al.* 2007).

Nevertheless, some generalizations appear possible. One apparently general finding of studies simulating species loss from empirical food webs or interaction networks is that the networks tend to be quite robust (in terms of secondary extinctions) to random loss of species, but much more vulnerable to loss of highly connected species (Solé and Montoya 2001, Dunne *et al.* 2002a, Memmott *et al.* 2004). Since skewed distribution in the number of links among species appears typical of empirical food and interaction webs (Berlow *et al.* 1999), this result emphasizes the importance of focusing conservation attention on highly connected species. A second apparent pattern emerging is that cascading extinctions seem to be most severe after loss of species at low trophic levels and to affect most severely those at higher trophic levels (Table 5.3). These patterns, however, are sensitive to variation in model assumptions and clearly preliminary.

5.3.5 Summary and conclusions

The complexity of interactions within food webs highlights both conceptual and empirical challenges to further progress. A key conceptual challenge is development of an objective, empirically based framework for defining interaction traits at the level of individual organisms that reliably predict trophic level, competitive ability, and compensation. For example, high trophic level

Table 5.3 Traits influencing community reorganization via secondary extinctions after loss of a focal species.

Study type	Trait(s) promoting secondary extinction		Effect of species richness	Reference
	Focal species	Secondary spp		
Model	–	Trophic specialization of predators (after plant or herbivore removal); trophic generalism of prey (after carnivore removal)	Richness increases secondary extinctions; connectance increases secondary extinctions	1
Model	Low trophic level (autotrophs)	–	Richness reduces secondary extinctions	2
Model	–	High trophic level	Richness reduces secondary extinctions (only in absence of stochastic variation)	3
Model	Low trophic level; strong interactions with consumers	High trophic level; weak interaction with resources; strong interaction with consumers	–	4
Model	High trophic level	–	Richness reduces secondary extinctions	5
Model	Highly connected; low trophic level	High trophic level (consumers generally)	Connectance reduces secondary extinctions generally (but increases loss of top predators)	6
Model	Low trophic level (with intraspecific competition)	High trophic level	Richness increases secondary extinctions in absence of consumer intraspecific competition, but decreases extinctions with consumer competition	7
Model; simulations of empirical food webs	–	High trophic level (consumers generally); trophic uniqueness	–	8
Simulations of empirical food webs	Highly connected	–	–	9
Simulations of empirical food webs	Highly connected	–	No effect of richness; connectance reduces secondary extinctions	10
Simulations of empirical food webs	Highly connected	–	–	11
Simulation of empirical plant-pollinator networks	Highly connected	–	–	12

1 Pimm (1980); 2 Borrvall et al. (2000); 3 Ebenman et al. (2004); 4 Christianou and Ebenman (2005); 5 Borrvall and Ebenman (2006); 6 Eklöf and Ebenman (2006); 7 Thebault et al. (2007); 8 Petchey et al. (2008); 9 Solé and Montoya (2001); 10 Dunne et al. (2002b); 11 Dunne et al. (2004); 12 Memmott et al. (2004).

has been identified as a correlate of extinction vulnerability in a range of taxa (Table 5.2), but trophic level is not a trait *per se* but an emergent feature of the focal organism's interaction with other species and its abiotic environment. Can we identify traits of individual organisms or species that index trophic level, and if so, how precisely? For example, morphological traits are routinely used by palaeontologists to identify trophic level in fossil assemblages (Dunne *et al.* 2008), and defensive chemicals in a plant may indicate its low vulnerability to grazers and thus low connectance in a food web. Similarly, are there measurable traits of individual organisms that index potential for biomass or numerical compensation? And does the degree of compensation differ systematically among taxa, by environmental conditions, or as a function of measurable individual-level traits? Can we identify traits of species that reliably predict their impacts on secondary extinctions, and their sensitivity to different drivers of extinction? Do species that are especially vulnerable to secondary extinction play important functional roles in communities? Even if the effects of traits are context-dependent, such information could still be useful if the trait–context interactions are consistent and predictable. Answering these questions will require a tighter integration between empirical investigations of species loss and interactions, and modelling of food webs.

5.4 Effect traits and ecosystem function

5.4.1 Background

Tremendous effort has been devoted to correlating plant traits with population and ecophysiological processes (effect traits), and with environmental gradients and responses to disturbance (response traits). For example, plant resource capture, productivity, and decomposition have been linked to a variety of morphological, chemical, and demographic traits (Table 5.4). This literature is extensive and we do not review it comprehensively (Hooper *et al.* 2002, Lavorel and Garnier 2002), but focus on two themes relevant to linking environmental change, through changes in biodiversity and community structure, to altered ecosystem functioning.

First, the nature of the linkage depends critically on scale. Several studies have shown that basic demographic and size traits of organisms scale allometrically (e.g. Enquist *et al.* 1999, Marba *et al.* 2007), and correspond closely to ecosystem processes, at global or cross-ecosystem scales. These include, for example, the allometric scaling of metabolism with body size in plants (Enquist *et al.* 2003). Such approaches have recently been extended to multitrophic ecosystems to predict the potential biomass and distribution of marine fishes on a global scale based on primary production and temperature (Jennings *et al.* 2008). These are results of central importance to linking traits mechanistically to ecosystem functioning, but it is unclear how such global-scale allometric relationships translate to the local scales of interest in environmental conservation and management. The scatter in global, log-log plots of trait versus process likely conceals large variance among species on local scales with narrower ranges of trait values and lower diversity. Thus, at local scales, prediction of ecosystem functioning may require finer resolution of traits.

A second, more serious complication is that many ecosystem processes are mediated by multiple traits. This poses the challenge of reducing the dimensionality of trait space, ideally by finding suites of consistently correlated traits. Accordingly, considerable effort has gone into methods for defining 'functional groups' of species (Hooper *et al.* 2002, Naeem 2002a, Naeem and Wright 2003, Petchey and Gaston 2006), yet many functional group classifications turn out to have limited usefulness at the level of local communities since functional traits can vary independently across species (Eviner and Chapin 2003, Wright *et al.* 2006). In at least some contexts, however, there are predictable suites of co-occurring traits that can be reduced to a small number of trait syndromes or 'strategies' that correlate with functional processes (e.g. Craine *et al.* 2002, Diaz *et al.* 2004). The existence of consistent suites of functional traits is also suggested indirectly by the strong differences among major clades of plants in, for example, decomposability of their foliage (Cornwell *et al.* 2008). Again, the power and utility of the

Table 5.4 A selective summary of plant traits influencing process rates in terrestrial ecosystems. Response shows the sign of the ecosystem response with increasing value of the trait.

Function	Trait	Response	Reference
Productivity	Specific leaf area (SLA)	+	1, 2, 3, 4, 5
	N concentration	+	2, 4, 5
	Plant size	+	2, 4
	Dry matter content	+	1, 5
	Rooting depth	+	1, 4
	Potential relative growth rate	+	6
	Shoot:root ratio	+	2
Decomposition	N concentration	+	4, 5, 7, 8, 9
	Taxonomy: dicots	+	9, 10, 11
	C:N ratio	−	4, 12, 13
	Lignin content	−	4, 11, 14
	Taxonomy: monocots	−	9, 10
	Specific leaf area (SLA)	+	5
	Leaf toughness	−	12
	Leaf dry matter content	−	5
Resource capture and utilization	Leaf N	+	15, 16
	Taxonomy: legumes	+	4, 17
	Specific leaf area (SLA)	+	3
	Root mass	+	16
	Plant mass	−	7
	N concentration	−	7

1 Lavorel and Garnier (2002); 2 Heisse et al. (2007); 3 Díaz and Cabido (1997); 4 Chapin (2003); 5 Garnier et al. (2004); 6 Vile et al. (2006); 7 Wardle et al. (1998); 8 Cortez et al. (2007); 9 Cornelissen and Thompson (1997); 10 Wardle et al. (2002); 11 Cornelissen (1996); 12 Perez-Harguindeguy et al. (2000); 13 Madritch and Hunter (2004); 14 Aerts (1997); 15 Diaz et al. (2004); 16 Drenovsky et al. (2008); 17 Hooper and Vitousek (1998).

relationships depend on scale. At regional or global scales, basic ecological processes can be clearly related to traits (e.g. Diaz et al. 2004, Díaz et al. 2007, Enquist et al. 2007). An important direction of research in this area involves the emergence of new approaches that allow interpretation of ecosystem responses within the context of multiple effect and response traits.

5.4.2 Empirical results

The most successful efforts linking organismal traits with ecosystem processes have been at regional to global scales. Recent syntheses confirm that certain syndromes of functional traits recur predictably across different habitats and unrelated floras. For example, measurement of a suite of 12 standardized traits, including leaf and canopy characteristics, phenology and life history traits, across 640 plant species from three continents revealed a globally consistent 'major axis of evolutionary specialization' reflecting the evolutionary trade-offs between rapid resource acquisition on the one hand and conservation and protection of tissues on the other (Diaz et al. 2004). Principle component analysis showed that the first axis, interpreted as a predictor of resource capture and use, was correlated with relative growth rate, leaf nitrogen content, and litter decomposition rate, as well as palatability to generalist herbivores (Fig. 5.3). Similarly, as mentioned in a previous section, grazing by vertebrates produces a characteristic change in the suite of traits dominating vegetation, although the details vary to some degree among regions (Díaz et al. 2007). Among animals, the simple trait of body size proves both general and useful on local scales, for example, the regular scaling of predator and prey body sizes

72 BIODIVERSITY, ECOSYSTEM FUNCTIONING, AND HUMAN WELLBEING

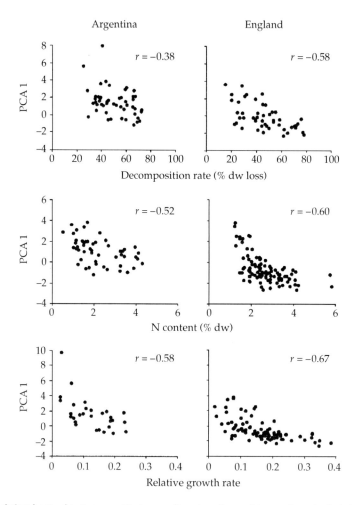

Figure 5.3 Syndromes of plant functional traits are consistent across disparate regions and taxa, and are closely tied to ecosystem processes. The figure shows similar correlations between Principal Component Axis 1 (defining plants with smaller, thicker, tougher leaves, and lower specific leaf area and three aspects of ecosystem functioning in two geographically and floristically different regions. After Diaz et al. (2004), used by permission.

across a wide range of systems (Brose et al. 2006). Refining such approaches for application at the scale of local communities would be a valuable step forward.

5.5 Empirical evidence linking realistic biodiversity change to ecosystem functioning

5.5.1 Background

The previous sections documented that species differ in extinction risk and that certain traits are associated with this variance in risk. We also showed that species differ in their effects on ecosystem functioning, and again that certain traits are correlated with this variance in function. These imply that the order in which species are lost from a community, the secondary extinctions that ensue, and the degree to which remaining species compensate in biomass, will have important consequences for how ecosystem function changes. This brings us back to the question of whether and how changes in function predicted by random-loss experiments approximate changes expected under more realistic (trait-based) extinction scenarios.

Studies addressing ecosystem impacts of realistic extinction scenarios have generally used one of two approaches. Most commonly, they have explored the plausible hypothesis that a certain trait (e.g. body size) predisposes species to extinction under a given driver, then used species trait data to simulate a corresponding extinction order (e.g. Solan et al. 2004, Bunker et al. 2005, McIntyre et al. 2007). A second approach has used empirical information on changes in assemblage composition along natural or anthropogenic environmental gradients as a proxy for extinction order, and asked how changes in trait distribution along the gradient influence ecosystem properties (Zavaleta and Hulvey 2004, Srinivasan et al. 2007). These approaches explicitly recognize that species are likely to go extinct in a particular order, as particular traits render a given species more or less sensitive to particular extinction drivers. Where the effects of environmental change (e.g. disturbance) are well characterized in time and/or space, position along the gradient allows reconstruction of the relative severity of disturbance and the actual sequence of species loss.

5.5.2 Empirical results

Several studies have contrasted ecosystem effects of realistic extinction scenarios and random-loss scenarios, and nearly all demonstrate that these have very different impacts on ecosystem functioning (Table 5.5). Recent simulation studies took advantage of existing data on the *per capita* functional impact of species to predict effects of extinctions on function when species with certain traits (e.g. large body size) or documented sensitivity to disturbance (e.g. intensive agriculture) were simulated to go extinct first. These include effects of terrestrial vertebrates on Lyme Disease risk (Ostfeld and LoGiudice 2003), benthic marine invertebrates on bioturbation (Solan et al. 2004), European grassland plants on biomass production (Schläpfer et al. 2005), beetles on dung burial, and bees on pollination rates (Larsen et al. 2005), neotropical trees on carbon storage (Bunker et al. 2005), perturbing lake communities on secondary extinctions (Srinivasan et al. 2007), and freshwater tropical fishes on nutrient recycling (McIntyre et al. 2007). In all seven studies, many – though not all – extinction scenarios showed substantially different patterns than those predicted if species were lost in random order. Similar results have been found in three experiments. First, grassland ecosystem functioning (resistance to invasibility) was more strongly related to diversity when the plant richness gradient simulated the observed nested pattern of species loss (Zavaleta and Hulvey 2004) than when plant assemblages were synthesized with a random loss scenario (Dukes 2001); in this case the invasion resistance provided by higher diversity was reduced more under the realistic extinction scenario (Fig. 5.4). Second, detritus processing by stream insects was more strongly related to diversity when insect richness gradients reflected realistic extinction scenarios than when the gradients were assembled at random (Jonsson et al. 2002). Finally, nitrogen uptake by tidepool seaweeds was positively correlated with seaweed diversity along a natural diversity gradient, but not a random-assembly gradient (Bracken et al. 2008).

These differences in the functional effects of realistic versus random extinction scenarios depend, in part, on the covariance between the traits that lead to high extinction risk and the traits related to high functional importance (Solan et al. 2004). If these two sets of traits are positively correlated, i.e. there is substantial overlap in the traits that mediate extinction vulnerability and effects on ecosystem processes, then functioning will decrease rapidly with only a few species lost from the community (Lavorel and Garnier 2002, Raffaelli et al. 2002). For example, in a simulation study of neotropical trees (Bunker et al. 2005) in which large trees disappeared first, there was a rapid decline in carbon storage because the remaining trees, even after increasing in density, did not have the traits (high wood density, height) required to store carbon maximally. At the other extreme, if the traits associated with extinction risk are negatively correlated with those important for function, then realistic extinctions will eliminate species of low functional importance and have less effect on ecosystem functioning, at least initially, than random extinctions. For example, in the study on productivity in European grasslands (Schläpfer et al. 2005), intensive farming led first to loss of the less productive species, and thus extinctions under this scenario had less impact on productivity than

Table 5.5 Summary of studies exploring consequences of non-random extinction sequences for ecosystem functioning 'Rand = random.'

Organism manipulated	Ecosystem process	Trait mediating extinction	Study type	Extinction driver	Change with declining diversity			Reference
					Random	Realistic	Real − Rand	
Grassland plants	Primary production (biomass accumulation)	Productivity, competitive ability	Experiment, simulation	Management (fertilizer and mowing)	−	0,+	Weaker	1
Grassland plants	Resistance to exotic invasion	Unspecified (from nested subset analysis)	Experiment	Disturbance (nested subset analysis)	0	−	Stronger	2
Forest trees	Carbon storage, (biological insurance)	Size, growth rate, drought-sensitivity	Simulation	Management, environmental change	0	0, −, 0	Similar, stronger, similar	3
Marine algae	Nutrient use	Unspecified (natural gradient)	Experiment	Unspecified	0	−	Stronger	4
Bees	Pollination	Body size	Simulation	Habitat loss/fragmentation	−	−	Stronger	5
Dung beetles	Dung burial (nutrient recycling, seed dispersal)	Body size	Simulation	Habitat loss/fragmentation	−	−	Stronger	5
Mammalian hosts of Lyme disease	Protection from Lyme disease risk	Body size, home range size, trophic level	Simulation	Habitat fragmentation	+	−	Stronger	6
Stream insects	Detritus (leaf) processing	Tolerance of acidification and organic pollution	Experiment	Pollution	0	−	Stronger	7
Tropical FW fishes	Nutrient regeneration	Body size, trophic level	Simulation	Fishing	−	−	Stronger	8
Freshwater organisms	Stability (resistance to secondary extinctions)	Unspecified (from nested subset analysis)	Simulation	Unspecified	−	0	Weaker	9
Marine benthic invertebrates	Nutrient regeneration	Body size, population size, sensitivity to disturbance	Simulation	Unspecified	−	−	Stronger (body size)/ weaker (rarity)	10
Three plant and three animal assemblages	Unspecified (response variable was functional diversity)	Leaf N content, rooting depth (plants); body size (animals)	Simulation	N deposition (plants), harvesting (animals)	−	−	Stronger	11

1 Schläpfer et al. (2005); 2 Zavaleta and Hulvey (2004); 3 Bunker et al. (2005); 4 Bracken et al. (2008); 5 Larsen et al. (2005); 6 Ostfeld and LoGiudice (2003); 7 Jonsson et al. (2002); 8 McIntyre et al. (2007); 9 Srinivasan et al. (2007); 10 Solan et al. (2004); 11 Petchey and Gaston (2002b).

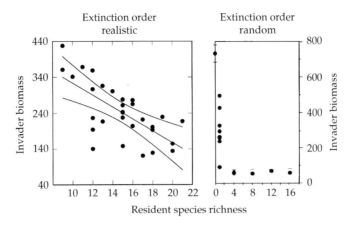

Figure 5.4 Evidence from field experiments that extinction impacts on ecosystem functioning (invasibility) depend strongly on the order of grassland plant species loss. Invader performance in experimental plots declined with native richness when the diversity gradient reflected real-world patterns (left, Zavaleta and Hulvey 2004, used by permission), but was unrelated to native richness when extinction order was determined randomly (right, Dukes 2001, used by permission). Invasibility is displayed as aboveground biomass of the alien starthistle, *Centaurea soltitialis*. Left panel shows the regression line and 95% confidence intervals.

random loss of species. Despite increasing realism in species loss trajectories, however, the few studies performed to date have considered only single forcing agents. If individual species have multiple traits that each expose them to different extinction drivers, it follows that in the presence of multiple drivers of extinction the ecosystem consequences of biodiversity loss may be much greater than current evidence suggests (Sala *et al.* 2000, Brook *et al.* 2008).

5.5.3 Summary

A nearly universal result of both the small number of experiments and the larger but still modest number of simulations that have compared random with non-random extinction scenarios (Table 5.5) is that the order of species loss has profound consequences for the direction, rate, and magnitude of change in ecosystem functioning with extinction. Importantly, studies of non-random extinction scenarios suggest that, more often than not, realistic extinction orders have larger effects on ecosystem responses than random loss sequences, and thus that random-loss designs may often underestimate the ecosystem consequences of diversity loss (Table 5.5, Duffy 2009). Another striking finding is that, even in very species-rich assemblages, strong skew in body size or other traits can result in major changes in ecosystem-level processes after a loss of a particular species. This is shown dramatically in extinction simulations of marine benthic invertebrates (Solan *et al.* 2004) (Fig. 5.1) and tropical freshwater fishes (McIntyre *et al.* 2007), in which biodiversity and ecosystem functioning relationships fall into two distinct domains depending on whether the dominant species is present or not.

5.6 Discussion

As the examples discussed here illustrate, the correlation between extinction traits and functional traits depends critically on both the extinction scenario and the function under consideration, so it is currently difficult to make global generalizations about the sign of the correlation (see Lavorel and Garnier 2002). The mechanism also matters: if niche complementarity drives diversity effects on function, then the variety of functional traits ('functional diversity') may be more important than any particular functional trait in determining ecosystem functioning. Simulations using empirical data from at least six assemblages (tested by Petchey and Gaston 2002b) suggest that functional diversity is indeed reduced more under realistic than under random extinction scenarios.

An important limitation of simulation studies comparing realistic and random species extinctions is that they require assumptions about how the remaining species in the community would numerically respond to the loss of species. These assumptions differ between studies. For example, Ostfeld and LoGiudice (2003) and Srinivasan et al. (2007) assumed no numerical compensation amongst remaining species, whereas Bunker et al. (2005) and Dukes (2001) assumed full compensation across all species. These assumptions are critical because, as discussed earlier, compensation matters. Species interactions also complicate predictions about the order of species loss in response to a stressor, because the loss of each species causes a reorganization of interactions and abundances in the food web. In the empirically based simulation studies discussed earlier, the order of species loss (and thus effects on ecosystem functioning) was based on species' direct vulnerability to the stressor (e.g. Solan et al. 2004), whereas it is clear that indirect interactions among species can change the sensitivity of species to stressors (Ives and Cardinale 2004). The covariance between functional response and effect traits is thus likely to be context-dependent in food webs of interacting species. Soberingly, dynamic species interactions such as compensation and cascading extinctions can result in a great range of possible outcomes of species extinction, with potentially vastly differing consequences for ecosystem functioning.

Another limitation of much trait-based research to date is its focus on single guilds or trophic levels (marine infauna, Solan et al. 2004, forest trees, Bunker et al. 2005, dung beetles, Larsen et al. 2005). Yet it is clear from observations, experiments, and simulations that the broader effects of extinction on ecosystem structure and functioning are often mediated by trophic interactions (Pace et al. 1999, Dunne et al. 2002b, Petchey et al. 2002b, Rayner et al. 2007). Thus, a critical frontier in conservation ecology involves expanding trait-based approaches for multi-trophic communities (Belgrano et al. 2005, Duffy et al. 2007). Indeed, the extent to which the last decade's high-profile research on biodiversity and ecosystem functioning can be applied to real-world conservation and management problems will depend critically on linking that work to realistic scenarios of extinction and invasion, and the reorganization of food webs that follows this turnover (Schwartz et al. 2000, Srivastava 2002, Srivastava and Vellend 2005). A focus on response and effect traits, and the covariance between them, is a promising way forward.

Finally, it is worth reiterating explicitly why we are interested in the links between biodiversity and ecosystem functioning in the first place. There are many arguments for concern over the declining state of biodiversity – from purely aesthetic and moral to nakedly commercial – but probably the most compelling reason is the role of diverse living organisms in creating and maintaining the complex life support system without which humanity would literally be doomed. The lack of public urgency regarding this problem can be traced in part to the limitations of classical economics in valuing both the benefits of living organisms and ecosystems that are outside traditional markets, and in estimating the costs of degrading natural systems (Chapters 17 and 18). Trait-based approaches can contribute to these practical challenges in several ways. First, they can give us a clearer prediction of how natural communities are likely to change under particular scenarios of environmental impact. Second, by discovering general links between environmental change, community response, and ecosystem functioning based on measurable traits, these approaches can reduce the dependence of policy implementation on specific taxonomic expertise, which is in short supply and declining. Third, defining the role of particular types of organisms (defined by traits) in mediating particular ecosystem services provides data necessary to parameterize economic models underlying policy instruments for sustainability (Chapter 17) generally, and is essential to moving from traditional single-species harvesting models to models that take account of the complex interactions typical of all real ecosystems (Chapter 19).

5.7 Conclusions and future directions

The field of extinction research is currently very active, and recent years have produced a rich body of data on the traits that predispose particular taxa to endangerment and extinction, and increasingly, their synergisms (Purvis *et al.* 2000c, Traill *et al.* 2007, Brook *et al.* 2008). These data can inform improved design of studies, experimental and otherwise, that explore the effects of specific environmental change drivers on community organization (including biodiversity) and its consequences for ecosystem processes. The importance of such studies is highlighted by the marked differences in outcomes of both experiments and simulations as a function of the specific extinction trajectories employed and, in simulations, the assumptions made about how species interact.

One otherwise complimentary reviewer of this chapter noted that 'Most of the generalizations that can be made might be guessed by an undergraduate fresh out of a good ecology class.' It is true that the vulnerability of species with large body sizes, 'slow' life histories, specialized habits, small geographic ranges, and low population densities are both well-established and self-evident. The bright side of this observation is that these strong generalizations offer correspondingly clear guidance on how to design tests and predictive models that better link realistic extinction to ecosystem functioning. As one example, the vulnerability of large animals high in the food chain, which also often are strong interactors, argues for a better marriage of approaches focusing on horizontal diversity within levels with predator–prey ecology and food-web modelling (Chapter 8; Duffy *et al.* 2007). A key research need is a better characterization, both theoretical and empirical, of the intrinsic traits and environmental factors that influence the degree of compensation among potentially competing species, and more generally, the strength of interactions of all kinds between species (Berlow *et al.* 2004). Another key research need is a focus on multiple traits and how they interact with multiple drivers of extinction. Early trait-based approaches to linking environmental change with ecosystem impacts have understandably focused on individual traits. However, it is clear that traits are often correlated with one another, often respond in concert to drivers of change, and can interact to influence ecosystem functioning in non-linear ways (Eviner and Chapin 2003). The same is true of extinction drivers (Sala *et al.* 2000, Brook *et al.* 2008). Thus, an important frontier is developing multivariate models of correlated trait change and of trait effects, and how they ripple out to influence functioning of ecosystems (Lavorel and Garnier 2002, Craine *et al.* 2002, Eviner and Chapin 2003, Poff *et al.* 2006.

Acknowledgments

We are grateful to Andy Hector for organizing the stimulating BioMERGE workshop in Ascona, Switzerland, from which this chapter arose, as well as the other members of the workshop. We also acknowledge Shahid Naeem and two anonymous reviewers for comments that improved the MS, and the National Science Foundation (OCE-0623874 to JED) and the Natural Sciences and Engineering Research Council of Canada (DSS) for support.

CHAPTER 6

Biodiversity and the stability of ecosystem functioning

John N. Griffin, Eoin J. O'Gorman, Mark C. Emmerson, Stuart R. Jenkins, Alexandra-Maria Klein, Michel Loreau, and Amy Symstad

6.1 Introduction

6.1.1 Overview

Concern that the rapid anthropogenic erosion of biodiversity (Wilson 1992) may undermine the delivery of ecosystem services (Daily 1997) has prompted a synthesis of community and ecosystem ecology over the last decade. Biodiversity-ecosystem functioning (BEF) research is central to this emerging synthesis, asking how biodiversity is related to the magnitude and stability of ecosystem processes. Isolating species richness (and diversity) effects from species composition has been a chief goal of BEF research. That species richness generally enhances the mean magnitude of a variety of ecosystem properties is now well established (Hooper *et al.* 2005; Balvanera *et al.* 2006; Cardinale *et al.* 2006a), but the effect of species richness on the stability of ecosystem properties remains equivocal (Hooper *et al.* 2005).

Early consensus (Odum 1953, MacArthur 1955, Elton 1958, Pimentel 1961) that diversity enhances various aspects of community and ecosystem stability was largely founded on qualitative observations and intuitive reasoning. This view was apparently contradicted by the results from rigorous mathematical treatments (e.g. May 1972) which showed that in a food web context stability was related to system properties, i.e. connectance, species richness, and interaction strength, and importantly, could be reduced by species richness. Recognition that stability has multiple definitions that can apply to all levels of ecological organization (Pimm 1984; see also Loreau *et al.* 2002) helped pave the way for a new, synthetic perspective that developed during the 1990s as attention to extinction's consequences for the stability of ecosystem functioning increased (Tilman and Downing 1994, Naeem and Li 1997). This BEF perspective recognized that fluctuating abundances of component species may not produce instability at the community or ecosystem level because compensatory reactions among species dampen fluctuations of aggregate abundance (Tilman 1996). The distinction between population and community-level variability was firmly drawn (Tilman 1996) and attention was focused on the latter through the BEF research agenda.

Within the BEF framework, experiments and theory explicitly relating to the effect of species richness on community-level aggregate properties (mainly biomass) have focused on variability through time in relation to background environmental variation (temporal stability) as well as on the impact (resistance) and recovery (resilience) of such properties to discrete, and often extreme, perturbations. In this chapter we review recent empirical studies examining the links between species richness and these three facets of stability (see Ives and Carpenter 2007 for a review of other types of stability). In addition, recognizing that explicit BEF experiments are limited in their spatial and temporal scope, we adopt a wider perspective by discussing how changes in biodiversity may undermine stabilizing properties of food webs and the ability of ecosystems to resist state-changes. Furthermore, using examples we emphasize that direct measurement of the stability of ecosystem services across gradients of human impacts can show how stability is influenced by human activities – both directly and indirectly via changes in diversity.

6.1.2 Theory linking biodiversity and stability

Theory has established strong links between species richness and the temporal variability of community-level properties and has provided a variety of possible explanatory mechanisms (reviewed by Cottingham *et al.* 2001, Loreau *et al.* 2002; see Box 6.2). Doak *et al.* (1998) argued that a reduction in community variability with increased diversity is an inevitable consequence of 'statistical averaging', i.e. the sum of many randomly and independently variable items is less variable than the average item. The strength of this effect depends on how the variances of populations scale with their means (Tilman *et al.* 1998), and the evenness of species' abundances and how their fluctuations are related (Doak *et al.* 1998, Tilman *et al.* 1998).

The importance of how species fluctuations are related is also reflected in another proposed stability mechanism that is closely related to the concept of statistical averaging: the 'insurance hypothesis'. This assumes that interspecific niche differentiation causes species to respond differently to the environment and that this differential response can produce compensatory dynamics among species, buffering the impact of environmental changes (McNaughton 1977, Walker 1992, Naeem and Li 1997). The insurance hypothesis depends upon functional redundancy: the concept that species within the same functional group may replace each other with no consequences for ecosystem function. The stabilizing role of this mechanism further depends upon the diversity of species' responses within functional groups (functional response diversity; Lavorel and Garnier 2002, Elmquist *et al.* 2003). Theoretical models of competitive communities formalized the insurance hypothesis and confirmed that species diversity can stabilize community properties in the face of changing environmental conditions (Yachi and Loreau 1999, Ives *et al.* 1999). This prediction also generally holds for simple, theoretical, multi-trophic systems (Ives *et al.* 2000; Thébault and Loreau 2005) and metacommunities (Loreau *et al.* 2003).

The resistance of aggregate community properties is also theoretically enhanced by diversity under the insurance hypothesis, as species tolerant to a pulse perturbation or directional change in conditions are more likely to occur in diverse communities (Walker 1992, Yachi and Loreau 1999). On the other hand, the resilience of aggregate community properties (the return rate following an equal reduction in the populations of all component species) is not incorporated under the insurance hypothesis, in which species populations are differentially affected by the environment or perturbations. The resilience of populations may in fact be reduced by high levels of diversity in competitive communities (Loreau and Behera 1999). This theory does not account for differences between species in growth rates, however, which may result in a positive relationship between species richness and return rate of community biomass (resilience) through a sampling effect (Steiner *et al.* 2005b). Theory has yet to explicitly address this possibility, however.

6.2 Empirical findings

Problems with the design and interpretation of experimental tests (Givnish 1994, Huston 1997, Fukami *et al.* 2001) together with a general shortage of such tests, have hampered the attempts of earlier reviews to assess the effect of diversity on temporal stability, resistance and resilience (Cottingham *et al.* 2001, Loreau *et al.* 2002, Hooper *et al.* 2005). Subsequent vigorous empirical research has recently been conducted in this area. This work has been generally less open to alternative interpretation, as experimental designs have evolved to limit several confounding factors previously identified (but see Wardle and Grime 2003). It has long been recognized that species composition can have strong influences on ecosystem properties. Isolating the role of species richness *per se* has been a chief goal of BEF research, and is the focus of empirical studies reviewed here.

6.2.1 Temporal stability

We first examine empirical tests of the general hypothesis that diversity enhances the temporal stability of community-level properties such as biomass or production, measured as the reciprocal of the coefficient of variation (1/CV). CV is an appropriate, and widely used, measure of variability because it is standardized to the mean, accounting for the tendency of variability to increase with the mean. Note that in the primary literature authors report either temporal variability (CV) or temporal stability (1/CV). For consistency, we interpret and discuss diversity effects from all studies in terms of temporal stability, such that positive effects are stabilizing (see Table 6.1). Eighteen separate papers published between 1994 and 2006 include a total of 22 cases in which community-level temporal stability was measured. These studies were conducted in a range of systems varying in scale from aquatic microcosms to natural forest stands. Overall, diversity stabilized community-level properties in 13 cases, had no significant effect in 8, and reduced stability in a single study.

We first deal with studies that have examined the temporal stability of community-level properties (principally biomass or its production) within a single trophic level in both manipulative and observational studies. We then consider studies that have manipulated the diversity of communities across multiple trophic levels and examined how such manipulations affect the temporal stability of biomass and ecosystem process rates.

6.2.1.1 Temporal stability of biomass within a single trophic level

Direct tests of the diversity–temporal stability hypothesis within a single trophic level have been mainly conducted within replicated grassland plots and microcosms. Recent grassland experiments in which diversity was directly manipulated show a stabilizing effect of diversity (Caldeira *et al.* 2005, Tilman *et al.* 2006b). These studies reported reduced stability of constituent populations but greater community-level stability with increased species richness, which is consistent with theory and previous experimental evidence (Tilman 1996).

Amongst the microcosm examples, only two of five studies supported the diversity–temporal stability hypothesis. Steiner *et al.* (2005a) found that the aggregate biomass of four zooplankton species in mixture exhibited greater stability than monocultures of the constituent species. Furthermore, Zhang and Zhang (2006a) found an overall stabilizing effect of algal species richness on community biomass. This effect was, however, context-dependent, in that it occurred only under conditions of low nutrient availability; no diversity effect was detectable under enriched conditions. Petchey *et al.* (2002) also did not find an effect of species richness on the temporal stability of community-level biomass, either under constant or fluctuating temperature conditions. In only a single case did diversity destabilize community biomass (Gonzalez and Descamps-Julien 2004).

The mixed outcomes of these experiments, and the lack of support for diversity–stability relationships in some cases, can be explained by a range of mechanisms that might obscure diversity effects:

1) A direct destabilizing effect of diversity on population level (growth rates: Gonzalez and Descamps-Julien 2004; biomass: Petchey *et al.* 2002) variances exists and caused some populations to vary more within diverse communities to the extent that they eclipsed the effect of stabilizing mechanisms.
2) Synchrony of species responses to environmental variability might have limited insurance effects of increased species richness.
3) Low evenness and hence high variance among population biomasses within communities could weaken the relation between species richness and community-level stability (Ives and Hughes 2002, Petchey *et al.* 2002).

A key challenge for future studies is to elucidate the source of variability among experiments and environmental contexts. Explicit consideration of the degree of functional response diversity (*sensu* Lavorel and Garnier 2002) represented by species within increasingly species-rich communities would be an important development (see e.g. Walker 1999) that could help explain effects of species richness more completely. Moreover, the degree of environmental heterogeneity through time will dictate the extent to which such functional response diversity can be realized (see Tylianakis *et al.* 2008) and should be explicitly considered in future studies.

Natural gradients in diversity are expected to be driven by external factors that may obscure the effect of diversity on stability and complicate interpretation (Ives and Carpenter 2007). However, examining natural patterns of diversity and

Table 6.1 Effect of diversity on the temporal stability of ecosystem properties.

Reference	Trophic level	Ecosystem function	Factor	Diversity gradient[a]	Ecosystem type	Time scale	Div levels	Type of div effect
Caldeira et al. (2005)	Primary prod	Community biomass		Exp. M	Grassland	3 yr	1 to 14	Positive
Dang et al. (2005)	Decomposer	Decomposition		Exp. M	Stream	28 d	1 to 16	Positive
Dodd et al. (1994)	Primary prod	Community biomass		Nutr. F	Grassland	42 yr	8 to 45	Positive
DeClerk et al. (2006)	Primary prod	Primary productivity		Nat. F	Forest	64 yr	1 to 4	None
Gonzalez and Descamps-Julien (2004)	Primary prod	Community biomass		Exp. M	Aquatic	64 d	1 to 6	Negative
Morin and McGradySteed (2004)[1]	Multi	Community CO_2 flux		Exp. M	Aquatic	42 d	3 to 9	None
Petchey et al. (2002)	Protists and bacteria	Community biomass	Constant temp.	Exp. M	Aquatic	6 w	2 to 8	None
Petchey et al. (2002)	Protists and bacteria	Community biomass	Fluctuating temp.	Exp. M	Aquatic	6 w	2 to 8	None
Romanuk et al. (2006)	Multi	Community biomass	Low nutrient	Exp. M (dilution)	Aquatic	5 w	1 to 8	Positive
Romanuk et al. (2006)	Multi	Community biomass	Med. nutrient	Exp. M (dilution)	Aquatic	5 w	1 to 8	None
Romanuk et al. (2006)	Multi	Community biomass	High nutrient	Exp. M (dilution)	Aquatic	5 w	1 to 8	None
Steiner (2005b)	Consumers only	Community biomass		Nat. F	Aquatic	5 m	11 to 24	Positive
Steiner et al. (2005a)	Consumers only	Community biomass		Exp. M	Aquatic	48 d	1 and 4	Positive
Steiner et al. (2005b)	Multi	Community biomass		Exp. M	Aquatic	22 d	1, 2, 4 spp (X 5 trophic groups)	Positive
Tilman (1996)	Primary prod	Community biomass		Nutr. F	Grassland	8 yr	1 to 26	Positive
Tilman et al. (2006)	Primary prod	Community biomass		Exp. F	Grassland	10 yr	1 to 16	Positive
Tylianakis et al. (2006)	Multi	Parasitism		Nat. F	Agricultural	16 m	1 to 4	Positive
Valone and Hoffman (2003a,b)	Primary prod	Community biomass	Quadrat scale	Nat. F	Grassland	11 yr	1 to 16	Positive
Valone and Hoffman (2003a,b)	Primary prod	Community biomass	Plot scale	Nat. F	Grassland	11 yr	1 to 16	None
Vogt et al. (2006)	Multi	Community biomass		Exp. M (dilution)	Aquatic	7 w	1 to 8	Positive
Zhang and Zhang (2006a)	Primary prod	Community biomass	High nutrient	Exp. M	Aquatic	77 d	1,2,4,6	None
Zhang and Zhang (2006a)	Primary prod	Community biomass	Low nutrient	Exp. M	Aquatic	77 d	1,2,4,6	Positive

[a] Exp.: experimentally created diversity gradient; Nat.: naturally occurring diversity gradient; Nutr.: gradient produced by different nutrient levels. F.: field study; M.: mesocosm/microcosm study.
[1] Reanalysis of McGrady-Steed et al. (1997)

stability can reveal inter-relationships between the environment, stability and diversity. Valone and Hoffman (2003a,b) used an 11-year time series of grassland plots that varied naturally in species richness to investigate the relationship between population and community-level temporal stability. In this system, population stability increased with diversity, perhaps because the natural, productivity-driven diversity gradient resulted in larger, and thus relatively more stable, populations at higher diversity (Valone and Hoffman 2003b). At the community level, the authors found a weak stabilizing effect of diversity at the small quadrat (0.25 m^2) scale but not at the larger plot (0.25 ha) scale (Valone and Hoffman 2003a). The incongruity could be due to the scale of biological interactions, or simply due to smaller samples sizes, and thus reduced statistical power, at larger scales. At an even larger spatio-temporal scale, DeClerck et al.'s (2006) analysis indicates that there is no relationship between a naturally occurring gradient of conifer diversity and the temporal stability of annual biomass production in Sierra Nevada forest stands. This may be explained by a low degree of species richness and highly correlated responses of species to environmental changes. Diversity–temporal stability relationships are expected to be highly variable across natural gradients depending on the environmental driver of species diversity, the functional response range among the species, and direct environmental influences on stability.

6.2.1.2 Temporal stability in multi-trophic communities

Broadly, studies investigating the effects of diversity on temporal stability in multi-trophic systems fall into one of two categories: 1) Those that examine diversity effects on the temporal stability of community-level (and often population) biomass; 2) Those that manipulate the diversity of a non-basal species and examine effects on the stability of a non-biomass ecosystem process, e.g. decomposition. We begin by addressing the former type of multitrophic level study.

Theory developed in the context of the BEF research field predicts that, despite the complexity added when considering multi-trophic interactions, biodiversity still acts as biological insurance for ecosystem processes (Ives et al. 2000, Thébault and Loreau 2005). Early empirical studies (Naeem and Li 1997, McGrady-Steed et al. 1997) showed a stabilizing effect of diversity on community-level properties but were difficult to interpret due to confounding factors (Huston 1997, Fukami et al. 2001). Indeed, Morin and McGrady-Steed (2004) re-analyzed data from their earlier publication (1997) and found that the previously reported effect was due to spatial variability among replicates. Recent BEF experiments, all conducted within microcosms, have provided a further, less controvertible, glimpse at the possible effects of diversity on stability in multi-trophic systems. Steiner et al. (2005b) showed that diversity increased the temporal stability of community biomass, whilst species composition best explained variability in population-level abundance. They invoke the positive selection effect, suggesting that dominance of species with high population stability could underlie much of the observed influence of diversity. It would be interesting to investigate whether the populations of species that dominate mixtures are generally more stable, since if there is a trade-off between resistance and productivity (Lepš et al. 1982), the opposite may be true.

Other mechanisms are proposed in studies based on a microbial rock pool system. In microcosm studies, Vogt et al. (2006) and Romanuk et al. (2006) also found a stabilizing effect of diversity on the aggregate abundance of the community, but here, greater population stability at higher species richness summed to produce greater community stability. The mechanism dampening population-level variability is not clear, but Romanuk et al. (2006) postulate that in pools with high levels of unused resources, populations will tend to fluctuate more, because niche complementarity (at higher diversity) reduces resource levels, thus stabilizing populations. Their finding that population variability was greater in high nutrient microcosms is consistent with such a mechanism (but contrasts with the findings of single trophic level studies which showed that increased productivity led to the loss of a diversity effect (Zhang and Zhang 2006b). In an observational study Kolasa and Li (2003) found that diversity increased the temporal stability of microbial rock pool populations, but only when increasing specialization, and therefore variability, of individual species with increasing diversity was statistically controlled. This study shows that opposing forces operating in natural communities may yield no net effect of diversity on population stability.

Two notable studies have considered how species richness affects the temporal stability of ecosystem functions other than biomass or its production in multi-trophic systems. Dang et al. (2005) tested the effect of fungal diversity on both the mean magnitude and temporal stability of decomposition. Whilst diversity had no effect on the magnitude, temporal stability increased in close correspondence to the null-model of statistical averaging. Furthermore, the outcome was robust to a range of environmental contexts. Tylianakis et al. (2006) examined the effect of parasitoid diversity on the temporal stability of parasitism of wasps and bees. Again, diversity enhanced the temporal stability of this ecosystem process, indicating that diversity may play an important role in stabilizing trophic control within complex food webs. Interestingly, the effect of diversity reported by both Dang et al. (2005) and Tylianakis et al. (2006) was non-linear, producing the most rapid decrease in variability at relatively low levels of richness. This is consistent with statistical averaging models (Doak et al. 1998; but see Box 6.2), but more studies, as well as theory pertaining to ecosystem processes other than biomass (Box 6.1), are needed to determine the generality of these results.

Overall, the very restricted number of studies limits our ability to assess the effect of diversity on temporal stability in multi-trophic systems. However, the possibility that diversity may increase population stability in these systems is intriguing and warrants further exploration, as it is contrary to theory and experiments conducted within single-trophic level, competitive communities and May's (1972, 1973) models. Theory for multi-trophic systems (Thébault and Loreau 2005) predicts that diversity may increase population stability under some conditions, for instance, when consumers are either specialists or generalists with a trade-off between niche breadth and attack rate, and their temporal niche differentiation is low. It is difficult to assess whether these mechanisms identified by theory explain the results of recent experimental and observational studies

---

Box 6.1 From abundance to functioning

The vast majority of diversity–temporal stability studies have used community biomass or its production as their measure of ecosystem functioning. Many of the theoretical mechanisms linking stability to diversity may equally apply to other community-aggregated properties, but this has seldom been tested. The stability of process rates represents an important divergence from recent theory; instead of variation in species' summed abundances forming the response variable, the *efficiency* of species mediating ecosystem functioning is also of interest. In this case, the density-mediated component directly linked to theory pertaining to community biomass stability remains, but a potentially density-independent 'efficiency' component is added. Furthermore, the specificity of the process measured is likely to impact results. For example, If the ecosystem function is the flux of a particular nutrient (Bracken and Stachowicz, 2006), there may be less functional redundancy than for a universal process such as primary production or decomposition. Since functional redundancy is a central tenet of the diversity–stability relationship, this suggests that, as functions become more specific, their stability will be increasingly associated with the population stability of the one or a few species mediating the function and less with the total diversity of the system. Where species vary in their contributions to multiple functions, functional redundancy will be further reduced if such multiple functions are considered concurrently (Gamfeldt et al. 2007).

---

because these studies did not test mechanisms. Other stabilizing mechanisms in food webs could be more prevalent as species richness and food web complexity increases, thus stabilizing populations (Section 6.3.1). Direct empirical evidence of such effects, and how their efficacy and prevalence varies with diversity, has yet to emerge. Detailed analyses of dynamic trophic interactions over a range of temporal and spatial scales would be necessary to demonstrate such effects in food webs.

6.2.2 The effects of discrete perturbations

6.2.2.1 Resistance
Tests of the insurance hypothesis could be garnered from early studies comparing stress resistance across successional diversity gradients. The findings of

Box 6.2 What causes diversity's effects on temporal stability?

Despite established theoretical mechanisms linking diversity and temporal stability, there is currently little consensus regarding the relative importance of mechanisms underpinning effects observed in empirical studies. A number of authors have examined components of temporal variability in order to gain insight into the mechanisms underlying observed effects of diversity. Temporal stability (S_T), measured as 1/CV of community biomass, can be expressed as (Lehman and Tilman 2000):

$$S_T = \frac{\sum \text{species biomasses}}{\sqrt{\sum \text{species variances} + \sum \text{species covariances}}}$$

The numerator in this equation captures the short-term or average effect of diversity on community biomass. All else being equal, an increase in average community biomass with diversity due to overyielding tends to increase community stability simply because average community biomass is used to scale the variances and covariances in the CV. Additional long-term stabilizing effects of diversity can result from reduced summed variances, reduced summed covariances, or both in the denominator. Reduction in summed variances has generally been interpreted as indicative of statistical averaging (Tilman 1999; Cottingham et al. 2001), whilst reduced summed covariances have generally been interpreted as indicative of compensatory dynamics owing to competitive release and/or differential response to environmental conditions (Tilman 1999; Petchey et al. 2002). But a fundamental problem with this statistical approach is that summed variances and summed covariances are strongly dependent on each other, and do not capture distinct biological mechanisms. Both reduced summed variances and reduced summed covariances are ultimately driven by the same mechanism, i.e. asynchronous species responses to environmental flutuactions (Loreau and de Mazancourt, unpublished manuscript).

Of the 18 studies discussed here (Table 6.1), nine included information on the statistical components of temporal stability; multiple contexts in two studies yield 13 experiments for consideration. Nine of these experiments reported a positive effect of diversity on temporal stability, whilst two did not detect an effect (Table 6.2).

Contrary to Tilman's (1996) suggestion, the negative covariance effect – the result of increasingly asynchronous population fluctuations with increasing diversity – is not a common phenomenon. Species richness resulted in increasingly negative summed covariances in just a single study (Petchey et al. 2002). Two experiments actually

Table 6.2 Studies reporting the statistical components of temporal stability, their reported effects of diversity on community stability (1/CV), summed covariances, summed variances, total community biomass, and the mean-variance scaling factor (z).

Reference	Trophic level	Stability	$\sum$Covariance	$\sum$Variance	$\sum$Biomass	z^*
Caldeira et al. (2005)	Single	Positive	Increased	Increased	Increased	>1
Petchey et al. (2002)	Single	None	Reduced	N/A	Increased	>1
Romanuk et al. (2006)	Multiple (low nutrient)	Positive	No effect	Increased	Increased	1
Romanuk et al. (2006)	Multiple (med nutrient)	Positive	No effect	Increased	Increased	0.83
Romanuk et al. (2006)	Multiple (high nutrient)	Positive	No effect	Increased	Increased	0.85
Steiner (2005b)	Single	Positive	No effect	Reduced	N/A	1.45
Steiner et al. (2005a)	Single	Positive	No effect	No effect	Increased	N/A
Steiner et al. (2005b)	Multiple	Positive	No effect	No effect	Increased	1.55
Tilman et al. (2006)	Single	Positive	No effect	Reduced	Increased	1.6
Valone and Hoffman (2003a)	Single	Positive	Increased	No effect	Increased	N/A
Vogt et al. (2006)	Multiple	Positive	No effect	Reduced	No effect	N/A
Zhang and Zhang (2006a)	Single (low nutrient)	Positive	No effect	No effect	No effect	1.74
Zhang and Zhang (2006a)	Single (high nutrient)	None	No effect	No effect	No effect	1.79

* z is a parameter in the equation relating CV for community biomass to total community biomass and the number of species in the community. The statistical averaging stabilizing effect only occurs when $z > 1$ (Tilman et al. 1998).

continues

Box 6.2 *(continued)*
revealed increasingly positive covariances with more diversity. If species are similarly influenced by environmental variability, species abundances will track environmental conditions in a correlated manner (Vasseur *et al.* 2006; Loreau and de Mazancourt 2008). Species may also respond similarly if changes in environmental conditions are extreme relative to the range of tolerances exhibited by the assemblage (see Allison 2004). Positive species covariances do not necessarily preclude positive net diversity effects on stability because any deviation from perfect correlation between species environmental responses can in principle stabilize aggregate community properties (Yachi and Loreau 1999; Ives et al. 1999).

Greater diversity yielded increased, reduced, and unaffected summed variances in four, three, and five studies, respectively. These mixed overall findings suggest that summed variances, just as summed covariances, depend on context-specific factors not universally linked to diversity, such as competitive interactions and how population abundance changes with diversity (see Valone and Hoffman 2003a), in agreement with recent theory (Loreau and de Mazancourt, unpublished manuscript).

The most consistent explanation for a positive diversity–stability effect is a combination of overyielding and asynchronous species fluctuations. Of the seven experiments that yielded a stabilizing effect of diversity and reported summed biomass with respect to diversity, overyielding (diversity and biomass were positively related) occurred in five. If overyielding occurs, variance can be smaller relative to the mean even in the absence of any changes in summed covariances or variances. Statistical averaging due to asynchronous species fluctuations most likely contributed to the temporal stability of aggregate properties in numerous studies — in fact, all seven studies that measured the scaling relationship between mean and variance reported values indicating that, even in the absence of changes in summed covariances, diversity would be expected to enhance stability (Doak *et al.* 1998; Tilman *et al.* 1998; Table 6.2).

Only two studies in multi-trophic systems measured the statistical components of temporal stability, and they found mixed results: whilst Vogt *et al.* (2006) invoke reduced population variability with diversity as a driver of community-level stability, Steiner *et al.* (2005b) found no such effect, instead crediting a form of the selection effect – the low population variability of dominant species with stabilization. With only two studies, comparisons between these and single-trophic systems are not possible. Their greater complexity allows for quite different patterns to emerge, however, as we discuss below in the context of food web ecology.

Overall, both theory and empirical data suggest that we have not yet started to disentangle the biological mechanisms that underlie the stabilizing effects of diversity on ecosystem properties. The statistical partitioning of summed species variances and summed species covariances, which was proposed for this purpose, has proved ineffective. New innovative approaches are needed to address mechanisms. One promising, but data demanding, alternative would be to test observed patterns of species temporal variations against a neutral model of community dynamics under the combined influence of density dependence, environmental forcing and demographic stochasticity (Loreau and de Mazancourt 2008).

such studies have been varied, with several suggesting positive (Hurd and Wolf 1974, Mellinger and McNaughton 1975, Lepš *et al.* 1982) and others suggesting a negative (Smedes and Hurd 1981, Berish and Ewel 1988) relationship between diversity and resistance. These studies should, however, be interpreted with caution, as species composition and life-history traits also vary during succession (Odum 1969), making the role of diversity *per se* ambiguous. Thus, a rigorous BEF experimental approach was developed during the 1990s to more explicitly test the hypothesis. Thirteen studies, including 14 experiments, three of which measured two ecosystem properties each, have used this approach (Table 6.3). Five experiments yielded a positive effect of diversity, whilst eight showed no effect and four showed a destabilizing effect.

Early empirical findings generally, but not completely, supported the supposition that diversity will increase the resistance of community-level properties to perturbation (Loreau *et al.* 2002). Diversity increased resistance to drought across a gradient of nutrient enrichment in experimental grassland plots (Tilman and Downing 1994), even after the confounding effect of fertilization was analytically controlled (Tilman 1996). Several subsequent studies also revealed a positive diversity–resistance relationship (Griffiths *et al.* 2000, Joshi *et al.* 2000, Mulder *et al.* 2001). Wardle *et al.* (2000a), however, emphasized the importance of composition, finding no effect of plant functional group richness on stability to drought in a greenhouse experiment.

Table 6.3 Effects of diversity on resistance to specific perturbations.

Reference	Stability property	Trophic level	Ecosystem property measured[a]	Diversity gradient[b]	Ecosystem type	Perturbation	Time scale	Div levels	Type of div effect
Allison (2004)	Resistance	Primary prod	Community biomass	Rem. F	Intertidal	Heating	21 m	1 to 3 FG	Negative
Caldeira et al. (2005)	Resistance	Primary prod	Community biomass	Exp. F	Grassland	Drought/frost	1 yr	1 to 14	None
DeClerk et al. (2006)	Resistance	Primary prod	Primary Productivity	Nat. F	Forest	Drought	64 yr	1 to 4	None
Griffiths et al. (2000)	Resistance	Multi	Decomposition rate	Rem. M	Pasture soil	Heating	1 yr	NA	Positive
Joshi et al. (2000)	Resistance	Primary prod	Community biomass	Exp. F	Grassland	Invasion	1 yr	1 to 32	Positive
Kahmen et al. (2005)	Resistance	Primary prod	AG Community biomass	Nat. F	Grassland	Drought	7 w	13 to 38	None
Kahmen et al. (2005)	Resistance	Primary prod	BG Community biomass	Nat. F	Grassland	Drought	7 w	13 to 38	Positive
Mulder et al. (2001)	Resistance	Primary prod	Community biomass	Exp. F	Bryophytes	Drought	5 d	1 to 32	Positive
Pfisterer and Schmid (2002)	Resistance	Primary prod	Community biomass	Exp. F	Grassland	Drought	8 w	1 to 32	None
Pfisterer and Schmid (2002)[1]	Resistance	Primary prod	Community biomass	Exp. F	Grassland	Drought	8 w	1 to 32	Negative
Tilman and Downing (1994)[2]	Resistance	Primary prod	Community biomass	Nutr. F	Grassland	Drought	2 yr	1 to 26	Positive
Van Peer et al. (2004)	Resistance	Primary prod	Community biomass	Exp. M	Grassland	Drought and heat	8 w	1 to 8	Negative
Wardle et al. (2000)	Resistance	Multi	Plant biomass	Exp. M	Grassland	Drought	14 m	2 to 7	None
Wardle et al. (2000)	Resistance	Multi	Decomposition rate	Exp. M	Grassland	Drought	14 m	2 to 7	None
Zhang and Zhang (2006a)	Resistance	Primary prod	Community biomass	Exp. M (HN)	Aquatic	Cold	6 d	1,2,4,7	none
Zhang and Zhang (2006a)	Resistance	Primary prod	Community biomass	Exp. M (LN)	Aquatic	Cold	77 d	1,2,4,8	none
Zhang and Zhang (2006b)	Resistance	Primary prod	Community biomass	Exp. M	Aquatic	Cold	105 d	1 to 5	Negative

[a] AG: above-ground; BG: below-ground
[b] Exp.: experimentally created diversity gradient; Rem.: gradient produced from selective removal of species; Nat.: naturally occurring diversity gradient; Nutr.: gradient produced by different nutrient levels. F.: field study; M.: mesocosm/microcosm study. LN: low nutrients. HN: high nutrients

[1] Resistance as measured as *absolute* biomass lost (see text)
[2] As reanalysed by Tilman (1996)

These earlier studies, as well as theory, did not consider pre-disturbance effects of diversity on resource use and community composition. Incorporating these effects yields several possible consequences stemming from selection effects and complementarity – the mechanisms that lead to positive relationships between diversity and the magnitude of ecosystem functioning. If the positive selection effect is in operation, fast-growing species, which tend to be more vulnerable to stress (Lepš et al. 1982), may dominate diverse mixtures, potentially producing a negative diversity–resistance relationship. Recent synthesis shows that while the positive selection effect is a common phenomenon in BEF studies, there are a substantial number of studies reporting negative selection effects (Cardinale et al. 2007). This raises the possibility that the above mechanism may be reversed in these cases, but empirical studies have yet to examine this possibility.

Pfisterer and Schmid (2002) postulate that complementarity indirectly rendered diverse grassland plots at the Swiss BIODEPTH site more vulnerable to experimental drought: the drought reduced the niche complementarity responsible for greater production in diverse communities. This raises the possibility that there may be a trade-off between a positive influence of diversity on the magnitude of ecosystem functioning and ecosystem stability. An important point is that Pfisterer and Schmid (2002) recorded a greater *absolute* reduction in biomass in more diverse plots, whilst there was no difference when resistance was measured *relative* to pre-drought biomass. Resistance is most meaningfully measured as a reduction in an ecosystem process relative to the pre-perturbation level; indeed, this is commonly practiced (Pimm 1984) and is consistent with measures of temporal stability.

Other experiments have revealed no effect of diversity on this metric of resistance, despite positive effects of diversity on pre-stress biomass (Wardle et al. 2000a, Caldeira et al. 2005, Zhang and Zhang 2006a). Furthermore, both Caldeira et al. (2005) and Zhang and Zhang (2006a) report that complementarity and selection effects were not modified by environmental stress. Finally, Van Peer et al. (2004) found a negative effect of diversity on resistance measured in relative terms.

The positive pre-stress relationship between diversity and biomass was diminished as the demand for water exceeded acquisition in species-rich communities.

In all of these experiments the positive diversity–resistance relationship predicted by the insurance hypothesis was absent, implying that diversity may not simultaneously increase the magnitude and resistance of ecosystem functioning. High community biomass could mean that each individual within a diverse community suffers greater resource limitation, as a finite resource supply is under greater demand; in effect, the disturbance size is greater for each individual. The shortfall between resource demand and supply may outweigh the effect of the increasing range of species' tolerances with greater diversity.

The implication of this hypothesis is that in the absence of a positive diversity–biomass relationship, evidence of the insurance effect will be more likely. There are insufficient studies to assess this rigorously, but three studies provide tentative support for this hypothesis. Across a natural diversity gradient in German grasslands, Kahmen et al. (2005) found that the resistance of belowground biomass to an experimental drought increased with plant diversity, but no effect on aboveground biomass was observed. Reference plots showed no relationship between diversity and pre-stress biomass. Mulder et al. (2001) and Hughes and Stachowicz (2004) similarly reported a positive influence of diversity on stress resistance in systems without a pre-disturbance diversity–biomass relationship.

6.2.2.2 Resilience

Loreau and Behera (1999) found that diversity and resilience may be negatively related within theoretical competitive communities. Based on the very few published empirical tests (Table 6.4), no consistent influence of diversity on resilience of community properties is evident. Studies of successional diversity gradients show negative relationships between diversity and resilience (Smedes and Hurd 1981, Lepš et al. 1982), but species' life-history traits probably played a confounding role here. Theoretical predictions (Loreau and Behera 1999) are supported by just a single BEF experiment (Pfisterer and

Table 6.4 Effects of diversity on resilience of ecosystem properties following a perturbation.

Reference	Stability property	Trophic level	Ecosystem property measured	Diversity gradient[a]	Ecosystem type	Perturbation	Time scale	Div levels[b]	Type of div effect
Allison (2004)	Resilience	Primary prod	Community biomass	Exp. M	Intertidal	Heating	21 m	1 to 3 FG	Positive
DeClerk et al. (2006)	Resilience	Primary prod	Primary Productivity	Nat. F	Forest	Drought	64 yr	1 to 4	Positive
Griffiths et al. (2000)	Resilience	Multi	Decomposition rate	Rem. M	Pasture soil	Heating	1 yr		Positive
Pfisterer and Schmid (2002)	Resilience	Primary prod	Community biomass	Exp. M	Grassland	Drought	8 w	1 to 32	Negative
Steiner et al. (2006)	Resilience	Multi	Community biomass	Exp. M (HN)	Aquatic	Non-selective	53 d	1 to 4	None
Steiner et al. (2006)	Resilience	Multi	Community biomass	Exp. M (LN)	Aquatic	Non-selective	53 d	1 to 4	Positive
Tilman and Downing (1994)[1]	Resilience	Primary prod	Community biomass	Nutr. F	Grassland	Drought	2 yr	1 to 26	None

[a] Exp.: experimentally created diversity gradient; Rem.: gradient produced from selective removal of species; Nat.: naturally occurring diversity gradient; Nutr.: gradient produced by different nutrient levels. F.: field study. M.: mesocosm/microcosm study. LN: low nutrients. HN: high nutrients
[b] FG.: Functional groups
[1] As reanalysed by Tilman (1996)

Schmid 2002), which showed lower resilience in high-diversity grassland plots nine months after drought. Tilman (1996) found no effect of diversity on resilience after analytically removing the confounding factors present in an earlier analysis (Tilman and Downing 1994). Conversely, resilience increased with functional group richness in intertidal seaweed communities (Allison 2004), with conifer species richness in the Sierra Nevada (DeClerck *et al.* 2006), and with diversity within five trophic levels in a microcosm experiment (Steiner *et al.* 2006). DeClerck *et al.* (2006) invoke complementarity, suggesting that during community recovery, resources are abundant, thus resource partitioning is possible along several niche axes – a postulation that perhaps deserves further theoretical and empirical consideration. Steiner *et al.* (2006), on the other hand, credit their result to the sampling effect, as resilient communities exhibited a reduction in evenness over time.

It is difficult to separate resilience from resistance. Indeed, most studies have not removed the legacy of resistance from measures of community recovery (Tilman and Downing 1994, Mulder *et al.* 2001, Pfisterer and Schmid 2002, Allison 2004). To achieve this, a non-selective mortality event must occur or be experimentally applied (Steiner *et al.*, 2006). In nature, however, resistance and resilience are inextricably linked, because the community recovers with its post-perturbation composition and abundance. Thus, whilst equally reducing abundances of populations to isolate resilience *per se* from the effects of the disturbance is of considerable theoretical interest, the relevance to real systems is questionable.

6.2.3 Summary of empirical progress

Numerous diversity–stability experiments have been published in the last half-decade, substantially improving our understanding of the relationship between diversity and various facets of stability at population and community levels. Whilst diversity was commonly found to enhance community-level temporal stability, the effect of diversity on resistance and resilience is more equivocal. There is some empirical evidence to suggest that positive diversity–productivity effects may preclude a stabilizing effect of diversity in the face of extreme perturbations – a trade-off that needs further mechanistic exploration. Multi-trophic studies have been limited in number and scope, mainly being conducted in microbial rock pools or laboratory microcosms. Nevertheless, the typical result that populations are actually stabilized by diversity in these systems is intriguing and represents a major distinction from classical theory (e.g. May 1973) and findings from some grassland experiments (Tilman 1996, Tilman *et al.* 2006b).

6.3 A broadening perspective

Whilst of undoubted value, controlled experiments are logistically constrained. Isolating the role of diversity *per se* from that of species identity has proven a formidable task requiring large numbers of treatments and replicates. This has limited experiments to tractable, closed, small-scale systems–predominantly grassland plots and laboratory microcosms. Hence the degree to which the findings from these experiments are applicable to larger, landscape scales, different systems, and the delivery of important ecosystem services is questionable.

Building a more complete understanding of the role of diversity in stabilizing ecosystem functioning in these broader contexts requires approaches that trade replication and control for studies conducted over larger scales and in complete systems. In this section we outline three areas that address this link. First, we discuss several recent developments in theory that demonstrate how properties of food webs can affect aspects of population and community stability. Second, we assess possible effects of diversity on stability in systems with multiple stable states, before finally highlighting the possible role of diversity in stabilizing the delivery of two key ecosystem services: pollination and yield from fisheries.

6.3.1 Lessons from food webs

Although BEF research has begun to examine the effect of diversity on the stability of multi-trophic aquatic systems, the tie between food web theory and BEF science is not yet strong. This is unfortunate, because recent developments in food web theory illustrate several stabilizing properties of food webs that may explain the inconsistency of results in BEF-stability studies so far. Moreover, how species loss affects these properties can provide insight on diversity's effects in systems too complex for controlled experimental treatments. It must be noted that the food web literature provides definitions of stability that often vary from those in the BEF literature, they generally focus on recovery times following perturbations to the system, but also encompass measures of temporal variability among component species. Additional aspects of stability, not addressed in BEF experiments to date, may also have implications for ecosystem functioning, and as such several are outlined here.

It is now widely acknowledged that weak trophic interactions confer stability (population-level resilience) in natural food webs, and therefore that food webs with lower mean interaction strength are more stable. However, the configuration of interactions is also important, as the destabilizing effects of strong trophic links can be dampened if those strong links are coupled to weak interactions (McCann 2000). Coupled weak and strong interactions can promote asynchronous population fluctuations of prey, stabilizing aggregate prey biomass as well as resource supply to a switching predator (McCann 2000). This weak interaction effect may have marked implications for the temporal stability of community biomass in food webs. Furthermore, it has been shown that weak interactions confer local stability to food webs when they occur in omnivorous food web loops (Neutel *et al.* 2002, Emmerson and Yearsley 2004). Simulation studies have shown that an analogous effect can be scaled up to fast and slow 'energy channels' within food webs generated through alternative energy sources (Rooney *et al.* 2006). Species loss will alter the number and configuration of stabilizing weak interactions, as well as the mean interaction strength (McArthur 1955, McCann 2000), potentially destabilizing populations and ecosystem functions.

In reality, food webs are not static structures fixed in time; they are dynamic, varying in structure seasonally and from year to year. How such dynamic topologies persist is poorly understood. In this context, food webs are flexible structures, constantly changing in species composition, structure, and dynamics (de Ruiter *et al.* 2005), yet most theoretical studies of diversity–stability relationships assume static patterns of trophic linkage (May 1972, Pimm *and* Lawton 1977, 1978, de Ruiter *et al.* 1995, Neutel *et al.* 2002). Adaptation is suggested as one mechanism from which food web flexibility arises (Kondoh 2006), with adaptive defences by prey and adaptive foraging by predators influencing the strength of trophic interactions. The flexibility provided by adaptive foraging should enhance community persistence, as predators capable of a foraging shift can maximize their net energy gain by switching away from a less profitable resource. Indeed, Kondoh (2003) demonstrated that the classic negative complexity–stability relationship of many theoretical studies is inverted when the effects of adaptive foraging behaviour are incorporated. There are obvious implications of this mechanism for the diversity–stability relationship, although they have yet to be explicitly explored in the context of BEF research. A reduction of species diversity through the loss of prey species will limit the prey-switching options of adaptively foraging predators.

Primary species loss can trigger secondary extinctions, further reducing diversity and its associated stabilizing effects on ecosystem functioning. The tolerance of a food web to species loss is also an important aspect of stability in its own right: robustness (see Loreau *et al.* 2002). The traits of the deleted species markedly affect the likelihood and extent of secondary extinctions (see also Chapter 5). For example, like keystone species and ecosystem engineers, the loss of a highly connected species (species with a high proportion of total possible trophic links realized) has been shown to have disproportionate effects on food web structure (Solé and Montoya 2001).

Robustness may also depend on the characteristics of the entire food web. The amount of connectance is important, as high connectance may delay the onset of an extinction threshold (Dunne *et al.* 2002b). Whether connectance and species richness are associated in empirical (not theoretical) food webs is unresolved, however. Whilst Dunne *et al.* (2002b) report no relationship, Montoya and Solé (2003) found connectance to be lower in species-rich webs. The distribution of trophic links between species, the degree distribution, also has consequences for robustness. Food webs with skewed degree distributions, i.e. numerous poorly linked species and a few highly linked species, are robust to random species deletion but sensitive to removal of the most connected species (Solé and Montoya 2001, Dunne *et al.* 2002b). This is because randomly deleted species are likely to be poorly linked, thus having minimal knock-on effects on others. Across 12 well-described food webs, Montoya and Solè (2003) show that the degree distribution becomes progressively more skewed with increasing species richness. Consequently, species-poor webs are less robust in response to random species loss, since most species are moderately well linked. Therefore, whether species-rich food webs are likely to be more robust to species loss depends on whether species loss is random with respect to species' connectedness (see also Chapter 5). Other key factors that determine the effects of diversity on cascading species extinctions and ecosystem functioning include the strength of intraspecific density dependence (Thébault *et al.* 2007).

Although the properties that affect food web stability are critical to understanding the stability of ecosystem functioning in complex (multi-trophic) systems, it must be noted that comparisons across food webs may give results that differ from changes within a single food web. It is unknown, for example, whether the degree distribution of a species-rich web will become increasingly centralized, mirroring species-poor webs, as diversity is eroded within it. If this happens, food webs losing species will become ever more sensitive to random species deletion, yielding a positive feedback that may exacerbate system collapse. Cross-fertilization of ideas between BEF and food web science will help answer this and other unknowns. Investigating how food web properties vary concurrently with species richness in natural and manipulated food webs will help to better integrate BEF science with food web theory.

6.3.2 Diversity–stability in complex, real-world systems

6.3.2.1 Systems with multiple stable states
In all explicit diversity–stability experiments discussed here and elsewhere (Cottingham *et al.* 2001, Loreau *et al.* 2002, Hooper 2005), stability has been considered with respect to a single 'stability domain.' Both theory and observations, however, have shown that many ecosystems can exhibit non-linear dynamics, switching between multiple stable states (Scheffer and Carpenter 2003). It is thus unclear the extent to which findings from the studies discussed here (Table 6.1) can be applied to multiple equilibrium systems. Ecological resilience *sensu* Holling (1973) describes the amount of disturbance a system can absorb whilst still remaining within the same basin of attraction and can be heuristically viewed as the size of a particular stability domain. Changing environmental conditions, compounded perturbations and/or species loss can reduce ecological resilience, increasing the probability of an abrupt 'catastrophic shift' to an alternative state. The insurance hypothesis has been incorporated into this view of ecosystems (e.g. Peterson *et al.* 1998, Gunderson 2000): assuming that biodiversity increases the range of responses to the environment (functional response diversity), a more diverse system is buffered against impacts of perturbations and resultant catastrophic shifts.

Despite a dearth of experimental studies, the idea that biodiversity begets stability in systems with multiple stable states seems to be widely accepted. Reviews of regime shifts and ecosystem resilience assume a strong connection between functional response diversity and resilience (e.g. Gunderson 2000) or loss of diversity and loss of function (Briske *et al.* 2006). The diversity–stability concept even seems to have influenced environmental management in certain systems. For example, maintaining or increasing landscape diversity in pastures and rangelands is encouraged or required by many land-use agencies and programs (e.g. Mason *et al.* 2003, Mitchell *et al.* 2005), often for maintaining or increasing biodiversity itself, but also for increasing resilience in the face of disturbance (Pellant *et al.* 2004, Drever *et al.* 2006). The question is whether evidence supporting these ideas exists in the systems to which they are applied. Rangelands and coral reefs, two systems that provide a number of ecosystem services for a large part of the Earth's population and exhibit dramatic instability (undergo state changes) serve as examples.

Range and pasture lands occupy approximately 20 per cent of the land surface of the globe in areas that are particularly susceptible to drastic ecosystem changes such as desertification and grass-to-shrubland conversion (Hodgson and Illius 1996). Consequently, they have been the focus of many conceptual developments regarding ecological thresholds and alternative state theory, which are closely tied to the concept of ecological resilience (e.g. Briske et al. 2006). Although there is a wealth of literature on these topics, and the entire basis of rangeland management is now shifting to this paradigm in some parts of the world (USDA-NRCS 1997), there has been little research directly testing the role of biodiversity in rangeland resilience, state changes, or other aspects of stability at the scales that are applicable to range and pasture managers.

Two studies that directly address the relationship between biodiversity and stability in actual production systems had contrasting results. In Australian sheep pastures with a range of plant species richness caused by various manipulations (grazing regime, fertilization) and environmental conditions (climate, soils), variability of herbage production over 3–4 years was either not significantly related or slightly negatively related to plant diversity or species richness (Kemp et al. 2003). On the other hand, in dairy cattle pastures planted specifically to compare plant species richness effects on herbage production, yield did not differ significantly among treatments of 2, 3, 6, or 9 species in years with normal precipitation, but the two-species pasture did have lower production than the others in a dry year (Sanderson et al. 2005).

The evidence from coral reefs is equally equivocal. A biogeographical comparison can provide a tentative insight into the possible role of species diversity in providing resilience on coral reefs. In the Caribbean, eutrophication from increased nutrient inputs, disease, and over-harvesting of herbivorous fishes has resulted in a phase shift from coral- to fleshy macroalgae-dominated reef communities (Hughes 1994). Although the suite of functional groups is similar between Indo-Pacific reef systems and Caribbean reef systems, the former have much greater taxonomic diversity within most functional groups, presumably making them less susceptible to such phase-shift-causing perturbations (Bellwood et al. 2004). However, taxonomic richness does not guarantee functional resilience. A different single species, which is relatively rare, was responsible for reversing an experimentally induced coral–algal phase shift in the same system by high consumption on the fleshy algae (Bellwood et al. 2006). The effect of this single keystone species in the latter study was a complete surprise, in that it was previously unknown to consume these algae. The identity of species, rather than richness *per se*, may thus have dominant effects on the resilience of coral reefs (see also Bellwood et al. 2003).

That this surprise occurred in one of the better-studied systems in the world clearly supports the precautionary approach in biodiversity conservation. These results also highlight the fact that high diversity does not guarantee high redundancy and the stability often associated with it. If all the redundancy is in one functional group, then diversity will not necessarily promote stability. This is particularly relevant in light of the perturbations that afflict ecosystems today. All ecosystems evolved with a regime of disturbance, and evolutionary processes likely led to functional effect redundancy within these systems because of differences in organisms' responses to this disturbance regime (e.g. Walker et al. 1999). Novel individual perturbations (e.g. new diseases), as well as new combinations of disturbance events, could be tapping into functional types where that response redundancy does not yet exist.

6.3.2.2 Diversity and the stability of ecosystem services

Several empirical studies have recently emerged that bridge the gap between controlled experiments and real-world applications, demonstrating how human-impacted ecosystems can be used to examine the roles of environmental change and biodiversity in the stability of ecosystem service provision. Here we discuss two ecosystem services, first pollination and second fisheries yield, as examples. Both of these examples suggest that species diversity can influence the stability of ecosystem service provision.

Pollination is one key ecosystem service that has received attention within agricultural landscapes. Kremen *et al.* (2002, 2004) showed that intensive farming practices and a reduced proportion of natural habitats negatively affects the diversity of pollinators and temporal stability of melon pollination. Sites with high pollinator species richness provided more stable pollination services over time than sites with low species richness because of asynchronous fluctuations in the populations of pollinators from one year to the next. The role of species richness in spatial stability of pollination was demonstrated in coffee plantations: greater pollinator diversity, which is affected by local (e.g. plant diversity, light availability) and regional (e.g. isolation from natural habitat) factors, reduced spatial variation in fruit set between coffee plants (Klein *et al.* 2003a, b; Chapter 14).

Fisheries provide an important source of food for much of the world's population, underpinning the diets and economies of many coastal communities in the world's poorer countries. Collating data from the world's fisheries, Worm *et al.* (2006) found that the proportion of collapsed fisheries in a region was negatively related to its fish taxonomic diversity. Furthermore, they discovered that the likelihood of recovery from fisheries collapse was positively associated with species richness across large marine ecosystems. Causality is difficult to infer from this correlative approach, but the results support the supposition that diversity increases both resistance to – and recovery from – over-exploitation. Those harvesting the fish benefit from greater diversity, as the reliability and abundance of total catches increases with diversity. Humans act as switching, wide-ranging predators, releasing stocks from predation as they become scarce (McCann 2000) and changing to a more abundant species, thus deriving a stable supply from numerous fluctuating resources (Worm *et al.* 2006). This mechanism may partly explain why diverse fisheries are less likely to collapse – it is more profitable to switch targets if more abundant species are available. The well-known collapse of the Newfoundland cod fishery could be attributed to a single target species, compared to the diverse portfolio of taxa exploited in tropical subsistence fisheries, for example. Further efforts to explore the role of species diversity in mediating various aspects of the stability of ecosystem service delivery will both inform the management of such services, and also contribute to our general understanding of diversity–stability relationships in real-world systems. Such studies will also help to integrate our understanding of the drivers of species diversity and its ecological effects.

6.4 Conclusions

Recent studies have yielded great progress in understanding diversity's effects on the stability of ecosystem functioning in increasingly complex systems. Clearly, however, there is still much work to be done to reconcile theory, experimental results, and observations from natural or human-altered systems. A key step towards this goal must be elucidating the mechanistic basis of diversity effects on aspects of stability in a range of systems; a challenge that requires greater integration of theoretical and empirical work (Box 6.2). Insights from a growing body of food web analyses and simulations may help to explain the findings of BEF stability studies in multi-trophic systems.

To increase the applicability of diversity–stability research the effects of realistic diversity changes on valuable ecosystem services must be investigated. Studies across gradients of anthropogenic impacts have great potential to address this need, as these gradients incorporate both local (habitat) and landscape factors responsible for shifts in diversity (Chapter 14). Long-term measures of diversity and related ecosystem services across these land-use gradients, combined with modelling and mesocosm studies based on the communities occurring across these gradients, would help to elucidate the effect of biodiversity on the stability of key ecosystem services and potentially shed light on the underlying mechanisms. Recent work in the field of ecological economics shows that stability adds additional economic value to ecosystem services in the form of insurance (Chapter 17), further underlining the importance of a thorough understanding of the effect of biodiversity on ecosystem functioning and associated services.

CHAPTER 7

The analysis of biodiversity experiments: from pattern toward mechanism

Andy Hector, Thomas Bell, John Connolly, John Finn, Jeremy Fox, Laura Kirwan, Michel Loreau, Jennie McLaren, Bernhard Schmid, and Alexandra Weigelt

7.1 Introduction

This chapter reviews the methods developed to investigate the mechanisms that generate relationships between diversity and functioning in biodiversity experiments. What do we mean by mechanism? An important recent advance in ecology and evolution has been the championing of mechanistic statistical models (Mangel and Hilbourn 1997). These statistical models are mechanistic in the sense that their parameters refer to biological processes that can be quantified, rather than to unmeasureable abstract concepts that often prove useful in purely theoretical models of ideas. Similarly, non-linear regression analysis is often described as 'semi-mechanistic' when parameters can be at least loosely related to biological processes (Pinheiro and Bates 2000). In many areas of science there are often multiple layers of mechanism underlying the phenomena of interest. As we will explain below, some of the models reviewed in this chapter could be termed fully mechanistic in that they can be built to include parameters that refer directly to ecological processes (e.g. predation rates), whereas some of the other methods could be termed semi-mechanistic in the sense that they can indicate the presence of ecological processes (e.g. 'complementarity effects') even if, as explained above, they cannot quantify the exact biological process that underlies these effects. To understand the motivation for the development of these methods we first review the debate over the mechanisms responsible for relationships between biodiversity and ecosystem functioning.

7.1.1 Background

Following a landmark conference in 1992, the study of the relationship between biodiversity and ecosystem functioning became a focused area of research. In the edited book that arose from that meeting (McNaughton 1993, Schulze and Mooney 1993) quoted the following from Chapter 4 of *On The Origin of Species*: 'It has been experimentally proved that if a plot of ground be sown with one species of grass, and a similar plot be sown with several distinct genera of grasses, a greater number of plants and a greater weight of dry herbage can thus be raised'. The quote concisely makes a prediction – that more diverse plant communities should be more productive – and indicates the underlying mechanism. Darwin contrasts one species with several distinct genera, implying that it is the ecological niche differences between species that underlie this effect. More extensive text from Darwin's *Natural Selection* (Stauffer 1975) clarifies that Darwin really was relating biodiversity to ecosystem functioning via what he termed the 'ecological division of labour' (Hector and Hooper 2002) when he wrote that, 'A greater absolute amount of life can be supported...when life is developed under many and widely different forms...the fairest measure of the amount of life being probably the amount of chemical composition and decomposition within a given period.' Following McNaughton many researchers have reproduced this quote and its popularity reflects the tendency of ecologists at this time to focus on these ecological niche differentiation mechanisms and the species 'complementarity' (as introduced by Woodhead 1906) that results from these differences. However, there is a second class of potential mechanisms that was under-represented in the early literature. The sampling effect hypothesis (Aarssen 1997, Huston 1997, Loreau 1998a, Tilman *et al.* 1997c) proposes that in biodiversity experiments randomly assembled diverse communities have a higher probability of containing and being dominated by the species which is most productive when grown alone. The selection effect (Loreau and Hector 2001) and dominance effect (Fox 2005b) are similar but more general effects that relate the relative abundance of species in mixtures to their performance when grown alone (see below).

Species complementarity is sometimes used with reference only to resource partitioning. However, both conceptually and in practice, it is often difficult to separate resource partitioning from facilitation and other ecological processes such as diversity-dependent differences in natural enemy impacts (Connell 1978, Janzen 1970, Root 1973, Zhu *et al.* 2000). This set of 'complementarity effects' and the alternative set of 'sampling', 'selection' or 'dominance effects' are not mutually exclusive (they often can and do occur together) and can produce very similar patterns. This means that when analyzing biodiversity experiments, process cannot be inferred from pattern alone. Distinguishing the contributions of these alternative classes of mechanism has become an important goal in the analysis of biodiversity experiments, since they can indicate what sorts of ecological processes have generated the observed pattern and are important considerations in interpreting the results of these experiments.

7.1.2 Limitations and needs

Some of the methods available to investigate mechanism in biodiversity experiments have limitations that mean they cannot be applied to many existing datasets. Two of the main limitations of these methods are that they require measurements of how species affect ecosystem functioning when grown alone (that is all species must be grown in monoculture) and the contributions of individual species to the ecosystem functioning of mixtures (e.g. the total productivity of a mixed community must be broken down into the contributions of individual species). For certain organisms and for particular functions, these requirements are often difficult or impossible to meet. Hence there is the need for a varied toolbox of methods for the analysis of mechanism in biodiversity experiments that can encompass all organisms and all ecosystem processes. We think that this suite of tools now exists and the aim of this chapter is to provide a comparative review and users guide to these methods. Given the number of methods and limited space, and the fact that all of the methods have already been described in the literature, we restrict the main text to general descriptions only (with detailed supplementary material provided by the authors of each method). The literature has mainly focused on productivity as an ecosystem function and our discussion does the same even though some of these methods can be applied to other ecosystem process.

7.2 Analysis of mechanism in biodiversity experiments

7.2.1 Transgressive overyielding

One of the simplest analyses that can be performed is a comparison of the functioning of mixtures relative to the best-performing single-species community (monoculture). Mixtures that perform significantly better than the best monoculture are said to transgressively overyield and transgressive overyielding has often been seen as the acid test for positive effects of biodiversity (Cardinale *et al.* 2006a, Cardinale *et al.* 2007, Hector *et al.* 2002b, Kirwan *et al.* 2007).

7.2.2 Overyielding: relative yields

A second approach is to look for overyielding more generally by comparison to the single-species performances of all of the species present in a mixed community. The concept of 'relative yields' (RYs) has been used for this purpose in plant ecology and agriculture since the mid-twentieth century (De Wit and Van den Bergh 1965, Harper 1977, Vandermeer 1989). The relative yield of a plant species is simply its biomass in mixture expressed as a proportion of its biomass in monoculture. Summing the relative yields for all species in a mixture provides a relative yield total (RYT). Relative Yield Total values greater than one show that increases in the abundance of some species in mixture have not been exactly compensated by decreases in others (as would be the case in a zero-sum game (Hubbell 2001)). Overyielding (as distinct from transgressive overyielding) is often taken to indicate resource partitioning but can also occur through facilitation and other mechanisms, such as the reduction of natural enemy impacts in mixtures (as described above). Loreau (1998b) devised a more general scheme of deviations from expected values that are closely related to the earlier relative yield-based measures. Overyielding-based methods usually assume a substitutive experimental design in which total density is held constant, independent of diversity, although the method can be adapted for additive designs too (where additive means that the total density is the summed total of the densities of all of the component species).

7.2.3 Additive partitioning methods

7.2.3.1 Two-way additive partitioning of biodiversity effects

The additive partitioning method (Loreau and Hector 2001) extends the relative yield approach described above to define an overall net biodiversity effect and to partition this into two additive components: a complementarity effect and a selection effect (see supplementary material). In a substitutive experiment, the net biodiversity effect (for a community formed from species started at equal densities) is simply the difference between the observed yield of the mixture and the average of the monoculture yields. The net biodiversity effect equals zero when individual plants grow equally well in monoculture and mixture. The complementarity effect is based on changes in relative yields (or rather, differences in observed relative yields versus their null expectation values) and is linearly related to RYT (but scaled to a null value of zero rather than 1). Complementarity effect values > 0 indicate positive effects of biodiversity on overyielding while values < 0 indicate interference competition. The other half of the partition is a covariance term that was inspired by the Price equation from evolutionary genetics (although as we explain below the additive partitioning method and Price equation are different). The selection effect measures the covariance between a species trait (e.g. monoculture biomass) and its performance in mixture. Positive selection effect values indicate that species with greater than average monoculture biomass perform better than expected in mixture, while negative values indicate the converse. While the Loreau–Hector additive partitioning method does not examine biological processes directly, it has allowed major advances in the debate over the mechanisms underlying the patterns found in biodiversity experiments (e.g. Loreau and Hector 2001, Cardinale et al. 2007).

7.2.3.2 Tripartite additive partitioning of biodiversity effects

One limitation of the additive partition is that it assumes, as do relative yields, that complementarity is distributed equally across species. This means that it may over- or underestimate total complementarity, some of which falls under the selection effect (Petchey 2004). The tripartite partition (Fox 2005b) is a modification of the additive partition of Loreau and Hector (2001) described above. The two versions share the same goal of identifying whether, and for what reasons, the functioning of a given mixture of species deviates from that expected under a simple null hypothesis (see supplement). However, the tripartite version partitions the difference between observed and expected function into three additive components: the dominance effect (DE), trait-independent complementarity effect (TICE), and trait-dependent complementarity effect (TDCE). The two versions are related as follows. The complementarity effect from the two-way additive partition corresponds exactly to the trait-independent complementarity effect from the tripartite version. However, the tripartite partition can be thought of as taking the original two-way split into complementarity and selection effects from Loreau and Hector (2001) and performing a further split by dividing the selection effect into the dominance effect and a trait-dependent complementarity effect (SE = DE + TDCE). Species with particular traits (monoculture yields) can do better than expected in mixtures either at the expense of other species (pure competitive replacement as quantified by the dominance effect), or not at the expense of other species (trait-dependent complementarity effect).

7.2.4 Applying the Price equation to biodiversity experiments

Relative yield-based approaches, and the related additive partitioning approaches, have primarily been used to aid the interpretation of substitutive experiments with plants or similar organisms. In such experiments, interest centres on whether the functioning of a diverse mixture of species deviates from that expected under the null model that intra- and interspecific interactions are identical. However, in many circumstances, interest centres on comparing the functioning of different sites directly with one another, rather than on comparing each to a null model. This may be because no appropriate null model exists, because information to parametrize a null model is lacking, or simply because the investigator wishes to consider all processes that cause ecosystem function to vary among sites rather than factoring out the effects of some processes by comparison to a null model. For instance, an investigator might be interested in how the functioning of a site has changed since a historical extinction event, or in explaining variation in function along a natural diversity gradient. The Price equation partition (Fox 2006, Fox and Harpole 2008) was designed for such cases.

The Price equation partition classifies the mechanisms that cause two sites (a 'pre-loss' site of higher species richness, and a 'post-loss' site of lower species richness) to differ in ecosystem function (see supplement). The Price equation partition assumes that the species at the post-loss site comprise a nested subset of the species at the pre-loss site, and that total ecosystem function comprises the sum of the separate contributions of individual species (Fox 2006). The assumption of a 'summed' ecosystem function covers primary productivity and many other functions, but does not cover many others (see Fox and Harpole 2008). The Price equation partition divides the difference in total function between two sites, ΔT, into three additive components. The species richness effect (SRE) is that part of ΔT attributable to random loss of species richness, independent of which species were lost. The species composition effect (SCE) is that part of ΔT attributable to non-random loss of species making higher- or lower-than-average contributions to total ecosystem function. The context dependence effect (CDE) is that part of ΔT attributable to between-site differences in the functional contributions of the species present at both sites (i.e. the functional contributions of these species are not constant, but rather are context-dependent).

The Price equation partition takes (and extends) the original Price equation developed in evolutionary biology to classify and partition the causes of evolutionary change in mean phenotype (Frank 1995, 1997, Price 1970, 1995) and applies it to the effects of changes in biodiversity on ecosystem processes. In evolution, the mean phenotype of an offspring population can differ from that of a parental population for two reasons: natural selection (covariation between parental fitness and parental phenotype), and imperfect transmission (factors, such as environmental change, that cause the phenotypes of offspring to deviate on average from those of their parents). In mathematical terms, natural selection is analogous to the Species Composition Effect. For instance, non-random death of (selection against) large-bodied individuals will reduce mean body size in the next generation, assuming body size is heritable and all else being equal. Analogously, non-random extinction of high-functioning species will reduce mean function per species, and thus total function, all else being equal. Imperfect transmission is precisely analogous to the Context Dependence Effect. For instance, if all offspring to have larger body sizes than their parents then, all else being equal, mean offspring body size will exceed mean parental body size. Analogously, if all species remaining at the post-loss site function at a higher level than they did at the pre-loss site, mean function per species, and thus total function loss, all else being equal, will be higher at the post-loss site (while functioning is usually thought of as declining with species loss it could also increase).

7.2.5 Classical statistical analysis of mechanisms

7.2.5.1 Random partitions design and analysis
Bell et al. (2005b) introduced a direct approach to the analysis that avoids calculating derived values (e.g. complementarity effects) which must then be statistically analyzed in a second stage. Their approach is a direct analysis of the primary data using normal least squares and general linear models. The Bell et al. approach includes several notable features of both the design and analysis.

The design takes a full species pool, N, and forms a diversity gradient by dividing by integer factors of N. For example, Bell et al. selected a pool of 72 study species (from 103 available species) so that their diversity gradient comprised the series 72 (72/1), 36 (72/2), 24 (72/3), 18 (72/4), 12 (72/6), 9 (72/8), 8 (72/9), 6 (72/12), 4 (72/18), 3 (72/24), 2 (72/36), 1 (72/72). In other words, the species pool was randomly divided in half, randomly divided into thirds and so on. The resulting communities were termed a 'partition' of the selected species pool (to avoid confusion note the use the word in a difference sense to the additive partitioning equations described above). This approach was then repeated using different random selections to produce different replicate partitions, that is replicate diversity gradients that divide up the species pool in different ways (for example, two replicate gradients would divide the species pool into two different half-pools rather than using the same selection of species). This approach ensures that, within each replicate partition, each species is present once at every level of diversity (a species is present in one monoculture, in one two-species community and so on).

The method fits a least-squares model to the data that includes terms for the species richness, the presence/absence (identity) of each species, and the composition of the community. The level of ecosystem functioning, y, is modelled as:

$$y = \beta_0 + \beta_{LR}x_{LR} + \beta_{NLR}x_{NLR} + \left(\sum_{i}^{s}\beta_i x_i\right) + \beta_Q x_Q + \beta_M x_M + e$$

where β_0 is the intercept, β_{LR} is the coefficient associated with linear richness (richness treated as a continuous variable), β_{NLR} is the coefficient associated with species richness treated as a categorical variable, the β_i's are the coefficients associated with the presence/absence of each species, β_Q is the coefficient associated with each partitioned species pool, β_M is the coefficient associated with each composition, and e is a normally distributed random variable. One important feature of this method of analysis is that, when it is used along with the experiment design described above, the non-linear richness and species identity terms are orthogonal (do not share sums-of-squares). Consequently, it is possible to parse some of the explained variation into either variation due to species identity or to variation due to non-linear richness. Another unique feature of the design is that the collective effects of species interactions can be captured by the non-linear richness term (β_{NLR}). This 'deviation from linearity' term provides an ensemble test for all species interactions combined.

7.2.5.2 The diversity–interactions statistical modelling approach
The diversity–interactions approach (Kirwan et al. 2007) is also a more direct application of classical statistical methods that has several similarities to the analysis conducted by Bell et al. (2005b) (see http://www.diversity-model.com/). The approach is based on a framework of statistical models whose coefficients reflect the effects of species identity and species interactions. The initial community compositions are described by the abundance of each species as a proportion of total initial abundance (M). The species proportions (P_i) are either planned experimental proportions or the relative abundances of species measured early in the experiment. The regression equations describe the ecosystem process response variable (y) as follows:

$$y = \sum_{i=1}^{s}\beta_i P_i + \alpha M + \sum_{i,j=1, i<j}^{s}\delta_{ij}P_i P_j + \varepsilon$$

Here, β_i (the identity effect of the ith species) is the expected monoculture performance of the ith species, α is the effect of overall initial abundance, δ_{ij} is a measure of the strength of interaction between species i and j, and ε is the residual term. The model is fitted using standard regression techniques. The sign of an interaction coefficient δ_{ij} indicates whether the interaction between species i and j has a synergistic or antagonistic effect on ecosystem function. The total contribution to ecosystem function of the interaction is $\delta_{ij}P_iP_j$ and also depends on the initial relative abundances of the two species. The response in a mixed community expected solely from monoculture performance is $y = \sum_{i=1}^{s} \beta_i P_i$. The net biodiversity effect in model (7.1) is $\sum_{i,j=1, i<j}^{s} \delta_{ij} P_i P_j$, the sum of all pairwise interactions among species. This diversity effect generalizes to a rich class of alternative models based on alternative assumptions about the strength of pairwise species interactions. For example, the strength of pairwise interactions may all be the same (identical values of δ_{ij}) leading to a diversity effect $\delta \sum_{i,j=1, i<j}^{s} P_i P_j$ that is related to evenness (Kirwan *et al.* 2007). Alternatively, there may be clear patterns among the δ_{ij} that reflect the traits of the species in the mixture (e.g. a functional group model that has a common interaction coefficient for all pairwise interactions between species from different functional groups). Interactions may also involve more than two species or more complex functions. Many of these alternative models are hierarchical to model (7.1) or to the model with a single interaction coefficient, which leads to straightforward comparisons of models to identify the most appropriate. For example, a pair of nested models with and without species interactions can be compared to test whether the ecosystem process response is determined only by species identity effects or by identity effects and species interactions. The complexity to which we can describe patterns of interaction, and the sensitivity to discriminate between alternative patterns, depend on the range and patterns of relative abundances that were selected in the experimental design. Many diversity–function experiments use communities with varying species richness, but equal relative abundances. Diversity–interaction models can be fitted to data from such a design. However, by including experimental communities that provide good coverage of the design space, such as communities dominated by one or a subset of species, we can test for more complex patterns of species interaction (Kirwan *et al.* 2007). Also, prediction of the diversity effect may be reliable over a wider range of communities in which all component species are not equally represented.

7.3 Discussion

7.3.1 Pattern

Meta-analysis of the results from the first decade of research in this area clearly shows a positive relationship between biodiversity and ecosystem functioning; a pattern which is consistent across trophic groups (producers, herbivores, detrivores, and predators) and present in both terrestrial and marine ecosystems (Balvanera *et al.* 2006, Cardinale *et al.* 2006a, Worm *et al.* 2006). However, in terrestrial ecosystems the relationship between biodiversity and ecosystem functioning is generally quickly saturating with increasing diversity (Cardinale *et al.* 2006a) suggesting that the effect of random biodiversity loss on ecosystem functioning will be initially weak but accelerating (Hector *et al.* 1999).

7.3.2 Transgressive overyielding

The low frequency of transgressive overyielding in the meta-analysis performed by Cardinale *et al.* (2006a) led them to suggest that the general positive relationship between biodiversity and productivity in their analysis was most likely due to sampling effects. The logic is that if complementarity is present it should increase the performance of the mixed community above that of even the best single species. However, as we show below, there can be widespread complementarity without transgressive overyielding. In other words, lack of transgressive overyielding does not mean lack of complementarity. On the contrary, it has been shown using the classical Lotka–Volterra competition model that stable coexistence can occur in mixed communities via niche complementarity without transgressive overyielding (Beckage and

Gross 2006, Loreau 2004). This can be simply illustrated as follows. Consider two species that differ in their productivities when grown alone so that the first is more productive than the second. Assume that these two species can stably coexist together through some form of resource partitioning (or equivalent form of niche differentiation). Complementary resource use will act to increase the productivity of the two-species mixtures above the level that would be expected if the two species were not complementary, that is if resource competition were a zero-sum game. However, this effect will be countered by the reduction in productivity caused simply by the replacement in the mixture of some of the more productive species by individuals from the less productive species. Transgressive overyielding will only occur when the increase in productivity due to complementarity is stronger than the reduction in productivity caused simply by the 'dilution' of the most productive species by the introduction into the mixture of individuals of less productive species.

The comparison of the yields of a variety of polycultures with particular monocultures selected *post hoc* also raises several statistical issues that complicate the test (Schmid *et al.* 2008). Furthermore, it is not clear how to best define transgressive overyielding in biodiversity experiments. The situation in an agricultural setting is clearer: for a farmer the question is whether a mixture can overyield the most productive monoculture (although even the agricultural reckoning is complicated by issues of monoculture and mixture production and price stability of components over time with varying climates and biotic challenges). However, outside of agriculture the choice is less clear because, in principle, every monoculture provides a potential benchmark for comparison (Hector *et al.* 2002a). The traditional agricultural test for overyielding is arguably the most natural test when the species with the highest monoculture yield dominates the depauperate communities. However, it is easy to imagine cases where traditional agricultural overyielding is not the only natural choice. One example occurs when the species that is highest yielding in monoculture is not highly abundant in the mixtures. Abundance is often taken as inversely related to extinction risk (small populations are often at greater risk of extinction) so that a species which is not highly abundant in the original full community may be one of the species which is lost as diversity declines. In this example, it is not clear that the species with the highest-yielding monoculture should be the benchmark for comparison since it may not even be present in the later depauperate community, let alone the dominant species (Hector *et al.* 2002a). As we discuss below (see: negative selection effects), in biodiversity experiments it is often the case that the species that dominate communities are not those that are most productive when grown alone (indeed they often have lower-than-average monoculture yields). In these situations the case for taking the species with the most productive monocultures as the benchmark for comparison is not clear. An alternative approach would be to take the monoculture value of the species that dominates mixtures instead since, for example, this is the species that would be expected to go extinct last based on population size arguments (Hector *et al.* 2002a).

7.3.3 Overyielding and the additive partitioning methods

Additive partitioning methods allowed the first attempts at identifying the relative importance of the different classes of mechanism underlying the patterns reviewed above (see also Schmid *et al.* this volume). Meta-analysis of plant biodiversity experiments reveals that almost all studies are driven by a combination of complementarity and selection effects but that overall complementarity effects are nearly twice as strong as selection effects (Cardinale *et al.* 2007). However, even though complementarity effects have a greater effect than selection effects they are not strong enough to cause mixtures to do significantly better than the best monocultures in most cases, as discussed above.

Another feature revealed by the additive partitioning method is the unexpected frequency of negative selection effects (e.g. counter to the predictions of the original sampling effect hypothesis). In the meta-analysis of additive partitioning results from experiments with plants 44% of studies showed negative selection effects. In other words, in nearly half of all experiments communities were

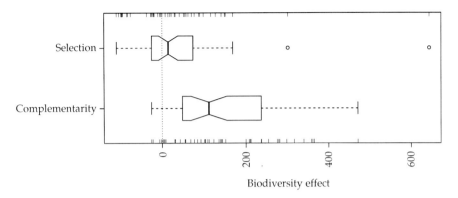

Figure 7.1 Box-and-whisker plot summary of the 44 studies with additive partitioning data reviewed in Cardinale et al. (2007). The box and whiskers show quartiles of the distributions of the selection (above) and complementarity (below) effects. The heavy central bar is the median and the notches on the boxes indicate an approximate $P = 0.05$ test for the median values versus zero. Points show two positive outliers for the selection effect values.

not dominated by highly productive species but by species with a lower-than-average monoculture biomass (Fig. 7.1). By looking only at the outcome of competition in mixtures plant ecologists have too often equated high productivity with competitive dominance. While this is often the case, there are many situations where less productive species are able to become dominant. The mechanisms by which they achieve this remain poorly understood and could be addressed by future research.

The key innovation of the tripartite additive partitioning method was the identification of the trait-dependent complementarity effect. Ecologically, a non-zero trait-independent complementarity effect occurs when the ecological mechanisms that mediate the strength of interspecific interactions relative to intraspecific interactions differentially affect high-yielding species, so that monoculture biomass covaries with traits that allow species to overyield in mixture (Petchey 2004). Fox (2005b) suggests that such covariation might reflect 'nested niches' (also called 'included niches'). In this view, species with 'larger' niches benefit from being planted in mixture, because their niches contain those of species with 'small' niches with 'room to spare'. However, species with 'small' niches experience equal niche overlap whether planted in monoculture or mixture. Therefore, species with larger niches grow better in mixture than in monoculture, but not at the expense of species with small niches. This leads to a positive or negative trait-dependent complementarity effect, depending on whether species with 'larger' niches have high or low monoculture biomasses.

The hypothesis that trait-dependent complementarity effects arise from 'nested niches', while trait-independent complementarity effects arise from non-overlapping niches, could be tested experimentally by manipulating the scope for niche differentiation. For instance, Dimitrakopoulos and Schmid (2004) planted monocultures and mixtures of plants in various depths of soil, and found an increasing selection effect with increasing soil depth. The complementarity effect (equivalent to the trait-independent complementarity effect from Fox 2006) increased with soil depth so it is possible that the increase in the selection effect was due, at least in part, to an increasingly strong trait-dependent complementarity effect in deeper soil. Increasing soil depth might be expected to increase the trait-dependent complementarity effect if some species with low monoculture biomass can produce only shallow roots, while species with high monoculture biomass can produce shallow and deep roots. Deep soil would allow species with high monoculture biomass to access a resource pool unavailable to shallow-rooted species, allowing deep-rooted species to attain high biomass in mixture, but not at the expense of shallow-rooted species.

To date, there have been few formal comparisons of the bi- and tripartite versions of the additive partitioning method. In other words, there have been few formal assessments of the contribution made by the trait-dependent complementarity effect. Our published (Fox 2005b) and unpublished results to date suggest that trait-dependent complementarity often makes a relatively minor contribution. One interpretation of the typically small magnitude of the trait-dependent complementarity effect is that niches are not usually 'nested'. That is, the ecological mechanisms that mediate the strength of interspecific interactions relative to intraspecific interactions typically do not differentially affect high-yielding species. This hypothesis could be tested by manipulating factors thought to mediate the ecological differentiation of species. It would also be interesting to look for the trait-dependent complementarity effect in circumstances in which it might be expected to be large (e.g. to examine the functioning of mixtures of generalist and specialist consumers).

As a general procedure we recommend analysts compare these related overyielding and partitioning approaches and, all else being equal, select the simplest one that describes the data well. For example, presenting the tripartite method will be essential when trait-dependent complementarity plays an appreciable role but presenting the selection and complementarity effect may otherwise suffice (in the limit, when trait-dependent complementarity effects are zero the dominance and selection effects are mathematically equivalent). In other cases, near equal monoculture biomasses make the selection effect covariance term trivial and relative yield totals or deviations from expected values (Loreau 1998b) provide a simpler alternative to additive partitioning (e.g. Vojtech *et al.* 2008).

7.3.4 The Price equation

The Price equation partition is a natural approach when interest centres on the effects of species loss from an initially diverse community, and the ecosystem function of interest comprises the summed contributions of individual species. The other major approach for comparing observed ecosystem function among sites is classical statistics (see Section 7.2.5). The Price equation partition and classical statistics can be viewed as trading off retention of information *vs* general applicability.

The Price equation partition requires knowledge of the functional contributions of individual species, and retains the information about which species were lost. Indeed, the reason for assuming that the less diverse site comprises a strict subset of the species in the more diverse site is so that information about which species were lost can be retained in a useful fashion (see Appendix B in Fox and Harpole (2008) and our supplementary material). By retaining this information, the Price equation partition defines terms (SRE, SCE, and CDE) that have a straightforward mechanistic interpretation independent of the details of study design. In situations where either the Price equation partition or classical statistical approaches can be applied, the investigator should carefully consider the question of interest in order to select the most useful approach.

7.3.5 The diversity–interaction statistical modelling approaches

One advantage of the application of these classical statistical methods to biodiversity experiments is that they avoid calculation of derived values (complementarity effects and so on) that must then be analyzed in a second stage. Furthermore, the methods do not require monocultures (as with additive partitioning), nor a full mixture (as in the Price equation), nor individual species contributions to the functioning of mixtures. Ideally a simplex design assures that species are grown in different combinations and at different relative abundances but species may be simply present (100 per cent) or absent (0 per cent). When species are simply present or absent the analysis of Bell *et al.* (2005b) can be seen as a special case of the approach of Kirwan *et al.* (2007). So, while the additive partitioning and Price equation approaches have mainly been applied to aboveground biomass production in plants these classical statistical approaches should be applicable to any ecosystem function (e.g. Sheehan *et al.* 2006). One advantage of classical statistical approaches is that they do not require knowledge of the functional contributions of individual species, and some classical statistical approaches also omit information about which species are absent from which sites. By omitting this information, classical statistics gains

Table 7.1 Overview of when the methods reviewed in this chapter can be applied depending on the information collected (whether or not individual species contributions to ecosystem processes can or has been measured) and type of experimental design (whether the relative abundance of species in communities is known or simply their initial presence or absence; and whether species mixtures and monocultures are all nested subsets of a single high-diversity community.

Information/Design	Transgressive overyielding	Random partitions	Diversity interactions	Relative yields and additive partitioning	Price equation partition
No individual species contributions Species presence/absence	√	√			
No individual species contributions Species relative abundance	√	√	√		
Individual species contributions	√	√	√	√	
Individual species contributions Nested communities	√	√	√	√	√

more general applicability, for instance to ecosystem functions that do not comprise the summed contributions of individual species, and to sites that do not comprise nested subsets of species. The general applicability of classical statistics allows a greater range of cross-study comparisons (Balvanera *et al.* 2006), although this generality comes with the risk of allowing statistically valid comparisons whose scientific interpretation is obscure (Fox and Harpole 2008). One cost of omitting information is that the interpretation of the terms of a fitted statistical model will necessarily depend on the details of the model and the study design. For instance, effects of species richness found by studies with different designs will have differing interpretations because of the other terms included in the respective statistical models. A final limitation of the classical statistical approach is that it cannot identify biological mechanism directly, but its ability to identify strong patterns among species interactions should direct the focus of more detailed explanatory research.

7.4 Conclusions and recommendations

Our main conclusion is that the range of techniques developed for the analysis of mechanisms in biodiversity experiments is now broad enough that we judge that some investigation of mechanisms should be possible for all studies published to date that examine effects of diversity within a single trophic level (plant biodiversity experiments for example; Table 7.1). This should enable a move from purely phenomenological studies to those that also address the underlying mechanisms. Wider application of the methods described here should help resolve the debate over mechanism that has continued largely due to the failure of many studies to address the underlying biological processes in an informative way (Cardinale 2006a, 2007). We finish with some words of warning and suggestions for future work.

Few, if any, of the methods reviewed here have been comprehensively explored. By that, we mean their behaviour has not been investigated with extensive simulation studies. This is a clear need for future work. Furthermore, most studies to date have tended to select one method and apply it in isolation. On the one hand, it is good that analyses should focus on the most appropriate method at the design stage. On the other hand, we are at a stage where it would also be interesting to run the different methods on the same dataset and to compare and contrast the results. In this way we can see where the different methods agree or disagree and demonstrate the advantages of one method over another in terms of what they reveal about the underlying biology. These approaches also need to be extended, or alternatives invented, that can deal with mechanism in multitrophic biodiversity experiments.

Finally, while our review emphasizes the ways in which analytical methods have tried to move closer to biological mechanisms, none of the methods described here measures the processes involved. Ideally the analyses described here will be supplemented by experimental approaches that directly quantify the processes involved in species interactions and which are not applied *post hoc* but are specified at the experimental design stage. These could include direct measures of natural enemy attack in monocultures and mixtures for example, or the use of isotope methods that can identify stocks and flows of resources. Only then will we be able to ask how well derived measures, like the biodiversity effects from the additive partitioning analyses and the statistical interactions from classical approaches, map onto biological interactions in diverse communities.

CHAPTER 8

Towards a food web perspective on biodiversity and ecosystem functioning

Bradley Cardinale, Emmett Duffy, Diane Srivastava, Michel Loreau, Matt Thomas, and Mark Emmerson

8.1 Introduction

One of the most common questions asked by researchers across a variety of scientific disciplines is 'How does the number of nodes connected together into a network influence the efficiency and reliability of that network?'. While social scientists and epidemiologists might think of 'nodes' and 'connections' as people interacting within a social network, computer scientists, neurologists, and civil engineers would instead think of servers connected together in a world-wide web, synapses connecting neurons in the brain, or hubs connecting to other hubs in a transportation network or telecommunications grid (Albert and Barabasi 2002, Newman 2003). Regardless of the particular study system, all of these individuals ask similar questions about how the number of nodes and connections among nodes influence the efficiency and reliability by which information, disease, energy, or matter is transmitted throughout that network.

Within the field of ecology, one of the oldest and most fundamental questions asked by researchers is 'How does the number of species interacting within a food web influence the efficiency and reliability by which energy and matter are transmitted through that web?'. Research on this topic can be broadly divided into two foci. Historically, much attention in ecology has focused on identifying those taxa that are the most influential nodes in a food web. For many years, it has been thought that some subset of species might represent 'hubs' of interactions and/or exhibit such strong interactions that they exert a disproportionate influence over food web dynamics. This idea has fueled much debate over the prevalence of omnivory in food webs (Polis and Strong 1996, Thompson *et al.* 2007, Yodzis 1984) and whether the increased number of feeding links that result from omnivory increases or decreases the stability of energy flow through a food web (McCann *et al.* 1998, MacArthur 1955). Identifying species that represent influential nodes has also been one of the primary goals in the search for 'ecosystem engineers' (Jones *et al.* 1994), 'keystone species' (Paine 1966, Power *et al.* 1996) or other types of 'strong interactors' (Wootton and Emmerson 2005) that might have cascading effects on the diversity and biomass of species at a variety of different trophic levels (Paine 1966, Carpenter *et al.* 1987, Elser *et al.* 1988).

In the 1990s, ecologists began to pursue a slightly different perspective on food webs. This perspective focused not on the cascading impacts of individual species, but rather on how the number of species that comprise any single trophic level might control fluxes of energy and matter. Research in this area was generally referred to as Biodiversity effects on Ecosystem Functioning (BEF for short), and was often justified on grounds that (1) loss of biological diversity ranks among the most pronounced changes to the global environment (Sala *et al.* 2000, Pimm *et al.* 1995), and (2) reductions in diversity, and corresponding changes in species composition, may alter fluxes of energy and matter that underlie important services that ecosystems provide to

106 BIODIVERSITY, ECOSYSTEM FUNCTIONING, AND HUMAN WELLBEING

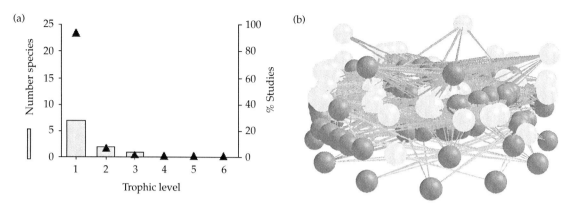

Figure 8.1 (a) Summary of the biological complexity of biodiversity-ecosystem functioning (BEF) studies performed to date. On the x-axis is the number of trophic levels included in different experiments. On the left-hand y-axis (plotted as grey bars) is the mean number of species per trophic level. On the right hand y-axis (plotted as triangles) is the percentage of studies that have included 1, 2, or more trophic levels. Note that 93 per cent of BEF experiments have focused on a single trophic level composed of a mean seven species. (b) An example of the complexity of a real, yet still relatively simple natural food web in a salt marsh (from Lafferety et al. 2007). Note that within this system there are dozens of species (nodes) and hundreds of feeding links (lines connecting nodes) among plants, herbivores, predators and parasites that span six or more trophic levels. Figure reproduced with permission from K. Lafferty. See Plate 4.

humanity (e.g. production of food, pest/disease control, water purification, etc. Daily 1997, Chapin et al. 1998). While the value of BEF research for conservation biology and management has been questioned by some (Schwartz et al. 2000, Srivastava and Vellend 2005), there is a more fundamental reason for the recent prominence of this topic. BEF is one of the few research topics in ecology that examines how biological variation *per se* acts as an independent variable to regulate key community and ecosystem-level processes (Naeem 2002b). Understanding the ecological consequences of variation among species has shown much potential to complement our historical focus on the ecological impacts of highly influential species.

Although the BEF paradigm has evolved considerably over the past 15–20 years and been increasingly applied to a variety of organisms and ecosystems, studies have continued to focus mostly on simplified 'model' communities. In fact, the typical experiment has manipulated an average of just seven species in an average of just one trophic group (Fig. 8.1(a)). Such minimal levels of complexity are far from the realities of natural food webs, where, even for some of the simplest communities, species interact within webs composed of hundreds of species spanning many trophic levels (Lafferty et al. 2006, Polis 1991, Martinez 1992). At present, it is unclear whether such oversimplifications are justified, or alternatively, whether they have led ecologists to potentially erroneous conclusions. However, what is clear is that a large body of research in ecology has shown that interactions of species across trophic levels can have cascading impacts that influence the diversity and biomass of organisms at numerous levels in a food web. At the very least, this suggests that the past focus of BEF on diversity within single trophic levels may be insufficient to quantitatively predict, and perhaps even qualitatively reflect, the ecological consequences of diversity loss.

In this chapter, we continue with the development of an idea that originated with other authors who have argued that, in order to understand how extinction alters the functioning of whole ecosystems, ecologists will likely need to merge modern paradigms of BEF with much more classic ideas in food web ecology that consider not only the functional role of diversity within trophic levels, but the interactions of species across trophic levels (Duffy et al. 2007, Bruno and Cardinale 2008, Petchey et al. 2004a). Our chapter is organized as follows. In Section 8.2 we briefly review five hypotheses about how fluxes of energy and matter through a food web might depend on the diversity of species comprising a web. Those hypotheses are divided into those that contrast diversity effects within different trophic levels versus those that focus on diversity effects across trophic levels. In Section 8.3 we outline the empirical support for or against these hypotheses, emphasizing that most are still unresolved and in need of testing. In the final Section 8.4, we outline just a few of the areas of research that we believe will be fruitful as ecologists move towards an integration of BEF into food-web ecology.

8.2 Five early hypotheses about multi-trophic biodiversity and ecosystem function

8.2.1 Diversity effects within trophic levels

8.2.1.1 Top-down effects of diversity grow increasingly strong at higher trophic levels

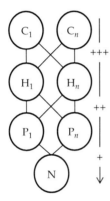

Early hypotheses proposed that species extinction from higher trophic levels was likely to have greater impacts on the functioning of ecosystems than extinction from lower trophic levels (Table 8.1). Duffy (2002) argued that three characteristics potentially make ecological processes more sensitive to extinction by consumers than plants: (1) because species at higher trophic levels have lower population sizes and are under stronger anthropogenic pressure than most wild plants, higher trophic levels face greater risks of extinction and higher rates of species loss; (2) consumer assemblages have lower overall richness and higher degrees of resource specialization, leading to less 'functional redundancy' and limited potential for surviving species to compensate for processes performed by lost counterparts; and (3) unlike plants, consumers often have impacts on processes that are disproportionate to their abundance or biomass. Duffy's (2002) paper was one of the first to call for a merger of BEF and food-web theory, and the hypotheses put forth in that paper were useful, in part, because they represented an alternative to those posed by a number of other authors. For example, some have argued that extinction at higher trophic levels may, in fact, have less impact on ecological processes than extinction at lower trophic levels. These arguments have usually been based on the idea that animals are more generalized in their use of resources than historically appreciated, either because the extent of omnivory and intra-guild predation has been underestimated (Rosenheim *et al.* 1995, Holt and Polis 1997, Polis and Holt 1992), or because animals can 'switch' among different prey species by moving across habitats (Polis *et al.* 1997, McCann *et al.* 2005). Resource generalization has been proposed to dampen the effects of consumer diversity on prey populations (Finke and Denno 2005, Snyder and Ives 2003).

8.2.1.2 Increasing diversity of a resource reduces the strength of top-down control by consumers

The majority of BEF studies performed to date have taken a 'top-down' perspective, meaning that they have examined how diversity within a given trophic level impacts the fraction of resources consumed, and production of biomass, by that focal trophic level. In contrast, diversity may also have 'bottom-up' effects on the dynamics of food webs, meaning that the diversity of resources may influence how efficiently those resources are consumed and converted into biomass by higher trophic levels (Table 8.1). At least three hypotheses have been proposed to explain how resource diversity might influence trophic dynamics: (2.1) the variance in edibility hypothesis argues that a more diverse prey assemblage is more likely to contain at least one species that is resistant to consumers (Leibold 1989,

Table 8.1 Five early hypotheses about multi-trophic BEF. What do the data say?

Section	Hypothesis	Key reference(s)	Section	Balance of evidence	Certainty	Key references
8.2.1	Diversity effects within trophic levels					
8.2.1.1	Top-down effects of diversity grow increasingly strong at higher trophic levels.	Duffy (2002)	8.3.1.1	The balance of evidence is not consistent with this hypothesis. Recent meta-analyses have found no difference in the direction or magnitude of diversity effects for groups of producers, herbivores, detritivores, or predators. In fact, there tends to be considerable generality such that decreases in species richness decrease the efficiency of resource capture and the amount of biomass produced by any given trophic group.	Medium to high	Balvanera et al. (2006), Cardinale et al. (2006a, 2007)
8.2.1.2	Increasing diversity of resources reduces the strength of top-down control by consumers.	Leibold (1989), Duffy (2002), Ostfeld and LoGiudice (2003), Root (1973)	8.3.1.2	The balance of evidence is consistent with this hypothesis. Summaries suggest that consumption of lower by higher trophic levels is reduced when a resource base is more diverse. Note, however, that most of the available data comes from studies that have not directly manipulated the richness of resources. Several controlled experiments have provided counter-examples, so the generality of this hypothesis remains unclear.	Low to medium	Andow (1991), Hillebrand and Cardinale (2004)
8.2.2	Diversity effects across trophic levels					
8.2.2.1	Top-down effects of consumer diversity oppose the bottom-up effects of resource diversity.	Holt & Loreau (2002), Thébault and Loreau (2003, 2005)	8.3.2.2	Recent meta-analyses suggest that the top-down effects of consumer diversity are qualitatively different than the bottom-up effects of resource diversity. However, these effects have not been opposing as suggested by this hypothesis. Note, however, that few studies have simultaneously manipulated the richness of species at adjacent trophic levels, so conclusions are tentative.	Low	Srivastava et al. (2009)
8.2.2.2	Diversity effects on biomass production and resource capture by a given trophic level are reduced in the presence of a higher trophic level.	Holt and Loreau (2002), Thébault and Loreau (2003)	8.3.2.1	The balance of evidence does not support this hypothesis. Of the few experiments that have manipulated species richness in the presence vs. absence of a higher trophic level, results are decidedly mixed. Analyses presented in this chapter further show no evidence that the effects of plant diversity on plant biomass differ for experiments performed in the presence vs. absence of herbivores.	Low	Mulder et al. (1999), Duffy et al. (2005), Wodjak (2005), and this chapter
8.2.2.3	Trophic cascades are weaker in diverse communities.	Strong (1992)	8.3.2.3	Experiments and data summaries to date have been equivocal and contradictory. At present, there is no clear reason to accept or reject this hypothesis.	None	Schmitz et al. (2000), Borer et al. (2005), Cardinale et al. (2003, 2006b), Wilby et al. (2005), Snyder et al. (2006), Finke and Denno (2005), Byrnes et al. (2006)

Duffy 2002); (2.2) the dilution hypothesis (Ostfeld and LoGiudice 2003), which has also been called the resource concentration hypothesis in the agro-ecology literature (Root 1973), suggests that specialist consumers become less efficient at finding and attacking their resource in a diverse prey assemblage; and (2.3) the balanced diet hypothesis suggests that a more diverse prey assemblage provides a more complete nutrition and, as a result, leads to higher consumer biomass (DeMott 1998). While hypotheses (1) and (2) predict that trophic efficiency will decrease as the diversity of resources increases, (2.3) predicts the opposite.

8.2.2 Diversity effects across trophic levels

8.2.2.1 *Top-down effects of consumer diversity oppose the bottom-up effects of resource diversity*

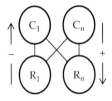

An important, but still unresolved issue is whether the overall impacts of diversity loss at adjacent levels are opposing or reinforcing, antagonistic or synergistic. Hypotheses (2.1) and (2.2) suggest that consumer diversity tends to enhance the flux of resources from lower to higher trophic levels, whereas resource diversity tends to reduce these fluxes. Collectively, these two hypotheses lead to a third hypothesis: that extinction of species from adjacent trophic levels will have opposing impacts on the flux of energy and matter through a food web (Table 8.1). This prediction has received some theoretical support from mathematical models showing that simultaneous changes in diversity from consumers and their resource leads to countervailing effects on total resource use and biomass production (Thebault and Loreau 2003, Thebault and Loreau 2005, Holt and Loreau 2002). Fox (2004b) provided a counter example in which he used Lotka–Volterra models to show that the joint response of prey biomass to prey and predator diversity is potentially more complex. While predator diversity generally decreases prey biomass, prey diversity can increase or decrease biomass depending on how different life-history trade-offs influence the coexistence of prey.

8.2.2.2 *Diversity effects on biomass production and resource capture by any focal trophic level are reduced in the presence of higher trophic levels*

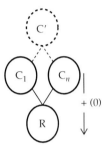

In their recent review, Duffy *et al.* (2007) used the terms 'horizontal' and 'vertical' diversity to distinguish between the richness of species within a trophic level and the richness of trophic levels that comprise a food web. They argued that one of the primary limitations in merging BEF with food-web theory is knowing how the impacts of divesity within trophic levels depend on the length of food chains (i.e. how horizontal and vertical diversity interact). The first step in overcoming this limitation is to ask how the diversity effects of any single trophic level are altered by the presence or absence of the next highest trophic level. Holt and Loreau (2002) used simple consumer–resource models to argue that the effects of plant diversity on nutrient uptake and plant biomass production are reduced in the presence of herbivores. This occurs because herbivory selects for dominance by poor plant competitors that are also the most tolerant to consumption by herbivores. Subsequent models by Thébault and Loreau (2003) also suggested that addition of higher trophic levels might qualitatively alter diversity–production relationships at lower levels; however, the direction of these impacts depends on both the nature of trade-offs between a plant's competitive ability and ability to resist herbivory, and on the degree of consumer specialization.

8.2.2.3 Trophic cascades are weaker in diverse communities

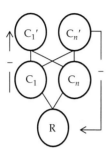

In his seminal critique of the empirical evidence for trophic cascades, Strong (1992) argued that cascades are 'a relatively unusual sort of food web mechanics... over the full range of ecological communities, evidence is that these cascades are restricted to fairly low-diversity places where great influence can issue from one or a few species.' He went on to suggest that trophic cascades are 'all wet,' meaning they occur primarily in aquatic ecosystems where communities are characterized by linear, low-diversity food chains. In contrast, he argued that terrestrial food webs are more reticulate and 'consumption is so differentiated in speciose systems that its overall effects are buffered.' The idea that diversity modifies the strength of trophic cascades can be broken down into at least two distinct hypotheses: (1) increasing the diversity of species comprising secondary consumers C' tends to decrease the strength of indirect effects on a basal resource R, and (2) increasing diversity of primary consumers C tends to decrease the indirect effects of C' on R. This latter hypothesis is very much an extension of hypotheses (2.1) and (2.2), as all of these rely on the assumption that an increasing diversity of resources tends to reduce the top-down impacts of consumers on food-web dynamics (Table 8.1).

8.3 What do the data say?

8.3.1 Diversity effects within trophic levels

8.3.1.1 Are diversity effects stronger at higher trophic levels? (Hypothesis 2.1.1)

Empirical evidence gathered to date does not appear to support the hypothesis that diversity effects are stronger at higher trophic levels. Balvanera et al. (2006) reviewed 103 studies in which they could examine 400+ correlation coefficients relating species richness to a variety of ecological processes. They found no evidence for differential correlations between diversity and any of the response variables at various trophic levels. Similarly, Cardinale et al. (2006a) collated data from 111 experiments that have manipulated species richness and examined how this aspect of diversity impacts the capture of resources and production of biomass. Their analyses compared four trophic groups: (1) microalgal, macroalgal, or herbaceous plants assimilating nutrients or water, (2) protozoan or metazoan herbivores consuming live algal or herbaceous plant tissue, (3) protozoan or metazoan predators consuming live prey, and (4) bacterial, fungal or metazoan detritivores consuming dead organic matter. They showed that, on average, experimental reduction of species richness decreases the standing stock abundance or biomass of the focal trophic group, resulting in less complete resource use by that group (Fig. 8.2). However, the standing stock of, and resource depletion by, the most diverse polycultures were indistinguishable from those of species that performed best in monoculture. Importantly, the authors could not detect any statistical difference in the magnitude of diversity effects among the four trophic groups.

Collectively, these meta-analyses suggest there is considerable generality in the way that the diversity of species impact resource capture and biomass production in food webs. The fact that Cardinale (2006a) and Balvanera (2006) both found that the BEF relationships did not change dramatically across trophic levels could imply that, if niche complementarity is the main mechanism driving these patterns, then the degree of niche complementarity could be similar across trophic groups. Identifying whether the mechanisms that dictate BEF relationships are the same across different levels of biological organization is a key next step in BEF research (a point we return to in Section 8.4.1). Although studies to date show considerable generality in diversity effects across trophic levels, we should emphasize that there still tend to be fewer absolute numbers of species at

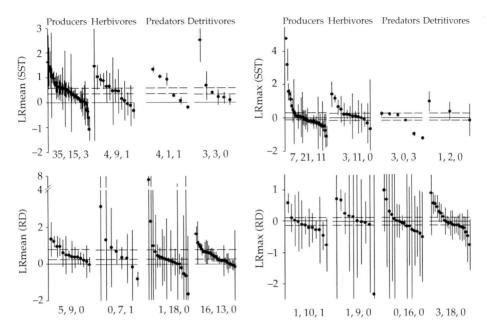

Figure 8.2 Summary of the results of experiments that have manipulated the richness of species in four trophic groups t (producers, herbivores, predators, and detritivores), and examined how richness impacts the standing stock abundance or biomass of t (SST – top graphs) or the fraction of resources depleted by t (RD – bottom graphs). The y-axes in all graphs give the diversity 'effect size', measured using two log ratios. LRmean (left graphs) compares SST and RD in the most diverse polyculture used in a study to the average of all monocultures. LRmax (right graphs) compares SST and RD from the most diverse polyculture used in a study to the species having the highest values of SST or RD in monoculture. Each data point is the mean effect size for all replicates in an experiment ± 95 per cent CI. Dashed horizontal grey lines give the 95 per cent CI for all experiments combined based on results from a mixed model ANOVA. Numbers below each figure are the number of studies that have shown significantly positive effects of diversity, no effect, or negative effects of diversity. Data are from Cardinale et al. (2006a).

higher trophic levels, and that these species tend to be disproportionately prone to extinction (a point we return to in Section 8.4.2). Thus, it is still reasonable to hypothesize that food webs can tolerate fewer extinctions at higher trophic levels before ecosystem functioning is altered.

8.3.1.2 Does resource/prey diversity weaken the strength of top-down control? (Hypothesis 2.1.2)
Empirical evidence gathered thus far is mostly consistent with the hypothesis that increasing prey diversity tends to reduce the impacts of consumers on prey. Andow (1991) tallied the results of 200+ studies of herbivorous arthropods and found that more than half of the herbivore species had lower population sizes on plant polycultures as opposed to monocultures. He argued that the resource concentration hypothesis, in which specialist consumers have a more difficult time finding their resource in a diverse prey assemblage, best accounted for the observed patterns. A summary of aquatic studies by Hillebrand and Cardinale (2004) tallied results from 172 experimental manipulations of herbivores and showed that consumption of algal biomass generally declined with increasing algal species richness. Although these patterns are consistent with hypothesis (2.2), some caution is warranted when interpreting these summaries, since the studies reviewed did not manipulate species diversity directly, and many potentially confounding factors were not controlled for. This caveat is particularly important when considering the mixed results from the limited number of experiments that have manipulated resource diversity directly. Several studies do provide evidence consistent with the variance-in-edibility hypothesis (Steiner 2001, Duffy et al. 2005), or for the dilution hypothesis (Keesing et al. 2006, Wilsey and

112 BIODIVERSITY, ECOSYSTEM FUNCTIONING, AND HUMAN WELLBEING

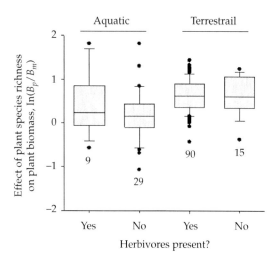

Figure 8.3 A summary of the impact of plant species richness on the production of plant biomass when herbivores are present or absent in experimental units. Data were taken from the summaries of Cardinale et al. (2006, 2007). The log ratio of plant biomass in the most diverse polyculture B_p to biomass in the average monoculture B_m was analyzed using a mixed model ANOVA with herbivores (y/n), ecosystem (aquatic vs. terrestrial), and their two-way interaction included as fixed effects, experiment accounted for as a random effect, and observations weighted by the inverse of their variance. Analyses indicate that the impacts of plant diversity on plant production do not differ when herbivores are absent vs. present ($F = 0.01$, $P = 0.92$), and that this conclusion is consistent among ecosystems ($F = 0.07$, $P = 0.80$ for interaction). These data and analyses should not be taken as conclusive evidence that herbivores do not impact plant diversity–biomass relationships since the studies summarized here differ in many ways that cannot be explicitly accounted for. However, these data can serve as a null hypothesis for experiments that explicitly manipulate plant diversity in the presence versus absence of higher trophic levels.

Polley 2002, Otway et al. 2005) where increasing diversity of resources leads to reduced consumption by higher trophic levels. Other studies provide support for the balanced diet hypothesis, showing that mixed diets of primary producers tend to enhance herbivore growth and biomass accumulation (Pfisterer et al. 2003, DeMott 1998). Thus, although the balance of evidence appears consistent with hypothesis (2.2), these conclusions should be considered tentative.

8.3.2 Diversity effects across trophic levels

8.3.2.1 Do top-down effects of diversity differ from bottom-up effects? (Hypothesis 2.2.1)

To date, studies that have simultaneously manipulated the richness of species at adjacent trophic levels are rare (Fig. 8.1(a)), and it is difficult to draw many general conclusions about the direction of top-down versus bottom-up effects of diversity in food webs. However, a recent meta-analysis by Srivastava et al. (2009) suggests that hypothesis (2.1) is not supported in detrital systems. These authors compiled the results of 90 experiments reported in 28 studies of detritivores to ask 'Do changes in consumer (i.e. detritivore) diversity have the same effect on rates of resource consumption as changes in resource (i.e. detrial) diversity?'. To address this question, they compared the top-down effects of consumer (detritivore) diversity on the consumption of dead organic matter (decomposition) to the bottom-up effects of resource (detrital) diversity on consumption of dead organic matter. Their meta-analysis indicated that reductions in detritivore diversity generally led to reductions in rates of decomposition, but changes in the diversity of detrital resources led to no detectable change in decomposition. The implication is that consumer, but not resource diversity, impacts consumption and energy flow in 'brown' food webs (detritus-consumer). However, an important point to keep in mind is that the resources studied by Srivastava et al. (2009) are 'dead,' meaning they are non-living resources that have no potential to show dynamic coupling to their consumers. A number of mathematical models suggest that diversity–function relationships could be qualitatively different when resources are 'living', such as in 'green' food webs (i.e. plant-based systems) where populations have the potential to respond to changes in the density of their consumers (Loreau 2001, Ives et al. 2005). The potentially important contrast between systems that have dynamic (living) vs. non-dynamic (non-living) is an issue that we return to in Section 8.4.1. For now, suffice it to say that we do not know whether the results of Srivastava et al. (2009) are specific to detrital systems, or whether they hold more generally.

8.3.2.3 Are diversity effects at one trophic level altered by higher levels? (Hypothesis 2.2.2.)

Only a handful of experiments have manipulated the richness of species in a focal trophic level and then simultaneously manipulated the presence/ absence of a higher trophic level. Mulder *et al.* (1999) varied plant diversity in the presence and absence of insect herbivores in a grassland plant assemblage. In the absence of herbivores, plant biomass increased with plant diversity, whereas when insects were present, they fed heavily on species with intermediate biomass, weakening the impact of plant diversity and biomass. Conversely, in a seagrass system, effects of herbivore richness on plant production were stronger in the presence of a higher trophic level (crabs) than in their absence (Duffy *et al.* 2005), which presumably occurred because of tradeoffs between species abilities to compete for resources versus resist predators. In other experiments, addition of a higher trophic level changed not only the magnitude but also the sign of the diversity–function relationship at the prey level (e.g. Hattenschwiler and Gasser 2005, Wojdak 2005).

We have been able to further examine hypothesis (2.2.2) by collating data from the meta-analyses of Cardinale *et al.* (2006a, 2007) for studies that have manipulated the richness of primary producers. We divided experiments into those that did versus did not allow herbivores access to experimental plots or pots, and then compared how plant diversity influenced plant biomass between the two types of studies. Although plant species richness generally increased the production of plant biomass, we found no evidence that herbivores alter the magnitude of plant diversity effects (Figure 8.3). This was true for studies performed in both aquatic as well as terrestrial ecosystems. Although these analyses are far from conclusive, when taken with the mixed results of experiments they suggest that widespread support for hypothesis (2.2.2) is presently lacking.

8.3.2.4 Are trophic cascades weaker in diverse communities? (Hypothesis 2.2.3)

Experiments and data summaries that have addressed hypothesis (2.2.3) to date have been equivocal and contradictory. Schmitz *et al.* (2000) performed a meta-analysis of 14 terrestrial experiments that manipulated higher predators and found evidence that the cascading effects of predator removal on plant damage were weaker in systems that had higher herbivore diversity. A more comprehensive analysis of trophic cascades measured in a variety of ecosystems found no evidence that variation in the strength of cascades was related to the richness of predators, herbivores, or plants (Borer *et al.* 2005). In contrast, a limited number of experiments have manipulated the diversity of predators at top trophic levels and shown that diversity can indirectly alter plant biomass by changing rates of herbivory. Cascading effects of predator diversity have been demonstrated in agricultural (Cardinale *et al.* 2003, Wilby *et al.* 2005, Snyder *et al.* 2006), salt marsh (Finke and Denno 2005), and kelp forest systems (Byrnes *et al.* 2006), and have been attributed to non-additive interactions (Cardinale *et al.* 2003, Cardinale *et al.* 2006b), omnivory (Bruno and O'Connor 2005), intraguild predation (Finke and Denno 2005), and changes in herbivore behavior (Finke and Denno 2005, Byrnes *et al.* 2006). Yet, the magnitude and direction of predator richness impacts on plant biomass and production have been inconsistent among studies (see Bruno and Cardinale 2008 for a review). Thus, although predator richness frequently has cascading impacts on food-web properties, it is difficult at this point in time to predict whether these cascading effects generally increase or decrease plant biomass. Therefore, at present, there is no clear evidence that can be used to accept or reject Strong's (1992) hypothesis that trophic cascades are restricted to low-diversity linear food chains.

8.4 Where do we go from here?

8.4.1 Detailing mechanisms: niche partitioning and life-history tradeoffs

William Dillard, founder and Chairman of Dillard's department stores, once said that the three most important factors for the success of a business are 'location, location, location.' Similarly, we believe that the three most important factors that will determine the success of the BEF paradigm will be our ability to identify mechanisms, mechanisms, mechanisms! Understanding the mechanisms that underlie diversity effects essentially requires that researchers return to several of ecology's classic questions about how niche partitioning and life-history tradeoffs allow species to coexist. Chesson (2000) provided what is perhaps the most elegantly organized summary of the mechanisms that allow

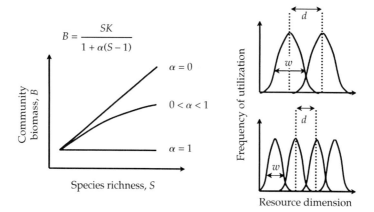

Figure 8.4 (a) Solutions to Lotka–Volterra competition equations showing how species richness affects community biomass production for differing levels of interaction strength. Note there is a positive, but decelerating relationship between B and S for all $0 < \alpha < 1$. This is an inevitable consequence of niche packing (insets) where the addition of species to a system with finite resource forces the average species to occupy a smaller fraction of resource space. Thus, the more species there are, the less each species contributes to resource capture and biomass production, on average.

coexistence. He showed that, for a wide variety of mathematical models, coexistence is ultimately determined by the balance of two interacting forces, which he called *equalizing* and *stabilizing*. Equalizing forces are those that minimize differences in the fitness of species, causing interspecific interactions to have weaker influence over population dynamics. Hubbel's (2001) neutral theory of biodiversity is the extreme case of an equalizing force where demographic parameters are assumed to be identical among species such that interacting with another species has the same per capita impact as interacting with a congener. Equalizing mechanisms are not mathematically stable and cannot allow long-term coexistence. Rather, equalizing mechanisms only serve to slow the inevitable outcome of species interactions. Thus, long-term coexistence requires some type of stabilizing force that involves niche differentiation in space or time. Regardless of whether niche differentiation occurs through partitioning of limited resources, shared predators, or some other dimension of a species niche, stabilizing forces all share the feature that they reduce interspecific relative to intraspecific interactions, leading to a per capita growth advantage of a species when rare.

The literature is ripe with models that examine how reductions in interspecific relative to intraspecific interactions regulate the impacts of species diversity on the production of single trophic-level systems (Loreau 2004, Tilman et al. 1997c, Ives et al. 2005, Cardinale et al. 2004). The discrete time Lotka–Volterra models of competition serve as an example (Cardinale et al. 2004), where the biomass of any species i in a local community can be described as

$$b_i(t+1) = b_i(t)\exp\left[r_i\left(1 - \frac{b_i(t) + \alpha \sum_{j\neq i}^{N} b_j(t)}{K_i}\right)\right] \quad (8.1)$$

K_i is the equilibrium biomass of i in the absence of competitors, r_i is the intrinsic rate of increase in biomass, and α is the ratio of inter- to intra-specific interaction. If species have similar carrying capacities and symmetric interactions, then all species have the same biomass at equilibrium, $b(\infty)$, and for any local community

$$b(\infty) + \alpha(S-1)b(\alpha) = K \quad (8.2)$$

From this, the total biomass of the community is

$$B(\infty) = \frac{SK}{1 + \alpha(S-1)} \quad (8.3)$$

For the extreme cases of $\alpha = 1$ or $\alpha = 0$, eq. 3 reduces to $B = K$ and $B = SK$, respectively, which shows that community biomass is independent of, or a linear function of richness (Fig. 8.4). For all other scenarios where $0 < \alpha < 1$, community biomass is a positive but decelerating function of species richness. Importantly, the curvilinearity of this function has nothing to do with how 'unique' or 'redundant' species are. Rather, the decelerating relationship is an inevitable consequence of packing more species into a finite niche axis. Even when all species are specialists with a unique niche, the contributions by any single species to resource capture and biomass production must decline as a function of richness (i.e. $b \propto 1/S$, Eqn 8.2), causing each increase in diversity to contribute smaller increments to resource capture and biomass.

Equation (8.3) predicts a rather straightforward set of relationships between species diversity and community biomass for any trophic group that is supported by a non-dynamic resource (e.g. plants assimilating inorganic resources, detritivores feeding on dead organic matter, etc.). One of the key questions as we extend BEF theory to multi-trophic systems is whether this same set of simple relationships holds true for systems where the resources are themselves dynamic. Interestingly, several authors have analyzed Lotka–Volterra models for both dynamic and non-dynamic resources and found that the effects of species diversity on community biomass are often qualitatively similar between one and two trophic-level systems (Thebault and Loreau 2003, Thebault and Loreau 2006, Ives *et al.* 2005, Fox 2004b). There seems to be just two general instances where new behaviors emerge in a multi-trophic system. The first occurs when dynamic resources, which have the potential to be overexploited in a multi-trophic system, are brought to extinction by their consumers. Overexploitation or extinction of resources by a diverse group of generalist consumers can yield humped-shaped diversity–biomass relationships in both predators and prey, which is a BEF relationship that is not found in single trophic-level systems (Thebault and Loreau 2003, Thebault and Loreau 2006, Ives *et al.* 2005). Second, there are certain types of life-history tradeoffs that can alter the shape and magnitude of a diversity–biomass relationship (Thebault and Loreau 2003, Thebault and Loreau 2006). For example, when resource species exhibit a tradeoff between their competitive abilities and their ability to resist or recover from consumption, this can moderate coexistence among prey (Holt *et al.* 1994) and dictate whether prey biomass increases or decreases with diversity (Holt and Loreau 2002, Thebault and Loreau 2003). Similarly, the tradeoff between the degree of resource specialization and assimilation efficiency of consumers has important implications for the BEF relationship. The diversity of consumers that pay no cost to generalism, i.e. that do not trade off their ability to consume a wide diversity of resources against their efficiency at consuming each of these resources, typically has a strong destabilizing effect on both population- and ecosystem-level fluctuations, whereas species diversity has a stabilizing effect on ecosystem-level fluctuations when consumers do have such tradeoffs (Thebault and Loreau 2005).

So are the consequences of extinction the same in single versus multi-trophic systems? Theory predicts that the answer entirely depends on the form of tradeoffs that mediate the coexistence of both consumers and their resources, and whether or not resources exhibit density dependent dynamics and overexploitation by consumers. What we need now are innovative experiments that manipulate the strength of consumer–resource interactions and/or the existence of tradeoffs that are presumed to underlie diversity effects in multi-trophic systems. Although such innovative experiments will no doubt be challenging, they have the potential to yield some of the most important new insights into the functioning of food webs.

8.4.2 Realistic scenarios of extinction

It is well established that species extinction is a non-random process. Throughout both geological and modern time, certain biological traits such as dispersal ability, generation time, body size, geographic range, and local density have proven to be correlated with extinction risk (McKinney 1997, Lawton and May 1995, Purvis *et al.* 2000a). Trophic position also appears to be correlated with extinction risk. In marine systems, extinction of fish species generally proceeds from the top of food webs downward (Pauly *et al.* 1998), which is partly due to human preferences for large-bodied fish, and partly because such fish have low resilience due to late maturity and slow growth (Myers and Worm

116 BIODIVERSITY, ECOSYSTEM FUNCTIONING, AND HUMAN WELLBEING

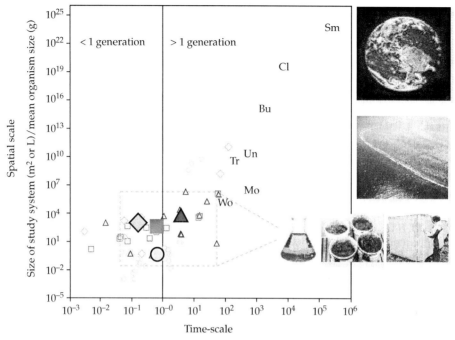

Figure 8.5 The spatial and temporal scale of biodiversity-ecosystem functioning experiments. The experimental duration (in days) and spatial scale (in m² or L) of experiments reviewed by Cardinale *et al.* (2006) were standardized to the mean generation time and body sizes of the focal organisms. Data were divided into four trophic groups: Plants = green circles, Herbivores = blue triangles, Predators = red squares, Detritivores = brown diamonds. The scale of each individual study is given by smaller symbols while the medians for each trophic group are shown as larger symbols. The box denoted by the dashed line gives the 10th and 90th percentiles for the scale of all experiments. For comparative purposes we show the scale of several natural extinctions: Wo = wolves from Yellowstone National Park, USA; Mo = Moa from New Zealand; Tr = Trout from Lake Superior, USA; Un = Unionid mussels from the lower Mississippi River, USA; Bu = Various species of butterflies in Europe; Cl = Loss of certain cladoceran zooplankton from Lake Superior, USA; Sm = Global eradication of the small pox virus. See Plate 5.

2005). In terrestrial systems, studies similarly report higher extinction probabilities for predators than their prey (Kruess and Tscharntke 1994, Didham *et al.* 1998b).

Non-random patterns of extinction can affect diversity–function relationships in at least two ways: via the functional traits lost, and via changes in community interactions. Initially, ecosystem function may be most affected by the functional traits of the species that preferentially go extinct (Srivastava and Vellend 2005, Lavorel and Garnier 2002). Positive covariance between extinction risk and the magnitude (Gross and Cardinale 2005) or uniqueness (Petchey and Gaston 2002b) of a species functional effects can exacerbate the impacts of species loss on ecosystem function (i.e. diversity–function effects are initially stronger for realistic extinctions than random extinctions). Predators may have high functional importance in food webs, first because of the strength of top-down processes in food webs (Duffy 2003), and second because predators may have traits that are additionally correlated with high functional impact (e.g. body size – Solan *et al.* 2004).

Following extinction of a species, diversity–function relationships are additionally influenced by the response of the surviving species to loss of a community member. Gross and Cardinale (2005) showed that the effect of species interactions amongst survivors depends critically on the mechanisms that underlie diversity–function relationships: niche partitioning, facilitation or the sampling effect each make very different predictions about how biased extinction scenarios differ from random extinction scenarios. In food web simulations, Ives and Cardinale (2004) showed that the coupling of directional extinction with species interactions can lead to unexpected changes in the functional importance of species. Although it is clear that non-random patterns of extinction can have very different implications than the random extinctions that commonly simulated in experiments, our ability to predict the functional changes that stem from non-random extinction – particularly the top-down effects of species loss on ecosystem function – is still in its infancy. After our need to characterize interaction strengths and interspecific tradeoffs (Section 8.4.1), our single biggest gap of knowledge stems from a lack of information about levels of covariance between extinction risk and species-specific impacts on rates of ecological processes at various trophic levels.

8.4.3 Environmental heterogeneity, patch dynamics, and scale

The typical biodiversity experiment performed to date has taken place in experimental units slightly larger than a five-gallon bucket, and has run for less than one generation of the focal organisms (Fig. 8.5). While there are noteworthy exceptions (Tilman *et al.* 2001, Hector *et al.* 1999), it seems safe to say that most of our inferences about biodiversity stem from experiments performed at spatial scales much smaller, and temporal scales much shorter than those at which species extinctions actually matter (also see Naeem 2001a for a more complete review). Overcoming this mismatch in scale is a daunting task, and the difficulties of performing large-scale, long-term experiments are why ecologists use simplified model systems in the first place (Srivastava *et al.* 2004). Nevertheless, ecologists have begun to make progress on these issues by incorporating the important ecological factors that co-vary with scale into their experimental designs (Cardinale and Palmer 2002, Dimitrakopoulos and Schmid 2004, Mulder *et al.* 2001) and accounting for them in meta-analyses of experiments performed at different scales (Cardinale *et al.* 2007).

The issue of scale is by no means unique to BEF research, nor is it specific to multi-trophic systems. There are, however, certain characteristics of multi-trophic systems that make it especially important that we deal more directly with the issue. Namely, dispersal as a process affecting species coexistence becomes particularly prominent at higher trophic levels where organisms are typically more mobile (at least, on the shorter time-scales of most experiments) and, therefore, have the ability to integrate information across a landscape and aggregate in response to the density of their prey. This is important because dispersal and aggregation across spatially distinct patches or habitat boundaries can translate into various forms of niche partitioning that stabilize competitive interactions and consumer–resource dynamics (Armstrong 1976, McCann *et al.* 2005). As it modifies coexistence, dispersal across patches or habitat boundaries can also qualitatively alter the BEF relationship (Mouquet *et al.* 2002).

Although most of the work that has examined how dispersal affects BEF relationships has focused on single trophic level systems, it is useful to quickly review here and then consider how these predictions might be extended to systems with dynamic resources. A wide variety of ecological models have highlighted the important role that dispersal plays in maintaining the diversity of communities (e.g. Island Biogeography Theory – MacArthur and Wilson 1967, 'mass' effects – Shmida and Wilson 1985, 'rescue' effects – Brown and Kodricbrown 1977). Historically, models of dispersal have been phenomenological, meaning they did not explain the existence of diversity based on first principles. Instead, these models assumed there was some 'magical' pool of species that coexisted at large scales via some unknown mechanism(s), and these species generated propagules that could subsidize local populations. The emergence of meta-community theory (Leibold *et al.* 2004) represented a major advance because these models acknowledged that everything in a propagule pool must ultimately come from the collection of patches or habitats that span a species range. Based on first principles, meta-community models predict both the causes and consequences of diversity at 'local' (organisms interacting as communities within patches) and 'regional' scales (patches of communities connected by dispersal).

One common form of meta-community models assumes that species coexist through tradeoffs in their abilities to compete in patches that have differing types or supply rates of resources (i.e. what Leibold *et al.* 2004 call 'species-sorting' models).

These models predict that at the scale of any local community, increasing the number of species in the meta-community serves only to ensure that species best adapted to a given patch will colonize and dominate that patch. This is the typical 'selection effect' of diversity (Loreau and Hector 2001, Huston 1997), which has been formalized as follows: assume that species can be ranked by their carrying capacities such that $K_{(m)}$ represents the species having the highest carrying capacity in any single patch, $K_{(m-1)}$ is the next highest, and so on. If competition among species is strong ($\alpha = 1$ in Eqn 8.1), only one species from the regional pool gamma will be present in a patch at equilibrium, and the biomass in a patch will be

$$B(\infty) = \frac{N_{col}}{\gamma} K_{(m)} + \left(1 - \frac{N_{col}}{\gamma}\right)\left(\frac{N_{col}}{\gamma - 1}\right) K_{(m-1)} \\ + \left(1 - \frac{N_{col}}{\gamma}\right)\left(1 - \frac{N_{col}}{\gamma - 1}\right)\left(\frac{N_{col}}{\gamma - 2}\right) K_{(m-2)} + \ldots \quad (8.4)$$

Equation (8.4) says that the amount of biomass produced in a patch at equilibrium is proportional to the probability, N_{col}/γ, that the species with the highest carrying capacity, $K_{(m)}$, will colonize the patch. If a patch is not colonized by the most productive species, then the probability that the second most productive species, $K_{(m-1)}$, will colonize and dominate the patch is $\left(1 - \frac{N_{col}}{\gamma}\right)\left(\frac{N_{col}}{\gamma - 1}\right)$. Note that as the number of species colonizing a patch increases, the probability that a patch becomes dominated by the most productive species in the regional species pool approaches unity. However, one key point is that for the selection effect to operate in the first place, species diversity must first exist in the regional colonist pool (i.e. γ must exist at the scale of a meta-community). But in order for diversity to be maintained in the regional colonist pool, species must exhibit some form of tradeoff that ensures they use resources in ways that are complementary across patches. This suggests that the same mechanisms that ensure complementary use of resources across patches in a region also produce species-specific selection effects at the scale of a local community (Cardinale et al. 2004).

Loreau et al. (2003) similarly showed that coexistence of species at a regional scale could maximize biological production at a local scale, and called this the 'spatial insurance' hypothesis of diversity (also see Chapter 10, where Gonzalez treats the issue extensively). The general idea of the spatial insurance hypothesis is that while one species may be sufficient to maximize production in any local community, the maximization of productivity across all patches in any heterogeneous landscape requires that a diversity of species exhibit niche differences at a regional scale. Meta-community models like that used to generate the spatial insurance hypothesis are important because they serve as a springboard from which we can address more pressing issues within the field of BEF research. From the perspective of basic theory, we need to extend meta-community models to consider how species diversity impacts the production of community biomass when consumers and their resources both move across a spatially heterogeneous landscape. We need to know what happens to BEF relationships when (1) resources have a spatial refuge from their consumers, (2) consumers and resources disperse at similar versus different rates, or (3) species exhibit spatially mediated tradeoffs, such as in their dispersal versus competitive abilities, or dispersal versus ability to resist consumption. At the same time, we need experiments that explicitly mimic the assumptions of different meta-community models, and then examine how diversity impacts the production of local and regional biomass for various mechanisms that allow consumer–resource coexistence. These advances are essential if we expect to predict the ecological consequences of extinction from real food webs where the norm is that species move across habitat boundaries and make choices about where to spend their time in order to maximize fitness.

8.4.4 Socio-economic impacts of food web diversity

After several decades of research, it has become apparent that loss of diversity from an ecosystem can have impacts on ecological processes that rival, if not exceed, many other forms of environmental change. Ecologists are now in a position to estimate the number of species required to maximize the removal of greenhouses gasses like CO_2 from the atmosphere, remove nutrient pollutants from streams and lakes that serve as drinking water, or to produce crops and fisheries. Indeed, it is now possible to make reasonably educated estimates of how diversity loss

Box 8.1 Socioeconomic impacts of predator diversity

One of the primary services that ecosystems provide to society is the biological control of insect pests. This service is estimated to be worth US$400 billion per year globally (Costanza *et al.* 1997). Although it has long been assumed that effective pest management requires a diversity of predators, parasites, and pathogens (collectively called 'natural enemies'), experiments designed to explicitly test this hypothesis have only recently begun. Two case studies highlight the range of results observed thus far.

A parasitoid female Aphidius wasp laying her egg in a pea aphid.

Case study 1: Predator diversity decreases pest populations

In a field experiment performed in Wisconsin, USA, Cardinale *et al.* (2003) manipulated the richness of three natural enemies of aphids (pea and cowpea) that are herbivorous pests of alfalfa. Two of the enemies – a ladybeetle and an assassin bug – were generalist predators that fed on both aphid species. The third was a specialist parasitoid wasp that attacks only pea aphids. They found that as generalist predators reduced the density of both aphids, the parasitoid wasp became more efficient at attacking the pea aphid. As a result, when all three enemies were together they reduced aphid populations to one-half of that achieved by any enemy species alone. This translated to a 51 per cent increase in the yield of alfalfa. Alfalfa is the fourth most widely grown crop in the USA with an estimated annual value of US$11.7 billion (source: US Department of Agriculture). In 2003 when this study was performed, alfalfa was selling for $150 per acre. The state of Wisconsin dedicates 3.5 million acres to the production of alfalfa. Assuming the results of this experiment can be generalized to Wisconsin, the economic benefit of predator diversity would be roughly US$525 million during a single harvesting cycle. In a typical year in the midwestern USA, alfalfa is harvested 3× per summer.

Case study 2: Predator diversity increases pest populations

In a second field experiment, Cardinale *et al.* (2006) manipulated the diversity of a different group of aphid predators, this time focusing on three species of ladybeetles that are all generalist predators. When the ladybeetles were placed together in field enclosures, they tended to compete with each other in a way that reduced their individual ability to capture prey. As a result, more diverse predator assemblages were roughly 60 per cent less efficient at controlling aphid populations than expected based on how each ladybeetle performed when alone. In this case, the antagonistic interactions among the predators led to a 17 per cent decrease in alfalfa yield. This result emphasizes that predator species can interact in ways that may have economic costs. A key challenge for ecologists is to determine the frequency of positive and negative interactions among predators that might help us evaluate the costs versus benefits of biodiversity.

translates into societally meaningful units – whether that be in dollars, health risks, carbon credits, or otherwise.

The socio-economic implications of biodiversity are perhaps most obvious from studies of higher-trophic levels, including those of pollinators, and of natural enemies that control pest populations. Invertebrate predators, parasitoids, and pathogens can be important promoters of top-down control in terrestrial food webs, helping to keep pests below economically damaging levels. This natural biological control of pests represents a valuable ecosystem service that is essential to sustainable production of food and fibre. Recent economic valuation of the services provided by insects suggests the value of biological control of native pests by natural enemies is $4.49 billion per year in the USA alone (Losey and Vaughan 2006), and > US $400 billion per year at a global scale (Costanza *et al.* 1997). While classical biological control tends to focus on the contribution of individual species of natural enemies, a growing number of studies suggest that the efficiency of biocontrol is often a function of non-additive interactions among multiple predators, parasitoids, and pathogens (Rosenheim 2007, Losey and Denno 1998, Snyder and Ives 2003, Snyder *et al.* 2006, Finke and Denno 2005, Cardinale *et al.* 2003, Cardinale *et al.* 2006b). Although it is not yet clear whether these interactions among enemies generally increase or decrease prey populations, it is clear that the economic impacts of natural enemy diversity can be substantial (Box 8.1).

Crop pollination is another ecosystem service centred on interactions across trophic levels. The global value of pollination services have been estimated at US$117 billion per year (Costanza *et al.* 1997) and in a recent review, Klein *et al.* (2007) concluded that fruit, vegetable, or seed biodiversity (i.e. richness, abundance, and distribution of multiple species of pollinators) in delivering this ecosystem service are often poorly quantified. Similar to evaluations of classical biocontrol, where the focus is on the action of one or few natural enemies rather than diversity *per se*, many of the economic valuations of pollination services consider the contribution of honey bees alone. Klein *et al.* (2007) report case studies for nine crops on four continents implicating a diversity of pollinators and revealing that agricultural intensification jeopardizes wild bee communities and their stabilizing effect on pollination services at the landscape scale. At the individual farm level, such natural pollination services can contribute significantly to annual income; a study from a coffee plantation in Costa Rica, for example, indicated native bee species account for $62,000, or 7 per cent of the farm's annual income

(Ricketts *et al.* 2004). At a more regional level, Losey and Vaughan (2006) calculate that native pollinators (mostly bees) may be responsible for > $3 billion of fruit and vegetables produced in the USA.

Although often less direct, changes in biodiversity and associated trophic structure have major implications for issues such as disease risk, with associated impacts on economics and human well being. For example, top predators are often the first species to disappear as habitat is destroyed and fragmented. As elaborated in Chapter 15, when predators are lost to ecosystems, their prey may increase in abundance, leading to increased transmission efficiency of zoonotic diseases such as Lyme disease (Ostfeld and Holt 2004, Dobson *et al.* 2006). While quantifying the benefit of biodiversity in terms of disease regulation and infected cases averted is clearly complex, many diseases such as malaria, tick-borne encephalitis, and West Nile fever have been shown to increase as biodiversity falls (Dobson *et al.* 2006, and Chapter 15).

8.5 Summary

The emerging paradigm of Biodiversity Effects on Ecosystem Functioning has shown great potential to augment ecology's historical focus on the causes of biodiversity with a much more contemporary understanding of its ecological consequences. Even so, BEF studies have, thus far, been limited to highly simplified 'model' communities that are nowhere near the trophic complexity of real communities. To overcome this limitation, it is now imperative that ecologists begin to merge the BEF paradigm with more classic ideas in food web ecology that detail how interactions among trophic levels that play out in space and time can constrain fluxes of energy and matter. Most hypotheses about the functional role of diversity within and across trophic levels are in their infancy, and they represent a rich opportunity for new work during the second generation of BEF experiments.

CHAPTER 9

Microbial biodiversity and ecosystem functioning under controlled conditions and in the wild

Thomas Bell, Mark O. Gessner, Robert I. Griffiths, Jennie R. McLaren, Peter J. Morin, Marcel van der Heijden, and Wim H. van der Putten

9.1 Introduction

Microbial communities have been and continue to be used widely to test basic ideas in ecology, and studies of the relationship between biodiversity and ecosystem functioning are no exception. Although microbial microcosm studies have been successful at shedding light on debates in biodiversity-ecosystem functioning research (Petchey *et al.* 2002), for many community ecologists, their *raison d'être* is limited to testing hypotheses with a degree of abstraction just slightly below that of mathematical models. The underlying idea is that simple microcosms, like theoretical models, are surrogates for what might be occurring in more complicated and, it is often implied, more interesting communities composed of larger organisms. In this context, the aim of most microbial microcosm studies is to create a highly simplified system to ensure unequivocal identification of patterns and causal mechanisms. In biodiversity-ecosystem functioning research, this has required experimental assembly of communities that are much less species-rich than the microbial communities in natural ecosystems.

Apart from the small coterie of 'microcosmologists' (Carpenter 1996), there is a separate fraternity of environmental microbiologists who typically publish in a different suite of journals and attend different conferences than ecologists interested in larger organisms, even though many of the research questions are similar. The focus of environmental microbiologists is to understand naturally occurring microbial populations and communities, so a familiar refrain is that microcosm experiments with constructed communities are much too simplified to have a bearing on what occurs in nature. In addition, many environmental microbiologists have argued that experiments using artificially assembled microbial communities are irrelevant at best (misleading at worst) because the experiments are conducted at inappropriate temporal and spatial scales. Although the dichotomy we describe between environmental microbiologists and microcosm microbial ecologists is a caricature, there is nevertheless a clear need for increased communication between these groups of researchers.

Largely due to the development of powerful and accessible molecular techniques, it is increasingly possible to ask ecological and evolutionary questions of natural microbial communities, even though it remains difficult to manipulate microbes in the field the way we can manipulate larger organisms. Without the ability to manipulate microbes in the field, causal mechanisms continue to be difficult to pin down.

The first purpose of this review is to give a glimpse of the considerable advances in microbial ecology, and to outline how microbial biodiversity affects the functioning of ecosystems in what is a rapidly expanding field of study. The second purpose is to contrast the results reported primarily in microbiological journals with those of ecologists who use microbes as model systems. We concentrate on non-pathogenic prokaryotic and eukaryotic microbial communities, leaving discussions of pathogenic microbial communities to Chapter 15. We do not discuss viruses in the current review, in part because of a lack of biodiversity-ecosystem functioning studies explicitly using viruses. However, given that many viruses appear to be host-specific and that they are important in controlling microbial populations and hence community structure in many ecosystems (Suttle 2007), viral diversity is very likely to have considerable repercussions for ecosystem functioning.

9.2 Microbial biodiversity and ecosystem functioning

9.2.1 Microbial biodiversity

One of the principal difficulties in analyzing the relationship between microbial diversity and ecosystem functioning is that it is notoriously complicated to describe the diversity of microbial communities. Most microbial groups lack sufficient morphological features to permit identification to the species level using conventional taxonomy or classical microbiological techniques (e.g. staining cell walls or using selective media). In addition, most of the microbial world eludes investigation under controlled conditions because of difficulties in culturing most microbial strains in the laboratory (Rappé and Giovannoni 2002). DNA-based molecular techniques have provided highly resolved descriptions of microbial communities (e.g. Venter *et al.* 2004, Sogin *et al.* 2006), but even current high-throughput sequencing approaches have not comprehensively surveyed a single ecosystem, and all techniques suffer from both sampling error and sampling bias (Fig. 9.1).

The first difficulty in assessing microbial diversity is that the number of individuals to identify is too large for any reasonable ecological survey (Hughes *et al.* 2001), although new sequencing technologies have the potential to provide reasonably good estimates of diversity in some species-poor microbial habitats in the foreseeable future (Quince *et al.* 2008). Even extremely unproductive environments such as drinking water will have thousands of microbial cells per cubic centimetre, and typical communities contain more cells than could possibly be identified individually using current technology. Techniques to assess microbial diversity therefore proceed by taking a number of sub-samples: first by taking only a subset of the cells in the community, then a subset of the DNA from those cells, and then, for example, amplifying a particular locus from the extracted DNA before sequencing a subset of the DNA that has been amplified (Fig. 9.1). Although it is true that surveys of non-microbial organisms do not identify all of the individuals in the examined community, such ecological surveys of larger organisms generally record a significant percentage. Surveys of microbial communities, in contrast, typically identify many orders of magnitude fewer individuals than are actually present. Currently, a clone library (i.e. a collection of clones of chromosomal DNA that can be used to quantify microbial diversity; Fig. 9.2) might comprise a few hundred individuals (perhaps hundreds of thousands of individuals in the near future), whereas the community would be likely to contain at least 10^5 cells per cubic centimetre (of soil or water), often several orders of magnitude more, and the community contains many thousands of cubic centimetres. Much less than 1 per cent of a community is therefore recorded for even the most extensive surveys of microbial communities. This creates a problem because many richness estimators rely on guessing at the distribution of abundance in the community (i.e. how many rare species have not been observed in the survey). Since such a small portion of the community is sampled, the distribution of abundance in microbial communities is still unknown. While it is possible to make educated guesses at a distribution of abundance and extrapolate numbers of species (Curtis and Sloan 2004, Quince *et al.* 2008), such estimates remain speculative. The new generation of sequencing technologies will curtail this problem to some extent, but even sequencing hundreds of thousands of individuals in a microbial community will still represent only a minuscule portion of the total community.

The second difficulty in assessing microbial biodiversity is that the sample is not a random subset of the larger community. In Fig. 9.1, each arrow represents a situation in which a sub-sample is taken. Each sub-sample is biased (i.e. non-random); for example, some sequences amplify more readily than others during a polymerase chain reaction

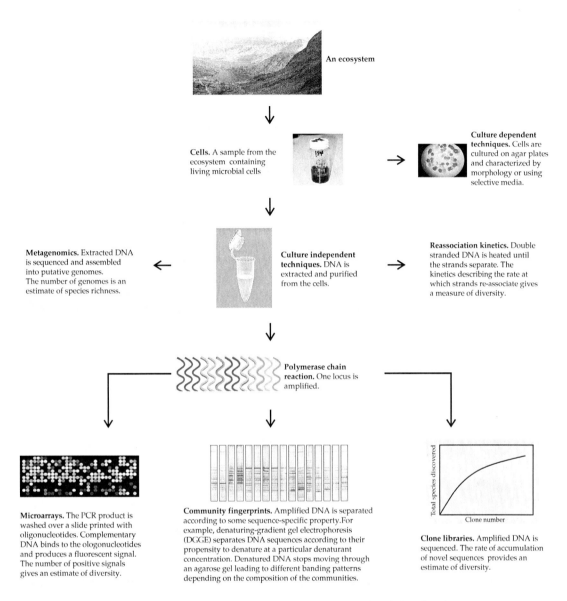

Figure 9.1 Popular methods for estimating microbial diversity. Microbial cells are too abundant and morphologically similar to survey using conventional ecological techniques. Rather, a series of subsamples are taken, where each arrow in the figure represents a subsample. Difficulties arise in estimating microbial diversity because the degree to which these are random subsamples is poorly understood.

(PCR) (Muyzer and Smalla 1998), which biases the final DNA copy number of each species compared to the original species mixture. Different initial copy numbers in different species can also significantly affect the final mixture of PCR products (Farrelly et al. 1995). Although the same issues of sample bias exist for larger organisms, the degree to which the sub-samples constitute non-random samples of the community remains largely unknown and an area of active study. It is only after these problems have been resolved that it will be possible to accurately assess microbial diversity.

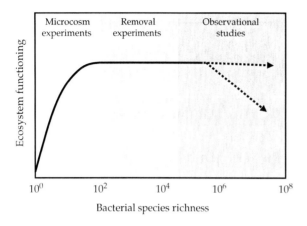

Fig 9.2 A schematic interpretation of microbial biodiversity-ecosystem functioning studies. Experiments that construct communities from culturable microbes have frequently found positive, saturating relationships between diversity and ecosystem functioning. These experiments have been conducted with at most 10^2 species. Removal experiments, where the diversity of natural communities is reduced, have shown mixed results, but there is often little effect of even significant declines in biodiversity on ecosystem functioning. For example, Wertz et al. (2006) reduced diversity from an estimated 8.3×10^6 bacterial species down to an estimated 8.4×10^2 species but found little reduction in a variety of ecosystem response measures. Observational studies of natural bacterial communities have found a variety of relationships, but there is frequently no clear pattern or a negative relationship.

The final difficulty with estimating microbial diversity is that there is ongoing controversy in defining what constitutes a microbial species (Achtman and Wagner 2008). As with larger organisms, species identity is relatively clear among microbial species that reproduce sexually. When reproduction is asexual, as is generally the case, the main alternative is to infer whether two individuals are of the same species if they share an arbitrary degree of DNA sequence similarity. Often, 97 per cent sequence similarity is taken as the threshold, although different cut-offs are used for different loci or for different taxonomic groups. In addition, many microbes frequently transfer DNA directly from an individual of one species to an individual of a different species, a process referred to as horizontal gene transfer. Because the biological species concept does not extend to the typical reproductive systems of most microbes, measures of 'biodiversity' in the microbial literature can have vastly different interpretations depending on the particular species definition that is employed. Although we use the term 'species' in the current chapter, it is evident that microbial biodiversity-ecosystem functioning studies will have to be interpreted in light of the species definition used, and conflicting results might simply result from differences in species definitions.

9.2.2 Microbial functions

Microbial communities are pivotal for the functioning of the world's ecosystems (Falkowski et al. 2008, van der Heijden et al. 2008). In aquatic environments, microbes often set the level of primary production and decomposition. In terrestrial ecosystems, microbial communities are important drivers of decomposition, thus converting organic matter into forms that are available for uptake by plants. As a result, microorganisms account for a major portion of ecosystem metabolism and biomass, on a global scale, accounting for a reported 50 per cent of the total protoplasm on Earth (Whitman et al. 1998). Microbes contribute to the regulation of Earth's climate by mediating fluxes of carbon dioxide, nitrous oxide, methane, and other greenhouse gases (Bardgett et al. 2008). The ability of microbial communities to metabolize even the most recalcitrant molecules is a testament to their enormous metabolic repertoire.

Microbes are also important in an applied context. For example, they are a key consideration for agricultural practices, particularly when agriculture relies on renewable resources and is employed to minimize environmental degradation. Microbes have been exploited as 'microbial pesticides'(Qaim and Zilberman 2003), including both naturally occurring and genetically modified pathogens that directly or indirectly reduce pest populations. Perhaps the best-known example is *Bacillus thuringiensis*, which produces a toxin that deteriorates insect mid-gut epithelial cells, thus halting larval insect foraging. There is enormous functional variation even within microbial species. In *B. thuringiensis*, for example, the variation in pathogenicity and host specificity is attributed in part to which

Table 9.1 Putting a price on microbial functions. Global economic valuations of some important microbial services. These data were compiled by FAO in 1998 and have not been adjusted to current market value.

Ecosystem service	Value (billion US$ per year)
Soil formation on agricultural land	500
Nitrogen fixation for agriculture	90
Pharmaceuticals of microbial origin	42.5
Synthesis of industrial enzymes	1.3

plasmids they carry (Rasko et al. 2005), but plasmid diversity in the context of microbial biodiversity-ecosystem functioning studies has been largely ignored. There is also the possibility that horizontal gene transfer is important in maintaining ecosystem functioning, which also is a process that is largely absent in larger organisms.

There are numerous other examples in which microbial diversity affects managed ecosystems and human life, the intimate involvement of microbes in so many processes making it difficult to put a monetary value on microbial diversity. Despite the growing recognition of the importance of microbial biodiversity, the significance of this biodiversity component is only beginning to be acknowledged in biodiversity policy debates. While most are not worried about microbial species going extinct, even small changes in the functioning of microbial communities have the potential for significant impacts on economy, welfare and other aspects of human life (Table 9.1). Perhaps as a consequence, all signatories to the World Trade Agreement now must adopt and implement patent laws for microorganisms and for biotechnology processes applied to living organisms, based on international tariff and trade rules (FAO 1998). This will undoubtedly have implications for the use and management of microbial diversity, as it becomes the legal property not only of the patent depositor, but also of the country of origin, regardless of where the organisms were collected. For microbial communities, it might not be imperative to conserve specific species or strains, but rather to conserve specific metabolic pathways, which may be the result of millions of years of evolution and might be lost due to chance conditions (Falkowski et al. 2008).

In summary, the importance of microbial communities in determining flows of energy and matter in ecosystems makes them of particular importance both for basic ecological investigations and in an applied context. Unlike with larger organisms, research on microbial diversity is not driven by conservation worries, but by the enormous benefits (and harm) to humans that could arise once the factors affecting microbial functions are well understood in terms of harnessing beneficial services and preventing harmful impacts. For example, even marginal increases in the ability of a microbial community to metabolize harmful constituents in sewage might have significant economic ramifications (Wagner and Loy 2002); algal communities have promise as biofuels or as food (Gross 2008); and altering microbial gut functioning might prevent obesity and therefore reduce medical costs (Backhed et al. 2004). In many of these areas, the key is to understand how differences in the structure and performance of microbial communities translate into differences in functioning, which is precisely the question being addressed by ecologists investigating the relationship between biodiversity and ecosystem functioning.

9.3 Functioning of microbial communities under controlled conditions

9.3.1 Model systems for biodiversity research

Microbial communities have a venerable history as model systems for studying questions in community ecology (Jessup et al. 2005). Their popularity stems in part from the ease with which it is possible to conduct experiments over many generations under controlled laboratory conditions. Such rapid generation times allow the communities to arrive at the equilibrium conditions upon which many mathematical models are based. In addition, controlled laboratory microcosm experiments allow for communities to be experimentally assembled according to a prescribed design by drawing different combinations of species from laboratory stocks of pure cultures. In contrast, studies of natural communities lack this opportunity, in part because the levels of diversity are too high to be tractable for this type of approach. The combination of short generation times and fine control of community membership has resulted in microbial microcosms becoming a reasonable 'halfway house between mathematical models and the full complexity of the field' (Lawton 1995).

Critics have argued that microcosm experiments lack generality, rely on overly simplified communities, or are experiments that are conducted at an inappropriate scale (Carpenter 1996, Jessup *et al.* 2004). Clearly, if the question under investigation is to understand how a specific ecosystem operates then there is no substitute for the real thing. But if the purpose is to understand more generally the processes that are important for how biodiversity affects ecosystem functioning, there is no intrinsic reason to choose one system over another, and it would seem sensible to use systems in which it is easiest to conduct the necessary experiments. With this line of reasoning, microbial communities appear to be the ideal model system for biodiversity-ecosystem functioning studies. Since much of this research was reviewed previously (Petchey *et al.* 2002), we limit the following sections to subsequent advances in the field.

9.3.2 Microcosm biodiversity-functioning experiments

9.3.2.1 Protistan food webs

Since the last review of studies of biodiversity and ecosystem functioning in eukaryotic microbes (Petchey *et al.* 2002) there have been a number of new developments. Having established that positive diversity-functioning relationships often exist in these microcosms (Petchey *et al.* 2002), the field has begun investigating how this relationship is modified by important ecological drivers, for example how systems of different diversity respond to experimental perturbations, to alterations in the amount of available resources, and to manipulating other factors which indirectly manipulate diversity, such as predator abundance.

Steiner and colleagues explored patterns of temporal variability in the presence or absence of perturbations in multitrophic food webs of differing diversity (Steiner 2005a, Steiner *et al.* 2006). Theoretical studies predict that increased productivity destabilizes populations (Rosenzweig 1971) but can stabilize total community biomass. In unperturbed food webs, greater diversity was correlated with reduced temporal variability in total community biomass, especially in more productive environments. Temporal variability in the biomass of individual populations was not affected by diversity, nor by productivity. Differences in the species composition among diversity treatments had no influence on the variability of community-level biomass, while composition did affect the variability of biomass at the level of individual populations. The effect of diversity, but not the destabilizing effects of productivity predicted by the paradox of enrichment, predominated in these microcosms.

In a separate experiment, diversity was manipulated and community biomass was predicted based on allometric relationships between organism size and abundance (Long *et al.* 2006). Microcosms with greater diversity tended to yield more total biomass than systems of lower diversity, and organism size played only a transient role in determining biomass. This result is consistent with positive effects of diversity on overyielding noted in other, very different systems (Loreau and Hector 2001, Loreau *et al.* 2001). Krumins *et al.* (2006) further explored the indirect impact of bacterivore diversity on rates of bacterially mediated decomposition of wheat seeds, and they found that bacterivores altered both the taxonomic composition and overall metabolic activity of the bacterial community.

Other studies have explored the impacts of various factors, including assembly history, productivity, and consumers on patterns of diversity. Fukami and Morin (2003) showed that the specific temporal sequence of species arrival had a profound effect on the form of relationships between microbial diversity and productivity, estimated as biomass accumulation (Fukami and Morin 2003). Hump-shaped, concave, or monotonically increasing patterns can all result simply from differences in the history of community assembly. Other work shows that different forms of predation, specifically that imposed by specialist versus generalist consumers, can either promote or eliminate positive relations between productivity and protist diversity (Jiang and Morin 2005).

Finally, a reanalysis of an earlier biodiversity-ecosystem functioning study (McGrady-Steed *et al.* 1997) in response to critiques of the design of early biodiversity experiments (Fukami *et al.* 2001, Loreau *et al.* 2001) confirmed that elevated biodiversity can reduce variability in some aspects of ecosystem functioning (Morin and McGrady-Steed 2004). The effect is not related to a reduction in the temporal variation in ecosystem functioning, but rather reflects reduced variation among replicate communities of similar diversity at any point in time. This is perhaps best interpreted as reduced spatial variation in functioning in more diverse systems.

9.3.2.2 Aquatic bacteria

Bacteria have provided some important advances in understanding biodiversity-ecosystem functioning relationships because of the relative ease with which it is possible to assemble communities and store strains for prolonged periods. For example, in the most species-rich microbial biodiversity-ecosystem functioning experiments to date, Bell *et al.* (2005b) used 72 naturally co-occurring bacterial species and found a positive relationship between species richness and community respiration in aquatic microcosms (Bell *et al.* 2005b). Both diversity and species identity were important in determining functioning in this study. A similar conclusion was reached when investigating the relationship between genotype diversity and ecosystem functioning using a single species of bacteria, *Pseudomonas fluorescens* (Hodgson *et al.* 2002), where evidence was found for a significant effect of functional (ecotype) and genotypic diversity on both productivity and invasion resistance as measured by the degree to which mixture yields exceeded the maximum monoculture yields of the constituent strains. There is therefore the opportunity for significant diversity effects even within species. In contrast, experiments using four bacterial species found little evidence that species richness affected ecosystem functioning (wheat seed mass loss) because species that contributed little to seed mass loss tended to become dominant (Jiang 2007). Similar experiments have suggested that a significant relationship between bacterial diversity and functioning might only become apparent in the context of the larger food web (Naeem and Li 1997, Naeem *et al.* 2000).

These outcomes might be determined by the method used to isolate bacteria from the environment. One recent study looked at mixtures of up to eight bacterial strains selected for their ability to grow on cellulose in monoculture (Wohl *et al.* 2004). By selecting only species that utilized a particular substrate (cellulose), and then offering those species only that substrate in spatially homogenous (shaken) microcosms, the study was designed to prevent resource use complementarity from being important. It was therefore not surprising that the observed influence of diversity was partly attributable to a selection effect. Interestingly, despite being selected on a single substrate, species richness appeared to be important in addition to the effects of selection, possibly due to facilitation among species. In contrast, facilitation is unlikely to have been driving the overall pattern of increase in functioning with increasing biodiversity in the Bell *et al.* (2005b) study or else an accelerating biodiversity-ecosystem functioning relationship (i.e. where the slope of the relationship increases with increasing biodiversity) would have been observed, although such an observation would have been hidden if facilitation was rare. This does not exclude the possibility that facilitation is common in microbial communities, but it might often be masked by the increasing degree of redundancy as species are added to the community.

One of the principle advantages of using bacteria in biodiversity-ecosystem functioning experiments is that short generation times allow simultaneous investigations of evolutionary and ecological processes. For example, a single generalist genotype of *Pseudomonas fluorescens* diverges into a number of distinct specialist ecotypes that inhabit the microcosm sides, air–water interface, and open water (Buckling *et al.* 2000). This niche differentiation leads to increased levels of functioning (Hodgson *et al.* 2002), so any processes that disrupt the extent to which diversification occurs (Brockhurst *et al.* 2006, Brockhurst *et al.* 2007) will also affect the level of ecosystem functioning. Thus, unlike with larger, longer-lived organisms, in microbial systems it is possible to examine the evolutionary processes directly and so too make hypotheses about which ecological mechanisms will be important in determining functioning (Fukami *et al.* 2007, Venail *et al.* 2008). If, as many have suggested, there is widespread functional redundancy in bacterial communities, bacterial diversity must be maintained in nature in the absence of niche differences, which is contrary to many theories of species coexistence. For the microcosms used by Bell *et al.* (2005b), there is evidence that the communities were neutrally assembled (Woodcock *et al.* 2007), leading to an independent prediction that any positive relationship between diversity and functioning in this system should be a consequence of the selection effect. The next generation of microbial experiments in this field is therefore likely to include studies that simultaneously investigate the causes and consequences of biodiversity.

9.3.2.3 Fungi

Fungal communities are important drivers of litter decomposition with potential for influencing carbon sequestration in soils and, for aquatic fungi, in downstream lakes, reservoirs and oceans. Consequently, there is great interest in understanding the degree to which the diversity and structure of fungal communities affect rates of decomposition. Varying the species richness of leaf-colonizing stream fungi (aquatic hyphomycetes) in microcosms had no effect on average leaf decomposition rates in communities with up to eight fungal species (Dang *et al.* 2005, Duarte *et al.* 2006), which implies a high degree of functional redundancy among these fungi. However, in a separate experiment, cultures containing two early fungal colonizers of leaves enhanced decomposition by 73 per cent compared to values expected from decomposition rates of single-species cultures (Treton *et al.* 2004). This outcome, in contrast to results from experiments using communities of several species (Dang *et al.* 2005, Duarte *et al.* 2006), is strong evidence of complementarity resulting in faster litter decomposition. In a similar vein, Bärlocher and Corkum (2003) reported a tendency towards faster decomposition with increasing fungal richness (1 to 5 species), although mixed communities never caused greater mass loss than the most effective species alone (Bärlocher and Corkum 2003). Raviraja *et al.* (2006) also found that both species richness and identities affected leaf mass loss in microcosms, although again the most effective fungal species degraded leaves faster than species mixtures. To date, all of these studies have been phenomenological in nature, so that the mechanisms behind and circumstances under which fungal diversity effects on litter decomposition emerge in aquatic microcosms are currently unknown.

There is also evidence that richness of aquatic hyphomycete communities can indirectly enhance decomposition through a positive effect on resource quality for invertebrate detritivores (Lecerf *et al.* 2005). Further more ecosystem processes other than litter decomposition (e.g. fungal biomass production) may be enhanced by diverse communities (Duarte *et al.* 2006). Lastly, even when average rates of decomposition are independent of species richness, variability of rates has been found to decline strongly with increasing fungal richness (Dang *et al.* 2005), as predicted from theoretical models (Doak *et al.* 1998) and observed in other types of systems (Tilman *et al.* 2006b, Lecerf *et al.* 2007). All else being equal, this should lead to higher predictability of litter decomposition rates when fungal communities in streams are diverse.

In microcosm experiments with culturable saprotrophic soil fungi, increased richness resulted in faster decomposition rates on grass litter (Deacon *et al.* 2006), forest soil (Wardle *et al.* 2004a, Tiunov and Scheu 2005) and powdered cellulose (Tiunov and Scheu 2005). Although species richness in these experiments was much lower than the number of species in the ecosystems from where the fungi were isolated, positive diversity effects on decomposition also emerged in an experiment that involved a rather high number of fungal species (43 taxa) (Setälä and McLean 2004). Nevertheless, communities with low species richness (six species) were as effective as the most diverse community in maintaining ecosystem functioning (Setälä and McLean 2004), indicating that fungal diversity effects on decomposition saturate at low levels, as has been found in many other circumstances.

9.3.2.4 Mutualistic microbe–plant interactions

Mycorrhizal fungi and nitrogen-fixing bacteria form symbiotic associations with approximately 80 per cent of all terrestrial plants, and therefore constitute an important pathway by which nutrients (e.g. inorganic nitrogen and phosphorus) are taken up by primary producers. Several studies have manipulated mycorrhizal diversity and measured plant performance. Results indicate that plant diversity, productivity and invasion success are responsive to either mycorrhizal diversity or mycorrhizal identity (van der Heijden et al. 1998, Stampe and Daehler 2003, Vogelsang et al. 2006). Similarly, manipulation of rhizobia in experimental dune grassland showed that presence of these nitrogen-fixing bacteria enhanced nitrogen capture (+85 per cent), plant productivity (+35 per cent) and plant evenness (+34 per cent) (van der Heijden et al. 2006). The diversity of bacterial symbionts was partly responsible for these effects because several of the legume species present in the microcosms formed host-specific associations with specific rhizobia. There is evidence that some of the effects of mycorrhizal diversity depend on the environmental context, depending on the species of plant that is infected and fertility of the soil (Jonsson et al. 2001). Recently, there have been some clues of how complementarity among mycorrhizal strains operates. In particular, some mycorrhizal families appear to protect against fungal pathogens, while other families enhance phosphorus uptake. As a consequence, plant growth is enhanced when both types of family are present (Maherali and Klironomos 2007).

9.4 Functioning of microbial communities in the wild

9.4.1 Diversity and functioning in the microbial wilderness

Since most microbes cannot be cultured, it is generally impossible to experimentally assemble communities from a library of constituent species to reflect the makeup of communities in natural ecosystems. The alternative to the culture-dependent approach is to conduct experiments or perform comparative analyses using exclusively culture-independent techniques to manipulate or describe microbial diversity and simultaneously measure ecosystem process rates *in situ* (Cavigelli and Robertson 2000) or in artificially created ecosystems (Bonkowski et al. 2001, Griffiths et al. 2004, Girvan et al. 2005). In both cases, differences in the microbial communities among replicates can be identified or created, and quantified. The advantage of this approach is that the results apply to whole microbial communities and not only to the subsets that can be cultured (Fig. 9.1).

In general, the difficulty with current culture-independent studies is that any genuine biodiversity effect tends to be confounded because of differences among species in their susceptibility to elimination during the biodiversity manipulation procedure. This is evident from the titles of the articles that use this kind of approach, where some studies claim to examine the impact of diversity (Cavigelli and Robertson 2000, Bonkowski et al. 2001, Muller et al. 2003), others investigate the impact of community structure (Franklin et al. 2001, van der Gast et al. 2003, Griffiths et al. 2004), and others focus on the impact of species composition (Cavigelli and Robertson 2000). Yet all of these studies performed precisely the same kind of manipulation; rare or susceptible species were eliminated. What these studies can demonstrate is whether either reductions in biodiversity or the elimination of particular species are correlated with changes in the magnitude or stability of functioning, but unambiguous attribution of effects to diversity changes is not possible.

9.4.2 Field experiments and observations

9.4.2.1 Soil microbes

Soil microbial communities play a pivotal role in determining rates of litter decomposition (Wardle 2002). One of the major hurdles is that microbial diversity in soils is extraordinarily high, with as many as tens of thousands of bacterial and fungal species in a single gram of soil. With so many species apparently competing for a limited set of resources, the predominant view is that many soil microbes are functionally redundant (Chapin et al. 1997, Wardle et al. 2004a, Deacon et al. 2006). There are opportunities for, resource partitioning especially on the most recalcitrant or exotic substrates. Lignin, for example, is most efficiently degraded by some basidiomycetous fungi commonly referred to as white-rot fungi. Although there may be pronounced differences in enzymatic capacities among microbial species to degrade other plant polymers, it appears that most have the metabolic machinery to break down the majority of the common substrates they encounter. Hutchinson's 'paradox of the plankton' (which asks how a large number of species can be maintained in a given community when they appear to be occupying similar niches) (Hutchinson 1961) remains unresolved for soil microbes. While the high levels of diversity might be explained in part by the intricate physical structure of the soil environment (Crawford *et al.* 2005) as well as the rich set of interactions among substrate availability, abiotic conditions (e.g. pH, temperature) and the specific species that are in the vicinity (Treton *et al.* 2004, Wardle *et al.* 2004a), it is difficult to envision how thousands or, for bacteria, many more species can occupy finely differentiated niches. The prediction is therefore that many soil species are functionally redundant.

There have been several experiments in which soil microbial diversity has been reduced through either fumigation or dilution. When a fumigation treatment was applied, Griffiths (2000) found that 'general' soil functions (such as total community respiration) were inversely correlated with biodiversity, while the relationship was positive for 'specific' functions such as nitrification (Griffiths *et al.* 2000). In contrast, Degens (1998) found that the reductions in diversity caused by fumigation caused a decrease in decomposition rate of simple organic compounds, although this relationship disappeared at high levels of soil moisture. A similar decrease in decomposition rate was found in soil microbial communities with reduced diversity achieved by dilution, although differences were only found between the highest and lowest dilution levels (Griffiths *et al.* 2001). Wertz *et al.* (2006) created differences in soil microbial diversity among replicate communities by diluting soil microbial suspensions before inoculation, after which they allowed soil microbial biomass to recover (Wertz *et al.* 2006). Their results suggest considerable functional redundancy, even for the specific processes they measured, such as ammonium transformation, which are carried out by relatively limited numbers of taxa (ammonia oxidizers, which carry out the first step of nitrification). Comparative studies that used natural differences in soil nitrifier communities found that the identity of the species in the soil affected rates of ammonia oxidation, and therefore concluded that there is potential for important effects of nitrifier community structure and diversity on nitrification rates (Cavigelli and Robertson 2000).

Manipulating the diversity of higher trophic levels in soil communities can also influence decomposition rate indirectly through their effect on bacterial and fungal populations (Cragg and Bardgett 2001, Hättenschwiler *et al.* 2005). For example, there was no increase in decomposition with increasing bactivorous nematode richness, even though some nematode species suppressed bacterial activity and diversity (De Mesel *et al.* 2006). However, species composition appeared to play a role in controlling decomposition when richness had no effect in collembola (Cragg and Bardgett 2001) or microbivorous nematodes (Mikola and Setälä 1998).

Overall, artificially reduced levels of diversity have not resulted in a clear effect on soil functioning. It is certainly the case that that functioning is sometimes decreased when biodiversity is reduced, but the reasons why this occurs remains equivocal (see 9.4.1 above). The predominant theme in the soil literature is that it is difficult to draw generalizations about the relationship between soil microbial diversity and processes (Raffaelli *et al.* 2002, Hättenschwiler *et al.* 2005). Relationships often appear to be idiosyncratic (Raffaelli *et al.* 2002, Hättenschwiler *et al.* 2005) although aspects of redundancy and keystone species effects are also evident (Setälä and McLean 2004). Clearly, these kinds of experiment remain in their infancy, and further investigations are necessary to construct an image of how the diversity and composition of soil microbial communities influence soil processes.

9.4.2.2 Microbe–plant interactions

Soil microbial communities interact directly and indirectly with plant roots as pathogens and symbionts (direct interactions) and decomposers and antagonists (indirect interactions) (Wardle *et al.* 2004a). In turn, plants manipulate the chemical environment surrounding their roots by secreting chemicals that favour particular microbial consortia, thereby altering the composition and diversity of the root-associated microbial communities (Wardle 2002). The positive and negative feedbacks between plants and microbial communities form an enormously complex network of interactions that is only beginning to be understood (Lepš et al. 2001, Scheu 2001, Bezemer et al. 2005, De Deyn and Van der Putten 2005, Bezemer et al. 2006).

There is some evidence to suggest that plant diversity plays a role in determining microbial diversity and ecosystem functioning (Hooper et al. 2005), but the importance of plant diversity compared with processes within soil microbial food webs remains uncertain (Kowalchuk et al. 2002). Although there are examples of effects of plant species richness on soil functioning (Zak et al. 2003), for the most part the effect of plant diversity appears to be plant species-specific (Cleland et al. 2004, De Deyn et al. 2004, Wardle et al. 2004a, Viketoft et al. 2005), and is often complicated because plant and microbial communities operate at different time-scales (Korthals et al. 2001, Hedlund et al. 2003) and there are complex interactions among soil microbial species (Milcu et al. 2006).

Influences of soil microbial diversity on plant performance or plant communities have been rarely examined, mostly owing to the difficulty of manipulating soil microbial diversity. Selectively excluding part of the soil microbial community requires severe manipulations (e.g. via fumigation, see above) before an effect on ecosystem processes can be observed (Bonkowski and Roy 2005). There have been some initiatives, such as Bradford et al. (2002), who manipulated size classes of soil organisms and concluded that effects of soil organisms from different size classes on plant productivity and plant community composition are relatively minor and that effects of larger organisms supersede those of the smaller organisms (Bradford et al. 2002).

Results from studies that have included soil microbes and soil invertebrates indicate that soil microbial effects on plant community performance may be overruled by effects of the invertebrates (Bonkowski et al. 2001, Bradford et al. 2002, Bezemer et al. 2005, Wurst et al. 2008). Overall, the effects of the decomposer community appear to depend more on community composition and relative species abundance than on diversity (Wardle et al. 2004a, Milcu et al. 2006).

9.4.2.2 Aquatic fungi

The relationship between fungal diversity and ecosystem functioning has also received some attention in aquatic ecosystems because of the importance of fungi in litter decomposition. Results from two surveys in streams suggest that the species-poor fungal communities presumably affected by forestry practices or water pollution do not result in altered leaf decomposition rates (Raviraja et al. 1998, Bärlocher and Graça 2002). However, the general paucity of such field data and the problem of drawing inferences about cause and effect from correlational data currently impede conclusive answers about the significance of aquatic fungal diversity for litter decomposition.

9.5 Synthesis

The experiments to date that have assembled microbial communities in microcosms point toward a positive relationship between diversity and ecosystem functioning in most cases, although this outcome is by no means universal. While much more research is required before firm conclusions can be drawn, the positive relationships observed primarily in microcosm experiments appear to be driven by a variety of mechanisms, including complementary resource use and selection effects. As with studies involving larger organisms, however, the task of identifying causal mechanisms is only just beginning.

While the experiments reviewed above have demonstrated that there is potential for positive biodiversity-ecosystem functioning relationships in microbial communities, they have not shown whether such a relationship is common in species-rich natural communities. In contrast with the results of manipulative microcosm experiments, non-manipulative surveys (which compare measures of ecosystem functioning associated with natural variation in microbial communities) and experiments (in which natural microbial communities are manipulated) have found inconsistent effects of diversity on ecosystem functioning. This raises the principal question in microbial biodiversity-ecosystem functioning studies of why apparent discrepancies exist between these two approaches.

There are real difficulties in attempting to bring together the two lines of enquiry that assemble species from their constituent species versus experiments or observations using natural communities. Currently unculturable bacteria cannot be directly manipulated, making it impossible in practice to conduct well-designed manipulative biodiversity experiments with truly natural microbial communities. Even though the ability to describe the diversity of these communities is rapidly increasing as new DNA sequencing technologies are becoming widespread, the approach with natural communities will remain comparative unless new means are devised to manipulate microbial species directly without the need for culturing. Such comparative studies can give clues to causal mechanisms, but they inevitably suffer from the fundamental problem of relating cause and effect in an unambiguous manner. Even if it were possible to manipulate natural communities directly, attempting to manipulate communities with thousands of species poses enormous practical challenges because of the exorbitant number of possible species combinations.

There are a number of possible reasons as to why discrepancies commonly occur between small-scale manipulations of species-poor cultivable communities and large-scale observations of natural microbial communities. Perhaps the most obvious possibility is that the two approaches focus on different parts of the same relationship. Manipulative experiments with assembled communities consider levels of species richness from 10^0 to at most 10^2 species, whereas field manipulations reduce levels of diversity from much greater than 10^2 down to 10^2 species (Wertz et al. 2006), and observational studies are comparing communities with $> 10^3$ species (Fig. 9.2). Since microbial studies have typically found a saturating relationship between diversity and ecosystem functioning, even large reductions of diversity in species-rich communities are expected to have little effect on ecosystem functioning. For example, Bell et al. (2005b) found that a generic measure of ecosystem functioning brought about by bacteria (community respiration) scaled with the logarithm of bacterial species richness. Consequently, the addition of the first 100 species to the community is expected to increase the level of community respiration by 200 per cent, whereas adding an additional 100 species would increase respiration rate by < 10 per cent. Thus, extrapolating curves from microcosm experiments such as those by Bell et al. (2005b) would predict that even very large reductions in the diversity of natural communities have negligible effects on ecosystem functioning.

A second possibility is that those species that can be cultured may be poorly representative of the larger community, in which case we should not expect the results of manipulative experiments to apply to microbial communities as a whole even if their diversity were as low as in the experiments. While it is certainly the case that libraries of cultivable microbes are unrepresentative of the larger community, and also that only an extremely small portion of the total microbial diversity has been studied in this manner, biodiversity-ecosystem functioning experiments have now been conducted with an array of cultivable organisms. In addition, it might be important to bear in mind that there is a flip-side to the 'cultivability' debate, which is that culture-independent techniques might be picking up primarily metabolically inactive cells. Depending on how community membership is defined, these inactive cells might not be community members of interest in the context of assessing effects on ecosystem functioning. If indeed diversity in field experiments is mostly reduced by affecting species in a resting state, it would not be surprising that even large declines in diversity fail to curtail ecosystem functioning, at least at a gross level and in the short-term scales. Evidently, metabolically inactive cells might play a key role in fluctuating or shifting environmental conditions, but it is unclear to what extent inactive cells become metabolically active. Thus, the question of how microbial diversity is measured or defined remains pivotal to discussions about the effects of microbial biodiversity on ecosystem functioning.

A third possible reason is that while microcosm studies favour fast-growing r-strategists, experimental manipulations of natural microbial communities are likely to eliminate preferentially rare species. Even when rare species have the potential to play a significant role they are least likely to contribute notably to ecosystem functioning (compared to common species) due simply to their rarity. Therefore, to some extent such experimental manipulations may lead to conservative estimates of microbial diversity effects on ecosystem functioning.

Advances in molecular techniques have revealed that there are many similarities in the community ecology of microbes and larger organisms, suggesting that there might also be similarities in the relationship between diversity and ecosystem functioning. These technological advances have not only increased our ability to do experiments, but also to survey the microbial wilderness the way we have explored wild communities of larger organisms for centuries. Currently, several studies have revealed that prokaryotes have been on Earth billions of years before eukaryotes and multicellular organisms, that microbes constitute enormous biomass, exhibit unique and extraordinarily varied biochemistry that drives biogeochemistry, have phylogenies and taxonomies that remain beyond our ability to resolve, and exhibit lateral gene transfer that is almost unheard of elsewhere in nature. Nevertheless, it appears that many of the patterns of diversity in microbial communities are broadly similar to those of larger organisms (Green *et al.* 2004, Horner-Devine *et al.* 2004, Bell *et al.* 2005a, Fuhrman *et al.* 2006, Martiny *et al.* 2006).

Establishing the generality of this conclusion in relation to microbial biodiversity-ecosystem functioning research is imperative. There is a great deal of excitement that findings in microbial biodiversity-ecosystem functioning research can produce both considerable academic advances in understanding how ecosystems operate, and practical advances in harnessing microbial communities or preventing microbial disease. However, the inconsistencies between the various approaches that we highlight in the current review underline the real need for a better understanding of the causes and consequences of changing microbial biodiversity before the potential of microbial biodiversity-ecosystem functioning studies can be fully realized.

CHAPTER 10

Biodiversity as spatial insurance: the effects of habitat fragmentation and dispersal on ecosystem functioning

Andrew Gonzalez, Nicolas Mouquet, and Michel Loreau

10.1 Introduction

Anthropogenic habitat destruction (e.g. strip mining or clear cutting of forests), conversion to agriculture (e.g. conversion of grasslands to croplands or rangelands, or conversion of forests to plantations), and fragmentation (e.g. dividing ecosystems inhabited by native species into parcels that are separated by inhospitable terrain) are generally considered the dominant drivers of biodiversity loss. The loss of inhabitable area is the predominant cause of population (Hughes *et al.* 1997) and species extinctions (Pimm *et al.* 1995). Isolation of fragments of habitat and edge effects associated with such fragmentation can cause further declines in both the number of species, changes in their relative abundance, and other aspects of biodiversity within remnant habitat patches (e.g. Andrén 1994, Fahrig 2003, Ewers and Didham 2006). Although other anthropogenic drivers (e.g. climate change, overexploitation, and the spread of non-indigenous species that adversely affect indigenous species) are growing in importance, it is clear that their impacts will be felt within the context of ongoing habitat loss. Indeed, strong synergies between habitat fragmentation and climate change are expected (Holt 1990, Travis 2003) and will likely compound the loss of biodiversity at local and regional scales.

The threat of widespread and rapid loss of biodiversity across most regions has prompted two decades of research on the impacts of biodiversity loss on ecosystem functioning and services. A number of controlled experiments have established that reduced levels of species diversity can impact community processes, such as biomass production and nutrient uptake (Cardinale *et al.* 2007), although data from unmanipulated plant communities suggest that these effects may be weaker or masked by other covarying factors in the environment (Grace *et al.* 2007, Hector *et al.* 2007). Overall, the beneficial effects of biodiversity in experimental conditions have been shown to saturate at relatively low to moderate levels, even when several functions are considered simultaneously (Hector and Bagchi 2007). The relevance of results from biodiversity and ecosystem function experiments, given the rapid saturation of biodiversity effects, has questioned their utility as a case for conservation biology and has led to calls for a broadening of empirical and theoretical perspectives within the field (Gonzalez and Chaneton 2002, Srivastava 2003, Srivastava and Velland 2005, Lawlor *et al* 2002).

Biodiversity effects on ecosystem functioning (BEF), though small or sometimes negligible in small-scale studies, may nevertheless be more significant at larger spatial and temporal scales (Yachi and Loreau 1999, Loreau *et al.* 2003, Cardinale *et al.* 2004). Typically, experimental BEF studies have been performed over small spatial and temporal scales, relative to the size, mean habitat range, and generation times of the organisms involved. Although these limitations are most acutely associated with studies of terrestrial plant and tree communities, experiments with aquatic systems can also have similar limitations. The results of BEF experiments, whether terrestrial or aquatic, although clearly valuable for establishing the effect of varying local biodiversity, cannot provide a complete understanding of the spatial processes affecting the relationship between biodiversity and ecosystem functioning at landscape (e.g. more than one ecosystem), regional, or global scales (Gonzalez and Chaneton 2002, Rantalainen et al. 2005, Srivastava and Velland 2005, Dobson et al. 2006).

The loss of biodiversity in fragmented landscapes has underscored the importance of viewing communities as 'open' structures dependent upon spatial fluxes from the surrounding communities in the region (Kareiva and Wennergren 1995, Leibold et al. 2004). Although the importance of dispersal for the maintenance of biodiversity is relatively well understood (e.g. MacArthur and Wilson 1967, Schmida and Wilson 1985; Loreau and Mouquet 1999, Amarasekare 2004, Holyoak et al. 2005, Mouquet et al. 2005) its importance to ecosystem functioning remains relatively unexplored (Kareiva and Wennergren 1995). For example, weak flows of individuals between habitats may have significant impacts on population production (Holt and Loreau 2002, Mouquet et al. 2002, Ives et al. 2004) and community stability (Huxel and McCann 1998, Loreau et al. 2003). From this perspective, a more complete understanding of the impacts of biodiversity loss on local ecosystem functioning requires a fuller understanding of dispersal-dependent mechanisms of biodiversity. At the regional, or metacommunity scales, the spatial components of diversity – both spatial variance in diversity among habitats or patches and turnover in composition from habitat to habitat or patch to patch – are significant determinants of ecosystem functioning at scales greater than the habitat or patch (Fukami et al. 2001, Bond and Chase 2002, Gonzalez and Chaneton 2002, Loreau et al. 2003a,b, Cardinale et al. 2004, Leibold and Norberg 2004). A framework for understanding how spatial processes mediate biodiversity-ecosystem functioning relationship is needed to improve our understanding and ability to predict the ecosystem consequences of biodiversity loss at larger scales.

In this chapter we will review several concepts that allow us to link local and regional scales of the biodiversity and ecosystem functioning relationship. First we will consider how the species–area relationship can link loss of habitat to delayed loss of diversity and ecosystem functioning in remnant fragments. Second, we will then show how spatial variance in biodiversity can affect estimates of regional functioning by non-linear averaging (Benedetti-Cecchi 2005). Finally we will use a metacommunity framework to formalize the *spatial insurance hypothesis* (Loreau et al. 2003). Throughout we will consider how each of these perspectives informs our understanding of the impacts of habitat destruction and fragmentation on biodiversity and ecosystem functioning and stability. We will conclude that our understanding of the relationship between biodiversity and ecosystem functioning is substantially altered when we incorporate the spatial processes required to link local and regional scales.

10.2 Fragmentation, species loss and functioning debts

Habitat destruction is not a uniform process and the end result is typically a mosaic of remnant fragments of habitat containing a subsample of the flora and fauna that occupied the formerly continuous habitat (Fahrig 2003, Ewers and Didham 2006). The loss of habitable area, increased isolation and increased edge effects associated with fragmentation collectively initiate a process of community disassembly (Diamond 1972), involving declines in both species abundance and diversity within remnant habitat fragments. Community disassembly following habitat fragmentation can be simplified to two processes operating at different time-scales: the first a relatively rapid sampling of the original diversity as habitat is lost and the second a longer-term process of decay or 'relaxation' in residual diversity from the remaining fragments. Ecologists have sought to calculate the extent of future species loss due to habitat destruction and fragmentation, but little attention has been paid to estimating the functional (i.e. biogeochemical or ecosystem process) effects of community disassembly in fragmented landscapes (e.g. Laurence et al. 1997, Larsen et al. 2005, Rantalainen et al. 2005)

Most experiments relating ecosystem function to changes in species diversity have so far adopted a 'static' approach. Ecosystem variables are usually measured across gradients of spatially interspersed diversity treatments, consisting of communities assembled at random from a given species pool (e.g. Tilman et al. 1997; Hector et al. 1999). Although these designs are useful for revealing diversity effects independent of species composition, they might not reveal the ecosystem changes that accompany diversity loss in fragmented habitats, where extinction is a non-random process dominated by non-equilibrium dynamics. In particular, the identity of species extinctions (e.g. rare versus dominant) and the timing of their occurrence may be variable and delayed (e.g. the *extinction debt* described by Tilman et al. 1994, Gonzalez 2000, Vellend et al. 2006). The functional effects of extinction debts in fragmented landscapes are unobservable in experiments if they use spatially structured gradients of fixed diversity levels as a surrogate for species loss following fragmentation.

Delayed losses of diversity due to habitat fragmentation should also generate a *functioning debt* – i.e. a delayed alteration in ecosystem attributes driven by the delayed decline and extirpation of species persisting in remnant patches (Gonzalez and Chaneton 2002). The possibility that extinction debts may be associated with functioning debts has received little attention to date. This is in part because the static and local approach of current experimental protocols precludes the study of these dynamic aspects of diversity loss. Although recent studies have addressed the problem of species loss as a non-random process through theoretical (Ives and Cardinale 2005) and statistical means (Solan et al. 2004) they have not the addressed the more complex dynamical issues arising from habitat fragmentation (Fahrig 2003). Extinction debts are threshold phenomena (Ovaskainen and Hanski 2002) that arise because species persistence depends upon the spatial configuration of the landscape; fragmentation affects landscape connectivity that alters local and regional colonization and extinction rates. The challenge now is to understand how this phenomenon of diversity loss affects ecosystem functioning.

For a long time the species–area relationship has been used to estimate the extent of species loss due to destruction and relaxation (the slow approach to a new equilibrium in species richness within the landscape, e.g. Brooks et al. 1999). The method involves increasing the exponent (z) of the species–area relationship to account for the disproportionate loss of species from small areas of habitat. If the original habitat area A_o, is reduced to A_n, we do not simply expect the original number of species to decline to S_n, but rather to S_n estimated with a new higher value of z. Here we use this approach to examine how habitat fragmentation will affect local ecosystem functioning.

Figure 10.1 depicts our conceptual model. We begin with the familiar species–area relationship, $S = cA^z$, which describes how diversity S scales with area A raised to the power z, where z ranges from about 0.15 for continuous tracts of continental habitat to about 0.25 for habitat islands. Two species–area curves can be drawn corresponding to 'before' and 'after' fragmentation. These curves can be used to estimate how many species there are in a given area before fragmentation by interpolating along the 'before' fragmentation curve ($z = 0.15$). The eventual loss of species from a fragment because of isolation (i.e. due to relaxation) can be estimated by switching to the lower, but steeper, species–area curve ($z = 0.25$). These changes in species richness can then be mapped onto the generally saturating function (Cardinale et al. 2006) describing the relationship between species richness and ecosystem functioning (e.g. productivity) to produce estimates of Δf, the delayed change in ecosystem functioning in a fragment due to the delayed species loss. This approach predicts that the smaller the fragment, the larger the reduction in ecosystem functioning (functioning debt) due to delayed species loss (Fig. 10.1).

Few data exist to test the validity of this approach for natural landscapes. A good starting point would be to use data from experimental model systems (e.g. Wardle et al. 2003b). Gonzalez and Chaneton (2002) noted the existence of a functioning debt following the delayed loss of species in experimental fragments of a bryophyte-based microecosystem, but they did not try to predict the extent of the functioning debt from the observed species loss. Although it may take considerable time for local extinction to occur, certainly experiments could be conducted in a grassland

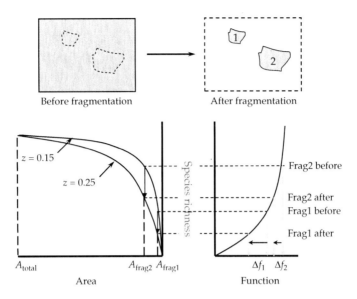

Figure 10.1 Predicting the delayed effects of habitat loss and isolation on local ecosystem functioning. The upper panel shows two fragments of differing area before and after habitat destruction (habitat is represented in grey). The lower left graph shows the two species–area relationships corresponding to before ($z = 0.15$) and after ($z = 0.25$) fragment isolation. The lower right graph shows a typical relationship between species richness and ecosystem functioning. The loss of species richness due to isolation is predicted by the increase in the slope of the species–area relation. This change in species richness due to isolation and community disassembly (frag 1 or 2 before) can be mapped onto the species richness-functioning relation (frag 1 or 2 after) to estimate the delayed change in function (Δf, the *functioning debt*).

setting to estimate the functional impacts of community isolation from the surrounding landscape (e.g. Robinson *et al.* 1992). Experiments in tractable systems, such as microbial microcosms, would be able to evaluate the different effects of fragmentation (e.g. isolation, edge effects, loss of area), and directly address how these causes of local extinction affect ecosystem functioning, although the laboratory context of microcosms often limits their utility for predicting phenomena in larger, more complex, less environmentally controlled ecosystems like grasslands or forests.

Others have recently pointed out the utility of the species–area relationship to address other impacts of area loss on ecosystem functioning (Tilman 1999a, Naelsund and Norberg 2006, Dobson *et al.* 2006), but have not used the approach to estimate the functional impacts of extinction debts. Dobson *et al.* (2006) pointed out that some functional groups (e.g. body size) and higher trophic levels are expected to have greater values for the exponent z, and thus differential sensitivity to area loss. An important consequence of this is that species loss due to habitat destruction should involve the top-down collapse of food webs (Holt *et al.* 1999). Although the evidence for this is equivocal (Mikkelson 1994, Holyoak 2000, Gonzalez and Chaneton 2002, Rantalainen *et al.* 2005), the functional consequences of such trophic collapse are likely to be great (e.g. Duffy 2003, Rantalainen *et al.* 2005, Rooney *et al.* 2006). Habitat destruction and fragmentation are the major causes of species loss in terrestrial ecosystems and more work is needed to establish how it affects ecosystem functioning (Kareiva and Wennergren 1995).

10.3 Spatial variance of biodiversity in fragmented landscapes

We noted that habitat fragmentation creates landscapes with many remnant fragments of variable size and species richness (Fahrig 2003). In the previous section we raised the problem of estimating the longer-term functional impacts of species loss within a given fragment. Here we address the problem of estimating the mean change in functioning across a set of fragments of variable size and species richness.

Within fragmented landscapes distributions of fragment size are often skewed with relatively few large patches distributed within a network of a large number of small patches (e.g. Keitt *et al.* 1997). This variation in fragment area translates into spatial variation in species diversity that may be exacerbated by a mix of deterministic and stochastic patterns of extinction and colonization across fragments (Wright *et al.* 2007). What are the consequences of spatial variance in diversity from fragment to fragment for estimating the change in mean ecosystem functioning at the landscape level? The answer involves spatial averaging. Benedetti-Cecchi (2005) recently pointed out that spatial variation in local diversity could significantly reduce estimates of mean ecosystem functioning in a fragmented landscape if the non-linear relationship between biodiversity and ecosystem functioning was not taken in account. This reasoning is based on Jensen's inequality (Jensen 1906, Ruel and Ayres 1999), the well known property that the expected value of a concave down function typical of biodiversity-functioning relationships, is lower than the function of the expected value: $E(f(X)) < f(E(X))$. Thus spatial variation in diversity (between communities) would, as a result of non-linear averaging across patches, produce lower than expected levels of ecosystem functioning. Drawing upon data from recent terrestrial plant experiments Benedetti-Cecchi (2005) found that, depending on the level of variance he assumed around the mean, Jensen's inequality could result in a 5–45% reduction in biomass production when compared to the case where spatial homogeneity in diversity was assumed. Although the range of this effect is large, it has generally been statistically significant.

Few biodiversity-functioning experiments have been conducted at the appropriate spatial scale to verify the importance of community-to-community variation in diversity due to non-linear averaging. In Fig. 10.2 we provide an example of how Jensen's inequality may lower the estimate of mean community production in landscapes with different distributions of habitat fragment size. Although we obtain a reduction in mean functioning due to Jensen's inequality, the effect is modest and corresponds to the low end of Benedetti-Cecchi's (2005) range. This is due in part because our use of the species–area relationship constrains the range of variation in species richness across fragments. However, Benedetti-Cecchi (2005) also used a larger range of variation (spatial variance) in species richness that may not be entirely realistic at the scale of local communities (Crawley and Harral 2001).

Our example raises empirical issues that merit further study. We assumed that the form of the BEF function is constant across the region of interest and thus the same function can be used to map diversity to function for any given community. Although this assumption probably breaks down across larger regions (e.g. Hector *et al.* 1999) we know remarkably little about spatial variation in BEF relationships at regional scales. We also assumed that the species–area relationship could be used as a first approximation to map the distribution of fragment size to the spatial variance in biodiversity. The general nature of non-linear averaging and the potential for large changes in spatial variation in diversity in disturbed landscapes suggest that this effect is likely to be sufficiently great that future studies should estimate it. Finally, we note that similar arguments can be made when estimating variation in mean ecosystem functioning when species diversity varies through time (Ruel and Ayres 1999), as it will in any fragmented landscape undergoing relaxation (Brooks *et al.* 1999, Gonzalez 2002).

10.4 Linking local to regional: the spatial insurance hypothesis

In the previous sections we have ignored dispersal and its role in driving spatial patterns of diversity and ecosystem process. The loss of biodiversity in fragmented landscapes forces a perspective that communities are 'open' structures dependent upon dispersal from the surrounding communities in the region (Kareiva and Wennergren 1995, Tilman and Kareiva 1997, Leibold *et al.* 2004). However, the way in which dispersal mediates the effects of diversity on ecosystem functioning has only recently been investigated theoretically (Holt and Loreau 2002, Mouquet and Loreau 2002, Loreau

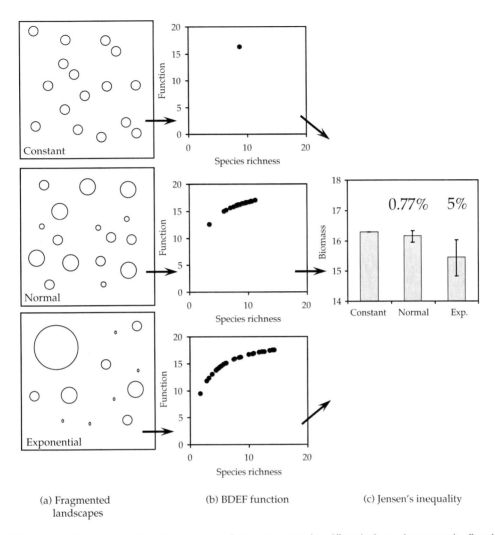

Figure 10.2 Spatial variation in species richness due to variation in fragment area across three different landscapes demonstrates the effect of Jensen's inequality. To calculate Jensens's inequality we assumed a nonlinear biodiversity-functioning relationship (a) of the form $aS/(S + b)$ where $S =$ species richness, $a = 20$, and $b = 2$. We then generated three types of landscape of 30 fragments each with equal total fragment area. In the first landscape every fragment has a constant area ($A = 5$ units), in the second each fragment area was drawn from a normal distribution ($\mu = 5$, $\sigma = 2$), and in the third the fragment area was drawn from an exponential distribution ($\mu = 5$). The fragments drawn here are for illustration only and are not to scale. We used the species–area relationship ($S = cA^z$, where $z = 0.3$) to generate species richness for each fragment in the three landscapes. Here we assume that the extinction debt has been paid and that z is at its equilibrium. (b) Values of species richness were then used to generate values for ecosystem function for each fragment (black dot for each fragment). We then compared mean biomass production per fragment (± 95% CI generated by Monte Carlo randomization, 1,000 iterations) for each landscape (c). The difference between the three landscapes defines the level of Jensen's inequality.

et al. 2003a and Loreau et al. 2003b, Mouquet and Loreau 2003, Cardinale et al. 2004, Leibold and Norberg 2004) and experimentally (Gonzalez and Chaneton 2002, Matthiesen and Hillebrand 2005, France and Duffy 2006b, Venail et al. 2008). We recently formalized the idea that dispersal mediates the effects of diversity of ecosystem functioning both directly and indirectly, an effect we have called the *spatial insurance hypothesis* (Loreau et al. 2003a).

The spatial insurance hypothesis is based on two mechanisms: (1) compensatory fluctuations between species (or functional groups) in the presence of spatio-temporal environmental heterogeneity, and (2) dispersal-driven spatial averaging of environmental heterogeneity. Under the first mechanism dispersal maintains biodiversity, which buffers ecosystem functioning against environmental fluctuations because functional compensations among species (or phenotypes) provide enhanced and more predictable aggregate ecosystem properties. Under the second mechanism dispersal directly buffers growth rates and inflates mean biomass production. This effect is based on the well-known principle that in the presence of dispersal, spatial variability in population growth averages arithmetically whilst in the absence of dispersal variability in growth averages geometrically (e.g. Ives et al. 2004). In general, the arithmetic mean will be greater than the geometric mean and this will translate into greater mean population biomass in the presence of dispersal. The extent to which spatial averaging due to dispersal translates into greater mean biomass will depend upon the extent of the increase in population growth rate and the strength of density dependence (Ives et al. 2004). Previous theoretical results suggest that the relationship between dispersal and mean biomass is unimodal with a peak at low to intermediate rates of dispersal (Holt 1993, Holt et al. 2005).

10.4.1 A source–sink metacommunity model: 'contemporaneous disequilibrium'

We reanalyze the source–sink metacommunity model of Loreau et al. (2003) with more species and by varying the input of nutrients within the metacommunity. We begin with a fragmented (patchy) landscape composed of a number of communities each experiencing variation in habitat quality (e.g. fluctuating temperature). We assume that each community experiences sinusoidal variation in the environment which varies out-of-phase across communities. We assume a set of species competing for a single resource and that species show dissimilar responses to the environment. Because they have different environmental optima each species will be the best competitor (defined by the rate at which they take up resource) in any given local community at different times. All species are assumed to be identical in all other aspects of their ecology (e.g. equal death and dispersal rates). Coexistence is not driven by temporal variation (i.e. resource partitioning through time) but is dependent on a spatial storage effect (Snyder and Chesson 2004), whereby dispersal allows species to persist by tracking spatio-temporal variation in environmental quality. In the absence of dispersal resource competition ensures that only a single species will persist with variable abundance in each community. Thus dispersal ensures local coexistence and environmental niche partitioning ensures regional coexistence (Mouquet and Loreau 2002).

The equations governing the metacommunity read:

$$\frac{dN_{ij}(t)}{dt} = [e_{ij}c_{ij}(t)R_j(t) - m_{ij}]N_{ij}(t)$$
$$+ \frac{a}{M-1}\sum_{k \neq j}^{M} N_{ik}(t) - aN_{ij}(t)$$
$$\frac{dR_j(t)}{dt} = I_j - l_j R_j(t) - R_j(t)\sum_{i}^{S} c_{ij}(t)N_{ij}(t)$$

$N_{ij}(t)$ is the biomass of species i (e.g. a plant) and $R_j(t)$ is the amount of limiting resource (e.g. a nutrient such as nitrogen) in community j at time t. The metacommunity consists of M communities and S species in total. Species i consumes the resource at a rate $c_{ij}(t)$, converts it into new biomass with efficiency e_{ij}, and dies at rate m_{ij} in community j. We assume that the resource is renewed locally through a constant input flux I_j, and is lost at a rate l_j. Species disperse at a rate a, dispersal is global, and propagules are redistributed uniformly across the landscape but do not return from the community they come from. We further assume that the consumption rates $c_{ij}(t)$ reflect the match between species traits and local environmental conditions as in Mouquet et al. (2002). Let the constant trait value of species i be H_i, which may be interpreted as its niche optimum along an environmental gradient, and the fluctuating environmental value of community j be $E_j(t)$. We assume that both species traits and environmental values vary between 0 and 1, and that a species' consumption rate is highest when the environmental value matches its niche optimum as measured by its trait value. Specifically, consumption rates are given by:

$$c_{ij}(t) = \frac{1.5 - |H_i - E_j(t)|}{10}$$

Ecosystem productivity at time t is the production of new biomass per unit time, average metacommunity productivity is thus:

$$\Phi(t) = \frac{\sum_{i=1}^{S}\sum_{j=1}^{M} e_{ij} c_{ij}(t) R_j(t) N_{ij}(t)}{M}$$

Here we study a metacommunity made up of 20 communities. Because various community properties (e.g. local and regional diversity and stability) are known to vary with rate of resource input we also examined the effect of varying the rate of resource input (system fertility) on metacommunity production and stability. In our simulations, we considered the following parameters: $e_{ij} = 0.2$; $m_{ij} = 0.2$; $I_j = $ (110 or 165); $\ell_j = 10$;. Environmental fluctuations follow a sinusoid with period T:

$$E_j(t) = \frac{1}{2}\left[\sin\left(Einit_j + \frac{2\pi t}{T}\right) + 1\right]$$

In contrast to our previous analysis (Loreau et al. 2003) we do not force the initial environmental conditions in each community so that each species is the best competitor in a different community: the $Einit_j$ were chosen randomly from a uniform distribution between ± 2π resulting in $E_j(0)$ being random between 0 and 1. This more realistic assumption results in fewer species having source communities, and demonstrates the robustness of our results to variation in initial conditions. We started the simulation with 20 species ($H_1 = 1$ and $H_i = H_{i-1} - 1/20$ for $i = 2$ to 20). Simulations at each dispersal value were repeated 50 times with different initial environmental conditions and results were averaged over the 50 repetitions. Each simulation lasted 800,000 iterations (Euler approximation with $\Delta t = 0.08$) with the period T chosen to be large enough ($T = 40,000$) so that there was rapid competitive exclusion in the absence of dispersal. Temporal mean diversity (local and regional) and temporal mean productivity and its coefficient of variation (CV) were calculated over the last 200,000 iterations.

10.4.2 Metacommunity dynamics

The dynamics of the metacommunity are strongly dependent upon the rate of dispersal. Dispersal permits local coexistence. Increasing dispersal, and resource input increases the level at which diversity is maintained both locally and regionally (Figs. 10.3 (a) and (b)). Local diversity attains a maximum value of seven species for $a = 0.07$, whilst regional diversity declines to a minimum for $a = 0.004$ and then returns to its maximum for $a = 0.1$. For values of dispersal greater than 0.1, local and regional diversity declines linearly. Species best adapted to the average environmental conditions at the metacommunity scale now prevail; recall that because the environment varies out of phase across the metacommunity the average condition is $E = 0.5$ to which intermediate species are best adapted. Species adapted to the extremes of environmental variation contribute progressively less to community productivity and eventually go extinct locally and regionally. Local and regional diversity are now at their lowest levels and the metacommunity has now been reduced to a metapopulation of a single species. Overall, increasing dispersal has strongly non-linear effects on the diversity of the source–sink metacommunity. Although diversity is causally dependent upon the rate of dispersal in this metacommunity it is useful to study the change in local and regional dynamics as a function of variation in species richness and as a function of dispersal rate.

10.4.3 Metacommunity productivity

Mean temporal productivity (the rate of biomass production per unit time averaged over the last 200,000 iterations) was affected by dispersal and attained its lowest level when dispersal was absent. Increasing dispersal resulted in a unimodal response in mean ecosystem productivity that peaked at dispersal rates of 0.01 (Fig. 10.3(c)). The position of this peak was not affected by the rate of

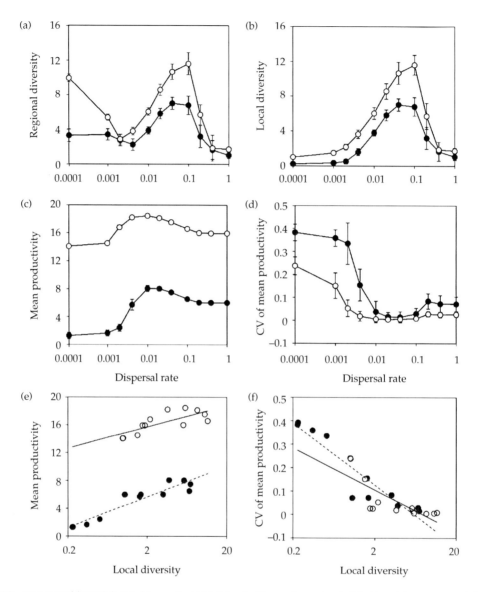

Figure 10.3 Mean regional (gamma) diversity (a), mean local (alpha) diversity (b), mean temporal productivity (c), and the mean temporal CV of mean productivity as a function of dispersal rate averaged over 50 simulations. Mean ecosystem productivity (e) and the mean temporal CV of mean productivity (f) as a function of dispersal rate. Variation between simulations is shown with standard error (a, b, c) and standard deviation (d). The symbols in each plot indicate different rates of resource input: filled circles $I = 110$; empty circles $I = 165$. All other parameter values are given in the text.

resource input, although average productivity increased significantly with increasing resource input, both in the presence and absence of dispersal. When productivity was plotted against local species richness (Fig. 10.3(e)) we obtained an increasing log-linear function that resembles the concave down relationship observed in many biodiversity-ecosystem functioning experiments. Increasing resource flow did not greatly alter the form of the richness–productivity relation; for a given level of species richness local productivity was increased roughly two-fold.

Increasing dispersal also enhances the productivity of each species directly through the

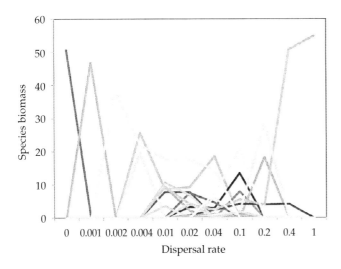

Figure 10.4 Example of how spatial averaging produces species-specific biomass peaks at different dispersal rates. We have averaged individual species biomasses over the last 250,000 time steps within a single community. We give values for different dispersal rates (each species is represented by a different grey line). Each species has a distinct mode, indicating the rate of dispersal at which each is most productive within this particular community. Parameter values are given in the text with $I = 140$.

phenomenon of spatial averaging. The effect of dispersal rate on productivity can be seen for all species in Fig. 10.4. The figure shows species-specific responses to increasing dispersal whereby the level of dispersal that maximizes mean productivity due to spatial averaging differs among species.

10.4.4 Metacommunity stability

We summarized the stability of ecosystem productivity with the coefficient of variation (CV = standard deviation/mean, e.g. Tilman 1999a, Hughes and Ives 2002). As local species richness increases the CV declines non-linearly (Fig. 10.3(f)). Increasing resource flow rate enhances this stabilizing effect of species richness (Fig. 10.3(f)). The CV shows a very different relation with increasing dispersal rate (Fig. 10.3(d)). Here we see an inverse unimodal relationship with the minimum CV obtained for intermediate dispersal ($a = 0.02$–0.1). Ecosystem productivity is strongly stabilized by dispersal and is very stable at intermediate levels of dispersal. Stability is still greater at high rates of dispersal than at very low rates, and this is again due to spatial averaging. Increasing resource flow further improves stability and reduces the CV across all levels of dispersal, but especially at the lowest rates.

Stability arises because of compensatory species dynamics within local communities (not shown).

This insurance effect of diversity is obtained at low levels of dispersal that maximize diversity. As above, spatial averaging also contributes to ecosystem stability by enhancing mean productivity (the denominator in the CV). However, low dispersal is not great enough to synchronize biomass growth across the metacommunity and eliminate compensation. This contrasts with the dynamics at high levels of dispersal ($a = 0.4$) where the dominant species is strongly synchronized across the community and spatial averaging dampens biomass fluctuations.

10.4.5 Discussion

The results of our metacommunity model reveal the importance of dispersal as a key process affecting the structure and function of patchy or fragmented landscapes. Two types of insurance effect were observed: an increase in the temporal mean of ecosystem productivity within and between communities, and a decrease in temporal variability. Dispersal rate mediates the strength of these insurance effects and links the diversity and functioning of communities from local to regional scales. Importantly our model allowed us to quantify and compare the indirect and direct roles of dispersal underlying the spatial insurance hypothesis: (1) dispersal indirectly affects the insurance effects of diversity by maintaining diversity in heterogeneous environments, and this effect is complemented by (2) the direct functional effects of dispersal due to spatial averaging. Furthermore, increasing the rate of resource flow enhances these effects. The relative importance of these two processes in fragmented landscapes will depend upon the type of coexistence mechanisms (dispersal-dependent or independent) and the degree of spatio-temporal heterogeneity and connectivity in the metacommunity. The assumptions of our source–sink metacommunity model are such that both effects of dispersal are important; we now discuss each in turn.

10.4.5.1 Diversity as insurance

Biodiversity can buffer ecosystem productivity against strong environmental variation as long as species with an appropriate environmental trait is present and able to grow under the prevailing environmental state (recently called 'response diversity' Elmqvist *et al.* 2003). Environmental niches of this type have been studied before in a single patch case and are thought to be a common mechanism by which diversity may buffer function over long time-scales (Chesson *et al.* 2002, Lehman and Tilman 2000, Gonzalez and De Feo 2007). When associated with dispersal and asynchronous environmental variation across communities, environmental niches can provide a robust mechanism for long-term species coexistence (e.g. a spatial storage effect, Chesson 2000).

The increase in the mean productivity with species richness in our metacommunity model stemmed from the compensatory dynamics between species through time. Longer-term studies of diversity and productivity in grasslands have revealed the importance of compensatory dynamics for the function and stability of grasslands (McNaughton 1985, Dodd *et al.* 1994, Bai *et al.* 2004). In our model the functional species trait (rate of nutrient uptake) was linked to the environment state. Because we assumed that species were evenly separated along the temporal environmental gradient, fluctuations in the environment altered the relative competitive ability of each species, which drove the strong compensatory dynamics we observed. Any environmental factor (biotic or abiotic) that imposes a shared source of mortality and disrupts phase synchrony in interspecific responses to the environment will reduce the insurance effects we show here (Ives *et al.* 1999). Although increasing interspecific synchrony will decrease the insurance effects evident in our model, it will conversely increase the species diversity necessary to generate the same level of insurance. Thus in spatially heterogeneous landscapes more species will be required to achieve the same level of buffering as environmental fluctuations become more phase-synchronized among communities.

Compensatory fluctuations in total community biomass also occurred in space across the metacommunity. Total metacommunity biomass was thus buffered by the out-of-phase variation in patch-to-patch community biomass. The propagation of this spatial insurance effect to the metacommunity level arose because of our assumption that the environment varied out-of-phase across local communities. Increasing the spatial autocorrelation in the environmental conditions will diminish phase difference amongst patches. At this scale phase synchrony will increase the number of patches required to achieve the same level of spatial insurance; two patches that fluctuate out-of-phase are sufficient to strongly buffer productivity, whereas a greater number of patches are required to generate the same effect as communities become increasingly in phase across the metacommunity. In general spatial autocorrelation decays with distance (Koenig 1999). The rate at which spatial autocorrelation decays with distance will determine the number of patches and the inter-patch distance required to maximize phase asynchrony, and ultimately, the scale at which spatial insurance effects are most pronounced. Spatially explicit versions of our model will be needed to address these important issues.

10.4.5.2 Dispersal and spatial averaging

Decreasing the rate of dispersal in the metacommunity has a significant direct effect on the mean and variability of ecosystem productivity and biomass. This direct effect of dispersal occurs independently of diversity and has been shown for single species metapopulation models (Holt 1993, Ives *et al.* 2004, Holt *et al.* 2005, Matthews and Gonzalez 2007). From these previous results the expected relationship between dispersal rate and mean biomass is unimodal with a peak at low to intermediate rates of dispersal. We have shown (Fig. 10.4) that this is indeed the case for most species in the metacommunity, the exception being the species that dominate the metacommunity at low and high rates of dispersal that show a U-shaped relationship between dispersal rate and mean biomass. Figure 10.4 also reveals differences between species in the level of dispersal that caused a peak in mean abundance; different species tend to contribute to the spatial averaging component of the spatial insurance effects as dispersal is increased. Overall, spatial averaging plays a relatively more important role in spatial insurance effects as diversity declines at higher rates of dispersal (Loreau et al. 2003).

Although theoretically spatial averaging is well understood there is relatively little empirical evidence. Good evidence for spatial averaging has been found in single-species microcosms (e.g. Ives et al. 2004, Gonzalez and Matthews 2007) and in a single multispecies microcosm experiment (Fig. 3 in Matthiessen and Hillebrand 2006). Clearly, much more empirical work is required to establish the importance of spatial averaging for the functioning and stability of fragmented landscapes.

10.4.5.3 Empirical tests of the theory

Recent years have seen several experiments address the effects of dispersal on the relationship between biodiversity and ecosystem functioning (Gonzalez and Chaneton 2002, France and Duffy 2006b, Matthiessen and Hillebrand 2006). The results of these experiments are broadly consistent with the spatial insurance hypothesis although none represents a precise test of the model we used.

Matthiessen and Hillebrand (2006) constructed laboratory metacommunities of benthic microalgae. They enhanced the rate of dispersal from the experimental 'regional pool' (aquaria) into the local communities (open-top, upright plastic tubes in the aquaria) by increasing the frequency at which the algae were scraped from the bottom of the aquaria and resuspended into the water column. As predicted by the spatial insurance hypothesis they found unimodal relationships between dispersal rate and local species richness and biovolume (a measure of primary production). When they used species richness as the predictor, variable biomass production showed the saturating, concave-down function typical of many biodiversity-ecosystem functioning experiments and similar to that shown in Fig. 10.3e. No attempt was made to study the community dynamics in this relatively short-term microcosm experiment.

Gonzalez and Chaneton (2002) used corridors to sustain dispersal from a large continuous block of habitat to satellites of isolated fragments of moss inhabited by a diverse community of microarthropods. In this experiment secondary production of this decomposer community was the ecosystem function of interest. Local extinction occurred in the isolated moss fragments after several months of delay, but not in those maintained by dispersal (corridor-sustained rescue effects). Fragments unconnected by corridors maintained two-thirds less secondary biomass than those connected by corridors. Thus dispersal seemed to buffer the ecosystem's capacity for organic matter processing in these experimentally fragmented habitats. However, dispersal rate was not controlled in a manner that could test the range of effects encompassed by the spatial insurance hypothesis.

France and Duffy (2006b) created experimental seagrass metacommunities in mesocosms and examined the effect of adding dispersal corridors on ecosystem functioning and stability. The dispersal corridors allowed mobile grazers to move from community to community, and thereby affect rates of grazing and primary production. Dispersal tended to decrease diversity (alpha and beta) and increased the temporal variability of local grazer abundance. Also dispersal tended to reduce spatial variability in grazer abundance, and enhanced grazer impacts on edible algae. Overall, results were considered inconsistent with the spatial insurance hypothesis, although because the main outcomes of the model are non-linear the theory also predicts destabilizing effects on functioning and stability; unfortunately, dispersal rates were not estimated during the experiment and so it is difficult to assess how they were affected by the presence of corridors. Importantly, there is a clear a mismatch between the assumptions of the theory that is based on dynamics of competition, and the multitrophic experimental system France and Duffy (2006b) used to test it. Further theory examining the diversity and functioning of food webs in metacommunities (e.g. Holt and Loreau 2002) is required to understand the results from this and other ecosystems with trophic complexity.

Although these experiments have tested various aspects of the spatial insurance hypothesis, none provides a complete test. The model assumes strong spatio-temporal heterogeneity that has highly dynamic affects on the outcome of competition. Under these assumptions dispersal has a strongly non-linear effect on diversity and ecosystem functioning. Future tests of the spatial insurance hypothesis will therefore require good experimental control of a number of factors – metacommunity size, dispersal rate, spatio-temporal heterogeneity – and the direct measurement of local and regional diversity and ecosystem functioning over extended periods of time.

10.5 Conclusions

This chapter has emphasized the spatial dimension of the relationship between biodiversity and ecosystem functioning. All the results we have raised have broad implications for the conservation and management of fragmented landscapes. First, we considered the link between local and regional processes using simple rules to scale diversity and function with area. Our results suggest that species loss due to habitat destruction may have delayed impacts on local function (a *functioning debt*). These results complement the idea that area loss will have the greatest functional effects as higher trophic levels suffer higher rates of extinction (Dobson *et al.* 2006). Spatial processes, just like multitrophic interactions, have the potential to generate complex non-linear effects on biodiversity and ecosystem functioning. Second, we indicated that spatial variance in species richness (perhaps caused by habitat fragmentation) should be taken into account when scaling the mean relationship between biodiversity and ecosystem functioning from local to regional scales.

We used a source–sink metacommunity model to examine various aspects of how spatial and temporal variability in the environment can be buffered by diversity. The metacommunity framework is a valuable way to link local and regional scales. The model predicts positive relationships between diversity, productivity, and stability. These results suggest that changes in landscape connectivity and fertility (resource flow) following anthropogenic fragmentation may alter both species diversity and ecosystem processes at local and regional scales. Because of the non-linear effects of dispersal, both increasing and decreasing landscape connectivity can either increase or decrease diversity and the temporal and spatial variability of (meta)ecosystem processes (Loreau *et al.* 2003b). The impact of reduced dispersal will depend upon the initial level of landscape connectivity and the dispersal ability of the organisms considered. Experiments addressing how scales of resource heterogeneity and dispersal interact to affect ecosystem diversity and stability are needed.

Recent work has established an important link between ecological and economic aspects of the insurance value of biodiversity (e.g. Armsworth and Roughgarden 2003, Baumgärtner 2007, and Chapter 18). Biodiversity has insurance value in economic terms because management decisions that alter biodiversity can affect the mean and variance of returns associated with an ecosystem good in a variable environment. Thus a risk adverse resource manager should optimize levels of biodiversity by, for example, adjusting the area and connectivity of a nature reserve (Armsworth and Roughgarden 2003), or sustaining pollinator services by maintaining pollinator-preferred habitat. This insurance value of biodiversity exists in addition to the direct and indirect (use and non-use) benefits normally associated with biodiversity. The theory outlined in this chapter stresses that knowledge of spatial processes across ecosystems will be essential if we are to understand the effects of fragmentation on the ecological and economic impacts of biodiversity loss.

Acknowledgements

A.G. and M.L. are funded by the National Science Engineering and Research Council of Canada, the Canada Research Chair Program, and a team grant funded by Le Fonds Québécois de la Recherche sur la Nature et les Technologies. NM is funded by the Programme National EC2CO and ANR Jeune Cheuchenr. The comments and suggestions of S. Naeem and two reviewers improved the manuscript.

PART 3
Ecosystem services and human wellbeing

CHAPTER 11

Incorporating biodiversity in climate change mitigation initiatives

Sandra Díaz, David A. Wardle, and Andy Hector

11.1 Introduction

Climate change mitigation through the sequestration of carbon (C), and the protection of biodiversity have captured the attention of scientists, governmental agencies, and the public in general in the past few years. This is justifiable in view of the formidable challenges posed by them to the long-term sustainability of the Earth's life support systems (Millennium Ecosystem Assessment 2005b, IPCC 2007).

Biodiversity and C sequestration in the biosphere have seldom been considered in an integrated way, either by international conventions or by the scientific community. Biodiversity considerations have been taken into account only marginally in international initiatives and agreements aimed at mitigating the ecological impacts of climate change. The most influential of these initiatives is the Kyoto Protocol to the United Nations Framework Convention on Climate Change (UNFCCC), which is intended to slow down the human contribution to increased atmospheric carbon dioxide concentration (http://unfccc.int/resource/docs/convkp/kpeng.pdf). This protocol was entered into force in February 2005 and has now been signed and ratified by 183 states. The Kyoto Protocol considers net C sequestration in the biosphere as one way to stabilize carbon dioxide levels in the atmosphere, and offers countries the opportunity to receive 'carbon credits' for enhancing sequestration. According to the definitions of the Marrakech Accord, climate change mitigation measures based on biological sequestration of C include afforestation, reforestation, revegetation, and forest, cropland and grazing land management (http://unfccc.int/resource/docs/cop7/13a02.pdf). However, when defining eligible C sequestration initiatives to be taken by different countries, the Kyoto Protocol explicitly excludes natural ecosystems already extant in 1990 as C sinks (http://unfccc.int/resource/docs/cop6secpart/l11r01.pdf). This is also the case with regard to the Clean Development Mechanisms (CMD, http://unfccc.int/resource/docs/2002/sbsta/misc22a04.pdf; see also Article 12 of the Kyoto Protocol) by which developed countries that emit C in excess of agreed-upon limits can obtain C offsets by investing in initiatives to sequester C and foster sustainable development in less developed countries. Here, only afforestation and reforestation qualify as eligible land use initiatives during the first commitment period of 2008–2012 (http://unfccc.int/kyoto_protocol/items/2830.php).

There is no mention of biodiversity in the main text of the Kyoto Protocol. The documents emerging from several meetings between 2001 and 2008 (Conferences of the Parties to the UNFCCC 7-13, and meetings of the Subsidiary Body for Scientific and Technological Advice, http://unfccc.int/meetings/items/2654.php) represent an advance in the sense that they incorporate biodiversity concerns. For example, the Marrakech (CoP-7), Milan (CoP-9) and Buenos Aires (CoP-10) accords, and the modalities for implementation of the CDM projects (CoP-11) explicitly state that LULUCF (land use, land use change, and forestry) and CDM initiatives must contribute to the conservation of biodiversity and sustainable use of natural resources, as well as to the promotion of C sequestration. Following the Montreal meeting (CoP 11), a request was issued to analyze the inclusion of avoided deforestation (Reducing Emissions from Deforestation and Degradation, or REDD) as part of the UNFCC activities in developing countries, either as part of the CDM next commitment period starting in 2012, or as a separate instrument designed specifically for this purpose. REDD are now an integral part of the 'Bali Road Map' (http://unfccc.int/resource/docs/2007/cop13/eng/06a01.pdf), which resulted from CoP 13. As in the case of the CDMs, the fact that the REDD initiatives should be compatible with the preservation of biodiversity is explicitly mentioned. These represent important steps forward, but biodiversity is still considered as a rather general 'side benefit' of carbon sequestration initiatives.

Academic publications (e.g. Kremen et al. 2000, Noss 2001, Niesten et al. 2002, Niles et al. 2002, Schulze et al. 2002, Sanz et al. 2004, Balvanera et al. 2005, Kremen 2005, Balvanera et al. 2006, Fearnside 2006b, Betts et al. 2008, Field et al. 2008) and assessment reports aimed to inform international conventions on the best ways to mitigate the effects of global change (e.g. Gitay et al. 2002, Díaz et al. 2003, Díaz et al. 2005, Stern 2006, Fischlin et al. 2007, Royal Society 2008) have stressed the importance of considering biodiversity, and analyzed the economic, social, and environmental costs and benefits of incorporating biodiversity-related criteria into C sequestration. However, in our opinion the fact that biodiversity not only has intrinsic value but could also enhance or reduce the effectiveness of C sequestration actions has not been sufficiently explored.

In this chapter we ask whether forest plant biodiversity, through its effects on ecosystem processes and especially on long-term C storage, is likely to have relevant consequences for the effectiveness of C sequestration. We first consider the theoretical background by which this could happen. Then we consider the available evidence. Finally, we make some recommendations based on this background and identify knowledge gaps and future research needs.

We refer to biodiversity as the number, abundance and identity of genotypes, populations, species, functional groups and traits, and landscape units present in a given ecosystem (Millennium Ecosystem Assessment 2005b, Díaz et al. 2006). In taking this broad approach, we consider species richness as just one component of biodiversity, and include other components, such as the identity and abundance of species and functional and structural traits, in our analysis, since recent syntheses (Díaz et al. 2005, Hooper et al. 2005, Díaz et al. 2006, Chapin et al. 2008) highlight the fact that composition is more important in determining ecosystem functioning than richness.

11.2 How can biodiversity affect C sequestration?

The success of C sequestration initiatives depends on how much C can be stored in the long term, which in turn depends on the net balance between C gain and C loss over long periods. It also depends on how important the C-sequestering ecosystem is perceived to be by the local stakeholders and the society at large, which in turn depends on the extent to which positive ancillary effects (such as preserved or enhanced ecosystem services other than C sequestration) can be obtained from it. This is because when stakeholders value the potential of an ecosystem to provide drinking water, food, aesthetic enjoyment, protection against natural disasters, and other services, they are more likely to protect its integrity, and therefore its C sequestration capacity, in the long term.

In this chapter we summarize the theoretical bases and some emerging evidence by which biodiversity as defined above could influence the overall success of C sequestration initiatives. We focus on path one of Fig. 11.1, and claim that biodiversity should be explicitly considered in the design of C sequestration initiatives.

It is common in international negotiations to use the term 'C sequestration' in a loose sense, to refer to the enhancement of both C stocks in and influxes into the biosphere through avoided deforestation, afforestation, reforestation, revegetation, and forest, cropland, and grazing land management. In the ecological sense, however, C sequestration refers to the maintenance or enhancement of C stocks in the biosphere. This is because large influxes can sometimes be accompanied by large effluxes, resulting in no net C accumulation. Net C sequestration occurs when the size and/or residence time of C stocks increases, due to a long-term positive balance between an ecosystem's C gains through

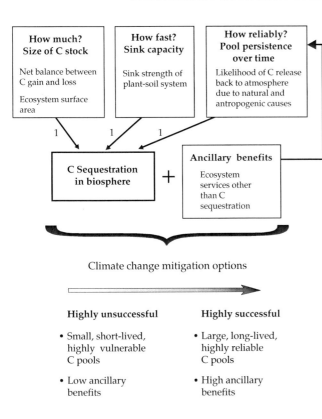

Figure 11.1 The success of climate change mitigation initiatives based on the biological sequestration of C depends on two main components: path (1), the amount and persistence of C sequestered in the plant–soil system; and path (2), the ancillary benefits provided by the C stock to humans. The positive effect of ancillary benefits is twofold. On the one hand, humans get extra benefits as well as climate change mitigation, such as regulation of water quality and quantity, soil fertility protection, traditional products, or cultural continuity ('win–win' options). On the other hand, the higher these benefits, the more likely the local communities are to preserve the C stock, thus increasing its long-term reliability.

net primary productivity and C losses through heterotrophic respiration and non-respiratory processes such as fire, harvest, and leakages of particulate, dissolved, or volatile C compounds (Catovsky et al. 2002, Schulze et al. 2002, Chapin et al. 2005, Schulze 2005). If biodiversity has the potential to affect C gain through productivity, or C loss through respiration and non-respiratory processes, then it follows that it should influence both the gross and the net C sequestration capacity of ecosystems. In this contribution, we use the term C sequestration (i.e. C storage) in the ecological sense, as a positive long-term change in, or maintenance of, C stocks. We refer to C influxes into the biotic system as C uptake or C capture.

Different theoretical backgrounds and some emerging evidence suggest that different components of biodiversity (species and genotype composition, number and spatial arrangement) differ in their potential to modify the magnitude, rate, and long-term permanence of the biosphere's C stocks and fluxes. Therefore, biodiversity consideration could be an integral part of the design and implementation of policy and management actions aimed at enhancing the long-term C sequestration capacity as well as the overall ecosystem-service value of primary, managed, and planted forests.

11.2.1 C sequestration predictions based on different theoretical approaches

How could biodiversity affect C sequestration in primary, managed, or planted forests? At present, there are three main theories leading to different predictions. These theories are the *neutral hypothesis*, the *mass ratio hypothesis*, and the *niche complementarity hypothesis*. We distinguish the neutral hypothesis from the other two because species differences play no role in it. Life history tradeoffs between species underlie both the mass ratio and niche complementarity hypotheses, but the first proposes that species influence ecosystem functioning according to their traits and in direct proportion to their relative abundance whereas the other also takes species interactions into account.

11.2.2 The neutral hypothesis

The Unified Neutral Theory of Biodiversity and Biogeography (Hubbell 2001) predicts that diversity can be maintained with random, neutral drift in species abundances so long as the evolution of new species can balance stochastic extinctions. Within the context of the links between biodiversity and C sequestration, the neutral hypothesis acts as a useful 'nothing happens' model. The neutral hypothesis assumes that individuals of all species have equal *per capita* probabilities of recruitment and mortality. On the surface the theory may seem to predict that all species are equal, but that is only the case for the recruitment and mortality rates, and functional traits are not explicitly considered. An attempt to reconcile neutral theory with niche theory proposes that species achieve equal *per capita* rates of recruitment and mortality by different resource allocation tradeoffs (Hubbell 2001: Chapter 10). However, the relative abundance of species is random with respect to their traits. If C storage is determined by the traits of species then under a neutral model, the sequestration capacity of forests will vary randomly over time along with neutral drift in the relative abundances of species.

11.2.3 The mass ratio hypothesis

According to the mass ratio hypothesis (Grime 1998), resource dynamics at any given time in an ecosystem strongly depend on the structural and functional characteristics of the dominant (i.e. most abundant) primary producers, and ecosystem functioning should be strongly affected by their life history tradeoffs. Therefore the total C stock of an ecosystem, its sink strength (the rate of change of the stock), and its residence time (the time that C will remain sequestered in the system) should strongly depend on the functional attributes of the dominant plants, as well as on climate and soil nutrients (Fig. 11.2). The traits of the dominants should strongly influence C uptake via net primary productivity and C loss via decomposition and disturbance. Fast acquisition of C per unit of leaf biomass or leaf area and long-term conservation of standing biomass are not expected to be maximized at the same time. This is because, across major taxa and biomes, there should be a tradeoff between a suite of attributes that promote fast C and mineral nutrient acquisition and fast decomposition, and another suite of attributes that promotes conservation of resources within well-protected tissues and slow decomposition (Grime 1979, Hobbie 1992, Cornelissen *et al.* 1999, Aerts and Chapin 2000, Díaz *et al.* 2004, Wright *et al.* 2004). The former, acquisitive, suite includes attributes such as leaves that are nutrient-rich, palatable, and short-lived, and often wood of low density. This suite is more common in light-demanding early-successional plants that act as pioneers after disturbance (Coley 1983, Pacala *et al.* 1996, Cornelissen *et al.* 1999, Ellis *et al.* 2000, Ter Steege and Hammond 2001, Laurance *et al.* 2006), and leads to shorter C and nutrient residence time in the ecosystem because of their short leaf lifespan and fast litter decomposition rates (DeAngelis 1992, Hobbie 1992, Aerts 1995, Wardle *et al.* 2004a). The latter, conservative, suite of traits includes leaves that are nutrient-poor, unpalatable, and long-lived, and often dense wood. This suite is more common in late-successional plants, which in forests include mostly disturbance-intolerant species (especially during ecosystem retrogression or decline, Walker *et al.* 2001, Wardle *et al.* 2004b); these species can increase C storage and mineral nutrient residence time as a result of their long leaf lifespan and slow litter decomposition rates. As a consequence of the existence of these suites of strongly associated attributes, there is a tradeoff at the ecosystem level between short-term C assimilation rate and long-term C storage. Within forest ecosystems, many forest types are successional mosaics where early- and late-successional patches coexist as a result of natural die-off events or, more commonly, small (e.g. tree fall) and large (e.g. forest fires) disturbance events (Denslow 1987, Crews *et al.* 1995, Pacala *et al.* 1996, Richardson *et al.* 2004). Early-successional and late-successional patches are dominated by acquisitive and conservative species, respectively, leading

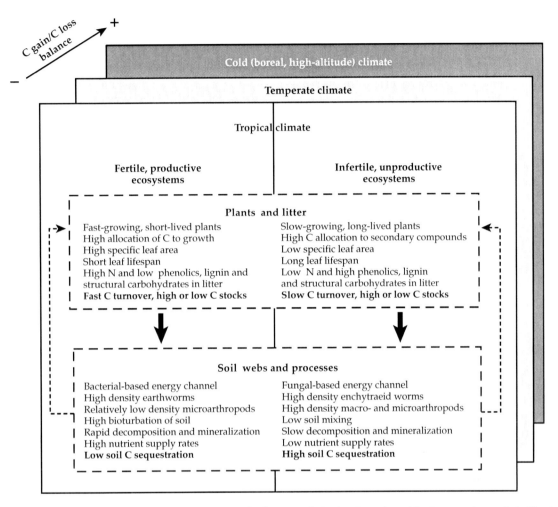

Figure 11.2 The traits of plants, especially dominant plants, strongly influence C and mineral nutrient cycling and thus C sequestration capacity in different ecosystems. Plant traits serve as determinants of the quality and quantity of resources that enter the soil and the key ecological processes in the decomposer subsystem driven by the soil biota. These linkages between belowground and aboveground systems feed back (dotted line) to the plant community positively in fertile ecosystems (left) and negatively in infertile ecosystems (right). C sequestration is highest in infertile conditions because decomposition is more impaired than net primary productivity by infertility and in colder conditions because decomposition is impaired more than net primary productivity by low temperatures (Derived from Wardle et al. 2004a).

to a differentiation in ecosystem processes between patches of different successional age. We should note here that while the mass ratio hypothesis describes the dominance of these strategies within patches, the landscape scale diversity between patches represents a form of niche complementarity (see below).

The structural and physiological traits of the dominant plants can also influence the probability of disturbances such as fire, wind-throw, and episodic herbivory, that are major avenues of C loss from ecosystems (Laurance 2000, Knohl et al. 2002, Lavorel and Garnier 2002, Chapin 2003, Pausas et al. 2004, Gough et al. 2008), and have important consequences for the long-term success of C sequestration initiatives. As well as this indirect effect through C sequestration capacity, the structural and phenological attributes of vegetation cover over large areas can affect climate directly. Functional traits of the dominants, such as leaf lifespan, growth form, root depth, and stomatal conductance affect albedo, roughness, and evapotranspiration. Through these biophysical feedbacks, the functional and structural composition of land patches can influence climate at the local, regional, and even trans-regional scale, depending on the land area covered by each vegetation type (Chapin et al. 2000a, Chapin et al. 2000b, Thompson et al. 2004, Chapin et al. 2005, Betts et al. 2008, Chapin et al. 2008). Recently, Körner (2005) has summarized the variety of functional traits in temperate and boreal tree species and their possible ecosystem-level implications, but a similar exercise has not yet been carried out for tropical and subtropical ecosystems.

Ecosystems consist of not just a producer but also a decomposer subsystem, and C sequestration is determined not just by ecosystem C gain (driven by net primary productivity, or NPP) but also by C loss (driven by decomposition). Thus, whether or not C accumulates in soils is driven to a large extent by the difference between C input to the soil (through litterfall, dead root production, and rhizosphere release) and C loss from the soil (through decomposition and respiration). Although decomposition at local (within-stand) scales is determined largely by litter quality (and hence the traits that drive litter quality), the linkages between aboveground (producer) and belowground (decomposer) communities are often relatively weak (Hooper et al. 2000, Wardle et al. 2004a, Hättenschwiler 2005). Thus decomposition rates need not respond to ecological gradients (e.g. succession, climate, diversity) in the same direction or to the same extent as does NPP. For example, decomposition is promoted by temperature more than is NPP, leading to reduced soil C sequestration at higher temperatures (Anderson 1991) and decomposition rates may decline across successional gradients while NPP is increasing, leading to rapid soil C accumulation (Wardle et al. 2004b). Further, plant species that produce high-quality litter may induce a 'priming effect' that accelerates the losses of native organic matter in the soil and thus promotes net ecosystem C loss (Jenkinson 1971). This may also explain why in some situations an increase in NPP is not matched by an increase in the amount of C stored in the soil (Fontaine et al. 2004), and may have important, though largely unrealized, consequences for soil C persistence and hence ecosystem C sequestration. Conversely, increasing domination of the plant community by plant species that are unproductive but contain high amounts of recalcitrant lignin and polyphenol compounds in their litter (such as can occur during ecosystem retrogression) can contribute to greater retention of C in the soil even when NPP is declining (Wardle et al. 2003a) (Fig. 11.2).

Tree species (or forest vegetation types) can differ markedly in the extent to which they promote sequestration of soil C (e.g. Jobbagy and Jackson 2000, Rhoades et al. 2000, Resh et al. 2002, Matamala et al. 2003, Russell et al. 2004), in a large part because they differ in their effects on the balance between C gain and C loss. For example, N-fixing trees will often accumulate more soil C than non-N-fixing trees (Resh et al. 2002). Systems dominated by slow-growing tree species that produce well-defended leaves (and hence poor litter quality) frequently promote substantial soil C accumulation relative to tree systems dominated by plants that grow rapidly and produce litter of high quality (Wardle et al. 2003a). The effectiveness of C sequestration initiatives depend on the magnitude and accumulation rate of soil C stocks, as well as the persistence of these stocks. Soil organic carbon (SOC) can be accumulated in short-lived pools, such as the microbial and labile pools (mean residence time of < 5 years), and long-lived pools in which SOC is protected by association to colloidal materials and the formation of stable microaggregates or recalcitrant compounds (mean residence time of thousands of years) (Lal 2005); tree species affect both of these pools. Dominant plant species have a clear influence on short-lived pools through root output and litter, and longer-lived pools through their litter quality (Wardle et al. 2003a), although their capacity to influence longer-lived pool is not always clear (Lal 2005, Jandl et al. 2007). Shallow rooting coniferous species tend to accumulate SOC in the forest floor, but they will sometimes accumulate less in deeper layers than comparable deciduous trees that often have deeper, more ramified roots. This is presumably in part due to the effective way in which root growth and subsequent root death can directly result in incorporation of organic matter inputs beneath the soil surface (Jobbagy and Jackson 2000, Trumbore 2000, Vesterdal *et al.* 2002).

The mass ratio hypothesis does not deny that less abundant species can sometimes play a major ecosystem role or face similar life history tradeoffs to those of abundant species (Grime 1998, Eviner and Chapin 2003, see below), but puts the emphasis on the functional composition of local *dominants* (Nilsson and Wardle 2005, Wardle and Zackrisson 2005). The niche complementarity hypothesis, in contrast, highlights the functional *differences* between coexisting species. These hypotheses are not mutually exclusive, and both processes can be operating in the same system (Loreau and Hector 2001, Fox 2005b, Potvin and Gotelli 2008). Many of the differences in life history traits reviewed above with regard to the mass ratio hypothesis may also be relevant to the discussion of niche complementarity that follows.

11.2.4 The niche complementarity hypothesis

This hypothesis is based on the idea that a greater range of physiological, structural, and phenological traits represented in the local community provides opportunities for more efficient resource use in a spatially or temporally variable environment (Trenbath 1974, Vitousek and Hooper 1993, Tilman *et al.* 1997c). This hypothesis is also compatible with the existence of trait tradeoffs, and indeed such tradeoffs are the basis for niche differences between species. But here there is less emphasis on the tradeoffs of the dominants as major drivers of ecosystem properties. When species show complementary niche differences it is likely – but not automatic (Hector 1998, Hector *et al.* 2002) – that a mixture of species may show greater overall resource uptake and rates of ecosystem processes than the same species grown in monoculture. Niche complementarity may relate to resource use, but mixtures may also perform better if rates of attack by natural enemies – either pests or pathogens – are higher in monocultures, in low-diversity patches, or near parent trees (e.g. Janzen 1970). Less abundant species are often minor players in ecosystem resource dynamics (Grime 1998) but may play an important role as a group, for example through ecosystem engineering (Jones *et al.* 1994), through keystone species effects (e.g. plant species that form mutualisms with nitrogen-fixing bacteria, Vitousek and Walker 1989), and through participating in complex indirect interactions (Eviner and Chapin 2003). Non-abundant species might be important in providing an insurance effect (a type of temporal niche complementarity) that helps sustain ecosystem functioning in the long term, particularly in a changing environmental context (Walker 1995, Walker *et al.* 1999, Yachi and Loreau 1999). There are few examples of insurance effects in the literature and it is therefore still too early for a formal assessment of their strength and occurrence.

The role of genetic differences between populations or genotypes of the same species in natural ecosystems has been little studied. In the case of herbaceous communities, Joshi *et al.* (2001) found that the performance of different genotypes was always best in the sites from which they were sourced, and Booth and Grime (2003) reported that communities composed of genetically uniform populations appear to be more variable in canopy structure, and to lose more species over time, than communities composed of genetically heterogeneous populations. Reusch *et al.* (2005) showed that genotypic richness of the cosmopolitan seagrass *Zostera marina* enhanced biomass production despite near-lethal water temperatures due to extreme warming across Europe. Crutsinger *et al.* (2006) showed that increasing population genotypic richness in the old-field herb *Solidago altissima* determined arthropod diversity and increased above-ground net primary productivity. However, it is difficult to know how general these patterns are, and whether they apply to woody ecosystems. Genetic variability among spatially separated populations of the same tree species has been shown to be an important driver of litter quality and ecosystem processes such as decomposition, herbivory and nutrient cycling (Treseder and Vitousek 2001, Whitham *et al.* 2003, Schweitzer *et al.* 2004, Schweitzer *et al.* 2005b), but experimental evidence on the effects of tree intraspecific genetic richness on ecosystem processes is still lacking (Hughes *et al.* 2008). Indeed, most of the evidence of the positive effects of high species and genotypic richness comes from the field of subsistence agriculture and forestry practiced by traditional peoples (Pretty 1995, Altieri 2004). This diversity is often lost during the process of selection for the production of high-yielding varieties. Therefore the possibility exists that the loss of inter- and intra-specific genetic variation could also lead to instability of plantations and other managed woody ecosystems in the face of a changing environment.

As for processes related to C loss, there are now a number of litter-mixing studies that collectively suggest that generally plant species composition of litter rather than its richness plays an important role in decomposition and nutrient cycling rates. Although the additive effects of species richness on litter decomposition cannot strictly be considered a niche complementarity effect in the sense of complementarity of resource use, they are discussed here because they involve 'richness-related' effects, as does the niche complementarity hypothesis. Litter mixing studies have found litter species richness to exert generally idiosyncratic or weak effects on litter mass loss (e.g. Wardle et al. 1997a, Bardgett and Shine 1999, Hector et al. 2000, reviewed by Gartner and Cardon 2004), while plant species richness has generally been found to exert weak or neutral effects on soil processes (Chapman et al. 1988, Hooper and Vitousek 1998). Further, it has been shown experimentally that addition of a greater richness of C substrates to the soil (such as might be expected in a more species-rich plant community) did not exert strong or consistent effects on C loss rates from soil, or on soil C storage (Orwin et al. 2006). However, in instances in which NPP is promoted by plant species richness, it is likely that decomposition rates would be less unresponsive, in which case greater C sequestration would be expected over time. The mechanistic basis through which plant richness might affect soil processes is relatively poorly understood. However, the available evidence suggests that plant species richness is not a powerful driver of soil decomposer richness (Hooper et al. 2000) and that decomposer richness is not a major determinant of soil process rates such as decomposition or nutrient supply rates for plants (Laakso and Setälä 1999, Setälä and McLean 2004, Hättenschwiler et al. 2005).

11.2.5 Where does the available evidence stand and what else do we need to know?

In summary, the predictions of these different hypotheses for the incorporation of biodiversity in C sequestration initiatives vary markedly. Taken to an extreme, the mass ratio hypothesis predicts that C storage would be maximized by planting a monoculture of the species with the combination of traits (stature, lifespan, timber density, decomposition rate, resistance to fire, wind-throw, and pests) that produces the highest species specific C storage for a given area. The niche complementarity hypothesis predicts that C storage will be impacted by interspecific differences among coexisting species, in terms of resource use and tolerance to biotic and abiotic factors. It also predicts that it may be possible to increase C storage by planting complementary mixtures of species, sets of species with known mutually facilitative effects, and/or ensuring that a mosaic of late- and early-successional patches is kept (e.g. Caspersen and Pacala 2001). Finally, the neutral hypothesis predicts that the C storage capacity of natural forests will vary randomly with stochastic shifts in species abundances. In plantations it may be possible to influence C storage by controlling the recruitment stage, for example by increasing seed or seedling input of species that are good at sequestering C but are poor recruiters.

The three hypotheses all stem from strong theoretical developments and are all supported by empirical evidence to varying degrees in forested systems. Most of the experiments from which this evidence is derived were not originally designed to test these hypotheses. Moreover, there is an important body of results of experiments specifically designed to test the effect of biodiversity (and most commonly species richness) on the functioning of grasslands (reviewed in Loreau et al. 2001, Díaz et al. 2005, Hooper et al. 2005) but there are few corresponding experimental studies in woody ecosystems, which may not necessarily behave in similar way to herbaceous ecosystems.

Table 11.1 provides an overview of recent studies of the role of different components of plant biodiversity in C gain and loss of forest ecosystems. They include primary forests, traditionally managed forests, and commercial and experimental plantations. Our synthesis, which is intended to be illustrative rather than exhaustive, reflects the scarcity of published studies involving woody plants. This is true for all continents, but particularly dramatic in Latin America, Africa, and Asia, precisely where most remaining high-diversity forests are located. There have been some studies that can be interpreted in the light of the mass ratio or niche complementarity hypotheses to varying degrees. As for the neutral hypothesis, we found no study directly linking it with the way in which biodiversity could affect ecosystem processes. According to the original authors' interpretation of their own results (Table 11.1, third column) there seems to be more support for the mass ratio hypothesis than for the niche complementarity hypothesis, in the sense that the authors conclude that composition (the presence of certain tree species) appears to play a more important role than species richness. However, compositional differences could arise from either mass ratio or niche complementarity effects or some combination of the two (Loreau and Hector 2001). Distinguishing the relative contributions of these two mechanisms will require future studies that are explicitly designed to discriminate among the two classes of causes. Evidence for relationships between species richness and stability of forests and plantations is mixed. It follows that particular attention should be paid to the identity of the species chosen for afforestation, reforestation and rehabilitation projects, with the actual richness of species planted taking second place. However, (1) positive effects on ecosystem functioning are often found in mixtures of two or more species compared to monocultures; (2) virtually all the reported studies were not specifically designed to distinguish between the three different hypotheses, and the patterns observed may fit more than one of them (e.g. Chave 2004, Volkov *et al.* 2005); and (3) mass ratio, niche complementarity, and neutral hypothesis mechanisms may all be acting simultaneously (e.g. Potvin and Gotelli 2008).

An experimental test of the neutral hypothesis through the removal of dominant species has recently been performed for intertidal communities (Wootton 2005), but the feasibility of this approach for use in other systems is unclear. Experiments to definitively establish the relative importance of the mass ratio and niche complementarity mechanisms for determining ecosystem properties in forests will ideally require the establishment of monocultures and mixtures of all component species under the same environmental and management conditions (e.g. Redondo-Brenes and Montagnini 2006, Potvin and Gotelli 2008). This may be practical for species-poor ecosystems (e.g. boreal forests), but it quickly becomes unfeasible if one is to incorporate even a fraction of the high richness of tree species characteristic of many tropical forests. We also emphasize that experimental approaches of this type are not the only way to formally test for the role of biodiversity in ecosystem functioning, and ideally the results of such studies should be considered alongside other approaches that have recently been employed to test how biodiversity affects forest C sequestration, such as simulation- and modelling-based approaches (Bunker *et al.* 2005), field removal experiments (Díaz *et al.* 2003, Wootton 2005), observational studies using well characterized gradients of plant diversity (Wardle *et al.* 2003a), and forestry projects that incorporate diversity components into their design (i.e. 'enrichment planting', e.g. Evans and Turnbull 2004). In the end, even being able to successfully distinguish between the relative importance of mass ratio, niche complementarity, and neutral hypothesis effects may not necessarily be crucial to the practical purposes of C sequestration, especially as these hypotheses are not all mutually exclusive. For example, experimenting with mixtures that contain non-random combinations of species (such as those that represent traditional mixtures), or maximize key ecosystem services like C sequestration plus food production, or are the most economically and socially feasible in each region, might make more practical sense than incorporating all the possible mixtures of component species within the experimental design.

Table 11.1 A summary (representative rather than exhaustive) of studies published during the last 13 years on the effects of different components of biodiversity on C sequestration through impacts on C gains or losses in woody ecosystems.

Ecosystem type and location	Main biodiversity component involved	Findings	Source
C gain			
Experimental plantations of fast-growing tropical tree species *Hyeronima alchomeoides*, *Cedrela odorata*, and *Cordia alliodora*; each species grown alone and with two perennial, large-stature, monocots (*Euterpe oleracea* and *Heliconia imbricata*)	Species and functional group richness	Ecosystem productivity and resource capture were increased when the monocots were grown with *C. odorata* and *C. alliodora*, but not with *H. alchomeoides*	Haggar and Ewel (1997)
Boreal forest trees and understorey vegetation on Swedish lake islands	Species and functional group richness	Species-rich islands less productive at large spatial scale (between islands) because more productive species dominate on less diverse islands; some evidence of greater understorey species richness promoting overall forest productivity within islands	Wardle *et al.* (1997), Wardle *et al.* (2003), Wardle and Zackrisson (2005)
Young plantations of four indigenous tree species: *Hieronyma alchorneoides*, *Vochysia ferruginea*, *Pithecellobium elegans*, and *Genipa americana*, growing in mixed and pure stands at La Selva Biological Station, Costa Rica	Species richness and identity	Total tree biomass production rate of the mixture was not significantly higher than that of the most productive monocultures	Stanley and Montagnini (1999)
Stand productivity in USA Forest Inventory and Analysis database	Species richness	Positive correlation between tree species richness and stand productivity, especially when comparing monocultures vs. mixtures of two or more species; variations in abiotic factors not considered	Caspersen and Pacola (2001)
Stand biomass in global forest dataset	Species richness	Forest stand biomass not associated with tree species richness	Enquist and Niklas (2001)
Experimental plantations of three native tree species, *Hyeronima alchoreoides*, *Cedrela odorata*, and *Cordia alliodora*, in monoculture and in mixtures with the palm *Euterpe oleracea* and the giant perennial *Heliconia imbricata*, in Costa Rica Atlantic lowlands	Species richness and composition and functional group richness	Tree species richness influenced ecosystem nutrient use efficiency in tree-only stands. Aboveground net primary productivity after four years was significantly higher in polycultures than in monocultures of *C. odorata*, and *C. alliodora*, but not in the case of *H. alchomeoides*. The presence of the additional life forms increased nutrient uptake and uptake efficiency, but only in some systems and years Although species and life forms exerted considerable influence on ecosystem nutrient use efficiency, this was most closely related to soil nutrient availability	Hiremath and Ewel (2001)
Stand productivity of boreal forests dominated by *Betula* spp., *Picea abies* and *Pinus sylvestris*, under the same environmental conditions and management	Species richness and composition	Mixtures of *Betula* spp. and *P. abies* more productive than *Picea* monocultures, but mixtures of *Betula* spp. and *P. sylvestris* were less productive than *P. sylvestris* monocultures; species richness effect significant only at early successional stages	Frivold and Frank (2002)

Long-term experimental comparison of different agroforestry systems in Brazilian Amazonia; peach palm (*Bactris gasipaes*) for fruit and heart-of-palm production, cupuaçu (*Theobroma grandiflorum*), and rubber (*Hevea brasiliensis*) planted in multistrata mixtures and in monocultures, also compared with adjacent primary rainforest and 14-year old secondary forest	Species richness and composition	Multistrata agroforestry system showed more accumulation of above- and belowground biomass than cupuaçu, rubber, or peach palm for heart-of-palm, but less than peach palm for fruit. Secondary forest accumulated 50%, and primary forest likely 500% more total biomass than the most productive plantation	Schroth *et al.* (2002)
Wood production in Catalonian forests with different degrees of species richness, dominated by *Pinus sylvestris* or *Pinus halepensis*	Species richness	In *P. sylvestris* forests wood production was not significantly different between monospecific and mixed plots. In *P. halepensis* forests wood production was greater in mixed plots than in monospecific plots. No significant effect of species richness when environmental factors were considered.	Vilà *et al.* (2003)
Experimental plantations of three native tree species, *Hyeronima alchoreoides*, *Cedrela odorata*, and *Cordia alliodora*, in monoculture and in mixtures with the palm *Euterpe oleracea* and the giant perennial *Heliconia imbricata*, in Costa Rica Atlantic lowlands	Species composition and functional group richness	Light particulate organic matter C and soil C:N ratio were significantly higher, and total soil organic matter C was slightly higher, under *H. alchoreoides*, as compared to under the other two tree species Functional group richness had a positive effect on total and light/particulate soil organic matter as compared to monocultures of *C. odorata* and *C. alliodora*, but not in the case of those of *H. alchoreoides*.	Russell *et al.* (2004)
Litter production in Catalonian traditionally managed forests	Species richness, species and functional trait composition	Litter mass larger in 2–5 species mixtures than in monospecific stand. In mixed forests, identity of trees determined whether litter stocks increase with tree diversity.	Vilà *et al.* (2004)
Simulation study of the magnitude and variability of aboveground C sequestration in 18 scenarios of tree species extinction within a species-rich tropical in Panama	Species richness and functional trait composition	Different trait-based scenarios (e.g. order of extinction determined by wood density, height, growth rate, drought tolerance, also a random extinction scenario) resulted in strong differences in magnitude and variability of C stocks	Bunker *et al.* (2005)
Long-term tree-planting experiment, established in 1955 in NW England; *Quercus petraea*, *Alnus glutinosa*, *Pinus sylvestris* and *Picea abies* planted in monocultures and in 2-spp mixtures	Species richness and composition	All mixtures involving *Pinus sylvestris* showed more growth in pure stands of either species; *A. glutinosa* mixtures not involving *P. sylvestris* did not outperform monocultures, *P. abies*/*Q. petraea* mixture showed less growth than monocultures	Jones *et al.* (2005)

(Continues)

Table 11.1 (*continued*)

Ecosystem type and location	Main biodiversity component involved	Findings	Source
Review of the 20th century forestry literature with emphasis on commercial trees in the temperate and boreal zones	Species and functional group richness and composition	Increased productivity in mixtures of species with different spatial, phenological or successional niches (e.g. *Larix/Picea*, *Quercus/Betula*, *Pinus/Picea*, *Pinus/Betula*). Some mixtures (e.g. *Picea abies/Betula pendula*) sustain production over a larger range of densities than monocultures and are thus more tolerant to risks	Pretzsch (2005)
Natural and seminatural forests, plantations and secondary woodlands in the Ecological and Forest Inventory of Catalonia (IEFC), including 95 tree species	Species richness, species and functional trait composition	Stemwood production increased from single-species to 5-species stands, but stand age and richness were negatively correlated. Species richness had a significant positive effect on stemwood production in stands dominated by sclerophyllous species (e.g. *Quercus*, *Arbutus*), and low-productivity conifer stands, but not deciduous species stands in humid or warmer climates	Vilà *et al.* (2005)
Experimental mixed plantations of native trees *Balizia elegans*, *Callophyllum brasiliense*, *Dipteryx panamensis*, *Hyeronima alchorneoide*, *Jacaranda copaia*, *Terminalia amazonia*, *Virola koschny*, *Vochysia ferruginea* and *Vochysia guatemalensis* in Costa Rican tropical rainforest. Monocultures were compared to 3-species mixtures, all of them consisting of one fast-growing sp., one slow-growing sp., and one legume, to keep functional richness as constant as possible	Species richness	Although some individual species were more productive in mixtures than in monocultures, none of the mixtures showed significantly higher growth or C storage than the monocultures of the most productive species involved in each mixture	Redondo-Brenes and Montagnini (2006)
More than 5000 permanent forest plots in the National Forest Inventory of Spain in the Catalonia region, including 51 tree species, growing in monocultures and in 2- to 5-species mixtures	Species richness, functional group richness and identity	Stemwood production was positively associated with tree species richness and with functional group identity (deciduous forests were more productive than coniferous or sclerophyllous forests). Functional group richness did not significantly explain stemwood production once the effects of environmental and structural variables were taken into account	Vilà *et al.* (2007)
Experimental plantations of native tropical trees representing a range of relative growth rate (*Cordia alliodora*, *Luehea seemannii*, *Anacardium excelsum*, *Hura crepitans*, *Cedrela odorata*, and *Tabebuia rosea*) in monocultures, and 3- and 6-spp. plots, in Central Panama	Species richness and composition	Plot biomass (estimated from basal area) did not differ between mixtures and monocultures or among mixtures. There was a significant species richness effect on growth, attributed to complementarity, in the 3-species mixtures as compared to monocultures, but there was no significant effect in 6-species plots. Mortality was strongly dependent on species identity, and independent of species richness. Overall, there was a positive complementarity effect (using the additive partitioning method of Loreau and Hector 2001) of species richness on plot biomass and a negative selection effect, resulting in no net species richness effect	Potvin and Gotelli (2008)

C loss

Boreal forest trees and understorey vegetation on Swedish lake islands	Species and functional group richness	Species-rich islands supported less soil respiration, microbial biomass and decomposition at large spatial scale (between islands), contributing to net C sequestration in the soil Some evidence of greater understorey species richness promoting these processes within large (but not small) islands Differences among islands in belowground processes and C sequestration are explicable by traits of dominant plant species but not species richness	Wardle et al. (1997), Wardle et al. (2003), Wardle and Zackrisson (2005)
Damage by beetle Phratora vulgatissima and rust Melampsora spp. on five Salix genotypes in monocultures and mixtures in regular and random spatial arrangements	Genetic richness and spatial heterogeneity	Mixtures showed less damage by rust and beetles than monocultures; no significant effect of structural design was detected, but the trend was for decreased damage in random configurations	Peacock et al. (2001), Hunter et al. (2002)
Microcosms experiments using litter of nine phenotypes of Quercus laevis in monocultures and in mixtures	Intraspecific phenotypic richness and composition	C and N fluxes within single phenotype treatments were significantly, but unpredictably, different from those of mixtures No effect of phenotype identity on soil bacterial or microarthropod communities	Madritch and Hunter (2002), Madritch and Hunter (2005)
Literature review of European forests (especially N Europe)	Species richness and/or composition	Different species and functional types differed in wind resistance; mixtures were not more stable than monospecific stands against windstorms	Dhôte (2005)
Literature review of decomposition rate of single-species litter vs. litter mixtures of several N Hemisphere tree species	Species richness and composition	Sometimes faster decomposition in mixtures; in other studies the effect was similar to that predicted from the decay rates of individual species and their relative contribution to the mixture; in two cases lower decay rate in mixtures; different mixtures involving Pinus or Quercus showed no consistent effect as compared to monocultures	Hättenschwiler et al. (2005)
Meta-analysis of 54 studies of insect herbivory on trees, with emphasis on temperate systems	Species richness and composition	Tree species growing in mixed stands overall suffer less damage by specialized herbivore insects than do pure stands; generalist insects showed a highly variable response	Jactel et al. (2005)
Heterobasidium annosum (butt rot) in pure vs. mixed stands under different climatic conditions (mostly N Europe)	Species richness	Incidence of H. annosum negatively correlated with tree species richness	Korhonen et al. (1998), as cited in Pautasso et al. (2005)
Cronartium ribicola rust and Phellinus weirii root rot in North American forests	Species richness and composition	Disease spread associated with certain host tree species, rather than with tree richness	Pautasso et al. (2005)
Literature review of boreal forests	Species richness and/or composition	Mixed stands not more resistant to fire than monospecific stands	Wirth (2005)

(Continues)

Table 11.1 (continued)

Ecosystem type and location	Main biodiversity component involved	Findings	Source
Review of 26 experimental studies on the effect of the diversity of trees in boreal forests on the damage by invertebrate and vertebrate herbivores and pathogen species	Tree species richness and composition, landscape heterogeneity	Species-rich stands not consistently less prone to pest outbreaks and disease epidemics than monocultures. Composition appeared to play a greater role than species richness *per se* Susceptibility to inspect pests decreased with increased isolation of stand within a forest mosaic of non-host species	Koricheva *et al.* (2006)
Experimental boreal forests of *Betula pendula*, *Pinus sylvestris*, and *Picea abies* in Sweden and Finland	Species richness and composition	Monocultures of *B. pendula* and mixed stands containing 25% of *B. pendula* and 75% of *P. sylvestris* showed higher defoliation by insects early in the season than *B. pendula* monocultures or 50–50 mixtures of *B. pendula* and *P. sylvestris*. No difference between monocultures and mixtures late in the season	Vehviläinen *et al.* (2006)
Experimental plantations of native tropical trees representing a range of relative growth rate (*Cordia alliodora*, *Luehea seemannii*, *Anacardium excelsum*, *Hura crepitans*, *Cedrela odorata* and *Tabebuia rosea*) in monocultures, and 3- and 6-spp. plots, in Central Panama	Species richness, species and functional trait composition	After c. 4 years from establishment, no consistent general effect of species richness was found on either litter production or decomposition. Litter production was significantly affected by tree species richness and identity, with the majority of intermediate-richness mixtures showing higher litter yields than expected based on monoculture. Litter decomposition also varied with species identity and functional attributes. High-richness mixtures decomposed at rates that were no different from expected on the basis of their component species. However, individual species changed their decomposition pattern depending on the richness of the litter mixture	Scherer-Lorenzen *et al.* (2007a)
Experimental decomposition of monocultures and mixtures of 2, 3, 4, and 5 dominant species of central Argentina mountain woodlands, representing a range of functional groups decomposition rates (*Acacia caven*, *Lithraea molleoides*, *Bidens pilosa*, *Hyptis mutabilis*, and *Stipa eriostachya*)	Species richness, species and functional trait composition	When up to five species were included, both species richness and functional composition showed non-additive, mostly positive effects on litter mixture decomposition. The synergistic effects of species richness were significant when the richness of the mixtures changed from 2 to 3–4 species. A greater positive effect was found in mixtures with higher mean nitrogen content and a higher heterogeneity in non-labile compounds. Litter mean quality and chemical heterogeneity were the most important factors explaining decomposability of mixtures	Pérez Harguindeguy *et al.* (2008)

11.3 Making the most of biodiversity in the design of climate change mitigation initiatives

The major hypotheses examined above, and the evidence available so far, indicate that the incorporation of biodiversity considerations has the potential to influence the magnitude and long-term persistence of C-sequestration initiatives. The leading role of the functional traits of locally dominant plant species is supported by strong evidence from a variety of ecosystems. However, considerably more experimental, observational, and modelling work is needed to elucidate many specific details, such as to what extent increasing the small-scale species richness of reforestation or afforestation actions can increase their ability to store C. Nevertheless, we believe that some practical recommendations can already be made based on the current level of knowledge.

- *Protecting primary forests is the best C sequestration option.* For obvious practical reasons, to date there is no published biodiversity experiment involving formal experimental manipulation of tree species richness beyond six species. However, primary forests usually have a larger number of species and a wider range of plant functional attributes than do planted forests. They also tend to be dominated by large-sized, slow-growing species that are conservative with resources. Therefore, under both the niche complementarity and mass ratio hypotheses, we expect them to maximize C stocks. Available evidence from the biodiversity and biogeochemistry literature supports this idea. Primary forest ecosystems represent the most important biological C sinks on the planet in terms of both quantity and likely stability through time (Buchmann and Schulze 1999, Valentini *et al.* 2000, Schimel *et al.* 2001, Schulze 2005, Luyssaert *et al.* 2008). With very few exceptions, they contain larger C stocks than younger forests in all biomes (Pregitzer and Euskirchen 2004, Schulze 2005). Recent studies suggest that C outputs and inputs in primary forests are frequently not at equilibrium, and that such forests are active, albeit sometimes small, C sinks (Schimel *et al.* 2001, Schulze *et al.* 2002, Sabine *et al.* 2004, Schulze 2005, Luyssaert *et al.* 2008). In temperate and boreal zones, forests contain large quantities of carbon and can continue accumulating it for centuries (Luyssaert *et al.* 2008). There is less empirical information for tropical forests, but their C exchange appears to be approximately balanced, or even slightly positive (Schimel 2007, Stephens *et al.* 2007). This points to a gross sink that compensates for emissions due to tropical deforestation and fires. Primary forests often show a lower uptake of C per unit time than do newly established plantations (Gower 2003) but on the other hand they sequester it for a longer time. Also, the process of land conversion, for example during the establishment of a new plantation, often releases very large amounts of C from the soil to the atmosphere (Valentini *et al.* 2000, Guo and Gifford 2002, Pregitzer and Euskirchen 2004). As a consequence, the net balance of C sequestered per hectare is usually more strongly positive in the case of primary forests than for new plantations, with the benefits from the latter being more transitory and uncertain (Schulze 2005). Primary forests are being destroyed at accelerated rates, especially in the African and Latin American tropics (Lambin *et al.* 2003, Fearnside and Barbosa 2004, Shvidenko *et al.* 2005). The amount of forested area lost is still impossible to match by plantation initiatives, and this is likely to continue to be the case for the next several decades. Plantations can also involve high monetary and environmental costs. For example, the monetary cost of sequestering 1 Mg of C by forestation and agroforestry activities has been estimated as being more than triple than that of sequestering the same amount by conservation of already existing forests (van Kooten *et al.* 2004). Another recent study shows that monospecific plantations of fast-growing trees in southern South America have strong negative impacts on water supply and soil fertility (Jackson *et al.* 2005). An additional reason to protect primary forests is that changes in the functional attributes of vegetation over large areas can affect climate directly through water and energy exchange (Chapin *et al.* 2008).

- *The maximization of short-term C sink strength is unlikely to be the best option for C sequestration in the longer term.* As explained in previous sections and illustrated in Fig. 11.2, the well-supported mass ratio hypothesis predicts that there is fairly a universal tradeoff between a suite of plant attributes that promotes fast C and mineral nutrient acquisition and loss ('acquisitive' syndrome), and another that promotes slower acquisition but long retention of resources within well-protected tissues ('conservative' syndrome). This suggests that a management regime that simultaneously maximizes rapid C uptake from the atmosphere and its long-term sequestration is unlikely to be found. This is directly relevant to C sequestration initiatives, since at any time a C-sequestering project is launched, a decision should be made in favor of one or the other side of the tradeoff (Aerts 1995, Caspersen and Pacala 2001, Noss 2001). For example, early-successional, light-demanding, fast-growing species should be selected when the goal is to maximize short-term productivity. However, C sequestration in the longer term will be greater in areas dominated by later-successional species that are slower growing but have denser timber, and whose litter decomposes more slowly. In view of this, high sink capacity in the short term should not be considered as the major criterion in reforestation/afforestation initiatives. In general, careful consideration of the species and genotypes chosen for each C sequestration project is needed (Lal 2004). There are strong ecological bases to suggest that fast-growing, genetically homogenous, easy-to-manage, widespread forestry species and genotypes (e.g. members of *Eucalyptus*, *Pinus*, and *Acacia* widely planted in South America, Africa, and East Asia) may not represent the most effective option in terms of long-term C sequestration. Also, the choice of species and genotypes with the appropriate attributes for local (present and projected) climatic and disturbance conditions (e.g. fire proneness, storm, or frost frequency) is very important. The same considerations apply to plantations that serve as sources of solid biofuel, although permanence is obviously less of an issue in that case.

- *Mixed forestry systems might be more stable in the face of environmental variability and directional change than monocultures, and they might sequester C more securely in the long term.* This recommendation is consistent with the niche complementarity hypotheses, as well as the results of several experiments in herbaceous communities. The evidence from forest ecosystems is still inconclusive, and long-term field-scale experimental, observational and theoretical studies are needed to rigorously test whether, how generally, and for how long increasing the number of genotypes, species and functional types can benefit afforestation, reforestation, agroforestry, secondary forest recovery and solid biofuel plantation initiatives. However, thousands of years of agricultural experience point to the use of polycultures as a promising precaution to buffer forest production throughout the year and also against environmental change and variability and pest and weed damage. Tree monocultures often, but not always, promote less SOC accumulation than primary or secondary forests (see Lal 2005, Jandl *et al.* 2007 for reviews). But even in cases where the amount of C sequestered by a monoculture is higher, the use of mixtures of more than one tree species may be a good alternative for small or medium-sized farms, especially in tropical and subtropical areas. This is because mixed plantations provide a wider range of products and opportunities. For instance, fast-growing and slow-growing species provide revenues in the short and long term, respectively; different species provide non-forest products such as fruit at different times of the year and thus improve food security and buffer market risks (Piotto *et al.* 2004, Montagnini *et al.* 2005). These ancillary benefits of mixed plantations and agroforestry systems increase the interest of local stakeholders in establishing and protecting forests and diminish incentives for changing to other land uses (Liebman and Staver 2001, Pretty and Ball 2001, Schroth *et al.* 2002, Piotto *et al.* 2004, Montagnini *et al.* 2005). Sometimes the recovery of the natural forest is limited by animal dispersal of propagules, soil moisture, and competition from herbaceous plants. Mixed plantations offer an alternative in these cases. For example, in Costa Rica, more individuals and species of native trees were found to regenerate in the understorey of mixed plantations than those under monocultures (Guariguata *et al.* 1995, Powers *et al.* 1997, Carnevale and Montagnini 2002).

- *Plantations established with the specific purpose of C sequestration or biofuel production can, and should, be compatible with biodiversity conservation.* It is vitally important that projects supported through the CDMs or other initiatives aimed at increasing C uptake do not come at the direct or indirect cost of clearing natural ecosystems, and that they maintain a high ecosystem-service value from the point of view of local communities rather than simply meeting the C credit priorities of external investors

(Niesten *et al.* 2002, Prance 2002, Fearnside 2006a). Niesten *et al.* (2002), Schulze *et al.* (2003) and Chadzon (2008) provide examples of forestry projects that, rather than decreasing pressure on natural ecosystems, may contribute to their destruction, in the name of the creation of C sinks. Agroforestry practices have the potential to store large amounts of C while at the same time protecting biodiversity. For example, Brandle *et al.* (1992) and Noss (2001) highlighted the potential of planted shelterbelts and riparian forests that store C and at the same time provide wildlife habitat and permanent regional vegetation connectivity. Modeling efforts by Bolker et al. (1995) and Pacala and Deutschman (1995) suggest that species-rich and spatially heterogeneous forests could have a C sequestration potential of up to 50 per cent more than monospecific, spatially homogeneous forests. As in the case of managed forests not specifically designed for C sequestration processes, high inter- and intraspecific genotypic richness, the inclusion of local genotypes, and the maintenance of a rich and heterogeneous landscape increases the value of plantations for local societies, and thus their willingness to protect them. This enhances their potential to preserve their long-term survival and C sequestration capacity (Prance 2002, Díaz *et al.* 2005). On the other hand, local communities have little to win and much to lose (e.g. traditional medicine, cultural and spiritual values, employment) from reliance on monospecific stands of fast-growing (and often introduced) tree species and varieties. The incorporation of what is 'valuable biodiversity' from the local community's point of view is essential for striking the right balance between biodiversity and C sequestration and for ensuring the long-term protection of C-sequestering plantations (Díaz and Cáceres 2000, Prance 2002, Saunders *et al.* 2002, Díaz *et al.* 2005, Canadell and Raupach 2008).

• *Decisions about the species and genotype richness and composition of protected or newly established plantations or agroforestry systems should be tailored to the local context.* It is important to keep an open perspective and to avoid mechanical application of general principles to individual projects without careful consideration of the resource base, prevailing disturbance conditions, scale of the project, and attributes of the organisms (including not only the planted species) and ecosystems involved. A practical way to increase our understanding of how, where, and why different biodiversity components affect the C-sequestration capacity of different ecosystems would be to incorporate an experimental component to climate change mitigation and agroforestry and forest rehabilitation initiatives (e.g. Ewel 1986, Montagnini et al. 2005, Scherer-Lorenzen *et al.* 2005b). Moreover, we are aware of a wealth of information being produced by the forestry sector, but this is not often reflected in the peer-reviewed literature. In this sense, the recent book edited by Scherer-Lorenzen *et al.* (2005a) has made a valuable contribution through making available a large body of difficult-to-access and diffuse literature from the forestry sector. A similar effort with specific focus on key regions (e.g. Latin America, Africa, Southeast Asia) including the wealth of information accumulated by governmental and non-governmental grassroots initiatives, would be valuable for helping find the best options for simultaneous C sequestration and biodiversity protection in primary, managed and planted forests.

11.4 Final remarks

In the past few years, the focus of international mitigation efforts seems to have shifted from cutting fossil fuel emissions to enhancing C sequestration, with the remarkable exception of some actions taken during the most recent COPs (see Introduction). The potential contribution of C sinks to climate change mitigation is clearly less important in terms of C released to the atmosphere, than that of decreasing emissions from fossil fuel burning (IGBP 1998, Prentice *et al.* 2001). Therefore, by no means do we believe that mitigation initiatives are a substitute for cutting fossil fuel emissions, however beneficial for the conservation of biodiversity they would be. That said, there is considerable potential for increasing the world's C stocks through management practices (Watson *et al.* 2000, Niles *et al.* 2002, Fischlin *et al.* 2007, Canadell and Raupach 2008). Considering the dramatic observed and projected consequences of climate change (IPCC 2007), we must exploit this potential to the largest possible extent. Equally important is making sure that C sequestration measures do not backfire in the long term, for instance by ensuring that their overall environmental costs do not offset their benefits.

On the basis of the findings summarized above, and in accordance with other authors (IGBP 1998, Schulze *et al.* 2002, Schulze *et al.* 2003, Fearnside 2006b, Luyssaert *et al.* 2008), we suggest that the conservation of natural ecosystems is the best C sequestration option available. Natural ecosystems, with their ability to simultaneously maintain C stocks, biodiversity, and ecosystem services, and their built-in capacity to cope with environmental change and variability, are the ultimate 'win-win' climate mitigation option. There is no substitute for the C-sequestration capacity of natural forests, nor any practical way to reproduce the biodiversity of some of them (Myers *et al.* 2000) or to substitute for the ecosystem services they provide (Millennium Ecosystem Assessment 2003, Shvidenko *et al.* 2005). There is evidence suggesting that their functional composition is changing and that they are losing species at an alarming rate due to land use change (e.g. Sala *et al.* 2000, Brook *et al.* 2003, Gaston *et al.* 2003), and climate change (Parmesan and Yohe 2003, Root *et al.* 2003, Lenoir *et al.* 2008). In view of this, probably the best long-term C sequestration option would be to encourage scientific and policy efforts that preserve their integrity.

In those areas where afforestation and deforestation will not come at the cost of destroying natural ecosystems (e.g. in degraded, not recently deforested areas, or areas where the forest is unlikely to recover naturally, Appanah and Weinland 1992, Montagnini *et al.* 2005), our findings strongly suggest that built-in biodiversity considerations will not only increase their overall ecosystem-service value (Millennium Ecosystem Assessment 2003), but also specifically enhance their long-term C sequestering capacity. In order to make a difference for mitigating the effects of global warming, the size, longevity, and reliability of biological C stocks are more important considerations than sink rates. Consequently, preserving the integrity of natural systems, and building diverse systems with a careful consideration of the most suitable dominant and subdominant species and genotypes, is probably the most appropriate way forward. This is not free of technical difficulties, but its long-term cost–benefit ratio appears low when all economic, social, and environmental factors are considered.

In view of this, the lack of biodiversity considerations in the main body of the Kyoto Protocol is unfortunate to say the least. Particularly worrying is the fact that in the first commitment period of the CDMs only afforestation and reforestation are included, considering that more than half of the world's forested area is located in developing countries and that they are facing accelerating deforestation rates (Lambin *et al.* 2003, Shvidenko *et al.* 2005). In our view, in order to reverse this trend, biodiversity considerations should be incorporated into C sequestration initiatives. In this sense, the request of some developing countries to incorporate the protection of tropical forests into the second commitment period of the Kyoto Protocol (http://unfccc.int/resource/docs/2005/cop11/eng/misc01.pdf), and the new international interest in avoided deforestation with explicit mention to biodiversity (e.g. REDD) are signs that the tide might be turning towards a more positive direction.

Acknowledgements

This chapter greatly benefited from input by D. E. Bunker, O. Canziani, and N. Pérez-Harguindeguy, and from critical review by M. Loreau. It is a product of Núcleo DiverSus (endorsed by DIVERSITAS and the Global Land Project). It has also benefited from fruitful interactions between its authors and the participants in the DIVERSITAS ECOServices Meeting 'Biodiversity and Carbon Sequestration' (7–10 September 2005, Danum Valley Field Centre, Sabbah). SD was supported by FONCyT, CONICET, Universidad Nacional de Córdoba (Argentina), the J. S. Guggenheim Memorial Foundation and the Inter-American Institute for Global Change Research (CRN II 2015, supported by the US National Science Foundation Grant GEO-0452325) while carrying out research leading to this chapter.

CHAPTER 12

Restoring biodiversity and ecosystem function: will an integrated approach improve results?

Justin Wright, Amy Symstad, James M. Bullock, Katharina Engelhardt, Louise Jackson, and Emily Bernhardt

12.1 Introduction

Twenty years ago, Bradshaw (1987) stated that ecological restoration should be an acid test of ecological understanding. In fact, the practice of restoration has developed more through trial and error than by the application of any scientific framework. Since Bradshaw's statement, restoration ecology has undergone a rapid increase in conceptual development and basic research, as indexed by the rising number of peer-reviewed publications (Young et al. 2005), the creation of the journal *Restoration Ecology* in 1993, and the recent publication of a number of edited volumes dedicated to exploring the conceptual underpinning of restoration ecology (Falk et al. 2006, Van Andel and Aronson 2006). In addition, meta-analyses of restoration studies (e.g. Pywell et al. 2003) are beginning to draw out some general patterns and relate them to broader ecological theory. In this chapter we contribute to this development by exploring the applicability of the biodiversity-ecosystem functioning (BEF) framework to restoration science.

Restoration ecology is the subdiscipline of ecology that informs the 'intentional activity that initiates or accelerates the recovery of an ecosystem with respect to its health, integrity and sustainability'(S.E.R. 2004). Like the broader field of ecology, restoration ecology is an integrative discipline, having drawn important influences from fields as diverse as applied sciences such as agronomy and engineering (Mitsch 1993); social sciences such as sociology (Geist and Galatowitsch 1999) and landscape architecture (Fabos 2004); and earth sciences such as soil science (Bradshaw 1997) and hydrology (Morris 1995); as well as the subfields of population (Rosenzweig 1987) and landscape ecology (Van Diggelen 2006). This diversity of influences has led to many different approaches and goals for restoration projects. However, given the primary focus on restoring the structure and function of ecosystems, the strongest conceptual basis for most of restoration ecology stems from community and ecosystem ecology (Ehrenfeld and Toth 1997, Palmer et al. 1997, Young 2000, Falk et al. 2006). Simply put, restoration ecology is typically interested in restoring either biodiversity, ecosystem functioning, or both.

Despite these overlapping areas of interest, little crossover is evident between restoration ecology and 'classical' biodiversity-ecosystem function research (Naeem 2006a). With few exceptions (Bullock et al. 2001, Callaway et al. 2003, Bullock et al. 2007), BEF experiments have not taken place in restoration settings. Although BEF research might inform restoration (Aronson and Van Andel 2005, Young et al. 2005, Naeem 2006a), thus far few concrete suggestions have been offered on how restoration ecology might benefit from a consideration of BEF research and theory or vice versa. In this chapter, we start by comparing community and ecosystem approaches to restoration and suggest how a BEF approach might differ from these. We then draw on BEF theory and empirical research to suggest three broad areas where understanding the links between biodiversity and ecosystem functioning could have significant impacts on the success of restoration: 'classical' BEF impacts (i.e. higher diversity leads to improved functioning); the effects of biodiversity on the stability of ecosystem functioning; and the effects of biodiversity on the provisioning of multiple ecosystem services.

For this chapter, we constrain the scope of what we consider to be restoration to management activities whose primary goal is to improve ecosystem services other than provisioning services, although improved provisioning services may result from the activities. Management activities whose primary goal is to improve provisioning services are considered in Chapter 13, which formally examines the role of BEF research in managed ecosystems.

12.2 Community, ecosystem, and BEF approaches to restoration

BEF research combines elements from community and ecosystem ecology. Consequently, Naeem (2006a) contrasted community and ecosystem approaches to restoration with an approach based on the BEF perspective. A community approach largely focuses on the restoration of the biotic components of an ecosystem – which species are present, their relative abundance, and their interactions (trophic, competitive, facilitative). It is often used when starting from essentially bare ground, as in tallgrass prairie or hay meadow plantings in former agricultural fields, and when the goal of restoration is to enhance the conservation value of a protected landscape – restoration for biodiversity's sake. This approach has evolved over time from assuming that restoration is essentially the speeding up of a linear approach to a specific, predictable, equilibrium state, to accepting the dynamic nature of communities, possible alternative stable states (Bullock *et al.* 2002, Hobbs 2006), and the influences of disturbance and dispersal limitation on diversity (Pywell *et al.* 2003, Walker *et al.* 2004).

This evolution may help overcome some of the problems encountered when using the community approach. For example, restoring the target community is often more difficult than expected, even in systems where community restoration has been practised for decades (Kindscher and Tieszen 1998). Furthermore, restoration of one part of a community has not always yielded restoration of the whole community. In both California grasslands and English meadows, plant composition and relative abundance of species have been restored to resemble the reference condition, but the soil microbial community has remained significantly different, implying that functions such as decomposition and nutrient turnover may not have been restored (Smith *et al.* 2003, Potthoff *et al.* 2005, Steenwerth *et al.* 2006, Pywell *et al.* 2007). Although greater understanding of factors influencing community assembly may overcome some of these problems, it may not overcome the common assumption of the community approach that functioning will be restored if the community is restored. This was explicitly studied in two constructed *Spartina* marshes. In both, vegetation composition, cover, and biomass of planted sites were similar to those of natural marshes within 18 months of planting, but in one case, the height structure failed to meet the needs of the endangered bird for which the marshes were constructed (Zedler 1993), and in another, soil organic matter and nutrient accumulation, denitrification rates, and tidal export of nutrients required much longer to resemble reference levels (Craft *et al.* 1999).

The ecosystem approach makes the opposite assumption, in that a habitat template that restores ecosystem processes (e.g. hydrology of a wetland) is created, but most or all species are left to colonize on their own in a process of self-assembly (e.g. Bradshaw 2000). The approach, which is characteristic of many early restoration efforts (Bradshaw and Chadwick 1980), may more accurately be described as 'rehabilitation' rather than 'restoration' (S.E.R. 2004). Some examples of this approach include constructing mechanical barriers in eroded gullies in an overgrazed rangeland in order to slow surface water flow and enhance water percolation into the soil (King and Hobbs 2006) and increasing meandering of a stream in order to reduce flash flooding and increase habitat complexity and sediment and nutrient retention (Rosgen 1994). Other activities using the ecosystem approach include ecological engineering and reclamation. Ecological engineering strives to achieve and maintain a specific ecosystem service within a relatively strict range, such as reducing nitrogen in wastewater to regulator-accepted standards via a constructed wetland. In contrast, reclamation strives to make a highly altered system serve some useful purpose by achieving an acceptable level of safety and aesthetics, such as stabilizing mine tailings with vegetation that can tolerate the harsh conditions but are not necessarily native to the site (S.E.R. 2004). Both use ecological processes and biotic components to achieve their goals, but these goals do not include creating a system whose functioning or composition resemble a reference ecosystem.

Although economically attractive, the 'Field of Dreams' approach of solely manipulating the physical environment ('If you build it, they will come'; Hilderbrand et al. 2005) is risky because dispersal barriers limit the colonization of desired biotic components and, even if these can be overcome, interactions with species not yet present (e.g. pollinators or mycorrhizae) may be necessary for establishment or reproduction of important species. Furthermore, initial seeding with native or exotic species that may grow quickly and provide good cover can prevent establishment of desired later successional species. For example, in old field sites in England, sowing of a few native grasses limited successful colonization by desired species (Pywell et al. 2002). In addition, projects using the ecosystem approach with no attention to the composition of the biotic component may achieve their physio-chemical goals, but the narrow focus of these goals may lead to long-term problems. For example, some of today's most troublesome invasive species were introduced as a rehabilitation measure in over-grazed, drought-stressed rangelands (Christian and Wilson 1999, Clarke et al. 2005, Schussman et al. 2006, Williams and Crone 2006). Finally, the approach does not take advantage of the multiple functionality and resilience potentially provided by systems with an actively restored biotic community.

In contrast, a restoration approach based on BEF theory and empirical results stresses the relationship between the biotic community and ecosystem functioning. Although the BEF perspective does not encompass all of the topics relevant to ecological restoration, it does cover the majority of a more general framework recently proposed for restoration ecology (King and Hobbs 2006). A BEF approach to restoration is based on the asymptotic relationship between biodiversity and ecosystem functioning. This relationship is what sets the BEF approach to restoration apart from the other two approaches. Restoration strives to restore an ecosystem to that relationship by removing anthropogenic inputs that maintain high functioning but low diversity in managed systems (e.g. fertilizer or pesticides) or by enhancing functioning in degraded systems by adding key sets of species (Naeem 2006a). Debate about the applicability of BEF theory outside of experimental settings (Huston 1997,

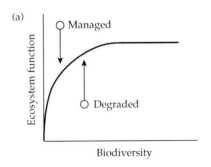 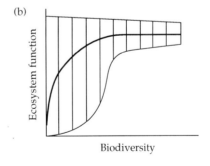

Figure 12.1 The BEF perspective for ecological restoration. (a) The fundamental assumption of the BEF perspective is that biodiversity affects ecosystem function, often in the manner depicted here, where adding diversity when diversity is low has a stronger effect on ecosystem functioning than adding diversity when diversity is high, until the point at which increasing diversity has no effect on functioning. Restoration (vertical arrows), in this perspective, is the restoration of this relationship in degraded systems, which have lower functioning than expected, or in managed systems, in which anthropogenic inputs produce functioning at a higher level than expected given the system's diversity. (b) The theoretical realm of possible ecosystem functioning in relation to biodiversity. The hatched area incorporates the assumption, supported by empirical evidence, that it is possible to have greater functioning with a low number of species than with the most diverse system, as well as the assumption that the range of variability in functioning decreases as biodiversity increases. The latter assumption has less support from empirical evidence (see text). Figure is adapted from Naeem (2006).

> **Box 12.1 Should the BEF restoration approach apply to all ecosystems?**
>
> Ecological restoration, like most things ecological, is inherently and frustratingly context-dependent. Restoration is being carried out in habitats as diverse as streams, rivers, forests, wetlands, grasslands, estuaries, deserts, and alpine habitats to achieve an equally diverse suite of goals. Given the diversity of contexts in which restoration occurs, it is appropriate to ask whether the BEF approach is relevant to all forms of restoration. We see great potential for the BEF perspective to improve the practice of ecological restoration in many cases. However, it is important to recognize that in systems where abiotic structure serves as the dominant control of ecosystem properties, the BEF approach may not be as relevant to successful restoration.
>
> Streams and rivers are an instructive case study of restoration in ecosystems that are strongly controlled by abiotic forces. River restoration is one of the most extensive forms of restoration in the USA, with at least $1 billion dollars invested annually (Bernhardt et al. 2005). A growing body of scientific literature has documented biodiversity effects on ecosystem function in stream ecosystems (e.g. Jonsson and Malmqvist 2000, Cardinale et al. 2002). Yet because the dominant taxa in river ecosystems are easily dispersed, short-lived, and small it is difficult to apply these BEF perspectives to river restoration projects. Seeding streams with the appropriate algae, macroinvertebrates, and fish is no guarantee that those organisms will establish. In terrestrial ecosystems, where vegetation itself provides much of the physical structure of the environment on which other organisms depend, it is easy to see how planting diverse native species assemblages can effectively 'restore' not only an ecosystem function (= productivity) but also key components of ecosystem structure (e.g. canopy architecture). This is not the case in rivers, where the physical template is primarily controlled by hydrology and geomorphology.
>
> However, we would argue that a BEF perspective can and should inform river restoration in two important respects. First, terrestrial vegetation can play very important roles in river ecosystems, and thus BEF approaches can directly inform riparian management aspects of river restoration. The majority of small stream ecosystems are primarily fueled by leaf litter inputs (Fisher and Likens 1973), and higher diversity litter inputs can lead to greater secondary production (Swan and Palmer 2006). Riparian trees themselves contribute large wood to stream channels which can play important roles as both habitat and biogeochemical hotspots in rivers (Wallace et al. 1995, Valett et al. 2002, Wright and Flecker 2004, Warren and Kraft 2006).
>
> Second, a BEF perspective should inform the evaluation of river restoration projects. Even when aquatic organisms cannot be directly introduced, monitoring changes in community composition following restoration activities can provide important insights into what is and what is not working. For example, the absence of shredding functional feeding groups from restored streams relative to reference conditions may indicate that organic matter dynamics have not been effectively restored, the presence of nitrogen-fixing blue–green algae may indicate that phosphorus loads are excessively high, or the absence of a diverse hyporheic meiofauna may suggest that groundwater and surface waters have not been effectively reconnected. While restoration practitioners may not be able to actively manage the diversity of all ecosystems to affect services, recognizing the links between biodiversity and ecosystem functioning that exist even in ecosystems strongly controlled by abiotic forces can lead to improved assessment of the success or failure of restoration projects.

Wardle 1999, Naeem 2000, Grace et al. 2007) has highlighted that restoring a highly diverse community does not necessarily guarantee a high level of functioning – environmental factors such as soil fertility and climate are also crucial determinants of ecosystem functioning (Huston and Mcbride 2002, Naeem 2002b). However, when environmental factors are held constant, the BEF approach suggests that greater biodiversity provides a high level of ecosystem functioning, although achieving the highest level of functioning does not require the restoration of the entire community (Lehman and Tilman 2000, Naeem 2006a). The rest of this chapter explores these implications for restoration in greater detail.

12.3 'Classical' BEF implications for restoration

Since its earliest inception BEF research has largely been focused on testing the hypothesis that the loss of diversity of species (or functional groups) leads to changes in ecosystem functions such as productivity

> **Box 12.2 Diversity of grassland plantings**
>
> Grassland restoration in the central portion of the USA illustrates the relatively low level of diversity used in some restorations compared to their reference systems and the implications.
>
> - Native northern mixed-grass prairie in western Nebraska and South Dakota has approximately 37–80 native plant species per 0.1 ha (Symstad *et al.* 2006), whereas the recommended seed mixtures for native rangeland plantings (> 0.1 ha) in this region have a maximum of 13–25 species and a minimum of four species (http://www.nrcs.usda.gov/technical/efotg/).
> - The federal Conservation Reserve Program (CRP) pays farmers to plant or maintain perennial vegetative cover, grassland being one allowable type, in areas that would otherwise be used for agricultural production. The program, which affects more than 10 million ha of grassland nationwide, rewards plantings that provide ecosystem services such as reduced soil erosion, increased wildlife habitat, water quality protection, soil salinity reduction, and carbon sequestration (Barbarika 2005). The diversity of these plantings is difficult to track, but typical values are 4–10 species.
> - The species planted in restorations like these are often dominants (e.g. warm-season bunchgrasses) that drive major aspects of ecosystem functioning (Camill *et al.* 2004), but low functional diversity, particularly the lack of nitrogen-fixing legumes in some plantings, may limit a restoration's functioning potential (Kindscher and Tieszen 1998).

or nutrient cycling (Naeem *et al.* 1994, Hooper and Vitousek 1997, Tilman *et al.* 1997b). Since these early studies, the field of BEF research has flourished, with over 100 published experiments testing this general hypothesis (Cardinale *et al.* 2006a). Contentious debates have ensued about the proper experimental design or how best to interpret the results of these experiments (Garnier *et al.* 1997, Huston 1997, Wardle *et al.* 1997a, Thompson *et al.* 2005, Wright *et al.* 2006), and considerable work still needs to be done to determine the mechanisms that might link diversity to ecosystem functioning. Despite these uncertainties, several recent meta-analyses have demonstrated that, on average, ecosystem functioning does increase with increasing numbers of species in BEF experiments (Balvanera *et al.* 2006, Cardinale *et al.* 2006a). For example, Cardinale and colleagues (2006a) found that diversity enhanced both plant productivity and nutrient uptake. These biodiversity effects can be tied to two ecosystem services that are often the focus of restoration efforts. It should be noted that while the most represented ecosystem in this meta-analysis were grasslands, these comprised only 34% of the studies, with the rest coming from a broad diversity of other terrestrial and aquatic ecosystems. Thus there is a growing body of work addressing BEF questions in other ecosystems such as streams (Jonsson and Malmqvist 2000, Cardinale *et al.* 2002), wetlands (Engelhardt and Ritchie 2001, Callaway *et al.* 2003, Sutton-Grier *et al.* In Review), forests (Bunker *et al.* 2005, Scherer-Lorenzen *et al.* 2005a), and marine systems (Duffy *et al.* 2001, Solan *et al.* 2004, France and Duffy 2006b, Worm *et al.* 2006). For most systems in which restoration is being conducted, potentially relevant BEF experiments have been conducted. Applying the findings from BEF research to restoration practices should be a natural extension of existing research. Indeed, a study that examined the consequences of restoring plant communities on ecosystem functioning in a California estuarine salt marsh demonstrated that increasing plant richness led to higher rates of nitrogen uptake and greater above- and belowground biomass (Callaway *et al.* 2003). Grassland and wetland restorations typically start from bare ground and try to recreate natural communities through the addition of seed. The seed mixes used are not particularly diverse relative to the natural ecosystems that serve as restoration targets (Box 12.2). Thus, many terrestrial restoration activities are operating in the region of diversity where varying species richness is most likely to have a significant effect on ecosystem functioning, since most ecosystem functions saturate at relatively low levels of species richness (Cardinale *et al.* 2006a).

However, the application of classical BEF research to restoration activities may still be limited given our current knowledge. First, while average ecosystem function has been shown to increase with increasing diversity (Balvanera *et al.* 2006, Cardinale *et al.* 2006a), meta-analysis of BEF experiments has also demonstrated that the highest performing species in monoculture produces a level of ecosystem function that cannot be distinguished from the level observed in the highest diversity treatment (Cardinale *et al.* 2006a). Second, due to the constraints of experimental design, BEF experiments do not always include low-diversity polycultures capable of outperforming the highest diversity treatment (Chapter 2). Thus, if the goal of restoration is to provide a maximum level of ecosystem services, arguments could be made for establishing a high-performing species to rapidly achieve a high level of functioning early in the restoration process. However, because these high-performers are usually dominant species, subordinate species may be more difficult to establish after the fact, as observed in tallgrass prairie plantings (Weber 1999), hay meadows (Pywell *et al.* 2002) and coastal marshes (Keer and Zedler 2002). This is particularly important because there is some evidence that the effects of greater diversity on functioning take some time to develop (Tilman *et al.* 2001, Cardinale *et al.* 2007), and because these subordinate species may contribute to stability of functioning.

12.4 BEF and stability of services in restoration

The question of whether biodiversity contributes to stability (e.g. resistance to disturbance, resilience after disturbance, and moderate range of variability through time) has been a topic in ecology for more than half a century (e.g. Macarthur 1955, May 1972). The recent development of the BEF perspective has provided resolution to some aspects of this question by suggesting that species that share functional effects traits (characteristics that affect ecosystem functioning in a specific way) often differ in their functional response traits (characteristics that determine how they respond to a specific perturbation). As a result, a relatively low number of species may provide a certain level of an ecosystem function in a constant environment, but if these species are adversely affected by a perturbation (e.g. drought, flood, fire, or herbivory), that level of functioning will only be maintained if species with a similar effect on functioning respond positively to this perturbation. For example, Eviner *et al.* (2006) identified a strong seasonality in the effects of individual species on N and P cycling in northern California grasslands. Their results suggest that when in mixture, the species' contributions to ecosystem functions would vary throughout the season, providing a mechanism for maintaining these nutrient cycles throughout the season. In addition, they investigated the relationship between a variety of plant traits (live tissue and litter chemistry and biomass, modification of bioavailable C and soil microclimate) and the ecosystem functions. The influence of individual traits on N mineralization and nitrification also varied throughout the growing season, illustrating how response traits (to seasonal changes in moisture and temperature) are not necessarily correlated with functional effect traits (Landsberg 1999, Lavorel and Garnier 2001, Hooper *et al.* 2002, Naeem and Wright 2003).

Several reviews describe this and other mechanisms in greater detail and show the strong theoretical support for the diversity–stability hypothesis (McCann 2000, Cottingham *et al.* 2001, Loreau *et al.* 2002, Hooper *et al.* 2005, Chapter 6), and empirical support for the hypothesis exists from experiments in systems relevant to ecological restoration. In plots in which plant species richness varied because of nitrogen manipulations, aboveground biomass was more resistant to, and recovered more fully from, a major drought in more diverse grassland plots in central Minnesota, USA (Tilman and Downing 1994). In the same system, but in plots in which plant species richness was directly manipulated, temporal stability (measured as temporal mean/standard deviation) of aboveground plant biomass over ten years increased linearly, from approximately 3.5 to 5.8, as planted species richness increased from one to 16 species, with the diverse plots having lower temporal standard deviations for a given mean biomass than the monocultures (Tilman *et al.* 2006b). Despite these encouraging examples, there are also counter-examples. More diverse plots had lower resistance of primary productivity to drought in a Swiss grassland experiment (Pfisterer and Schmid 2002); individual species, rather than species richness, affected biomass stability in constructed or natural wetlands (Rejmankova *et al.* 1999, Engelhardt and Kadlec 2001); and rocky shore intertidal communities with the greatest diversity were most severely affected by heat stress (i.e. had low resistance) but were the quickest to recover from the stress (i.e. high resilience) (Allison 2004). A recent meta-analysis of a large number of studies confirmed this inconsistency – averaged across experiments, more diverse systems were more resistant to nutrient perturbations or invasions, but diversity had either a neutral or negative effect on resistance to drought, response to warming, and variation in response to long-term environmental variability (Balvanera et al. 2006).

Maintaining ecosystem services within a reasonable range of variability is an important component of restoration projects focused on restoring functioning. The balance of evidence is currently not very strong that biodiversity plays a large role in determining this variability. However, relatively few field-scale studies have investigated this question, and there is no evidence that diversity strongly negatively affects stability, particularly over the long term, so the cost of increasing the diversity of a restoration is likely only the direct cost associated with adding that diversity.

12.5 Biodiversity and the restoration of multiple ecosystem functions

Restoration activities are not typically conducted with the goal of restoring a single ecosystem service. Rather, there is an implicit understanding that 'healthy' ecosystems provide a large number of services (Duraiappah and Naeem 2005), and that restoration can serve to increase multiple ecosystem services (N.R.C 2001, Bernhardt et al. 2005). For example, the restoration of Iraq's Mesopotamian marshes has been assessed based on five separate ecosystem functions: productivity of the dominant plant *Phragmites australis*, redox status, hydrologic function, salinity, and bird diversity (Richardson and Hussain 2006). Similarly, the US National Resource Council suggested that five major functions of wetlands need to be considered in the construction or restoration of wetlands: hydrologic functions, water quality functions, support of vegetation, support of fauna, and soil functions (N.R.C 2001).

This demand for multiple ecosystem services from a single restoration may be the strongest argument for incorporating greater biodiversity into ecological restorations. As was discussed above, high diversity plots in BEF experiments do not, on average, significantly outperform the highest performing monoculture (Cardinale et al. 2006a). However, what is not clear from this analysis is whether a single species can maximize the provisioning of multiple ecosystem functions. Evidence from a biodiversity, carbon dioxide, and nitrogen manipulation experiment in a Minnesota grassland context (Reich et al. 2001, Reich et al. 2004) suggests that this is not the case. In this experiment, the total above- and belowground biomass produced by a species and the amount of inorganic nitrate left in the soils after plant uptake were strongly correlated in the first year after planting (Fig. 12.2(a)). Such a result would suggest that a single species might be capable of both fixing high amounts of carbon and improving water quality by reducing nitrogen export. However, in subsequent years of the experiment, this correlation broke down (Fig. 12.2 (b,c)), making it more difficult to suggest a single species maximizes both ecosystem services. In fact, Hector and Bagchi (2007) found that maximizing seven ecosystem functions required between 8 and 16 species at eight grassland sites across Europe. In a recent review, Gamfeldt et al. (2008) found that multifunctional redundancy (i.e. the degree to which multiple species could sustain multiple functions) was generally lower than single-function redundancy (i.e. the degree to which multiple species could sustain a single function).

Species effects on ecosystem functions are functions of key morphological and ecophysiological traits (Engelhardt and Kadlec 2001, Eviner and Chapin 2003, Diaz et al. 2004), and BEF research is increasingly focused on how particular traits interact to determine the effects of diversity on ecosystem functioning (Eviner and Chapin 2003, Naeem and Wright 2003, Solan et al. 2004, Bunker et al. 2005). While attempts to determine which traits are most important in regulating particular ecosystem functions are still in the early stages, it seems reasonable to assume that different processes might be affected by different combinations of traits. Given this assumption, the extent to which single species can maximize multiple functions will depend on the extent to which the traits responsible for regulating the ecosystem functions of interest are

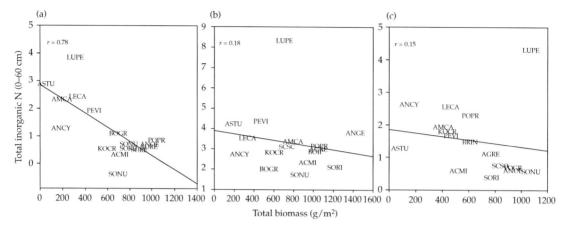

Figure 12.2 Relationship between two ecosystem functions, total above- and belowground biomass (a surrogate for carbon storage) and total extractable soil inorganic nitrogen (a surrogate for nitrogen removal from groundwater) from monocultures of different species in control plots (ambient levels of atmospheric CO_2 and nitrogen) of the BioCON BEF experiment from (a) 1999, (b) 2000, and (c) 2001. Species codes of the monocultures are: *Achillea millefolium* ACMI; *Agropyron repens* AGRE; *Amorpha canescens* AMCA; *Andropogon gerardi* ANGE; *Anemone cylindrica* ANCY; *Asclepias tuberosa* ASTU; *Bouteloua gracilis* BOGR; *Bromus inermis* BRIN; *Koeleria cristata* KOCR; *Lespedeza capitata* LECA; *Lupinus perennis* LUPE; *Petalostemum villosum* PEVI; *Poa pratensis* POPR; *Schizachyrium scoparium* SCSC; *Solidago rigida* SORI; *Sorghastrum nutans* SONU. Note that the scale changes on the axes between years.

correlated. Several recent large-scale analyses have found significant correlations between several important traits in plants (Diaz *et al.* 2004, Wright *et al.* 2004, Reich *et al.* 2006) and animals (Brown *et al.* 2004). These correlations yield suites of traits (e.g. those contributing to rapid acquisition of resources vs. those that contribute to conservation of resources in well-protected tissues in plants) that typically occur together in organisms. The uniformity of these suites across broad taxonomic groups and geographical gradients argue for the existence of fundamental tradeoffs in organismal development and life history. Whether or not these tradeoffs hold up within local species pools, where selection might push for diversification along trait axes (Grime 2006, Ackerly and Cornwell 2007), and for the traits actually responsible for ecosystem functions of interest in restoration (Eviner 2004), remains an open and important question.

12.6 The economics of BEF in restoration

For BEF research to be useful for ecological restoration, ecosystem functions must be related to the ecosystem services desired as the outcome of restoration. A key issue in performing this translation is that BEF research often does not assign a specific worth to a level of ecosystem functioning other than a vague 'more is better' for primary productivity or nutrient capture (Srivastava and Vellend 2005). In contrast, in restoration, the relative worth of various levels of ecosystem services must be assessed and agreed to by many stakeholders when any but the simplest restoration project is commenced (Fig. 12.3). Provisioning services, such as forage production, can be easily converted into currency value (Bullock *et al.* 2007). However, while Worm *et al.* (2006) showed dramatic increases in tourism-related revenue following the closure of fisheries in marine protected areas, which they attribute to increases in diversity, this conversion is not as easy for other types of ecosystem services resulting from ecological restorations. For example, results from BEF research might predict grams of carbon fixed per square metre, grams of nitrogen removed through denitrification, and grams of the greenhouse gasses N_2O and methane produced by different combinations of species used in a wetland restoration project. Until markets exist that allow translation of these ecosystem processes into a common currency, determining which mixture of species maximizes benefits while minimizing costs is, at best, just a guess. The growth of carbon trading markets (Bonnie *et al.* 2002) and early

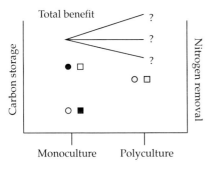

Figure 12.3 Hypothetical restoration scenario involving two ecosystem services: carbon storage (circles) and nitrogen removal (squares). Managers could choose between planting a monoculture of species 1 (filled symbols) that provides high levels of carbon storage but low levels of nitrogen removal, a monoculture of species 2 (open symbols) that provides low levels of carbon storage but high levels of nitrogen removal, or a mixture of both species that provides intermediate levels of both carbon storage and nitrogen removal. The total benefit of the ecosystem services provided by these different management choices (straight lines) is unknown and will depend both on the shape of the relationship between diversity and these two functions and on the relative economic weight placed on the two services.

attempts at nitrogen trading schemes such as the EPA's Water Quality Trading Program (active in areas of 10 states in the USA) suggests that economic valuations of at least some of the non-provisioning ecosystem services provided by restored ecosystems are being developed. An even more difficult task is deriving a common currency for services such as protecting higher trophic levels or maintaining hydrologic services, although tools for evaluating different options in the face of such uncertainties do exist (Lynam et al. 2007).

12.7 Recommendations

We argue that understanding the relationship between biodiversity and ecosystem functioning is important in enhancing restoration success in many ecosystems and that restoration activities can serve as powerful tools for exploring some of the central BEF questions. However, information exchange between the scientific community and restoration practitioners is too little and too slow for BEF research to be relevant in adaptive management of degraded ecosystems. Ecologists are recognizing that performing policy-relevant science involves more than simply publishing in high-profile academic journals (Palmer et al. 2004). This point is reinforced by a recent survey of stream restoration practitioners that showed that less than 1 per cent of over 300 restoration projects had specifically been informed by results published in scientific journals (Bernhardt et al. 2007). So what can be done to improve the situation?

For scientists, we make a few suggestions. First, continue basic BEF research, but keep in mind the information needed by restoration practitioners. Deeper understanding of how functional traits acting alone and in combination affect ecosystem functions that are related to ecosystem services is particularly important. Although general relationships between diversity and functioning explain why the restoration of biodiversity is important for restoring ecosystem functioning, restoration practitioners ultimately need to know which specific species combinations to restore to have confidence in the outcome of restoration projects. However, BEF theory, while explanatory, is not yet predictive (Hooper et al. 2005) and is therefore not yet ready to be embraced by the management community. Furthermore, experimental perturbations in BEF experiments, and monitoring of experiments over long time periods in which the environment fluctuates naturally, will help resolve what role biodiversity plays in stabilizing ecosystem services. However, because species and communities respond to environmental fluctuations in seemingly idiosyncratic ways, investigations that tie these two themes together, by seeking patterns in traits that determine species' responses to environmental variations, will yield the most relevant information for restoration.

BEF research focusing on microbial diversity is particularly crucial because many of the most critical ecosystem services are underpinned by microbial processes (e.g. the nutrient transformations that improve water quality). Our understanding of both how plant and animal diversity affect microbial community structure and how microbial diversity directly affects ecosystem functioning in real systems is relatively weak (Fierer et al. 2007, Jackson et al. 2007). To date there has been little focus on the importance of restoring microbial communities or how to go about doing so (Hasselwandter 1997). Restoration of the microbial community in meadows was not achieved by a simple soil microbe addition technique (Pywell *et al.* 2007), but experiments have shown that microbial community composition is related to both the diversity of the restored plant communities and to particular 'facilitating' plant species (Smith *et al.* 2003, Bardgett *et al.* 2006). A deeper understanding of the controls and consequences of microbial diversity is an area where BEF research is well-poised to make important contributions to restoration ecology.

Another aspect of diversity that may be critical to restoration success, but has not been as extensively studied in BEF research, is the importance of genetic diversity. Many ecosystems, including many that are important targets for restoration, are dominated by a single species that controls ecosystem function, e.g. giant kelp in kelp forests, Ponderosa pine in many forests of the Western US, seagrasses such as *Zostera marina* in shallow estuarine systems, and *Spartina* in intertidal zones. A growing body of work is showing that genetic diversity within a species can be an important regulator of ecosystem function (Hughes *et al.* 2008). Work on *Zostera*, an important species in estuarine restoration, showed that clonal diversity measurably affects ecosystem processes or the stability of those processes (Hughes and Stachowicz 2004, Reusch *et al.* 2005). Williams (2001) demonstrated that reduced genetic variation in *Zostera* used in restoration efforts resulted in decreased population growth and individual fitness. For both cottonwood (*Populus fremontii* x *augustifolia*) (Schweitzer *et al.* 2005a) and apsen (*Populus tremuloides*) (Madritch *et al.* 2006), genotypic richness can affect decomposition rates and nutrient cycling. Clonal diversity of *Solidago altissima* were shown to affect, not only primary productivity, but the diversity of higher trophic levels (Crutsinger *et al.* 2006). Given these compelling examples, research on the consequences of genetic diversity of species commonly used in restoration is likely to yield benefits both to basic science and restoration.

Third, to make the results of any BEF research applicable to practitioners, the connection from biodiversity to function to services needs to be stronger. For example, although the positive effects of plant diversity on above- and belowground biomass production in grassland experiments have been vaguely related to forage production, carbon storage, and soil erosion, little attention has been paid to whether the nutrient content of the biomass is sufficient for the purported foragers (see Bullock *et al.* 2007 for an exception) and direct measurements of soil movement or long-term C sequestration are rare (though more common in wetland studies). Given the need of many restorations to restore multiple services, this connection needs to be made simultaneously for multiple functions and their related services. Addressing these issues is already one of the central thrusts of the next generation of BEF research (Naeem and Wright 2003) and should not require significant changes to how we proceed with our science beyond the functions and properties measured in typical BEF experiments.

Fourth, BEF researchers must evaluate the relevance of experiments to real-world conditions. There is some conflict between the typical BEF experiment, which is carefully designed to disentangle the effects of individual species, functional traits, and diversity *per se* on the functions measured, and restoration projects, which are concerned with achieving a desired result with available materials and limited financial resources. Restoration practitioners may look at the high density of expensive species (many forbs, for example) planted in some BEF experiments and question the applicability of the results to their work: does the diversity effect require this relatively high input of normally subordinate or rare species? They might also question the relatively controlled situations of the field experiments: would the results be the same if vertebrate herbivores were not excluded from the plots, or if colonizing species were not removed? Finally, a practitioner would never think of restoring a *Spartina alterniflora* marsh without *Spartina alterniflora*, but many BEF studies include treatments analogous to this situation. Of greater interest to a restoration practitioner would be the question of how strong the diversity effect is when the only portion of diversity manipulated is the subordinate and rare species. These subordinate species may be essential to certain key functions which define the success or failure of the restoration. For example, grassland restoration in the UK to meet national Biodiversity Action Plan targets requires the presence of food plants of certain declining butterflies and other insects, but these plants are often uncommon and particularly difficult to establish (Pywell *et al.* 2003). Thus, BEF researchers seeking to provide answers for restoration practitioners and other natural resource managers need to ensure that they design their experiments with these issues in mind.

Concurrently, restoration practitioners and ecologists can potentially contribute significantly to strengthen BEF research. For them, we stress the utmost importance of monitoring and reporting short- and long-term effects of various restoration projects and practices. In the US, the federal government pays private land owners millions of dollars each year to plant and maintain perennial grasslands through the Conservation Reserve Program. A landowner's proposal is more likely to be funded if s/he plants a higher diversity of species (U.S.D.A 2006), but follow-up on the establishment success and environmental benefits of individual plantings is rare, and only recently have comparisons among the ecosystem services provided by plantings of different diversity levels been explored. In the USA, a major source of funding for stream and wetland restoration is associated with mitigation of wetland losses under the Clean Water Act (N.R.C 2001). Guidelines for successful restoration vary from state to state, but typically involve recommended species lists and some assessment of total vegetation cover. Some states are currently developing improved vegetation assessments that include repeated measurements of the cover of all planted and unplanted species. In the case of North Carolina, this improved monitoring scheme has been developed in coordination with the Carolina Vegetation Survey to ensure that monitoring data can be directly incorporated into an existing vegetation database that is actively being used in ecological research (Lee *et al.* 2007). As a member of the European Union, the UK Government pays out many millions of pounds a year in funding restoration on farmland under the Environmental Stewardship scheme (http://www.defra.gov.uk/erdp/schemes/es/).

This scheme exemplifies the biodiversity or ecosystem service dichotomy. Certain activities aim to restore a service, such as sowing field margins with a mix of plants designed to provide pollen and nectar for bees and butterflies or to provide winter seed for birds. In these cases the plant mixture does not resemble any seen in (semi-) natural systems and often contains non-natives. Other activities are focused on biodiversity, such as the sowing of specific plant mixtures into bare arable land or species-poor grasslands to restore particular target species-rich grasslands. Certain authors have criticized the effectiveness of European agri-environment schemes (e.g. Kleijn *et al.* 2006). This is partly because these authors have confused the aims of different activities (e.g. expecting the bird-seed mixtures to restore rare arable weed communities), but also because monitoring of outcomes has been poor. The limited monitoring (e.g. Critchley *et al.* 2004, Feehan *et al.* 2005) of vegetation has shown some success, but suggests that better targeting and more precise methods are required. This is leading to planning for more extensive and repeated monitoring programs and consideration of how biodiversity and ecosystem service aims might be integrated (e.g. for soil conservation).

Clearly, BEF research still has many avenues to explore before the majority of questions surrounding the relationship between biodiversity and ecosystem functioning are answered. Just as clearly, restoration ecology still has a long way to go before the results of reconstructing an ecosystem can be as predictable as the construction of a bridge or even a space station. Although not all restoration projects are suited to answering basic science questions regarding the relationship between biodiversity and ecosystem functioning, they are crucial to making BEF science applicable to real-world situations. Direct partnerships between researchers testing BEF concepts and restoration practitioners are crucial for ecological restoration to live up to its potential as an acid test for BEF ecology.

CHAPTER 13

Managed ecosystems: biodiversity and ecosystem functions in landscapes modified by human use

Louise Jackson, Todd Rosenstock, Matthew Thomas, Justin Wright, and Amy Symstad

13.1 Introduction

There is growing agreement that the long-term sustainability of terrestrial landscapes that provide ecosystem services, such as food, fibre, and timber, will be increased by the conservation of existing biodiversity and the adoption of biodiversity-based practices (Collins and Qualset 1999, Kates and Parris 2003). The broad concept of sustainability is characterized by a set of complementary goals: optimizing production of food, fibre, and forest products while protecting the resource base and social wellbeing. Biodiversity is increasingly recognized as a key component for the sustainability of managed ecosystems for future generations. For example, in the Millennium Ecosystem Assessment (MEA 2005), biodiversity is viewed as an important coping strategy against agricultural risks in an uncertain future. However, some would argue that this strategy should be viewed as 'received wisdom' rather than substantiated proof of process (Wood and Lenné 2005). Indeed, while humans have always relied on biodiversity for provisioning services within managed ecosystems, understanding how biodiversity influences the ecological functions that affect these services, such as pollination, pest control, nutrient cycling, and water purification, is incomplete. Yet with further research, the provision of multiple ecosystem functions (multifunctionality) by different sets of species (Hector and Bagchi 2007), may become one of the most viable tools for managing ecosystems that produce food, fibre, and a range of other ecosystem services (Jordan et al. 2007).

Human utilization of the Earth's ecosystems is progressing at a rapid rate (MEA 2005). This has resulted in enormous loss of biodiversity, both in ecosystems that are managed for provisioning services and in neighbouring unmanaged ecosystems that are impacted by these management activities (Bawa et al. 2007). For example, the increase in cropland (250 per cent) and pasture (440 per cent) in the last 300 years has resulted in the loss of approximately 30 per cent of forests and 40 per cent of steppe worldwide (Lambin et al. 2003) (Fig. 13.1). At present, 10 per cent of the world's terrestrial land base is used for intensive, high-input agriculture, 15 per cent is associated with low-input agriculture, and 40 per cent is in agricultural mosaics used for other extractive purposes, such as grazing (Mooney et al. 2005, Wood et al. 2000). With the world's population of 6.7 billion people projected to grow to 9 billion by 2050, increasing demand for resources will continue to lead to major changes in land use and conversion of natural biomes to managed ecosystems. Irrigated and pasture lands are both expected to double in area by 2050, with a net loss of 10^9 hectares of wildlands worldwide (FAO 2003), thereby increasing the global pressure on biodiversity in natural ecosystems. To keep pace with population growth, it is expected that more land will be converted to agriculture and further intensification of already converted land will increase through greater reliance on non-renewable,

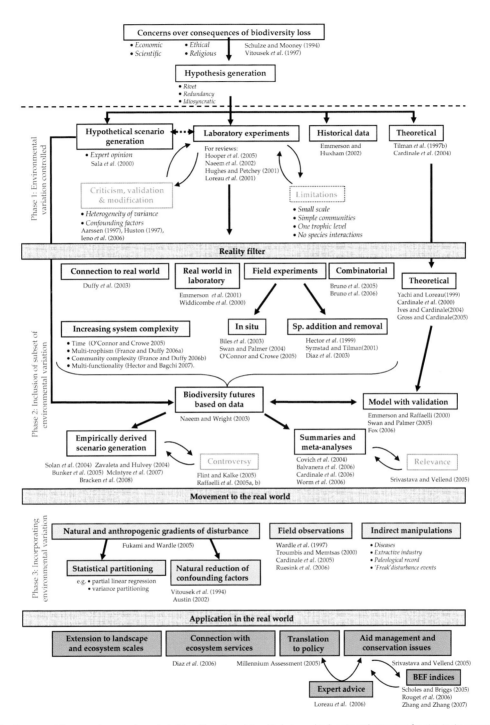

Plate 1. Summary of the research approaches adopted to address the relationship between biodiversity and ecosystem function in the peer-reviewed scientific literature. Modified from Godbold (2008). See page 32.

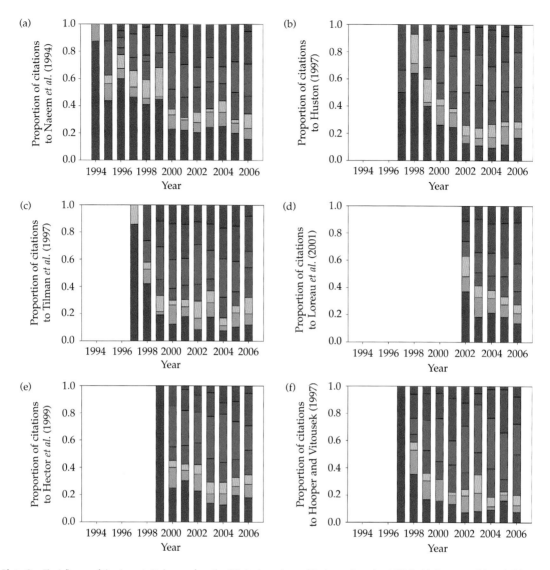

Plate 2. The influence of the six most cited papers from the BEF database since publication to December 2006 for (a) Naeem *et al.* (1994), (b) Huston (1997), (c) Tilman *et al.* (1997), (d) Loreau *et al.* (2001), (e) Hector *et al.* (1999) and (f) Hooper and Vitousek (1997). These publications influenced subsequent publications by contributing to discussion (red), development or reinforcement of theory (orange) or methodology (yellow), or by initiating or informing laboratory experiments (green), field manipulations (turquoise), or field observations (blue), or they were used to underpin practical applications in the real world (pink). Expanded from Benton *et al.* (2007). See page 38.

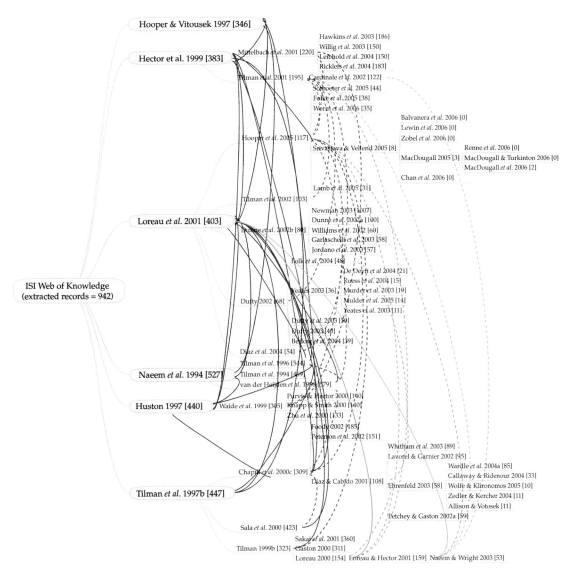

Plate 3. The rate, form, and connectivity of information flow within the BEF community and to multiple disciplines from the six most highly cited papers (=BEF-6) within the BEF database. For each generation of publications, the five most highly cited publications citing the previous generation were determined and linked, either directly (solid lines) or indirectly (dotted lines) to the BEF-6 via other highly cited publications. Line colour indicates the generation sequence (blue → red → green → orange). Publications not included in the BEF database are presented as a citation. The number of cites (from publication until December 2006) are indicated in square brackets. References listed are available in the electronic appendix. See page 39.

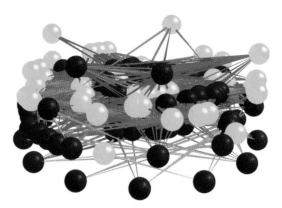

Plate 4. An example of the complexity of a real, yet still relatively simple natural food web in a salt marsh (from Lafferety et al. 2007). Note that within this system there are dozens of species (nodes) and hundreds of feeding links (lines connecting nodes) among plants, herbivores, predators and parasites that span six or more trophic levels. Figure reproduced with permission from K. Lafferty. See page 106.

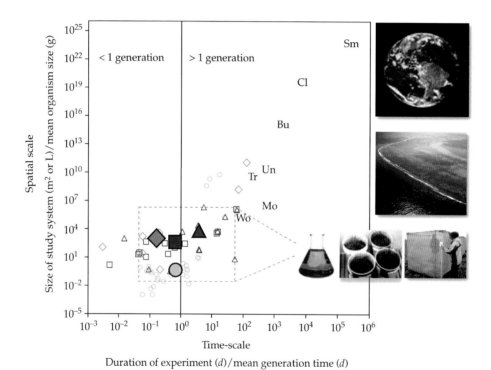

Plate 5. The spatial and temporal scale of biodiversity-ecosystem functioning experiments. The experimental duration (in days) and spatial scale (in m^2 or L) of experiments reviewed by Cardinale et al. (2006) were standardized to the mean generation time and body sizes of the focal organisms. Data were divided into four trophic groups: Plants = green circles, Herbivores = blue triangles, Predators = red squares, Detritivores = brown diamonds. The scale of each individual study is given by smaller symbols while the medians for each trophic group are shown as larger symbols. The box denoted by the dashed line gives the 10th and 90th percentiles for the scale of all experiments. For comparative purposes we show the scale of several natural extinctions: Wo = wolves from Yellowstone National Park, USA; Mo = Moa from New Zealand; Tr = Trout from Lake Superior, USA; Un = Unionid mussels from the lower Mississippi River, USA; Bu = Various species of butterflies in Europe; Cl = Loss of certain cladoceran zooplankton from Lake Superior, USA; Sm = Global eradication of the small pox virus. See page 116.

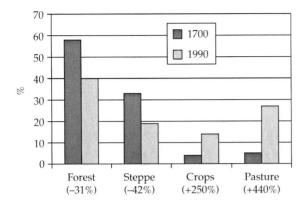

Figure 13.1 Global historical (1700) and current (1990) percentage of potentially arable land as a function of land use type, and the percentage change in different types of land use during this period (listed below each land use type on the x-axis). Steppe refers to natural steppe/savannah/grassland ecosystems, while pasture refers to ecosystems managed for livestock production. Based on data compiled by Lambin et al. (2003).

off-farm inputs unless more sustainable paradigms are developed (Clay 2004).

Incorporating biodiversity-ecosystem function (BEF) relationships in efforts to increase the sustainability of managed ecosystems requires combining biophysical and social sciences (Jackson et al. 2007, McNeely and Scherr 2003). Involving end users when determining the context and design of research is widely recognized as an approach that ensures research outcomes are relevant for decision makers. This is especially true for ecological processes and biodiversity assets that determine the wellbeing of resource-limited people, for whom poverty restricts their land management options. A landscape perspective provides better insights on management options than solely ecosystem- or plot-based research (Robertson and Swinton 2005, Tscharntke et al. 2005). These criteria are difficult for most ecological studies to achieve. How, then, do ecologists conduct meaningful research on biodiversity and its functions in managed ecosystems so that ecological processes are better understood and outcomes are directly conveyed to land users and other stakeholders?

This chapter addresses this central question considering three essential components. First, we examine the effects of management and intensification processes on biodiversity in managed ecosystems, as well as the possible impacts of measures aimed at mitigating negative effects. Second, we review what is known about the BEF relationships within managed ecosystems and highlight challenges for future research to deliver insights that embrace real-world complexity at temporal and spatial scales that are relevant to management. Finally, acknowledging the influence of social and economic factors in managed systems, we consider the need to place ecological insights within an interdisciplinary context to link BEF research to end users.

13.2 Management intensification as a driver of biodiversity loss

Management intensification is defined as the process by which managed systems increasingly rely on the use of non-renewable or purchased inputs that are indicative of higher energy and capital costs per unit land area (e.g. synthetic fertilizers, pesticides, and hybrid seeds). Here, substitution of mechanization and fossil fuels for human labour is considered to be part of intensification, although this might be viewed differently within a social science context. Intensification is a huge source of biodiversity loss within managed ecosystems and in adjacent wildland ecosystems (Lambin et al. 2003, Mooney et al. 2005, Wood et al. 2000). Agricultural intensification leads to a loss of diversity of crop varieties, animal breeds, and associated agroecosystem biodiversity (Brush 2004, Jackson et al. 2007) such as farm birds (Butler et al. 2007, Green et al. 2005), insects (Wilby and Thomas 2002), and soil biota (Loranger et al. 1998, Wardle et al. 1999). Additionally, intensification puts wild biodiversity at risk through gene flow from domesticated varieties to wild species (Thies and Devare 2007), exposure to potentially virulent pathogens such as avian flu, and adverse effects of agrochemicals on non-target species (McLaughlin and Mineau 1995).

Since the full set of ecosystem functions are rarely accounted for, understanding the range of impacts of intensification is difficult. Moreover, society has rarely mandated that the full social value of biodiversity conservation be assessed (Pascual and Perrings 2007).

13.2.1 Meta-analysis: biodiversity across landscape gradients

To better understand the effects of intensification, we conducted a meta-analysis of field studies across ecosystem types in agricultural landscapes. Studies were located using a keyword search in 'Web of Science' (Thomson Scientific), using the search phrases "biodiversity or species richness and (ecosystem function or ecosystem service or productivity or yield or pollination or nutrient cycling)", or "biomass and (agricultu* or agricultur* intensificat*)". We also followed up on other references cited within these studies. Studies were excluded if they used: inference of biodiversity across gradients from models; measurement of effects at a certain distance from a targeted ecosystem rather than within specific ecosystems; cursory descriptions of land use; or, data that were difficult to attribute to taxa or functional groups (e.g. restriction length fragment polymorphism for soil microbial studies).

Out of 50 papers examined, these search criteria yielded a total of 26 studies and 109 ecosystems that ranged from mature, climax forests to intensively managed cropland (Table 13.1). The taxa included birds, reptiles, amphibians, insects, soil microbes, and vascular plants, and spanned many biomes and continents. Studies contained two to nine ecosystems (mean = 4.2, sd = 2.1).

Classification of ecosystems along an intensification gradient used a multivariate clustering approach, based on three parameters described for each ecosystem within each study. These parameters were:

1) Land use: a descriptive classification of the land form and commodity into one of five categories (forest, grassland, perennial crop, annual crop, or fallow)
2) Successional level: a snapshot of the temporal (past, present, and potential future) successional stage into one of four categories (monoculture, polyculture, abandoned/fallow, or late successional vegetation)
3) Management: an integrative parameter subjectively assigned according to the degree of management intensity. This degree was based on a combination of fertilizer and pesticide practices, disturbance, and energy source (human labour vs. mechanization). Here we categorize mechanization as a form of intensified management due to the use of fossil fuel. Categories were high, medium, and low.

Clustering analyses used two different multivariate statistical methods: Partitioning Along Metroids (PAM) in R (R Development Core Team 2006) and Hierarchical Ward Clustering in JMP (SAS Institute 1995). We chose these two methods because of the slightly different algorithms with which they partition observations into clusters. These clustering techniques divided the data set into four clusters that were arranged along a gradient of agricultural intensification: (1) Forest, (2) Grassland/Pastures/Abandoned, (3) Complex Agroecosystems, and (4) Intensified Agroecosystems. The first two categories represent ecosystems dominated by natural vegetation and processes, while the two latter categories are both arable ecosystems, but differ in the management type and source of inputs. Complex Agroecosystems were managed with a greater emphasis on renewable inputs, rotations, reduced tillage, polycultures, or hedgerows, and the category contained a wide range of management types. Contrasting examples classified in this category are 'traditional polyculture coffee' (Armbrecht and Perfecto 2003, Armbrecht et al. 2005) and 'organic wheat' (Holzschuh et al. 2007).

PAM clustering resulted in 27 ecosystems in Forest, 62 in Grassland/Pasture/Abandoned, 44 in Complex Agroecosystem, and 43 in Intensified Agroecosystem, while Hierarchical Ward clustering resulted in 28, 62, 27, and 59, respectively. The two methods thus gave fairly similar results, although they varied in how systems were placed within the two agricultural categories. The PAM clusters were chosen for further analyses because of the robust nature of this methodology for smaller data sets and the high proportion of the variation (80 per cent) that was explained by the first two components. Each system's cluster designation was examined for accuracy. Three ecosystems were moved to a more

Table 13.1 Studies included in the meta-analysis, including ecosystem types, target species, the basis for assertions linking diversity to ecosystem functions, and the number of systems utilized in the meta-analysis.

Reference	Management category				Target species					Ecosystem Function[1]	Number of Systems[2]
	Forest	Pasture/Grassland/Fallow	Complex agroecosystem	Intensified agroecosystem	Birds	Reptiles/Amphibians	Insects	Vascular Plants	Soil Microbes		
Armbrecht and Perfecto (2003)	x						x			3	3
Armbrecht et al. (2005)	x			x			x			2	4
Baur et al. (2006)	x	x					x	x		2	4
Bullock et al. (2001)		x						x		3	2
Carney et al. (2004)	x	x		x					x	3	3
Debras et al. (2006)			x	x			x			1	5
Fedoroff et al. (2005)	x	x	x	x				x		2	6
Gabriel et al. (2006)			x	x				x		2	4
Genghini et al. (2006)			x	x	x					2	3
Gillison et al. (2004)			x	x				x		1	7
Glor et al. (2001)	x		x	x		x				1	9
Gordon et al. (2007)	x		x	x	x					3	5
Harvey et al. (2006)	x	x		x	x		x			2	6
Holzschuh et al. (2007)			x	x			x			3	2
Hutton and Giller (2003)		x					x			1	3
Joyce (2001)		x						x		1	3
Kremen et al. (2002)			x	x			x			3	3
Loranger et al. (1998)	x	x		x			x			1	6
Perner and Malt (2003)		x		x			x	x		1	7
Philpott et al. (2006)			x	x			x			3	2
Rundlof & Smith (2006)			x	x			x			2	4
Sall et al. (2006)									x	3	2
Snelder (2001)	x		x					x		3	2
Steenwerth et al. (2003)		x	x	x					x	1	9
Wardle et al. (1999)			x	x			x			1	2
Wilby et al. (2006)		x	x	x			x			1	3

[1] 1 = Cited, 2 = Not cited, 3 = Tested
[2] Not all systems or treatments in a published paper were included. When many were similar, the extreme examples were used. For studies with a temporal gradient, the last time point was used.

appropriate cluster when a unique set of circumstances was not captured in the simple criteria given above. Baur et al.'s (2006) extensive hay ecosystem was reassigned from Complex Agroecosystem to Grassland/Pastures/Abandoned due to use of minor management inputs. The *mogote* hilltop (Glor et al. 2001) was reassigned from Forest to Complex Agroecosystem since it was 'relatively undisturbed'. A hayfield (Steenwerth et al. 2003) that was classified as Grassland/Pasture/Abandoned was reassigned to Complex Agroecosystem because it had recently been tilled and seeded (Table 13.1).

To determine the effect size statistic (ES_s) of biodiversity across this agricultural intensification gradient, the mean species richness for each ecosystem, s, within each study, i, was extracted (Osenberg et al. 1999). From these means, a proportional effect size statistic for each ecosystem within the study, i, was calculated as the log of the ratio of mean species richness for the ecosystem, $\bar{t}_{is}$, divided by the species richness of the ecosystem with the absolute maximum species richness, m_i:

$$ES_{i,s} = \ln\left(\frac{\bar{t}_{is}}{m_i}\right)$$

For each of the four management categories, the means and standard errors of this effect size statistic were calculated, and t-tests were used to compare these means. Pairwise t-tests were computed for every combination ($P \leq 0.05$) (Fig. 13.2). A Tukey's HSD test for multiple comparisons was not significant, probably due to the large differences in sample sizes. Thus there is a risk that the means of the management categories are actually equal, but the less conservative t-test results at least show major trends in the data.

For all of the management categories designated by the PAM clusters, the mean ES_s was within a range of –0.37 to –0.65 (Fig. 13.2). This indicates that the mean species richness for a given ecosystem represented about 50 to 75 per cent of the maximum observed for any ecosystem in that study, regardless of the type of taxa surveyed. Biodiversity tended to be lowest in the Intensified Agroecosystems, but was not significantly different from the more Complex Agroecosystems, based on t-tests comparing the mean ES_s. The arable ecosystems tended to have lower species richness compared to ecosystems that were dominated by natural vegetation such as forests or grasslands, but only the Intensified Agroecosystems were significantly different from the natural ecosystems.

Even in Complex Agroecosystems, which usually had lower use of pesticides, herbicides and fertilizers than Intensified Agroecosystems, agricultural disturbance may be one of the main reasons for the trend towards a decline in biodiversity. For example, hilltops in a national park in the Dominican Republic had many more lizard species than were ever found in any agricultural habitats, and even abandoned agricultural areas still contained only 69 per cent of the region's species (Glor et al. 2001).

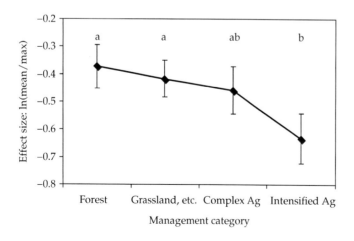

Figure 13.2 Response of species richness to management intensification quantified by meta-analysis of field studies across four ecosystem types in agricultural landscapes, showing the mean effect size statistic for each ecosystem, based on the mean species richness for an ecosystem within a site in relation to the maximum observed richness in the study (see text). Ecosystem types were determined by Partitioning Along Metroids (PAM), and four clusters were identified: Forest – Grassland/Pastures/Abandoned – Complex Agroecosystems – Intensified Agroecosystems. Studies included many types of taxa and biomes (Table 13.1). Effect size means that share lower-case letters do not differ significantly ($P \leq 0.05$) based on pairwise t-tests; error bars are standard errors.

Few studies utilized enough land use types in close geographic proximity to detect small-scale responses to intensification. Two exceptions, however, showed different patterns. In six landscape types, ranging from mature forest to arable cropland in a national park in Central France, plant biodiversity was highest in meadows and pastures, which were managed under intermediate disturbance intensity (Fédoroff et al. 2005). This reconciles well with the intermediate disturbance hypothesis, which proposes that biodiversity is highest when disturbance is neither rare nor frequent, such that both highly competitive and more disturbance-adapted species can co-exist (Connell 1978). In contrast, in Sumatra, one of the world's top five biodiversity 'hotspots' (Myers et al. 2000), plant biodiversity increased progressively from simple monoculture farming systems through more complex agroforests to late secondary and complex forests (Gillison et al. 2004). In this case, the diversity of the closed canopy rain forest surpassed any of the disturbed systems. Both of these regions have a long history of management (arable fields, pastoralism, and forest harvest in France, and rotating slash and burn agriculture in Sumatra (Palm et al. 2005)), but the intensity of management pressure in France may have been more consistent in space and time. This may have caused a relatively higher loss of species dependent on unmanaged systems, as well as more selective pressure for evolution of species adapted to intermediately disturbed meadows and pastures. More studies like these are needed to understand this discrepancy in the effect of management intensity on landscapes differing in their inherent species richness.

Organic farms, which were usually classified as Complex Agroecosystems, generally had higher biodiversity than conventional farms. This was true for birds (Genghini et al. 2006), flower-visiting bees (Holzschuh et al. 2007), dung beetles (Hutton and Giller 2003), and butterflies (Rundlof and Smith 2006). Reviews specifically addressing the organic–conventional difference have shown that organic farms characteristically do harbour greater biodiversity than conventional farms (Gabriel et al. 2006, Hole et al. 2005). However, since many of the studies in the meta-analysis encompassed a much broader range of ecosystem types, these differences apparently are small within the context of a greater landscape spectrum. A comparison between sets of organic and conventional farms showed that up to 37 per cent of the beta diversity of plant species was explained at the farm scale and up to 25 per cent at the regional landscape scale (Gabriel et al. 2006), suggesting that the landscape matrix affects biodiversity, and that field management only partly explains species richness.

The trends for differences among management categories (Fig. 13.2) suggest that practices such as agroforestry and complex rotations support greater biodiversity than does intense agricultural management, but they still create enough disturbance that species diversity tends to be reduced compared to pasture/grasslands or forests. One explanation may be that arable land use favours broad-niched, ruderal (i.e. 'weedy') species that become ubiquitous, even in situations where greater spatial heterogeneity in the past may have supported species adapted to unique managed and unmanaged habitats (Grau et al. 2003).

13.2.2 Mitigation options

Complex Agroecosystems tended to have higher levels of biodiversity than the Intensified Agroecosystems in the same agricultural landscape (Fig. 13.2). If this is true across a wider range of landscapes than examined here, then at least some forms of wildlife-friendly farming may produce reasonable levels of agricultural commodities (although measurements of agricultural productivity are absent from our analyses) and still conserve a certain level of biodiversity. The wildlife-friendly farming concept proposes that certain practices, such as lower agrichemical inputs, reduced tillage, and polycultures instead of monocultures, would reduce impacts on non-target biota and could allow the expansion of low-intensity production into wildlands to meet food demands without a concomitant loss of biodiversity (Donald 2004, Green et al. 2005).

In contrast, the alternative approach to biodiversity conservation in agricultural landscapes, land sparing, is to increase intensification while leaving larger areas uncultivated for wildlife habitat. Land sparing requires further intensification (e.g. more agrichemical inputs and monocultures) on lands currently under low-input, extensive production, but promises little change in the area of cultivated lands, either locally or globally, to meet food demands. This may cause wildlife populations to decrease in and around farmland, but may ultimately reduce the human impacts on biodiversity overall (Green *et al.* 2005, Mooney *et al.* 2005). Intensification, however, which requires more water, nutrient, and agrochemical inputs, has effects outside the allegedly fixed area that is being cropped. Our meta-analysis suggests that landscapes under arable production have already lost substantial biodiversity; further intensification and homogenization across landscapes may result in even greater biodiversity losses (e.g. Matson and Vitousek 2006, Rand *et al.* 2006, Tscharntke *et al.* 2005). Another issue is that set-aside lands can be put back into production, for example, the Conservation Reserve Program in the USA has recently allowed grazing on some lands, as cattle feed is in short supply due to artificially high corn prices driven by interest in biofuels.

The concepts of wildlife-friendly farming and land sparing may both be too simplistic in terms of their potential tradeoffs between agricultural production and biodiversity conservation for other ecosystem services. Many regions are experiencing both rapid expansion and intensification of agriculture across large areas (Morton *et al.* 2006), suggesting that neither pathway plays out as hypothesized when the configuration of ecosystems in these agricultural landscapes is changing rapidly. Intensification often results in the complete loss of fragments of natural vegetation, especially when the efficiency of agricultural production increases with the volume of goods (i.e. economy of scale) so that land sparing does not actually materialize. For example, historical accounts of agricultural intensification often demonstrate that transformation is rapid and ubiquitous across a landscape, such as the clearing of native vegetation for grain and livestock production after the Gold Rush in California (Vaught 2007). Consequently, land sparing may be quite difficult to achieve at the landscape level when a 'boom' cycle is under way to intensify production.

The challenge is to increase agricultural production in a way that utilizes biodiversity for sustainable alternatives to non-renewable and environmentally damaging inputs, and that increases the positive benefits for wild species, which in turn provide ecosystem services in human-dominated landscapes. This is the essence of the 'ecoagriculture' concept (McNeely and Scherr 2003). Strategies thought to increase both agricultural productivity and save biodiversity of wild species are often hampered, however, by lack of inventories or local knowledge about regional biodiversity, as well as a lack of information on multiple ecosystem services provided by habitats targeted for conversion. Without such information, it is difficult to determine what the impacts of different management options might be.

13.3 Biodiversity-ecosystem function relationships in managed ecosystems

Landscape gradients allow evaluation of biodiversity and ecosystem services in relation to land use. In the same meta-analysis as described above, we tallied BEF relationships for each of the studies by recording if (1) there was a citation given for a relevant ecosystem function or service in the discussion section (Cited), (2) if no citation was made for a relevant ecosystem function or service (e.g. the main interest was in conservation or the intrinsic value of biodiversity), and an ecosystem was inferred by the authors (Not cited), or (3) if the study tested ecosystem function or services (Tested) (Fig. 13.3). Pollination/predation, productivity, stability/resilience, and nutrient cycling were considered as the categories of ecosystem functions and services. Since more than one ecosystem function or service was sometimes given, the total tally was greater then number of studies analyzed. For the 'Not cited' category, we subjectively differentiated the studies into possible ecosystem function categories based upon the target species and their ecological roles, either as inferred or discussed. For example, ants in coffee systems may impact predation and pollination.

Only one-third of the studies actually tested an ecological function of biodiversity (Fig. 13.3). Most commonly, measurements were made of pollination and predation, followed by productivity and nutrient cycling. Stability and resilience were never tested, yet they were cited as potential functions in several papers in which no tests of ecological functions were made. The lack of information on

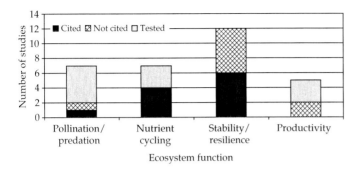

Figure 13.3 Abundance of studies in the meta-analysis invoking Tested, Cited, or Not Cited relationships between biodiversity and ecosystem function. Tested studies relied on actual measurements, Cited studies relied on literature to support the relationship, and Not Cited studies inferred a relationship.

ecosystem functions and services along intensification gradients is evident from these studies; this poses a major challenge for the ecological research community, as well as for policy makers.

13.3.1 Relevance of controlled experiments to managed ecosystems

Many of the recent advances in the understanding of BEF relationships have come from fairly narrow-scale, controlled experimental frameworks, such as grassland experiments where plant species and functional group combinations are assembled in homogeneous environmental conditions and the response variable is usually standing biomass stock (Hooper et al. 2005 and other chapters in this book). The extensive, low-input, species-rich assemblages used for this research make it difficult to infer implications for ecosystem services in managed ecosystems, although there are exceptions (e.g. biofuel production potential from low-input, high-diversity mixtures of native grassland perennials (Tilman et al. 2006a)).

A novel approach to designing and analyzing BEF relationships examined the effects of adding a small number of grass and legume species to species-poor, intensively managed grasslands across 28 European field sites, using different relative species abundances (Kirwan et al. 2007). The results showed transgressive overyielding (i.e. higher performance of mixtures compared to the highest-yielding monoculture) in three of the four geographic locations (Fig. 13.4), a positive effect of species evenness on production, and a negative effect of diversity on the abundance of unsown species. Thus small increases in plant diversity, as well as sowing rates and adaptive management to maintain evenness levels, could increase production and reduce the propensity for invasions of undesirable species in agricultural grasslands.

13.3.2 Research approaches for biodiversity and ecosystem functioning in managed ecosystems

In a distinctly different set of approaches, a growing number of ecological studies in managed ecosystems have developed methods to show BEF associations based on functional traits of organisms (Naeem and Wright 2003). For example, a set of analytical tools was developed to rank species according to their ecosystem function, and to determine those species that contribute disproportionately to function relative to their abundance (Balvanera et al. 2005). Another approach was to use indices of key features of plant biodiversity and functional types that relate to land use types and their associated ecosystem services, such as above-ground carbon storage and soil nutrients (Gillison et al. 2004). Other studies have used landscape gradients to infer how changes in biodiversity relate to management practices and ecosystem functions such as productivity or soil carbon storage (Bullock et al. 2001, McCrea et al. 2004, Steenwerth et al. 2003). In one such study, arthropod diversity did not recover after many years following disturbance, even though a typical indicator of recovery from disturbance, soil carbon, increased (Loranger et al. 1998) (Fig. 13.5). An ambitious and more controlled approach was to plant genetically diversified rice crops in all the rice fields in five townships in China; rice blast disease was > 90 per cent less

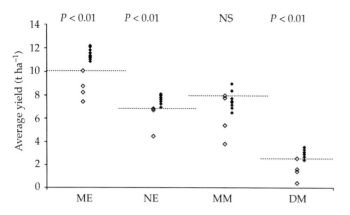

Figure 13.4 Evidence for transgressive overyielding, in which performance of overyielding mixtures exceeds that of the highest-yielding monoculture. Mean yield is shown over 28 intensive grassland sites in four locations (mid-European (ME), north European (NE), moist Mediterranean (MM), and dry Mediterranean (DM)) where two grass species and two legume species were sown into species-depauperate intensively managed systems. For each location, a unique group of four species was sown at different densities to create 11 mixture communities (closed symbols (◆) and four monocultures (open symbols (◇)). From Kirwan et al. (2007). Requested by Jackson et al. from Kirwan et al. (2007). Copyright permission granted.

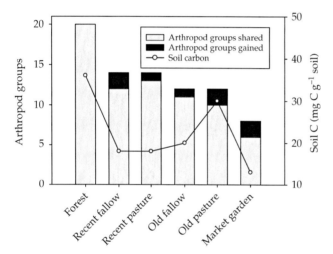

Figure 13.5 The diversity of arthropods decreased along a gradient of agricultural intensification (pesticides, tillage and weed control) along a land use gradient in Martinique (French West Indies). Arthropods were classified into 22 groups (19 orders and three larval groups). In the old pasture, diversity remained low despite higher accumulation of soil C. Soil C was measured in the 0–10 cm depth. From Loranger et al. (1998). Copyright permission granted.

severe in mixtures than in monocultures (Zhu et al. 2000). While these research contexts have direct implications for biodiversity in agriculture and managed grasslands and forests, they usually cannot demonstrate the mechanisms by which changes in species composition and richness affect specific ecosystem functions (Srivastava and Vellend 2005).

Some of the factors that complicate field studies in managed ecosystems beyond those of manipulation experiments are redundancy of species or functional groups, trophic complexity, and spatial heterogeneity (Naeem and Wright 2003). The challenge is to find approaches that test hypotheses about the mechanisms by which biodiversity affects ecosystem function while providing management relevance. Such approaches often depend on fortuitous 'experiments' that involve spatial or temporal gradients within managed landscapes. For example, to determine the role of flower-visiting bee diversity on the fruit set by coffee, Olschewski et al. (2006) chose a set of agroforestry farms near a national park in Indonesia, such that distance to the forest was not correlated with shade density. Transitioning to a new management scenario with increased biodiversity (e.g. conventional to organic production (Smukler et al. 2008)), provides another approach for identifying correlated changes in biodiversity and ecosystem functioning. These types of approach benefit from the participation of land managers, to ensure that sites and situations are well-paired, management records are accurate, and that ecological interpretation is relevant to actual situations.

13.3.3 Example: functions of biodiversity for pest control

Pest control provides examples that show the complex BEF relationships in managed ecosystems. In their meta-analysis, Balvanera *et al.* (2006) found that increasing biodiversity increased resistance to consumption and to invasion by exotic species. Landscape heterogeneity also contributes to higher invertebrate diversity and to the effectiveness of natural enemies. In a literature review, natural enemy populations were higher and pest populations lower in complex landscapes comprising cropland intermixed with mid- and late-successional non-crop habitats vs. simple landscapes (Bianchi *et al.* 2006). However, data to confirm impacts on actual crop damage or yield were generally lacking.

The patterns and relationships between different measures of diversity and heterogeneity within agroecosystems and pest control functions are not consistent. In one study, complex landscapes generally had higher rates of parasitism and parasitoid diversity compared with simple landscapes composed primarily of cropland (Menalled *et al.* 2001). However, this pattern was not consistent across all sites, so no clear effect of landscape complexity on parasitism could be determined. Other studies, such as Weibull *et al.* (2003), indicate that while species richness generally increases with landscape heterogeneity at the farm scale, changes in diversity do not clearly affect natural pest control. In simple landscapes, increasing non-crop habitat can have an impact on biological control but in structurally diverse landscapes that already have a high proportion of non-crop areas, increasing the proportion of non-crop areas further has little effect (Thies and Tscharntke 1999). Although more complex landscapes can support increased parasitism, they also tend to be subject to increased aphid colonization, with no net gain in biological control compared to simpler landscapes (Thies *et al.* 2005).

A growing body of literature has explored the effects of multiple natural enemies on predator–prey interactions. These studies indicate that as enemy diversity increases, so does the potential for a range of density- and trait-mediated interactions within and across trophic levels. For example, higher diversity of enemies increases the chance of antagonistic interactions among natural enemies that generate sub-additive effects of enemy diversity on consumption rates (Finke and Denno 2006, Rosenheim *et al.* 1995). In contrast, synergistic interactions between enemy species can generate super-additive effects with overall prey consumption rates greater than the sum of the single species consumption rates acting independently (Cardinale *et al.* 2003). Furthermore, the expression of these different diversity effects can be dependent on ecological context, with factors such as habitat complexity (Finke and Denno 2006) and extent or diversity of prey resource (Prasad and Snyder 2006, Wilby *et al.* 2005) strengthening and/or weakening the emergent properties of even the same assemblages of natural enemies.

These potentially complex, and even idiosyncratic, effects of predator diversity on prey consumption appear to sit at odds with the key findings from the recent meta-analysis of Cardinale *et al.* (2006a). This comprehensive evaluation of empirical BEF studies indicated that an increase in species richness increases, on average, the extent of resource utilization regardless of trophic level. The mechanism best explaining this trend was the 'sampling effect' whereby diverse communities are more likely to contain and be dominated by a functionally significant species. By implication, this suggests that species identity is more important than any emergent effects of biodiversity that might arise through, for example, resource-use partitioning or species interactions. The evidence from numerous multiple enemy studies, however, indicates the potential for neutral and even negative effects of diversity (i.e. not only positive), and also clear significance of intra- and inter-specific interactions in addition to species identity effects.

Mechanistic understanding of BEF relationships is needed before it can be concluded that species identity effects override trophic effects. First, the necessary data required to confirm that the species that does best in monoculture actually dominates function in the polyculture (i.e. the sampling effect) are generally lacking in the majority of studies. Moreover, for most BEF studies, the diversity–function relationship for a given function begins to saturate after just one or two species (Cardinale *et al.* 2006a), but many more species may be required to deliver maximum and multiple functions. At higher levels of diversity, complementarity effects and other emergent properties of biodiversity deriving from trophic complexity are likely to play a role. First, just one or two species of natural enemies may contribute 50 per cent of pest control function, but this may not be sufficient to deliver enough control to maintain a pest below an economic threshold level; in the context of practical pest control, we are interested in maximum function, not just the shape of the curve. Second, a common approach is to examine predation rates using simple food webs comprising a limited number of species. These species modules allow for interactions between the particular species present (identity effects) but do not necessarily test for diversity effects *per se* (e.g. Finke and Denno 2004). Additionally, many studies are short-term and consider intra-generational effects only, yet recent modelling studies indicate that short-term experiments and studies that ignore alternate resources can fail to capture the long-term significance of interactions such as intraguild predation (Briggs and Borer 2005). In a similar way, a characteristic of most controlled BEF experiments is that they restrict complexity (e.g. no alternative prey, no higher order predators or parasites, limited spatial or temporal extent), and therefore reduce the potential for expression of properties of diversity beyond identity.

Even if we consider just niche partitioning and complementarity, the ecological context and extent of each study has an important influence. This issue was illustrated recently by Wilby and Thomas (2007) using a hypothetical example (Fig. 13.6). Essentially, the range of available spatial and temporal niches, and the extent of the process under study influence the potential for complementarity. Very different relationships occur between pest control functioning and predator diversity if we consider consumption of one aphid species in one field in one year compared with consumption of all insect herbivores across several crops and multiple years. This argument, consistent with the insurance hypothesis, identifies a need for BEF research at appropriate temporal, spatial and process scales, and cautions against extrapolation of existing results until scaling relationships are adequately understood (Wilby and Thomas 2007).

13.3.4 Insurance value of biodiversity in human-dominated landscapes

Temporal stability, resistance, and resilience are three types of stability conferred by biodiversity (Griffin *et al.* 2009). At present, the potential for biodiversity to provide ecological resilience (i.e. the capacity to recover from disruption of functions and the mitigation of risks caused by disturbance (Holling 1996, Swift *et al.* 2004)) is compelling, but evidence at the landscape level is limited. In a long-term controlled BEF experiment, higher plant diversity decreased the temporal variance in productivity (Tilman *et al.* 2006b), supporting the ideas that biodiversity can be used to buffer against high variance in productivity (Yachi and Loreau 1999), and that high productivity in monocultures can only be consistently maintained through intensification (Naeem 2006a). Yet some have suggested that the functional significance of biodiversity may be most profound at larger spatial and temporal scales, by providing insurance value, especially when dispersal abilities of organisms allow for immigration within the landscape (Loreau *et al.* 2003, Swift *et al.* 2004). Species or phenotypes that appear to be functionally redundant for a specific ecosystem process at a given time may diverge in response to environmental fluctuations, thereby stabilizing the aggregate ecosystem function through time.

Multifunctionality is another compelling argument for greater biodiversity in managed ecosystems. Mixtures that simultaneously minimize crop environmental stress, contribute to pest control, provide traditional foods or medicines, and lead to greater market potential for a range of agricultural commodities exhibit the insurance value of biodiversity in managed landscapes (Perfecto *et al.* 2004, Swift *et al.* 2004). The multifunctionality of biodiversity has not been studied rigorously in this context. However, it appears to operate in traditional farming communities that depend strongly on landraces, as well as in situations where farmers select a large number of crop taxa not only to increase their direct marketing potential, but also to minimize the risks from various types of crop damage or failure.

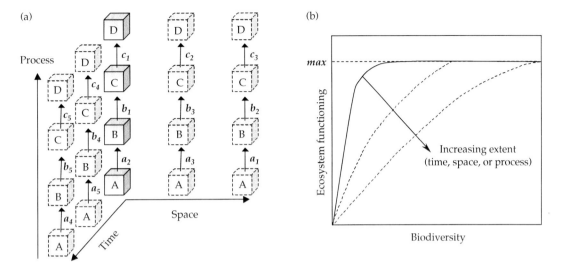

Figure 13.6 The influence of process, spatial extent, and temporal extent on biodiversity-ecosystem functioning relationships. (a) The role of elements of biodiversity (shown as lower-case letters with subscripts) in the transition between states (boxes) within a process. For a given position in time and space (i.e. a particular ecological context), the number of elements of biodiversity necessary to fulfill ecosystem function depends on the extent of the process in question. If the extent of the process of interest is simply the transition from state A to state B, then just one biodiversity element (such as a species) is required (e.g. a_2). If, however, a process involves transition from state A to state D then three elements are required for full function (e.g. a_2, b_1, c_1). As the temporal or spatial extent of a study is expanded, then the number of elements necessary for maximal functioning increases further, since different elements are likely to be most efficient at different points in space and time. For the full extent of the process A to D across all combinations of time and space, a maximum of 27 elements is required for full function, based on these assumptions. (b) The increase in biodiversity required for maximum function as the extent of process increases. Time or space changes the shape of the biodiversity–ecosystem functioning relationship from the baseline set of ecological conditions. From Wilby and Thomas (2007). Copyright permission granted.

13.3.5 Cross-ecosystem exchanges of biodiversity and their functions

The traits that permit species to move and persist across ecosystem boundaries will affect the ecosystem services that are provided by heterogeneity in the landscape (Gonzalez et al. 2009). Tscharntke et al. (2005) suggest three main possibilities of exchanges of organisms between crop and non-crop habitats: (1) exchange among the non-crop habitats as long as the landscape matrix allows the specialists in these habitats to disperse and persist, (2) non-crop habitat as a source of generalist species that invade cropped ecosystems, with effects on pollination, biological control, or pest infestations, and (3) crop habitat as a source of generalist species that invade the non-crop ecosystems, and affect biotic interactions in natural habitats.

Some information exists on the traits that permit species to move and persist across ecosystem boundaries, and on the ecosystem services that result. A landscape mosaic with agroecosystems in close proximity to fragments of natural habitats can increase pollination (Greenleaf and Kremen 2006b, Kremen et al. 2002) and pest regulation (Bianchi et al. 2006). Predatory wasps and parasitoids that act as natural enemies of crop pests are more abundant near field edges than in field centres, and in more complex landscapes with higher proportions of forest, hedgerows, grassland, fallows, and wetlands (Bianchi et al. 2006). Higher diversity of bees and wasps, especially parasitoids, are observed in agroecosystems close to rainforests, probably because their populations build up in the forests due to lack of disturbance (Klein et al. 2006). This can result in higher fruit set near forest remnants, and higher income for the farmer (Ricketts 2004). But in some areas in Nicaragua and Indonesia potential crop revenues of forest land-sparing exceed the value of coffee pollination from these adjacent forests, and thus the higher income derived from conversion from forests to coffee plantations drives land use change, even when some compensation for pollination services is available (Olschewski *et al.* 2006). At the landscape level, forest conversion may outweigh the immediate financial benefits of conserving biodiversity and its ecological functions (Lambin *et al.* 2003).

There are also spillover edge effects in the other direction (Rand *et al.* 2006). For example, many generalist insects that serve as natural enemies of agricultural pests feed on alternative hosts in nearby woodland and grassland fragments, with largely unknown effects on food webs. This is most likely to occur when adjacent ecosystems differ in productivity and/or when the timing of temporally variable resources shifts across ecosystem edges.

Fragments of natural vegetation along wetland margins are another example of cross-ecosystem exchanges. They filter agricultural nutrients and pollutants, reduce erosion, improve water quality, and provide habitat for higher trophic level species (Lovell and Sullivan 2006). Multispecies riparian buffers with herbaceous and woody species can result in higher total productivity, soil carbon sequestration, and nitrogen immobilization than monoculture plantings (Hill 1996, Marquez *et al.* 1999, Rowe *et al.* 2005).

There are situations where landscape-level heterogeneity and its associated biodiversity may cause agricultural problems. Some crop diseases only overwinter if there are hosts in the vicinity, leading to enforcement of crop-free periods and mandatory elimination of host species. In the Salinas Valley of California, one of the most intensively managed agricultural areas in the USA, lettuce mosaic virus infects both cultivated lettuce (*Lactuca sativa* L.) and its wild relative, prickly lettuce (*Lactuca serriola* L.), which is a ruderal species along field margins and roadsides. One solution was the legal designation of a month-long period when lettuce cannot be present in any field, and another was the eradication of prickly lettuce (Jackson *et al.* 1996) which entailed killing other annual species, including natives. Another example from the same region is the virulent strain of *E. coli* on spinach, which killed three people in 2006, and matches the DNA of *E. coli* in cattle, wild pigs, and stream water of a nearby grassland (Bailey 2007). A strong movement is now developing in this region to eliminate natural habitats, and landscape heterogeneity, around vegetable fields, in order to prevent future outbreaks.

In an example from the Great Plains of North America, restoring native bison to areas now dedicated to cattle ranching has been suggested to increase biodiversity and ecosystem health – a viable economic alternative in a region with declining human population density (Callenbach 2000, Popper and Popper 1987). This notion originally met with substantial resistance from many cattle ranchers, who felt that bison would transmit disease to their cattle or that bison were biologically and economically inferior to cattle (Popper and Popper 2006). These examples demonstrate that the general lack of information on landscape-level interactions between wild and agricultural biota can create strong human reactions that can ultimately result in decisions that may not favour biodiversity conservation or ecosystem services.

13.3.6 Indicators for the functions of biodiversity in managed ecosystems

Given the rapid loss of biodiversity, there has been a strong effort by various agencies to develop indicators of biodiversity (OECD 2003) often with the implicit assumption that increased biodiversity confers higher ecological functioning. There has been less emphasis on indicators that demonstrate the function of biodiversity, although some examples are given below. To date, three primary approaches are applied: correlations based on functional types, use of specific indicator taxa, and multitaxon studies. This is an emerging area of much needed science, and involves a great deal of ecological complexity.

Correlating trait-based functional types and specific functions is one approach. In Sumatra, Indonesia, a biodiversity 'hotspot', plant species diversity is lowest in intensive non-shaded coffee systems and increases through more shaded and complex agroforests to late successional secondary forests and closed canopy mature forests (Gillison *et al.* 2004). The number of plant functional types was shown to be correlated with soil carbon, nutrient availability, aboveground carbon, biomass (based on allometric equations), land use, and plot age since disturbance.

Birds serve as excellent indicators of biodiversity (Donald 2004), and species composition is known to be linked to agricultural intensification (Green *et al.* 2005). In orchards and vineyards in Northern Italy, diversity of birds, especially insectivores, were highest in organic systems, and pest management practices (number of pest control treatments and pesticides used) explained much of the variation among systems (Genghini et al. 2006). Using this type of knowledge about responses of different species to changes in management practices, Butler et al. (2007) developed a trait-based risk assessment framework to predict the effects of environmental change on bird biodiversity. For six components of agricultural intensification, they examined the potential foraging and habitat responses of different bird species, and assigned a risk score, with the end result of identifying particularly harmful management practices (e.g. that genetically modified herbicide-tolerant sugar-beets would impact the ecological habitat of 39 bird species). This framework could be expanded to include the response of specific functions, such as pollination, seed dispersal, and predation, on either desirable or less desirable species.

Multitaxon studies of biodiversity draw inferences about multiple ecosystem functions and conservation value. For example, the species richness and abundance of forest birds was positively associated with shade cover, epiphyte abundance, and canopy height in Mexican coffee agroecosystems, but no such associations occurred for understory dwelling mammals, which were captured in only two of 15 study sites (Gordon et al. 2007). Individual taxa perceive the landscape in different ways, and indicators must reflect these differences. For example, different forms of tree cover provide habitat for birds, bats, and insects in a pastoral landscape in Nicaragua (Harvey et al. 2006). In a study of subalpine hay meadows in Romania, each stage of succession held the maximum species richness for one of four taxonomic groups (plants, gastropods and diurnal and nocturnal *Lepidoptera*), indicating that all stages had high conservation value (Baur et al. 2006). There has been little work to assess the ecological functions of different functional groups of taxa and the tradeoffs involved in managing for each within a given ecosystem or landscape.

The same issues of complexity that have been the main theme of this section hinder the development of indicators for specific functions resulting from particular aspects and levels of biodiversity in managed ecosystems. Species richness clearly is too simplistic to be used as an indicator of ecosystem functions in actual ecosystems and landscapes, despite the typical results of controlled BEF studies (Cardinale et al. 2006a).

13.4 Linking biodiversity, ecosystem functions, management, and end-users

As indicated in the introduction, mounting population pressure is increasing the demand to maintain and enhance the productivity of agricultural systems. To do this in a sustainable way without impacting a range of non-provisioning ecosystem services represents a significant ecological challenge, and must be set within a socioeconomic context of market forces, social and economic policy, and human values (Robertson and Swinton 2005), and must consider the tradeoffs for different end-users, from farmers to policy makers (Jackson et al. 2007).

Broad assessments of the significance of ecosystem services suggest considerable value of regulating and supporting services within managed systems. For example, natural pest control services have been valued at > 400 billion USD per year globally (Costanza et al. 1997), with estimates specifically for cropland of 100 billion USD per year (Pimentel et al. 2004). Indeed, it has been suggested that 99 per cent of the populations of pests and diseases are controlled by their natural enemies (de Bach 1974). While such figures provide a measure of the overall economic significance of services provided by biodiversity, there is a general lack of detailed field-level assessments of particular services, the mechanisms behind them, and their economic value.

One recent study in New Zealand attempted to address this gap by using experimental manipulations to assess the value of natural pest control services in organic vs. conventionally managed arable production systems (Sandhu et al. 2008). The approach used artificial prey to quantify background predation rates coupled with knowledge of prey economic thresholds to calculate an economic value of biological control, using estimates of 'avoided costs' and 'total avoided costs' of insecticide applications. Avoided costs are the monetary savings derived from not needing to apply insecticides to control specific pests, whereas total avoided costs also include a component for external costs, such as the environmental impact of insecticide application. Background biological control of pests was severely reduced, four- to ten-fold on average, in 14 organic fields compared with 15 fields under conventional management. The mean economic value of this biological control in the organic fields was estimated at 55 USD ha^{-1} yr^{-1} (avoided costs) and 22 USD ha^{-1} yr^{-1} (total avoided costs) for both aphid and carrot rust fly. In contrast, little economic value of natural pest control services occurred in the conventional fields due to extremely low predation rates. However, it should be noted that the economic value of biocontrol varied across fields and positive economic values could not be demonstrated for all organic fields.

In this example, the avoided costs provide a measure of private benefit (i.e. benefit to the farmer) while the total avoided costs provide some measure of additional public benefit due to externalities beyond the farmer's field. Variation creates further complexity in valuation. Benefits varied with local ecological context, and positive effects were measurable in some field settings but not in others. Similar spatial variability in biodiversity and pest control has been reported previously (Menalled *et al.* 2001, Weibull *et al.* 2003), as both management intensity and landscape structure shape local arthropod communities (Bianchi *et al.* 2006, Schweiger *et al.* 2005). The overall conclusion is that generating clear guidelines for farmer and policy makers end-users is difficult, yet adoption of biodiversity-based practices may still be advocated as a potential component of economic and environmental risk reduction.

13.4.1 Impacts of scale on managing for biodiversity and its ecosystem functions

Context and the scale of adoption or implementation alter the impacts of management practices. This issue was explored recently by Griffiths *et al.* (2008), who used conceptual models to describe how a management strategy aimed at manipulating biodiversity affected an associated ecosystem service when the strategy was adopted at different scales. Non-crop habitats within agricultural landscapes were manipulated to increase natural enemy diversity and abundance and increase natural pest control. Three scenarios were considered. First, benefits were assumed to be independent of scale (Fig. 13.7(a)).

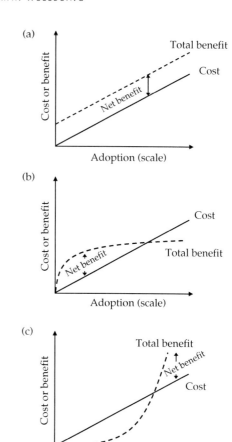

Figure 13.7 Costs and benefits of natural biological control as a management strategy that promotes an ecosystem service, and effects of the level of adoption of that strategy. (a) A simple linear relationship whereby effect on biocontrol is independent of level of adoption and hence, net benefit remains constant. (b) A non-linear relationship where effectiveness is greatest at small scales such that net benefit decreases as adoption increases. (c) A contrasting non-linear case where impact increases with adoption. Here adoption needs to exceed some threshold level before a net benefit is derived. From Griffiths *et al.* (2008). Copyright permission granted.

Here, increased adoption led to an increase in total cost and total benefit of the approach, but net benefit remained constant. In the second scenario, benefit was assumed to saturate with increasing adoption (Fig. 13.7(b)). Such a pattern could arise if, for example, the effects of the habitat manipulation derived largely from local redistribution of natural enemies. At small scales (a single field), natural enemies might be drawn from adjacent habitats, thereby boosting pest control, but at larger scales (multiple fields), this effect diminishes as the species pool for redistribution becomes exhausted. A technique might look promising initially and favour early adopters, but then fail as it becomes implemented more widely. The final scenario assumed that the benefit of a technology increased with scale of adoption (Fig. 13.7(c)). This could occur if habitat manipulation increased natural enemy richness and abundance as more habitat was created. With this pattern, an individual farmer might have little incentive for adoption unless the technology was implemented widely. Such negative scale or transition effects will likely hinder adoption whenever farmers try, or partially adopt, a practice in order to evaluate its performance. Since scaling effects are not well-understood, one of the key challenges for BEF research is to determine appropriate temporal (season to multiple seasons) and spatial (field to farm and landscape) scales to define optimum management strategies that link in a meaningful way to policy and market mechanisms.

13.4.2 Incentives for managing for biodiversity and its ecosystem functions

In most cases, people decide to utilize and conserve biodiversity because there is a private socioeconomic advantage, including rewards obtained from land stewardship (Pascual and Perrings 2007). As an example, in Austria, across a range of commodities and farming styles, plant species richness was not correlated with agri-environmental subsidies per hectare (Schmitzberger et al. 2005), suggesting that stewardship was not limited by finances but rather by awareness and mentality. In general, agri-environmental schemes have not been particularly successful in protecting farmland biodiversity (Kleijn and Sutherland 2003). The impact of financial support on agrobiodiversity conservation is still poorly understood, but it may need to be specifically targeted to well-defined ecosystem functions to be successful (Gottschalk et al. 2007).

Governmental backing of intensive land use can result in loss of biodiversity and ecosystem services, such as agricultural subsidies for high-input crops (e.g. corn for ethanol production in the USA). Tropical forest regions can rapidly undergo deforestation when there is sudden change to more ubiquitous, intensive systems, often occurring with the knowledge and implicit support of local and federal governments (Geist and Lambin 2002). An example of the latter comes from the Veracruz state of Mexico, where immigration to oil-producing areas led to intensive urban expansion accompanied by a 1400 per cent increase in croplands and grasslands for cattle raising in 30 years; these increases were at the expense of mature and secondary forests (Garcia-Romero et al. 2004). The loss of ecological functions from the biodiversity in these later successional ecosystems was apparently not a high priority for policy makers.

Social institutions are needed to promote incentives that ensure that biodiversity will be conserved to its fullest potential of functional benefits over the long term. An example is the wide price swings for coffee in the international market. Price increases result in land clearing or intensification, while price crashes lead to widespread abandonment of coffee plantations, and the end result is biodiversity loss (Donald 2004). Low-input, floristically diverse, shade-grown coffee tends to be associated with higher biodiversity of many taxa. Due to lower input costs compared to intensified, unshaded coffee, low-input systems can be very profitable in Mexico, and are less likely to be converted to cattle pasture, sugar cane, or housing developments when the price of coffee falls (Gordon et al. 2007).

Poverty also has profound implications for biodiversity in managed ecosystems. When there is uncertain tenure over land, limited or no access to credit, and poor market linkages, there may be little capacity to manage for biodiversity and its ecological functions. When social programs increase credit through micro-lending, and provide value addition to agricultural and forest products, there is greater potential to create viable enterprises that both conserve biodiversity and improve human wellbeing (Bawa et al. 2007, de Clerck et al. 2006).

In human-dominated landscapes, high levels of biodiversity are more likely to be maintained for reasons of intrinsic ('option' or 'bequest') values or utilitarian ('direct use') than for other functional values (Swift et al. 2004). Biodiversity at the farm/plantation and plot levels is most likely to be maintained for solely utilitarian values. The challenge for ecologists is to provide meaningful research on ecological functions of biodiversity at different scales so that it is valued, used, and conserved by society, even as the current trend for management intensification continues.

13.5 Conclusions

This chapter has focused on management intensification and the opportunities for conserving biodiversity and ecological functions in ecosystems situated in landscapes that have been modified by human land use, and the ways that researchers can address ecological processes in managed ecosystems for multifunctional outcomes. Three general messages emerge. First, although there is much recent progress in describing species diversity in managed ecosystems and landscapes, the functions of different components of biodiversity still remain unknown. Second, process-oriented research on functions of biodiversity may be most effective if it focuses on unique traits or relationships between species, especially if this can serve to develop indicators for the multiple functions of biodiversity. This will be more beneficial than a research focus on the effects of increasing species richness on one or a few ecosystem functions. Finally, since human decision-making plays a large role in land use in managed ecosystems, researchers must involve land managers and other stakeholders to grasp both the ecological and socioeconomic factors that determine biodiversity, its functions, and trade-offs between different management scenarios.

Although much of earlier BEF research has focused on basic processes in community ecology, delaying the so-called application phase for many years, this chapter indicates that there has been much recent creativity and activity toward understanding these processes in relation to ecosystem services, particularly in agricultural landscapes. Given the rapid rate at which biodiversity loss is now occurring, and the prognosis that climate change will further hasten these losses, much greater understanding of biodiversity and ecological functions in managed ecosystems is necessary to steer policy decisions toward more sustainable outcomes.

CHAPTER 14

Understanding the role of species richness for crop pollination services

Alexandra-Maria Klein, Christine Müller, Patrick Hoehn, and Claire Kremen

14.1 Introduction

Flower visitation and pollination by animals are the first steps and therefore critical to the sexual reproduction of many plant species (Kearns *et al.* 1998), including up to 90 per cent of Angiosperms in tropical rainforests (Bawa 1990). Flower visitors are insects or vertebrates like hummingbirds or bats visiting flowers for receiving nectar and/or pollen, and the visitation can, but does not necessarily, result in pollination, the successful fertilization of the flower. Recent estimates showed that 35 per cent of global food production directly consumed by humans comes from crops that benefit from flower visitation, primarily by bees and other insects (Klein *et al.* 2007). Many of these crop plants increase fruit or seed quantity from 5–50 per cent when flower visitors are present under experimental conditions, and a small number of crops ($n = 13$) do not set fruit at all if pollinators are absent. Flower visitors can be also important for our dairy and meat industries, since many forage crops, like alfalfa, clover (Delaplane and Mayer 2000), and soybeans produced for cattle forage (Chiari *et al.* 2005) benefit from insect pollination for seed production. Flower visitors are also often important in modern agriculture to mediate gene flow between varieties in hybrid seed production (Greenleaf and Kremen 2006a, Van Deynze *et al.* 2005). Many wild plant populations have been shown to be limited by access to flower visitors (Ashman *et al.* 2004).

Intensification of land use by tilling, irrigation, and fertilizer and biocide use, along with declining habitat heterogeneity from increased field sizes, monoculture plantings, and the modification of natural habitats by humans, significantly impacts pollinator communities across a variety of scales (Klein *et al.* 2007, Kremen *et al.* 2007, Ricketts *et al.* 2008). Local (field-level) alterations that impact pollinators and pollination include changes in: (1) the abundance and distribution of flower resources (Holzschuh *et al.* 2007, Klein *et al.* 2002, Williams and Kremen 2007, Tylianakis *et al.* 2008), (2) species richness of flowering plants (Ebeling *et al.* 2008, Ghazoul 2006, Potts *et al.* 2006), (3) availability of nesting sites and materials (Shuler *et al.* 2005, Kim *et al.* 2006), (4) light levels (Klein *et al.* 2003a, b), and (5) temporal continuity of floral resources (Greenleaf and Kremen 2006b). Examples of landscape-level factors affecting bee communities are the proportion and quality of natural or semi-natural habitats in the landscape (Greenleaf and Kremen 2006b, Kim *et al.* 2006, Kremen *et al.* 2002, 2004), the distance of natural or semi-natural habitats from crop systems (Klein *et al.* 2003a, b, Ricketts *et al.* 2004, Ricketts *et al.* 2008), and the presence of mass-flowering resources (Herrman *et al.* 2007, Westphal *et al.* 2003). Therefore, resource availability and the area and isolation of natural habitats are all important for the conservation of pollinator species richness, which provides a key service to crop production.

In this chapter, we discuss what is known about the relationship between flower-visiting species richness and pollination services. Following the introduction (Section 14.1), the relative importance of species richness, individual numbers, and combined effects in providing services to their mutualistic plant partners are described (Section 14.2). Next, the mechanisms underlying the role of species richness in ensuring insect pollination (Section 14.3), and the evidence for and against *sampling effects* (Section 14.3.1), *niche complementarity* (Section 14.3.2), and *functional facilitation* are discussed (Section 14.3.3). Since the pollination of crop species is embedded within plant–flower visitor (PFv) webs including both other crop and wild plant species, the utility of analyzing quantitative plant–flower visitor interactions for managing and restoring crop pollination services is discussed (Section 14.4). This section starts with a paragraph about web structure and characteristics (Section 14.4.1), to explain how general PFv characteristics buffer species extinction and integration (Section 14.4.2). The role of higher trophic levels in PFv webs is discussed, and in Section 14.4.3 ideas are summarized of how PFv webs provide information for crop pollination services. Section 14.5 presents the current knowledge concerning the consequences of current and projected pollinator declines for global crop production and wild plant populations. The chapter ends with outstanding research gaps in the understanding of how flower-visiting species richness contributes to pollination services (Section 14.6).

14.2 The importance of flower-visitor species richness and individual numbers for pollination services

Pollination services strongly depend on the total of all individuals in a community, the so called aggregate abundance of flower visitors (Morris 2003, Vázquez *et al.* 2005). The European honey bee, *Apis mellifera* L., for example, is the world's most important crop-pollinating species because their high sociality and large colonies yield large populations that are readily managed and moved in and among agricultural fields (Free 1993, McGregor 1976), ensuring fruit set and increasing fruit production and yield (e.g. Chiari *et al.* 2005, Roubik 2002, Stern *et al.* 2001). Although honey bees can provide pollination services for the majority of crop species (Klein *et al.* 2007), in some areas crops can be solely pollinated by species-rich communities of wild non-honey bees (Kremen *et al.* 2002, Winfree *et al.* 2008).

Consequently, under honey bee-scarcity scenarios, species richness of wild species may ensure adequate crop pollination services (Klein *et al.* 2003a, Winfree *et al.* 2008). Given the economic importance of bee pollination (Gallai *et al.* 2008), wild bee species richness will be critical in ensuring pollination services in the face of declining honey bee numbers.

Species richness of flower-visiting communities may be important for several reasons, four of which have been recently studied. First, more species-rich communities frequently have higher aggregate abundance and therefore contribute more to pollination services (Kremen and Chaplin-Kramer 2007, Larsen *et al.* 2005). With agricultural intensification, both flower-visiting species richness and aggregate abundance decline, because compensation by pollinator species that tolerate agricultural intensification is lacking (Larsen *et al.* 2005, Winfree and Kremen 2008). With lower species richness and individual numbers, pollination services become inadequate, as was shown for watermelons that were only sufficiently pollinated on the low-intensity farms that supported the richest bee communities (Kremen *et al.* 2002, 2004). Similarly, intensively cultivated coffee was found to be pollinated inadequately, except for plants growing close to forest fragments that supported the richest bee community (Ricketts *et al.* 2004). The ubiquity of the positive relationship between flower-visitor species richness and individual numbers across agricultural systems (Kremen and Chaplin-Kramer 2007), and the possible lack of density compensation in these systems (Winfree and Kremen 2008) suggest that high species richness is often necessary to provide sufficient pollination services through aggregate abundance of flower-visiting species.

Second, temporal turnover between years is extremely high in bee communities and this turnover affects crop pollination services (Herrera 1988, Pías and Guitián 2006, Williams *et al.* 2001). Species richness on watermelon farms with high flower-visitor species richness remained high from year to year, but the identity of bee species changed (Kremen *et al.* 2002). Temporal stability of the pollination service between days within a season (measured as the inverse of the coefficient of variation), also increased on low-intensity farms (Kremen *et al.* 2004) that supported higher species

Table 14.1 Similarity in flower-visiting communities among four different crops blooming in Yolo County, California. Almond blooms in early spring (February), while the other three crops are summer-flowering (May–August). Jaccard index of similarity is presented. Data from Kremen et al. 2002, 2004, Greenleaf and Kremen, 2006a, b, and unpublished data.

	Almond	Tomato	Watermelon	Sunflower	Pollinator species richness
Almond	1				14
Tomato	0.313	1			7
Watermelon	0.179	0.219	1		32
Sunflower	0.231	0.171	0.32	1	34

richness. Thus high species richness may ensure that flower visitors and pollination services are available to the plants over all days and years, but more studies are needed to test if temporal stability of pollination services is related to bee species richness or individual numbers.

Third, spatial preferences for different flower-visiting species using flowers located on different parts of the plant (Lortie and Aarsen 1999, Hambäck 2001) or in different locations in the field (Klein and Kremen unpublished data) can also lead to the necessity of high species richness for adequate pollination services. Spatial stability of pollination service between plants within sites along a land-use gradient (measured as the inverse of the coefficient of variation) can also be influenced by flower-visiting species richness (Klein et al. 2003a), but more experimental tests are needed before conclusions can be drawn about the generality of this relationship between spatial characteristics in pollination stability and flower-visiting species richness and individual numbers. More aspects of spatial stability and spatio-temporal species turnover will be discussed in Section 14.3.2.

Fourth, flower-visiting species richness is essential when different crops are grown within the same cultivation system. In California, four well-studied crops (watermelon, sunflower, tomato, and almond) share only 17–33 per cent of the species that visit and potentially pollinate them (Table 14.1). Almond blooms in February, March (winter to spring), while the other crops are all summer-flowering. Therefore, it is not surprising that the bee community was almost entirely dissimilar between almond and the other crops, while watermelon, sunflower, and tomato shared a substantial proportion of their visiting bee species. Flower visitors to tomato comprised a subset of the much richer visitor communities to watermelon or sunflower, which shared about 30 per cent of their visitor community. Despite blooming at the same time, watermelon and sunflower clearly attracted different flower-visiting species, because they differ in flower morphology and pollen and nectar availability, and because of the dominance of specialists on sunflower, which is the only one of these four crops native to North America. Similarly, in New Jersey, wild bee communities visiting watermelon versus tomato crops were found to be distinct as identified by non-metric multidimensional scaling (an ordination method to analyze compositional differences among communities). The community differences were visible even when controlling for differences between farm sites and the differential attractiveness of the two crops (Winfree et al. 2008). Similarly, Meléndez-Ramirez et al. (2002) showed a low species overlap of bees (similarity and shared species) visiting different cucurbit crops (cucumber, melon, pumpkin, watermelon) in Yucatán, Mexico.

Fifth, different species respond differentially to disturbances such as land-use intensification (response diversity, Elmqvist et al. 2003), potentially buffering the provision of pollination services across land-use gradients. On watermelon fields in New Jersey and Pennsylvania where bee communities provide a relatively consistent level of pollination services across a land-use intensity gradient (Winfree et al. 2008), differential responsiveness of bee species to land-use intensity may be a mechanism for achieving stability across the gradient (Winfree and Kremen 2008). In California, where pollination services decline dramatically across the landscape with increasing intensification (Kremen et al. 2002), differential responsiveness among bee species may dampen the effect of increasing land use, which would otherwise be even more dramatic (Winfree and Kremen 2008).

Table 14.2 Spatial variation (measured as CV, Coefficient of Variation) of insect-pollinated fruit set among highland (*C. arabica*) and lowland (*C. canephora*) coffee trees per site across a gradient of pollinator species richness in different coffee agroforestry systems (Klein *et al.* 2003a,b, Klein *et al.* 2008). (A) Results of simple regressions; (B) selected glm (generalized linear model) that explains the highest variance when including functional group richness, species richness, individual number and all interation terms. Models were selected according to lowest AIC and highest P-values. AIC (Akaike Information Criterion) and significance level $^{(*)} p < 0.1$; $^{*}p < 0.05$; $^{**}p < 0.01$; $^{***}p < 0.001$ of different Models were fitted. Analyses were done in R, version 2.6.2.

Response variable	Explanatory variable	AIC
Coffea arabica		
A Simple regressions		
CV bee pollinated fruit set	Functional group richness	190.7***
"	Species richness	195.9***
"	Individual number	195.9**
B Selected glm model		
CV bee pollinated fruit set	Functional group richness + species richness + individual number + (functional group richness × individual number) + (species richness × individual number)	190.1**
Coffea canephora		
A Simple regressions		
CV bee pollinated fuit set	Functional group richness	142.2*
"	Species richness	143.4$^{(*)}$
"	Individual number	144.2
B Selected glm model		
CV bee pollinated fruit set	Functional group richness + (species richness × individual number)	141.5*

Although species richness is often correlated with aggregate abundance in bee communities (Kremen and Chaplin-Kramer 2007), data on three crop species suggests that species richness may be even more important than aggregate abundance. The data were collected along land-use gradients in Indonesia differing in habitat and landscape management to 'experimentally' manipulate bee species richness and individual numbers in real world ecosystems. Flower-visiting observations and pollination experiments were conducted for three species representing three different breeding systems: (1) lowland coffee, *C. canephora*, a mainly self-incompatible species, (2) highland coffee, *Coffea arabica*, a mainly self-compatible species (Klein *et al.* 2003a, Klein *et al.* 2008); (3) and pumpkin, *Cucurbita moschata*, a monoecious species, meaning that female and male gametes are spatially and temporally separated (Delaplane and Mayer 2000). For all three crop species flower-visitor species richness was more important to explain pollination services when factoring out for the effect of flower-visiting individual numbers (Hoehn *et al.* 2008, Klein *et al.* 2008). Hence the studies demonstrate that bee species richness is essential for overall pollination services. Consequently, the following question arises: what are the mechanisms behind the relationship of species richness and pollination services?

14.3 Mechanisms for the effects of flower-visiting species richness on providing pollination services

Three mechanisms known to be important in BEF patterns may explain why diverse flower-visiting communities can function better than depauperate communites. First, under *sampling (selection) effects* (Loreau *et al.* 2001), high species richness increases the likelihood of random inclusion within the flower-visiting community on a given crop of an efficient and effective pollinating species. Second, under *niche complementarity* (Loreau *et al.* 2001), high species richness improves pollination services because species differ in their foraging behavior and may complement one another by visiting and pollinating different flower species, or in the spatial or temporal distribution of visits within or among flowers of a species. Third, under *functional facilitation* (Cardinale *et al.* 2002) a given species has a positive effect on the functional capability of other species. For example, interspecific interactions during foraging can lead to higher frequencies of cross pollination (Klein *et al.* 2008).

All three mechanisms assume that different flower-visiting species differ in their morphology, physiology, behavior, or other traits, so that high species richness maximizes resource use and thus pollination success, as discussed in more detail below.

14.3.1 Sampling (selection) effect

The sampling effect asserts that in experimental designs with varying species richness, increasing richness results in a higher chance of including a species that makes a disproportionate contribution to ecosystem functioning (Loreau 2000). In this case, the species in the species pool that is the most efficient (e.g. produces the most seeds per visit or deposits the most high-quality pollen) in pollination would dominate the pollination service and mask minor benefits by other species in multi-species communities whenever it was present. Due to the difficulty of experimentally manipulating flower-visiting communities, little evidence for or against the role of random sampling effects in pollination function is available. In natural experiments along land-use intensity gradients, Larsen *et al.* (2005) found that the most species-rich communities visiting watermelon were the only ones contained the most effective flower-visiting species for pollination. However, they showed that this was not due to random sampling effects, but rather to non-random local extinction processes.

The three crop studies from Indonesia (Hoehn *et al.* 2008, Klein *et al.* 2008) indicate that not sampling effect but niche complementarity seems to be the prodominant mechanism for high quality and quantity in pollination services. This prediction is based on the finding that functional group richness explained the pollination service better than species richness *per se*. In all three studies, functional group richness was classified using *a priori* groups of morphological and behavioural pollinator traits.

Generally, sampling effects might be more important in more specialized plant–flower visitor relationships, where a few visiting species, maybe due to proboscis length, are much better adapted to flower morphology than others and thus more effective as pollinators. In such cases a more species-rich community has a higher chance of including the most effective pollinator species by random chance.

14.3.2 Niche complementarity and spatial stability

The theory of complementarity between species assumes that high species richness increases functional effectiveness by increasing the efficiency of resource use over space and time. Therefore, contributions of many species will increase resource use with each additional, individual species through distinct resource partitioning, until a plateau is reached. For example, a diverse flower-visiting community whose members exhibit species-specific or functional group-specific use of spatio-temporal niches during flower visitation should lead to more efficient resource use. By exploiting different spatio-temporal niches (e.g. different parts of the flower, tree or field, different times of the day), different flower-visiting species might maximize their ability to obtain pollen and nectar resources (Herrera 1988). Resource use for flower visitors consists of gathering pollen and nectar and this action simultaneously pollinates the plant. More efficient resource use through complementarity could thus lead to better pollination services for the plants, for example, a greater number of flowers per plant are pollinated, or a better distribution of pollen on the stigma is obtained.

For pumpkin (Hoehn *et al.* 2008), characterizations of the species-specific differences in spatio-temporal pattern of flower visitation were also made in high and low-bee species richness systems. In the high bee species richness sites, pumpkins reached their maximum in seed numbers (hand-pollinated flowers served as reference) (319 ± 105 standard deviation), whereas in the low-bee species richness sites, the number of seeds per pumpkin was significantly lower (179 ± 44). Comparing the spatial (flower-visiting height of the plant) and the

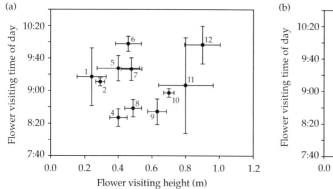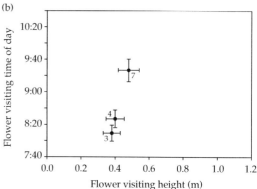

Figure 14.1 Flower height and forage time preferred by the most abundant flower-visiting bee species of planted pumpkin plants in two different habitat types: (a) grassland with high bee species richness and (b) forest with low bee species richness in Central, Sulawesi, Indonesia. Each habitat type was replicated with four plots in spatially separated sites. Experimental pumpkin patches of 2 × 5m were planted in each plot and liquid fertilizer was applied every other week (for more information see Hoehn et al. 2008). Arithmetic means ± standard errors are given. Numbers are presenting following bee species: 1 – *Nomia concinna*, 2 – *Lasioglossum* sp., 3 – *Apis cerana*, 4 – *Xylocopa dejeani*, 5 – *Nomia fulvata*, 6 – *Ceratina cognata*, 7 – *Trigona* sp., 8 – *Amegilla* sp., 9 – *Xylocopa confusa*, 10 – *Lasioglossum halictoides*, 11 – *Apis dorsata*, 12 – *Xylocopa nobilis*.

temporal (flower-visiting time of the day) activity of the flower-visiting species in both systems, it could be shown for the high species richness sites, comprising 10 species, that these species covered the entire spectrum of spatio-temporal niches. This increases the probability that flowers at any height were visited and flower visitors were active during the receptive period of the flowers (Fig. 14.1(a)). In contrast, the three bee species found visiting the low species richness sites did not cover all the spatio-temporal niches (Fig. 14.1(b)). In other words, in the low-bee species richness sites, spatial and temporal niches become unoccupied when bee species are lacking.

Another study on strawberries (Chagnon et al. 1993) demonstrates complementarity due to spatial factors. Here the crucial pollination trait was the species-specific behaviour of the bee interacting with the floral parts. Strawberries depend on pollen dispersion over the whole stigma to develop completely formed fruits. Small wild bees pollinated predominantly at the base of the stigma, while large honey bees were active on the centre of flowers, pollinating the apical part of the stigma. Absence of one of the two functional guilds causes deformed fruits, due to incomplete resource use by only one functional group. This example demonstrates that within-flower foraging behaviour can be strongly body size-dependent. Also, in the pumpkin system, species-specific flower-visiting behaviour (time of visitation and height and duration of floral visit) was strongly related to the body size of the visiting species (Hoehn et al. 2008). Therefore, morphological traits can be responsible for differences in spatio-temporal floral resource use.

Additionally, Table 14.1 shows the spatial variation of insect-pollinated fruit set among coffee plants per site across different agroforestry systems comprising different bee communities. Simple relationships between insect-pollinated fruit set and functional group richness or species richness or individual numbers showed that for the spatial stability aspect, functional group richness seemed to play an important role. Nevertheless, models including interaction effects between functional group richness, species richness, and individual numbers explained fruit set similar to simple models using species richness alone (Table 14.1). Here, statistical limitations were reached, because of the collinear nature between functional group richness, species richness, and individual numbers in the real world (as opposed to experimental) study systems.

These case studies comprising species with different plant breeding systems show that high flower-visiting species richness including different functional groups of flower visitors can lead to optimal (high levels with enhanced stability) pollination services for agricultural systems. Many studies show that flower-visiting communities are structured by behavioural niche differentiation, due to body size, circadian rhythm (as a result of differences in temperature tolerance), competition hierarchy, sociality, and other species-specific behavioural traits (Bishop and Armbruster 1999, Pinkus-Rendon *et al.* 2005, Stone 1994, Stone *et al.* 1999). Hence different functional traits of a species rich flower-visiting community can be important because niche complementarity between species maximize both resource use and pollination services provided.

14.3.3 Functional facilitation

A study by Greenleaf and Kremen (2006a) showed that wild bees contribute to sunflower pollination indirectly via facilitation. Hybrid sunflower seed production depends on animal-mediated pollen transfer from male (pollen- and nectar-producing) to female (only nectar-producing) parents. Although honey bees have low sunflower pollination efficiency per-visit, they are the only commercially available pollinators. Much higher pollination per visit efficiency of honey bees was achieved in areas with higher wild bee individual and species numbers. Encounters between honey bees and wild bees increased the likelihood that honey bees switched between female and male plants, thus increasing per-visit pollination efficiency. Otherwise, individual honey bee foragers tended to forage either on male plants for pollen or on female plants for nectar, and rarely transferred pollen between plants. Another potential facilitative mechanism could be dispersal of clumped pollen by honey bees of pollen deposited on female sunflowers by the wild bees. In the sunflower case, both individual and species number of the wild bee community significantly contributed to increasing the per visit pollination effectiveness of honey bees (Greenleaf and Kremen 2006a). However, in examples where only a few species are involved in interspecific interactions, species identity and individual numbers might be more important than species richness *per se* (Jonsson and Malmqvist 2003).

14.4 Plant–flower visitor interaction webs in crop pollination systems

Crop pollination studies are based on a given crop and its pollinating species, although crop plants may be influenced by interactions of their flower visitors with nearby wild plants, forming networks of dozens to hundreds of species. To date, mutualistic network approaches have only been applied in non-crop systems in order to understand ecological and evolutionary processes, but their application to crop systems may provide underlying information for managing and restoring flower-visiting species richness and crop pollination services. In Section 14.4.1 plant–flower visitor interaction webs (PFv webs) are shown and specific characteristics of the general structure highlighted, followed in Section 14.4.2 by a discussion of the importance of PFv web characteristics for species extinction and integration. Afterwards the role of higher trophic levels for PFv webs is described (Section 14.4.3). Section 14.4.4 summarizes some ideas of how quantitative PFv webs provide information to manage and restore flower-visiting species richness for crop pollination services.

14.4.1 PFv web structure and characteristics

Fully quantified plant–flower visitor webs provide the clearest description of the flower-visiting community structure, since they express the species individual numbers and the frequency of interactions among species (e.g. Forup *et al.* 2008, Memmott 1999, Fig. 2). Within such webs, the integration into the community of certain plants, like very rare species or an abundant flowering crop, or a managed pollinator, such as honey bees, can be studied (Gibson *et al.* 2006, Memmott and Waser 2002). Here, four PFv webs collected in an agricultural landscape in Dorset, UK, are shown to demonstrate the interactions between bees, crop, and wild plant species of surrounding flowering strips for bean (Fig. 14.2(a)), oilseed rape (Fig. 14.2(b)), and lupine (Fig. 14.2(c)). An additional PFv web in calcareous grassland is shown to visualize the differences between crop PFv webs and a PFv web of a semi-natural habitat without crop species (Fig. 14.2(d)). The webs of the crop fields are ordered from relatively poor

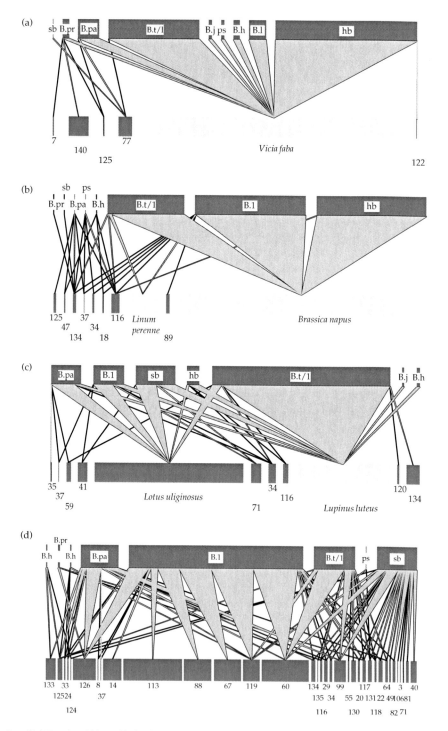

Figure 14.2 Quantified PFv webs in (a) bean, (b) oilseed rape, (c) lupine, and (d) calcareous grassland. The area sampled ranged between 10 and 13 ha and included a 10 m margin of flowering herbs around each crop field. Quantification was done by counting the specific interactions between a plant and a bee species over a sampling time of two to three hours on a sunny, windless day, except in the case of the grassland, where sampling was done repeatedly over the period of floral availability. All flowering plants were counted within one metre quadrates that were placed in a stratified manner over the area to estimate the

plant-flowering strips of the bean field to relatively rich plant strips of the lupine field. The oilseed rape field was adjacent to a linseed field which was at the end of its flowering period during sampling. The most abundant generalist bees were bumble bees, *Bombus terrestris/lucorum*, which were strongly attracted to the flowering crop, while more specialized bumblebees, such as *B. pascuorum*, *B. lapidarius*, *B. hortorum* and solitary bees appeared to prefer alternative flowering resources in the flowering strips. There was little attraction for solitary bees to either bean or oilseed rape, while lupine was visited frequently by leafcutter bees. In comparison, more evenly distributed interactions among plants and bees and a species-rich local community of solitary bees were found in calcareous grasslands where bees specialized on many different plant species (Fig. 14.2 (d)). Honey bees as generalist foragers were strongly attracted to mass-flowering crop and were not encountered in the grassland (non-crop) site.

The four quantitative PFv webs visualize the interaction between crop and wild plants and their flower-visiting species and can be used to identify whether the use of flowering strips adjacent to a crop field attracts several species to forage and interact via facilitation and complementarity to benefit the crop or divert flower-visiting species from the crop plants, which could lead to reduced flower visitation.

PFv webs are characterized by their highly nested structure. This indicates that flower-visiting specialists mainly interact with generalized plants that are visited by a whole array of other flower-visiting species. Specialist plants tend to be visited by generalist insects (Bascompte *et al.* 2003, Bascompte *et al.* 2006, Blüthgen *et al.* 2007, Montoya *et al.* 2006, May 2006). This nestedness leads to an asymmetry in plant–flower visitor interactions with strong dependence in one direction often linked to weak dependence in the reverse direction (Thompson 2006, Vázquez and Aizen 2004). Furthermore, mutualistc networks are dominated by a few extreme generalists, meaning only a few species build the core of the nested web and are much more connected than would be expected by chance (Jordano *et al.* 2003).

14.4.2 Species extinction and integration in PFv webs

Habitat loss or alteration, species extinction and addition of invasive species are the main factors disrupting PFv webs. Given the current, accelerated biodiversity loss, it becomes important to understand how whole networks of interacting plant and animal species are affected by extinction events and especially how extinction events might affect crop pollination services. The characteristic structure of PFv (nestedness and asymmetry) provides important alternative routes when species disappear and thus a certain insurance against biodiversity loss (Bascompte *et al.* 2006, Memmott *et al.* 2004).

Simulated progressive extinction models are currently only available for non-crop flower visitor webs. Memmott *et al.* (2004), for example, simulated flower-visiting species extinction to explore the resulting extinctions in plants depending on flower visitors for reproduction. The extinction scenarios

Figure 14.2 *(continued)*

abundance of floral resources per square metre. Bee counts were then extrapolated to the square metre unit. The upper represents bees and the lower bars plants. Each bar depicts a species and the length of the bar contains information about its abundance. The wedges depict interactions between plants and pollinating bees. The width at the basis of a wedge contains information about the interaction strengths. Honey bees (hb) did not occur in the grassland but were abundant on all three crop species. Bumble-bee groups consisted of *Bombus terrestris* and *B. lucorum* (B.t/l: pooled because they were difficult to distinguish in the field), *B. lapidarius* (B.l), *B. pascuorum* (B.pa), *B. pratorum* (B.pr), *B. hortorum* (B.h), *B. jonellus* (B.j) and parasitic *Bombus* species (ps; former *Psythirus*). Solitary bees (sr) were pooled because straight identification in the field was difficult. The flowering crop plants were labelled by name and bars were depicted in grey; the other plant species were: 3 = *Achillea millefolium*, 7 = *Anchusa arvensis*, 8 = *Anthyllis vulneraria*, 14 = *Bellis perennis*, 18 = *Buddleja davidii*, 20 = *Campanula glomerata*, 22 = *Campanula rotundifolia*, 24 = *Carduus nutans*, 29 = *Centaurea nigra*, 33 = *Cirsium acaulis*, 34 = *Cirsium arvense*, 35 = *Cirsium palustre*, 37 = *Cirsium vulgare*, 40 = *Crepis capillaris*, 41 = *Crepis paludosa*, 47 = *Digitalis purpurea*, 49 = *Echium vulgare*, 55 = *Eupatorium cannabinum*, 59 = *Galeopsis tetrahit*, 60 = *Genista tintoria*, 64 = *Helianthemum nummularium*, 67 = *Hippocrepis comosa*, 71 = *Hypochaeris radicata*, 77 = *Lamium purpureum*, 81 = *Leontodon autumnalis*, 82 = *Leontodon hispidus*, 88 = *Lotus corniculatus*, 89 = *Lotus uliginosus*, 99 = *Odontites verna*, 106 = *Pilosella officinarum*, 113 = *Ranunculus repens*, 116 = *Rubus fruticosus*, 117 = *Scabiosa columbaria*, 118 = *Senecio jacobaea*, 119 = *Serratula tintoria*, 120 = *Silene dioica*, 122 = *Sinapis*, 124 = *Stachys officinalis*, 125 = *Stachys selvatica*, 126 = *Succisa pratensis*, 130 = *Teucrium scorodonia*, 131 = *Thymus polytrichus*, 133 = *Trifolium pratensis*, 134 = *Trifolium repens*, 135 = *Ulex europaeus*, 140 = *Veronica persica*.

were either random extinction of flower-visiting species, extinction from least linked to most linked species, or *vice versa*. The webs were surprisingly robust towards flower-visiting species extinctions unless the most generalist (most linked) species were removed first. In this case, the extinction trajectory of plant species was tightly correlated to that of declining flower-visiting species richness following a relatively rapid (essentially linear) decline. Furthermore, some morphological features, such as body size, may make certain flower-visiting species more prone to extinction than others. Hence in systems where generalist flower-visiting species are large bodied, extinction would follow the worst-case scenario and pollination function would disappear as quickly as flower-visiting species (Larsen *et al.* 2005).

The opposite of extinction through habitat loss is the integration of alien species into existing mutualistic networks. In some ways, this occurs in crop systems when sudden appearance of abundant floral resources (the flowering crop) will change the behaviour of the flower-visiting species. Resident plants may either facilitate (increase) or compete with (decrease) visitation to the introduced (crop) plant. Experimental studies on such multi-species interactions typically have looked at the role of invasive plants on wild plant reproduction and have show various results from negative (Chittka and Schürkens 2001, Ghazoul 2002, Larson *et al.* 2006), to positive (Johnson *et al.* 2003, Moller 2004), to neutral (Aigner 2004). Community studies that investigate the impact of alien plant species on the whole mutualistic network show substantial integration of these plants into native PFv webs, with generalist flower-visiting speices most likely to visit introduced plant species (Memmott and Waser 2002, Morales and Aizen 2006). Studies on the effects of one alien plant on an array of native plants find mixed results that depend on the species identity of native plants (Moragues and Traveset 2005) and on the stages of invasion (Aizen *et al.* 2008b). Despite increased visitation frequencies in plots with a highly rewarding invasive plant, pollination of native plant species is not necessarily facilitated by the co-occurrence of these high-rewarding alien plant species, possibly because of the large amount of alien pollen they receive (Larson *et al.* 2006), as was shown by quantified pollen transport webs (Lopezaraiza-Mikel *et al.* 2007). By analogy, these studies suggest that flowering crop species will be visited by flower-visiting generalist species and will integrate reasonably well in existent interactions webs.

14.4.3 Higher trophic levels in PFv webs

A further aspect that can be examined using the food web approach is to elucidate the role of higher trophic levels such as natural enemies of pollinators like predators, parasites, and parasitoids. Because of habitat losses and modifications through agricultural practices, pollinator communities may become more susceptible to some dominant enemies (Klein *et al.* 2006). A well-documented case is the dramatic increase of managed (Downey and Winston 2001) and feral (Kraus and Page 1995) honey bees infested by the *Varroa* mites and the subsequent declines or losses of many colonies. PFv webs analyzed to date, however, exclude predators, parasitoids, and diseases. Higher trophic levels increase the connectedness and nestedness of food webs (Lafferty *et al.* 2006), but food web properties are affected by more than just the species present. Habitat modification through agriculture also affects food web properties. These properties may negatively affect species richness, composition of communities, and the diversity and strength of interactions. Tylianakis *et al.* (2007), for example, used 48 quantitative parasitoid–host (bee and wasp) webs and showed, although overall species richness were not affected by land-use intensity, that web properties, such as connectance, linkage density, compartment diversity, and interaction evenness, were all affected by land-use management. This can be explained by the high vulnerablility of the interaction structure to the presence, identity, phenology, physiology, behaviour, and diversity of different species. Hence, parasitoid–host interactions might show changes before species loss becomes apparent (Tylianakis 2008).

14.4.4 Quantitative PFv webs and the management of crop pollination services

In summary, quantitative food web analysis visualizes the structure and links between trophic levels and provides more detailed views of entire communities when assessing management effects on crop-flower visitor interactions. Nestedeness, asymmetry, and interaction strength are PFv characteristics that might indicate stability in crop pollination services, meaning insurance for flower-visiting species losses and effects of added invasive species.

Quantitative analyses can be also used to assess crop pollination services of multi-cropping systems or to evaluate management or restoration plans using flowering strips. Here web analyses give detailed information to help selecting appropriate wild flowering plant species, which do not divert flower-visiting insects from the crop plant and provide additional resources for them. Conversely, quantitative webs can also help to understand whether crop flower resources benefit certain wild flower-visiting insect species and/or their enemies, species of the higher trophic level. In this respect, statistical analyses using replicated networks to show, for example, degrees of specialization and generalization of the enemy species, the flower-visiting species, and plant species and the number of shared species in different management conditions will help to understand management effects on the flower-visiting community and its services for crop and wild plants (Forup et al. 2008, Tylianakis et al. 2008).

14.5 Consequences of pollinator decline for the global food supply

Both wild and managed pollinators have suffered significant declines in recent years. Managed *Apis mellifera*, the most important global source of crop pollination services, have been diminishing around the globe and particularly in the USA, where colony numbers are now at < 50 per cent of their 1950 levels (NRC 2006). In addition, major and extensive colony losses have occurred over the past several years in North America and Europe, possibly due to diseases and other factors (Cox-Foster et al. 2007, Stokstad 2007), causing shortages and rapid increases in the price of pollination services (Sumner and Boriss 2006). These recent trends in honey bee health illustrate the extreme risk of relying on a single bee species to pollinate the 75 per cent of crops that rely to some degree on insect visitation.

At the same time, although records are sorely lacking for most regions, comparisons of recent with historical records (pre-1980) have indicated significant regional declines in species richness of major pollinator groups and the plants they pollinate (bees and hover flies in Britain, bees alone in the Netherlands) (Beismeijer et al. 2006). Despite the alternative routes in PFv webs when species disappear, a recent meta-analysis on effects of habitat fragmentation on plant reproduction demonstrated that the reproductive success of self-incompatible (i.e. typically pollinator-dependent) but not self-compatible plants is strongly negatively influenced by habitat fragmentation, and that effects on both pollination and reproductive success were strongly correlated (Aguilar et al. 2006). These results strongly implicate pollinator loss as a major causative factor for the susceptibility of plant reproduction to one major disturbance, habitat fragmentation.

Large reductions in species richness and individual numbers of bees have also been documented in regions of high agricultural intensity in California's Central Valley (Kremen et al. 2002, Klein and Kremen unpublished). Traits associated with bee and hoverfly declines in Europe included floral specialization, slower (univoltine) development, and lower dispersal (non-migratory) species (Beismeijer et al. 2006). In the most well-known taxon, bumble bees in Europe, declining species are long-tongued specialists on Fabaceae, and their increasing rarity may be due to the loss of unimproved grasslands that are rich in their food resources (Goulson et al. 2008). Specialization is also indicated as a possible correlate of local extinction in flower-visiting communities studied across a disturbance gradient in Canada. It could be shown that communities in disturbed habitat contained significantly more generalized species than those associated with pristine habitats (Taki and Kevan 2007). Large-bodied bees were more sensitive to increasing agricultural intensification in California's Central Valley, and ominously, bees with the highest per-visit pollination efficiencies were also most likely to go locally extinct with agricultural intensification (Larsen et al 2005).

Thus, in highly intensive farming regions such as California's Central Valley that contribute comparatively large amounts to global food production (e.g. 50 per cent of the world supply of almonds), the supply of wild bee species is lowest in exactly the regions where the demand for pollination services is highest. Published (Kremen et al. 2002) and recent studies (Klein and Kremen, unpublished) clearly show that the services provided by wild bee flower-visiting species are not sufficient to meet the demand for pollinators in these intensive regions. These regions are instead entirely reliant on managed honey bees for pollination services. If trends towards increased agricultural intensification continue elsewhere (e.g. as in Brazil, Morton et al. 2006), then pollination services from wild species are highly likely to decline in other regions, based on a recent comprehensive analysis across 16 crops on five continents that showed how wild bee pollinators decline in species richness and individual numbers with distance from natural habitats (Ricketts et al. 2008). At the same time, global food production is shifting increasingly towards production of pollinator-dependent foods (Aizen et al. 2008a), increasing our need for managed and wild pollinators yet further. Global warming, which could cause mis-matches between flower-visiting species and the plants they feed upon, may exacerbate pollinator decline (Memmott et al. 2007). For these reasons, more serious shortages of pollinators in the future may indeed be faced.

A recent global assessment of the economic impact of pollinator loss (e.g. total loss of pollinators worldwide) estimates our vulnerability (loss of economic value) at (153 billion or 10 per cent of the total economic value of annual crop production (Gallai et al. 2008), a much larger figure than previous global estimates (e.g. Costanza et al. 1997). Although total loss of pollination services is unlikely to occur and to cause widespread famine, they potentially have both economic and human health consequences. For example, some regions of the world produce large proportions of the world's pollinator-dependent crops. Such regions would experience more severe economic consequences from the loss of pollinators, although farmers and industries would undoubtedly quickly respond to these changes in a variety of ways, passing the principle economic burden on to consumers globally (Gallai et al. 2008, Southwick and Southwick 2002). Measures of the impacts on consumers (consumer surplus) are of the same order of magnitude ((195–310 billion based on reasonable estimates for price elasticities, Gallai et al. 2008) as the impact on total economic value of crop production.

Nutritional consequences may be more fixed and more serious than economic consequences, due to the likely plasticity of responses to economic change. The 75 per cent of crop species that are pollinator-dependent supply not only up to 35 per cent of crop production by weight (Klein et al. 2007), but also provide essential vitamins, nutrients, and fibre for a healthy diet (Gallai et al. 2008, Kremen et al. 2007). The nutritional consequences of total pollinator loss for human health have yet to be quantified. However, food recommendations for minimal daily portions of fruits and vegetables are well known and already often not met in diets of both developed and underdeveloped countries.

14.6 Conclusions and future directions

Pollination is a critical step in the reproduction of many plant species and can significantly reduce primary production in natural communities and crop fruit set and quality in agricultural systems if not provided or provided in adequate supply. Some landscapes promote high flower-visiting species richness and frequency offering adequate crop pollination services because they offer season-around foraging and nesting resources (Kremen et al. 2002, Winfree et al. 2008). Simple landscapes with low proportion of natural and semi-natural habitats can negatively affect flower-visiting species richness and frequency, which sometimes affect fruit set (Ricketts et al. 2008). Consequently, in intensively managed agricultural landscapes where pollinator-dependent crops are grown, farmers are managing bees for pollination services, as for example in alfalfa, almond, watermelon, and sunflower production in California. The effect of isolation from natural habitats on crop pollination in 23 studies is summarized by Ricketts et al. (2008), but more studies using fruit set and production data are needed to produce key synthetic results.

Can high numbers of just one species like the honey bee substitute for flower-visiting species richness? Californian's farmers are producing high yields and the honey bee pollination strategy seems to be successful. The examples discussed in this chapter indicate that in some situations managing for one pollinator species will not only result in higher pollination insurance but also result in optimal pollination success. Different mechanisms lead to these effects, such as niche complemantarity for resources, facilitation among species, or sampling to reach highest pollination frequency.

What further research is needed for a holistic understanding of the role of flower-visiting species richness for crop pollination? To predict in which crop-flower-visiting system and under which conditions species or functional group richness can be a limiting factor for pollination services, research in experimental settings to separate for the effects of functional group richness, species richness, and individual numbers *per se* are needed. Flower-visiting community composition needs to be experimentally set up to separate out the effects of aggregate abundance and species richness on pollination services. Such experiments are frequencly carried out for plant communites to understand plant productivity. For pollination services, it is important to understand how much a species-rich, but individual-poor flower-visiting community can buffer pollination services loss of an abundant species like the European honey bee.

Mechanisms of species richness–pollination services relationships are complex, particularly when considering whole PFv webs including different trophic levels. These complex interactions outside the crop-flower visitor network itself have received little attention to date. Quantitative PFv webs can be used to (1) assess crop pollination services of muti-cropping systems or to evaluate crop pollination management or restoration plans using flowering strips, (2) and help to understand whether crop flower resources benefit certain wild flower-visiting insect species or their enemies.

Studies in real systems need to include the complex spatio-temporal multitrophic interactions among pollinating species, their mutualistic partners, and their enemies. Ideally, all production limiting factors and services per crop production system should be studied to calculate the exact value of pollination services to crop production (Bos *et al.* 2007). As many farmers will only adopt restoration activities on their properties when recognizing an economic advantage, opportunity cost for restoration and conservation programs needs to be calculated (Ghazoul 2007). Estimating the costs and benefits of restoring pollinator habitat for increasing wild pollination services are also needed to assess the economic potential for conservation-incentive markets for pollination services.

Box 14.1 Economic value of bees as crop pollinators – the case of alfalfa seed production

Commercial pollination services are mainly provided by managed honey bees through a standing and organized market between beekeepers and farmers. At the global scale, alfalfa seed production depends on the services of managed honey bees (mainly in California) and other managed species like alkali bees and leafcutter bees (mainly in Canada, western USA), and managed bumble bees (in Turkey), but also on diverse wild insect communities, because seed production and quality is extremely low (4 per cent) if pollinators are excluded (Cecen *et al.* 2008).

Alfalfa or Lucerne, *Medicago sativa* L., is the most important forage crop in many parts of the world (FAOStat 2006). Economic markets for pollination services for the Californian alfalfa industry developed between 1949 and 1951 (Sumner and Boriss 2006). Commercial pollination for alfalfa started with honey bee pollination experiments in 1947 followed by increased renting of honey bees for alfalfa seed production. At the same time, research to use alkali bees and leafcutter bees for alfalfa seed production was carried out, as it was noted already in 1940 that a high diversity of insect species are visiting alfalfa flowers. Some solitary bee species were also noted to forage under different environmental conditions (Bohart 1947, Olmstead and Wooten 1987, Mueller 1999). Even for California with the honey bee as the cheapest and most important pollinator, farm advisers recommended a combination of leafcutter bees and honey bees to reach high seed production (Mueller 1999). Experimental testing, however of whether the honey bee alone or a combination with other bee species results in higher seed set or quality is missing.

Although bumble bees and solitary bees access the alfalfa blossoms in a more effective way, resulting in higher per visit seed numbers relative to honey bee visits (Delaplane and Mayer 2000, Cecen *et al.* 2008), honey bees are dominant in alfalfa seed production. In California, farmers usually prefer short-term honey bee colony rental because these portable pollination units are cheap and beekeepers usually take the responsibility to provide strong units during bloom.

Continues

Box 14.1 (*Continued*)

The establishment of solitary bee nests is expensive and farmers need additional training for successful long-term establishment. Losey and Vaughan (2006) estimated the value of wild bees (excluding pollination by managed bees) for alfalfa seed production. They assigned an annual amount of $5.45 million to wild pollinators. This is 5 per cent of the annual average US gross yield for alfalfa seed production, assuming that 5 per cent of this production is provided by wild bees. This upper-bound value represents the contribution of wild pollinators to gross production. Such a calculation does not consider future adaptive responses by farmers and consumers to pollinator decline (Muth and Thurman 1995). Such responses, like adopting alternative crops or farming techniques that reduce the dependence of farmers on pollinators, will lower the value of pollination services from wild pollinators, as would consideration of net rather than gross revenues (Olschewski et al. 2006).

Another approach to assess pollination services by managed bees is to estimate how many bee workers will be needed to fully pollinate a given area of crop production (Robinson et al. 1989). Here, such a lower-bound value based on production and cost per area of alfalfa seed production in Montana (USDA-NASS 2008) is calculated for pollination services provided by rented honey bees and by managed leafcutter bees (table below). Mean honey bee colony rental numbers and mean renting cost for the blooming period of alfalfa were used to calculate farmers' costs for renting honey bees (Sumner and Boriss 2006). The cost to manage leafcutter bees on alfalfa-growing land is available and was calculated for a management period of 15 years (British Columbia Ministry of Agriculture and Food 1998). Management costs of leafcutter bees to pollinate alfalfa are similar to those of honey bee renting during alfalfa bloom. This is only the case when managing leafcutter bees over a longer time period, as the main expenses are covered in the first year of leafcutter bee establishment (bee cocoons, nesting blocks, nesting shelters). In other words, only long-term management of solitary bees is economically reasonable.

The average annual value (gross yield) of the total alfalfa seed production in Montana is $4,729,400 (USDA-NASS 2008). The state-wide farmer's cost and value for the pollination services of honey bees was calculated to be $1,220,550 and for leafcutter bees $1,394,620 (see table below). Thus pollination service values (farmer's cost) represent about 25.8–29 per cent of the value (gross yield) of this crop.

Relying on a single pollinator species is a risky strategy and this chapter highlights that in at least some cases different species can complement each other to achieve the best pollination results. An equivalent economic evaluation approach to estimate the pollination services provided by wild flower-visiting communities is to calculate replacement costs for honey bees by wild insects (Kremen et al. 2007). For this approach a calculation of the costs to restore agricultural landscapes for crops to get full pollination services by wild insects is needed, but to date not available.

Annual value of bee pollination of alfalfa seed production in Montana, based (a) on farmer's mean annual cost of renting honey bees; or (b) managing leafcutter bees, to fully pollinate alfalfa for seed production based on 1997–2007 production and yield data.

	Area in production (ha)	Stocking rate (hives/ha or bee cells/ha)[1]	Rental, management cost ($/hive, $/10,000 cells)[2]	Cost/ha	State-wide cost ($)
Honey bees	5150	7.9	30	237	1,220,550
Leafcutter bees	5150	74,100	36.6	270.8	1,394,620

[1] average number calculated by Delaplane and Mayer (2000)
[2] HB= average annual value considering the years of 1996–2005 (Sumner and Boriss 2006); LCB = average annual value for bee management of 15 years (British Columbia Ministry of Agriculture and Food 1998).

CHAPTER 15

Biodiversity and ecosystem function: perspectives on disease

Richard S. Ostfeld, Matthew Thomas, and Felicia Keesing

15.1 Introduction

Experimental and comparative studies increasingly reveal that certain ecosystem functions are maximized in highly diverse ecological communities. Key among the ecosystem functions correlated with high biodiversity are rates of nutrient cycling, primary production, and resistance to disturbances such as drought (see Chapters 1 and 2). Despite considerable interest among ecologists in this dynamic area of research, we suspect that non-ecologists are unaware of or unimpressed by these functions served by high biodiversity. A 2007 survey shows that considerably more respondents in the European Union consider moral reasons stronger than utilitarian or economic reasons for protecting biodiversity (http://www.ec.europa.eu/public_opinion/archives/flash_arch_en.htm). In a 2002 survey in the USA (http://www.biodiversityproject.org/resourcespublicopinion.htm), utilitarian and economic functions of biodiversity, with the exception of human health, are not even mentioned. It would appear that the ecosystem functions typically addressed by biodiversity scientists are not among the principal concerns of many citizens when considering the consequences of biodiversity loss, even though there are increasing arguments for why we should be concerned (the Millennium Ecosystem Assessment (2005a) see Chapters 17–19). Globally, the Convention on Biological Diversity has been signed or ratified by the vast majority of the world's nations, committing them to preserve biodiversity in order to achieve sustainable development, but it has had little traction. Thus, there is a disconnect between public perception and governmental and scientific understandings.

The public, however, is interested in disease, both newly emerging and resurgent diseases, and it is here that the role of biodiversity may resonate with citizens. Recent studies have shown that high biodiversity can strongly reduce rates of disease transmission, and consequently that biodiversity loss can exacerbate disease risk. These studies themselves are diverse in the aspects of biodiversity under study, in the taxa of pathogens, hosts, and vectors involved, and in the mechanisms postulated. For example, high genotypic diversity within a species reduces transmission of bacteriophages that attack bacteria (Dennehy et al. 2007), microsporidians that attack *Daphnia* water fleas (Pulkkinen 2007), and fungi that attack rice plants (Zhu et al. 2000). On the other hand, high *species richness* reduces transmission of fungal pathogens of herbaceous plants (Roscher et al. 2007), viral pathogens of birds and humans (Ezenwa et al. 2006), helminth parasites of snails and vertebrates (Kopp and Jokela 2007), and bacterial pathogens of humans and other mammals (Ostfeld and Keesing 2000a, LoGiudice et al. 2003). Reduced structural diversity in tropical forests as a result of deforestation has been associated with higher risk of exposure to malaria (Vittor et al. 2006). High diversity can reduce disease transmission by reducing encounter rates between infected and susceptible hosts, and by regulating the abundance of species that are important for pathogen persistence, among other mechanisms (Keesing et al. 2006). Knowledge that biodiversity influences human health and that of valued wildlife and plants is likely to heighten interest among both scientists and non-scientists in the importance of biodiversity.

In this chapter we explore some of the major issues that have shaped recent explorations of the biodiversity-ecosystem function relationship and apply these concepts to infectious diseases. These issues include: (1) the shape of the association between biodiversity and ecosystem function; (2) the relative importance of species biodiversity *per se* (e.g. species richness or evenness) vs species composition; (3) the relative importance of species biodiversity *per se* versus diversity of functional groups or relevant life history traits; (4) whether functions are performed better by the most diverse communities or by monocultures of the species with highest performance of that specific function; (5) how natural sequences of species loss under environmental change (community disassembly) vs. random sequences imposed experimentally influence ecosystem function; and (6) the importance of diversity at organizational levels other than (host) species in influencing ecosystem function. Because the empirical basis for exploring these particular issues is limited, we focus on describing each issue as it pertains to infectious disease and speculating on what might be found in future studies.

15.2 Shape of diversity–disease curves

When the performance of an ecosystem function increases linearly with increases in biodiversity, three related conclusions can be drawn. First, each species contributes approximately equally to the ecosystem function. Second, functional redundancy among species is weak. Third, the loss of biodiversity will result in reduced functioning, irrespective of initial diversity. In contrast, the performance of an ecosystem function can increase asymptotically with increasing biodiversity, implying that individual species contribute unequally, that some redundancy occurs (particularly in highly diverse communities), and that loss of biodiversity has a stronger impact on ecosystem functioning when the initial community is species-poor than when it is species-rich (Schwartz *et al.* 2000 and Chapter 1).

The shapes of curves relating some measure of disease dynamics (e.g. risk or incidence) to species richness have only rarely been measured or predicted, but the scientific basis for the relationships between biodiversity and these properties of ecosystems are the same for functions such as energy flow and nutrient cycling; each species contributes in some way, uniquely or redundantly, to risk or incidence. For West Nile virus incidence rates in humans (Ezenwa *et al.* 2006), and both prevalence and severity of rust fungal infection in perennial ryegrass (Roscher *et al.* 2007), biodiversity–disease curves were negative and apparently linear. In contrast, for Lyme disease risk (measured both as density of infected ticks and as tick infection prevalence), increases in biodiversity resulted in curvilinear decreases to an asymptote (Schmidt and Ostfeld 2001, Ostfeld and LoGiudice 2003). Dizney and Ruedas (in press) show a strongly asymptotic, negative relationship between mammalian diversity and prevalence of infection of deer mice with hantavirus. Other studies linking biodiversity and disease dynamics cannot address this issue because of insufficient variation in diversity; when comparisons are restricted to communities with one versus two or three species, curve shape cannot be estimated.

We suspect that, in some cases, the shape of the curve relating biodiversity to disease risk might be more complex than simple linear or asymptotic functions. In particular, for arthropod vectors or parasites with complex life cycles, some lower threshold level of diversity might be necessary for the disease to exist. This would occur if the vector or parasite requires several host species to fulfil its life cycle. Beyond this lower richness threshold, however, one might expect a negative relationship between diversity and disease transmission, leading to a unimodal distribution. We are not aware of specific tests of this hypothesis.

15.3 Diversity (richness) versus species composition

When individual species have strongly disparate effects on an ecosystem function, then the species composition (identity and relative abundance of all species) of a community is likely to be more important than the number of species or the quantitative value of a diversity metric such as the Shannon index or Simpson index. Much debate has surrounded the issue of how to contend with this issue (see Chapter 4). For researchers interested specifically in effects of species richness, the potential for species to have unequal effects is sometimes considered a complication to be minimized by carefully designed experiments. Often, these researchers will generate experimental communities by randomly selecting species and creating replicate communities at each level of diversity, in order to account for sampling artifacts. Such a design is intended to reduce the probability that the random inclusion of species with particularly strong effects is responsible for the relationship between diversity and function (see Chapters 1 and 2). However, these randomly assembled communities might not represent real communities of interacting species. To contend with this possibility, researchers concerned with how natural variation in diversity causes natural variation in ecosystem functioning must attempt to constrain species composition to mimic natural patterns and determine the relative importance of richness versus composition.

We are aware of only one study that explicitly contrasted species richness and community composition in affecting disease dynamics. In this study, LoGiudice et al. (2008) assessed the causes of variation in the proportion of nymphal blacklegged ticks infected with the Lyme disease spirochete among 49 forest fragments in three northeastern USA states. They found that species richness was significantly, but weakly, negatively correlated with nymphal infection prevalence. However, when they predicted nymphal infection prevalence using a model that specifically incorporated species identity, their results were significant and strong. Qualitatively similar results were obtained in studies of foliar fungal pathogens on grassland plants (Knops et al. 1999, Mitchell et al. 2002, 2003). Although severity of fungal disease was negatively correlated with species richness, the effect of diversity on disease severity was indirect; plots with higher diversity had lower density of the most susceptible grass hosts, and disease severity was positively correlated with host density (Mitchell et al. 2002). As a consequence, it appears that community composition had a more direct impact on disease dynamics than did species diversity *per se*.

15.4 Species diversity versus functional or trait diversity

In many ecological communities, some species are strongly similar in their performance of particular functions (e.g. rate of resource depletion or of net primary production). As a consequence, researchers have begun to classify species into functional groups (e.g. grouping herbaceous plant communities into legumes, non-leguminous forbs, and grasses) and ask whether diversity of functional groups better explains ecosystem functions than does species diversity (e.g. Naeem and Wright 2003, Reich et al. 2004). Functional group classifications often are based on taxonomy or on physiology directly relevant to the ecosystem function of interest (i.e. they represent 'complementarity' of functions). In most cases, functional group diversity performs better than species diversity in explaining ecosystem functioning (Diaz and Cabido 2001, Hooper et al. 2005, Reich et al. 2004). However, the predictive precision of standard classifications of species into functional groups rarely is significantly higher than that of random classifications (Wright et al. 2006 and Chapter 4), and the most appropriate algorithms for classifying species into functional groups are not yet clear.

Comparisons of species diversity versus diversity in functional traits have not, to our knowledge, been made for studies of the effects of biodiversity on disease dynamics. Potentially, one could categorize hosts of a pathogen according to their abilities to support pathogen population growth and transmission (i.e. reservoir competence). Alternatively, one could categorize species in a community by trophic level relative to the predominant hosts for the pathogen. Such a categorization would allow researchers to ask whether diversity of natural enemies of the host, or of resources for the host, reduces or enhances disease transmission (Ostfeld and Holt 2004, Holt 2008).

Increasingly, there is a move to abandon functional groups and focus on species traits, or the characters of species that represent what they do, but disease ecologists are only now evaluating this approach. For example, host quality, reservoir competence, body size, longevity, reproductive rate, and other traits would cleanly identify species into their roles in particular processes, which seems far more useful than attempting to group species into discrete categories when most species overlap in what they do (see Chapter 4). Studies of plant traits abound, but the study of traits of heterotrophs lags behind, and of pathogens and their hosts, even further behind (see Chapter 20).

15.5 Most diverse versus single highest functioning

Niche theory predicts that natural communities will consist of species that do not overlap completely in their suite of functions. One consequence of this expectation is that a diverse assemblage of species will tend to perform ecosystem functions more efficiently than will a monoculture, even if the species comprising the monoculture is more efficient than other single species. In other words, the functional complementarity arising from polycultures can cause them to out-perform all monocultures. Selection effects, however, in which one or a few species have disproportionate impacts on ecosystem functioning, may dominate and minimize the influence of complementarity. While there is strong evidence for complementarity in some studies, other experimental studies comparing performance of diverse communities to monocultures suggest that selection effects are stronger than complementarity (reviewed in Cardinale et al. 2006a). On the other hand, this pattern may also be due to limitations in the existing empirical studies and their interpretation. Cardinale et al. (2007) demonstrated that most of the reported plant-based studies actually do exhibit complementarity, or at least do not support the sampling effect alone. The resolution of this issue is important because the consequences for both the biodiversity-ecosystem function debate and the conservation of biodiversity are so profound.

Whether, or under what circumstances, monocultures can perform better than polycultures in influencing disease dynamics has not, to our knowledge, been assessed. Numerous examples of disease outbreaks in crop and livestock monocultures provide evidence that monocultures can have extremely high rates of disease transmission (reviewed in Keesing et al. 2006). But for certain disease systems, one might expect low diversity to be beneficial. For example, in the Lyme disease system in eastern North America, a small number of small mammals – white-footed mice, eastern chipmunks, short-tailed shrews, and masked shrews – are responsible for feeding and infecting a large percentage of the tick vector of the bacterial pathogen (Brisson et al. 2008). A large suite of other mammals and birds feed many ticks, but infect very few of them, and are considered 'dilution hosts' (Ostfeld and Keesing 2000). The dilution hosts differ in the degree to which they reduce the proportion of ticks infected, with grey squirrels being the most potent of the group (Ostfeld and LoGiudice 2003, Ostfeld et al. 2006). Simulation models indicate that a 'community' of dilution hosts consisting solely of grey squirrels would reduce disease risk to lower levels than would a naturally diverse community of dilution hosts (R. S. Ostfeld, unpublished). One caveat to generalizing from this result is that the performance of an ecosystem function, such as diluting disease risk, by a particular species is often contingent on the composition of the remaining community. For instance, shrews can act as a dilution host in some (species-poor) communities, but as an amplification host in other (species-rich) communities (Ostfeld et al. 2006). This is an example of context dependency that is well described in other studies of biodiversity-ecosystem functioning (Chapter 4). Therefore comparisons of performance in monoculture vs. polyculture must be made with caution. Even if monocultures of high-performing species outperform polycultures in diluting disease risk, applying this knowledge to real-world situations seems difficult or impossible. Returning to the Lyme disease example, replacing a diverse assemblage of vertebrates with a monoculture of squirrels seems neither practical nor desirable, even though it would reduce disease risk. Of course, these issues highlight the importance of defining what we mean by function. If the only function of interest is reducing risk of Lyme disease, then one non-permissive species might suffice, but if we are concerned with other diseases (i.e. other functional roles), then it is likely that greater diversity would be required.

15.6 Natural versus random sequences of community assembly or disassembly

A common approach to assessing the consequences of variation in biodiversity for ecosystem functioning is to assemble ecological communities by drawing species at random from a species pool and correlate functioning with species richness. A less common approach is to remove species from intact communities, which is more appropriate in mimicking local extinction, but is far less tractable (Symstad *et al.* 2001, Diaz *et al.* 2003). The main purpose of such experimental designs is to specifically assess the importance of species diversity *per se*, while avoiding the potential for species diversity, species composition, and other factors to co-vary. Of course, such experiments can also focus on functional diversity or both, and often the two are correlated (Naeem 2002). Natural communities, however, appear not to be assembled or disassembled randomly (Zavaleta and Hulvey 2004, Bracken *et al.* 2008). More commonly, some species are widespread and highly resilient to disturbances, while others are more restricted in occurrence and sensitive to disturbances. As a consequence, some species will tend to occur in most or all communities from rich to poor, whereas others will tend to exist only in species-rich communities. Studies in which species are added or removed in random order cannot provide insight into the importance of natural sequences of species addition (community assembly) or loss (community disassembly) (e.g. Fukami and Morin 2002). Moreover, given that many directly transmitted diseases exhibit density-dependent transmission, the nature of the methodologies commonly used to assemble experimental diversity treatments is likely to confound simple richness effects. Additive designs, which maintain similar abundance of individual species in single and multiple species assemblages, act to increase overall density as richness increases, whereas substitutive designs, which maintain the same overall abundance in single and mixed assemblages, act to reduce density of individual species as richness increases. However, it is important to note that sequences of species assembly or disassembly are often unpredictable. For example, extinction debt theory argues that dominant species are sometimes highly sensitive to forces of extinction. Tests of theory will require approaches where communities are assembled in both random and orderly fashions.

In a recent study, Ostfeld and LoGiudice (2003) used a simulation model to disassemble virtual vertebrate communities either in random sequence or in sequences suggested by natural disassembly rules. The roles of each species in contributing to Lyme disease risk had been empirically parameterized (LoGiudice *et al.* 2003), allowing the aggregate contribution of each community to be estimated. Random selection of species for removal resulted in a strong, positive correlation between species diversity and disease risk. This result occurred because one species – the white-footed mouse – has a particularly large positive impact on Lyme disease risk, and more diverse communities had a higher probability of including white-footed mice. But when communities were disassembled under rules with some empirical support, the opposite pattern was obtained – lower diversity caused increased disease risk. This result occurred because white-footed mice exist in all communities irrespective of diversity (LoGiudice *et al.* 2008, Nupp and Swihart 1996), and so less diverse communities had mice but few other hosts capable of diluting the impact of mice (Ostfeld and LoGiudice 2003).

A key question in understanding the effects of biodiversity loss on disease dynamics, or any other ecosystem function, is whether commonness, ubiquity, or resilience are correlated with the performance of the function of interest. For disease systems, if dilution hosts tend to be common and resilient and reservoir hosts rare and sensitive, then decreases in diversity will reduce disease risk because species-poor communities will be characterized by many dilution hosts and few reservoirs. However, if reservoir hosts are common and resilient and dilution hosts tend to occur only in more species-rich communities, then decreases in diversity will strongly increase disease risk. These correlations have not been adequately addressed in zoonotic disease systems, although it appears that common and ubiquitous species tend to be the most competent reservoirs for zoonotic pathogens (Ostfeld and Keesing 2000b). Potential evolutionary explanations for the correlation between ubiquity and high reservoir competence have not yet been explored. Two possibilities are that: (1) pathogens that are likely to experience a broad range of hosts (e.g. because they are transmitted by a generalist vector) might tend to specialize on hosts they are likely to encounter frequently; and (2) hosts whose life history traits promote commonness and resilience (e.g. rapid reproductive rate and short life span) might tend to tolerate infections (Jolles, Keesing, and Ostfeld, in litt.).

15.7 Other levels of biodiversity

Researchers assessing relationships between biodiversity and ecosystem functioning tend to focus on species diversity (and especially species richness), but biodiversity at other levels – genotypic, habitat type, landscape – can also potentially affect ecosystem functioning. These levels of diversity are only beginning to be studied in disease systems. Zhu *et al.* (2000) found that the incidence and severity of rice blast (a fungal pathogen) was considerably lower in agricultural fields with higher diversity of rice cultivars. Dennehy *et al.* (2007) found that two natural mutants of *Pseudomonas* bacteria were able to reduce prevalence of infection with bacteriophages compared with monocultures of the wild type. Neither mutant genotype was a viable host for the phage, and genotypic polycultures strongly diluted phage infection, sometimes to extinction.

Diversity at the genotypic level may also influence disease ecology. For example, Pulkkinen (2007) determined that the presence of > 1 genotype of *Daphnia magna* strongly reduced infection prevalence of a focal genotype with a parasitic microsporidian, *Glugoides intestinalis*. Another example is that of aphids and their fungal pathogens. Studies on susceptibility of pea aphids to a fungal pathogen have indicated significant genetic (between-clone) variability in fungal resistance (Ferrari *et al.* 2001). This suggests that a community with low genetic diversity comprising just the most resistant clone would result in the greatest reduction in disease risk. However, follow-up studies revealed significant Genotype × Environment interactions such that resistance ranking of individual clones changed with temperature (Blanford *et al.* 2003; Stacey *et al.* 2003). In this case, resistance under just one set of environmental conditions could be achieved with low genetic diversity, but to maintain resistance under environmental variation would require multiple clones. Further studies have now suggested that resistance to fungus is at least partly conferred by the presence of certain secondary bacterial symbionts within the aphids (Scarborough *et al.* 2005). How the symbionts differ between clones and how function is affected by environment is not clear, but the example illustrates the potential for considerable complexity in diversity–function relationships, even with just one host–pathogen combination.

Biodiversity can also be a factor *within* hosts. Recent studies have shown the potential importance of host coinfection with multiple disease agents. These mixed infections are common and have the potential to dramatically alter population dynamics and evolution of a particular disease agent (Cox 2001).

At the other end of the biodiversity spectrum, Vittor *et al.* (2006) found that decreases in tropical forest structural diversity that accompany deforestation strongly increase abundance and human-biting rates of the mosquito *Anopheles darlingi*, which is the primary vector of malaria in tropical America. Similarly, increases in yellow fever (Brown 1977), leishmaniasis (Sutherst 1993), and Ebola virus (Walsh *et al.* 2003) have all been linked to gross changes in habitat diversity and accompanying human encroachment on tropical forests increasing contact with key disease organisms and/or vectors. The exact nature of the diversity changes most potent in affecting disease transmission in these systems remains to be determined.

15.8 Looking forward

Ecologists have developed strong empirical and theoretical foundations for understanding the relationship between biodiversity and ecosystem functioning. Key issues include the shape of the relationships between biodiversity and specific ecosystem functions, the mechanisms underlying these relationships (e.g. the importance of species diversity per se versus species composition, functional traits), and how different patterns of community assembly affect functioning. Debates about these issues provide important insights for advancing our understanding of the effects of biodiversity on infectious diseases.

Considerable evidence is accumulating that biodiversity can strongly reduce disease transmission and risk (Dobson 2005, Rudolf and Antonovics

2005, Keesing *et al.* 2006, Dobson *et al.* 2006, Begon 2008, Molyneux *et al.* 2008). However, as we describe above, the issues of shape, mechanisms, and assembly patterns are only beginning to be addressed. Although we have limited our discussion to the effects of biodiversity on disease risk, we should also point out that diseases can strongly influence biodiversity.

Diseases play a fundamental (but largely unappreciated) role in shaping communities and mediating interactions between and across trophic levels. We cannot hope to fully understand the consequences of biodiversity change for almost any function without increasing our understanding of the role of disease organisms in community-level processes. Diseases can play important roles in natural regulation of pests and pathogens, provide new tools controlling them (e.g. Thomas and Read 2007), and can strongly influence the extent to which exotic plants and animals become damaging invaders in the introduced environment (Torchin *et al.* 2003; Mitchell and Power 2003).

Understanding the full relationship between biodiversity and disease risk will require exploration of both ecological and socio-economic factors (Fig. 15.1). Such exploration is particularly important in understanding the relationship between biodiversity and disease emergence (as opposed to tranmission dynamics of extant diseases). Risk is defined as the product of 'hazard × likelihood'. For emerging diseases, which are often accompanied by 'spillover' from a few principal host to secondary hosts, we suggest that 'hazard' might be represented by the prevalence of a particular pathogen and 'likelihood' represented by the probability or frequency of exposure to the pathogen. We further suggest that ecological change, such as loss of biodiversity or change in host species composition, is likely to have the greatest influence on the nature and magnitude of the hazard. On the other hand, many socio-economic factors (such as urban encroachment, land-use change, hunting, travel, and trade) broadly determine the chances of exposure, given a particular hazard. If we are to understand and manage disease risk, we need improved understanding of not just the ecology of infectious diseases, but also the social and economic contexts.

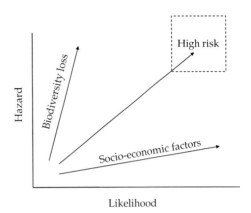

Figure 15.1 Conceptual model defining risk of an emerging infectious disease or disease spillover. Risk is the product of 'hazard × likelihood'. Ecological changes such as loss of biodiversity can affect the prevalence of a pathogen or parasite increasing disease hazard. However, transmission of the disease into a new population of hosts depends on frequency of exposure. For spillover of a zoonotic disease into a human population, likelihood of exposure is strongly determined by socio-economic factors.

Another research frontier is determining whether increases in vertebrate species diversity might increase the total burden of zoonotic disease. This might be expected if more vertebrate species simply add more zoonotic pathogens capable of infecting humans. Thus, even if risk of any particular disease declines with increasing vertebrate diversity, the total risk from all zoonotic pathogens might increase. Although such a relationship might exist, the evidence to support it is weak to non-existent. Interestingly, Jones *et al.* (2008) found that greater numbers of infectious diseases of humans have emerged in the temperate zone, where vertebrate diversity is lower, than in the tropics, where vertebrate diversity is higher. Although a detection bias partially accounts for this pattern, a simple correlation between vertebrate diversity and numbers of emerging diseases does not appear to exist.

Finally, one of the most important areas of biodiversity-ecosystem function research involves making it relevant to real-world problems, a main focus of this volume. As noted earlier, ecosystem functions like net primary productivity, nitrogen mineralization, and trace gas efflux do not resonate with the public at large; nor do they readily translate into ecosystem services. Diseases, however, matter tremendously to humans, are well studied, and the economic costs and consequences of disease are well known. The role of biodiversity appears to matter in virtually all its levels, from genotypic to population to community to structural, and a variety of mechanisms are at play. In a world where both biodiversity is changing dramatically and diseases are emerging and resurging, understanding the role of biodiversity in the ecology of diseases is arguably one of the most important areas in biodiversity-ecosystem function research.

CHAPTER 16

Opening communities to colonization – the impacts of invaders on biodiversity and ecosystem functioning

Katharina Engelhardt, Amy Symstad, Anne-Helene Prieur-Richard, Matthew Thomas, and Daniel E. Bunker

16.1 Introduction

Biodiversity is becoming increasingly homogenized as dispersal barriers are broken down (McKinney 2004) and as species respond to global warming through range shifts (Moen et al. 2004, Thuiller et al. 2006b). Such homogenization is leading to a net decrease in global biodiversity (Sax and Gaines 2003) with alarming consequences for the world's life support systems. Locally, however, addition of species through natural and anthropogenic causes can result in idiosyncratic, unpredictable, and sometimes latent changes in biodiversity. As found by studies reviewed throughout this volume and elsewhere (e.g. Hooper et al. 2005), changes in biodiversity may translate into measurable changes in ecosystem processes and the delivery of services to human societies. Thus, the addition of novel species can have impacts on the processing of energy and matter in ecosystems, either directly or indirectly, by affecting biodiversity.

Given the growing number of species additions to ecosystems worldwide, we would like to know whether it is possible to assess the risk of a large and potentially cascading and irreversible impact of a novel species on biodiversity and ecosystem functioning so that the riskiest species can be targeted for management. This requires linking two burgeoning fields in ecology: invasion ecology and biodiversity-ecosystem functioning research. Both fields are concerned with the loss of species, changes to ecosystem functioning, and measuring species traits to predict community and ecosystem-level impacts. However, whereas biodiversity-ecosystem functioning research has generally not allowed immigration from a regional species pool (but see Chapter 10), invasion ecology inherently focuses on open communities. Biodiversity-ecosystem functioning research focuses on the relationship and feedbacks between biodiversity and ecosystem functioning; invasion ecology generally focuses on one or the other. Both fields are therefore ripe for cross-fertilization.

In this chapter, we explore the consequences of opening communities to colonization and the new establishment of one or more species that strongly interact with resident species and the environment. We focus on the factors that determine the colonization, establishment, and impact of novel species on community properties and ecosystem functioning. We then use a risk analysis based on a bioeconomic framework to illustrate approaches to invasive species management that would be appropriate for different types of species. Most of our examples involve terrestrial plant communities, as these are the most commonly studied systems in the invasion ecology literature (Bruno et al. 2005), but we also draw from other trophic levels and systems. To avoid confusion in terminology regarding colonizers versus invaders, we follow the nomenclature of Davis and Thompson (2000; Box 16.1). Although our focus in this chapter is on the community and ecosystem-level impacts of the colonist, a large body of literature exists on the invasibility of the receiving community (Box 16.2).

Box 16.1 Terminology

Inconsistent and imprecise use of terms referring to species entering a community confuses the science of invasion ecology. To avoid this problem, we adhere to the terminology recommended by Davis and Thompson (2000), who emphasized dispersal rate of a colonizing species, its uniqueness to the region, and its impact on its new environment.

In this scheme, colonizing species are split into eight categories based on *dispersal type* ('diffusion' between adjacent or nearly adjacent environments or 'saltation' between widely distant environments), *uniqueness* to the region ('common' to the region or 'novel') and *impact* on the environment ('low' or 'high'). Davis and Thompson (2000) then lump these eight categories into three classes.

Regardless of the dispersal type or impact on the new environment, all colonizers common to the region are considered *successional colonizers*. Species with low impacts on the new environment that are novel to the region are *novel, noninvasive colonizers* regardless of the rate at which they disperse. These species comprise the bulk of exotic or non-native (terms we use interchangeably with 'novel') species, since most species reaching a new environment fail to establish, and most of those that establish do not become pests (Williamson and Fitter 1996). Finally, only those species that are novel to the region and have a large impact on the new environment, whether they disperse via diffusion or saltation, are *novel, invasive colonizers*, or true *invaders*.

Box 16.2 The receiving end

Establishment success of a potential invader is not only influenced by the traits of that species but also by the environmental characteristics of a site. Thorough reviews of the factors that influence a community's invasibility are published elsewhere (Lodge 1993; Levine and D'Antonio 1999; Richardson and Pysek 2006; Fridley et al. 2007). Here, we briefly highlight three of these factors particularly relevant to the focus of this chapter: diversity of competitors at a site, the presence of enemies, and the disturbance regime.

Diversity of competitors: A generation of ecologists grew up with the idea that diversity confers resistance to invasion through greater competition for resources: more species = more resources being exploited, therefore fewer resources available for a new species (Elton 1958). This idea remained largely untested until the advent of biodiversity-ecosystem functioning research in the early 1990s. Although the standard practice in biodiversity-ecosystem functioning experiments is to prevent unplanned species from entering the experimental units, when this practice has been relaxed, the diversity–invasibility hypothesis has largely been confirmed. In fact, the relationship between diversity and invasibility is one the most consistent effects across experiments (Balvanera et al. 2006). However, results have been contested by the fact that in surveys of natural and semi-natural communities, exotic richness is often positively correlated with native richness, suggesting that conditions conducive to high native diversity are also conducive to high exotic diversity (e.g. Levine 2000). The resolution of this discrepancy comes largely from considering the scale at which the relationship is considered. Since competition for resources occurs between neighbouring individuals, it would be expected, and has been confirmed by some studies, that the relationship between native and exotic species would switch from negative at the neighbourhood (0.1–1 m^2) scale to positive at larger scales (Brown and Peet 2003).

It must be recognized, however, that competition is not the only interaction that controls establishment success. Plant community diversity has been observed to modify the composition of invertebrate herbivore communities (Koricheva et al. 2000, Symstad et al. 2000) and this may influence the establishment success of a colonizing species (Prieur-Richard et al. 2002). Also, previous invasive species may facilitate the establishment of other species (Hacker and Dethier 2006, Kondo and Tsuyuzaki 1999) by modifying environmental conditions such as soil biotic community, pH, nutrients, oxygenation of water or sediments, thus rendering the habitat more hospitable to colonizing species.

Enemy release: One essential advantage for species colonizing a new region is the absence of some or all of the enemies that attacked them in their home range (Lodge 1993). This, of course, is the basis for biological control – the introduction of host-specific enemies (e.g. herbivores, pathogens, parasites) from an invasive species' region of origin (Butler et al. 2006, Simberloff and Stiling 1996). The role of natural enemies that are native to the invaded community in the establishment success of novel species is unclear, although examples do exist that illustrate the control of exotic species by parasites or herbivores native to the receiving community (e.g. Soldaat and Auge 1998).

Disturbances: Disturbances have unanimously been shown to favour plant invasions (Hobbs and Huenneke 1992). Disturbances simultaneously increase resource availability and decrease competition from resident species, allowing establishment of species that would not be able to enter an undisturbed community. A wide range of disturbance types have been shown to facilitate invasions: soil disturbance (e.g. Kotanen 1997, Burke and Grime 1996); nutrient enrichment (e.g. Burke and Grime 1996); modifications of fire regimes (e.g. D'Antonio and Vitousek 1992); and grazing by non-native herbivores (McIntyre and Lavorel 1994, Landsberg et al. 1999). Interaction between several disturbance types frequently leads to the highest rates of invasion (Hobbs and Huenneke 1992).

16.2 The relationship between biodiversity and ecosystem functioning

Biodiversity effects on ecosystem functioning are common, although the magnitude and direction of effects can vary across ecosystems, trophic levels, response variables, experimental designs (see Chapter 2), and, as we discuss in this chapter, communities that are relatively open or closed to immigration from regional or global species pools. Mechanisms that explain the relationship between biodiversity and ecosystem functioning can be placed into two main categories that differ fundamentally in how species interact. The *sampling effect* (or selection or dominance effect; see Chapter 7) occurs when greater diversity increases the probability that a highly competitive species is present and gains dominance in a community. Such species typically use more resources and produce more biomass than the average species and are therefore associated with higher levels of ecosystem functioning (e.g. higher productivity and higher resource use). An inverse sampling effect (Loreau 2000, Engelhardt and Ritchie 2002, Jiang et al. 2008) is a special case when the competitively dominant species does not have the greatest effects on ecosystem functioning, as can be the case when resource depletion is not the primary means for interspecific competition. In contrast, the *complementarity effect* (or niche differentiation effect; see Chapter 7) arises from resource partitioning among species. In this case, species-rich communities use more resources and are more productive than species-poor communities, up to a saturating point (Cardinale et al. 2006a), because species are using resources differently. Since both mechanisms are necessary for the maintenance of biodiversity (i.e. competitive asymmetries are common in communities *and* species partition resources through tradeoffs), they operate simultaneously in most systems. Quantifying the contribution of each mechanism to a biodiversity effect can be challenging and has received considerable attention in recent years (see Chapter 7).

Communities are not static, however, and the relative importance of these opposing community-structuring mechanisms may change as species immigrate, go locally extinct, and change in abundance. Hence the relationship between biodiversity and ecosystem functioning may shift between sampling and complementarity effects depending on the age of the community, the strength of interspecific interactions, species presence, and the *per capita* contribution of each individual within a community to ecosystem processes. Some biodiversity-ecosystem functioning (BEF) studies report that the effects of biodiversity on ecosystem functioning grow increasingly positive through time (Tilman et al. 2001, Jonsson 2006, Fargione et al. 2007). In nitrogen-limited grassland systems, for example, this effect is explained by a shift from the selection effect to the complementarity effect. In this case, complementarity in resource use increased the input and retention of nitrogen through time (Fargione et al. 2007). However, others have found that the effect of biodiversity on ecosystem functioning grows weaker over time (Bell et al. 2005b), especially in systems where facilitative interactions early in the development of the community are replaced by competitive interactions (Cardinale and Palmer 2002). Still others report that a positive BEF relationship is only transient (Hooper and Dukes 2004, Fox 2004a), possibly because interference competition allows monocultures to outperform polycultures in the long term (Fox 2004a).

While instructive in understanding how the BEF relationship changes through time as constructed ecosystems mature, these long-term studies of closed communities do not address how the BEF relationship changes during succession and/or invasion, which are processes that are inherent to open communities that allow immigration. Specifically, what happens to the BEF relationship when new species are allowed to immigrate?

16.3 Impacts of colonizing species on the biodiversity-ecosystem functioning relationship

Immigration and extinction processes, which are a function of biogeographic, environmental, and biotic constraints (Naeem and Wright 2003), strongly determine the structure and composition of all biological communities. A species will be absent from a community if it cannot disperse to a site, survive in the new abiotic environment, and successfully reproduce in the presence of the resident biota. Understanding the potential impact of a colonizing species on local biodiversity and ecosystem functioning will therefore depend on four factors. These are (1) the species' likelihood of colonizing a new area, (2) the species' likelihood of establishing a viable population and increasing in abundance, (3) the response of resident communities to the species' presence, and (4) the functional traits of the colonizing and resident species that determine biodiversity's effect on ecosystem processes. Functional traits (see Chapter 4) are quantifiable biological properties of species that affect how species respond to the biotic and abiotic environment through changes in the distribution and abundance of organisms ('response trait') and that affect ecosystem processes ('effect trait'; Lavorel and Garnier 2002, Naeem and Wright 2003, Engelhardt 2006).

16.3.1 The likelihood of colonization

The likelihood that a species will enter a new location is a function of the species' current range with respect to that location, characteristics of the landscape surrounding that location, and the organism's dispersal ability within that landscape.

Among plants and invertebrates, wind, water, and vertebrate dispersal modes, as well as small seed or body size and high propagule output, are associated with dispersal ability (Fenner and Thompson 2005). For birds and mammals, large body size and carnivorous diet type allow organisms to disperse long distances (Sutherland *et al.* 2000, Jenkins *et al.* 2007).

If some species traits confer greater dispersal ability and allow a species to reach new locations, are they effective predictors of the invasiveness of a species? For plants, for which the greatest body of literature is available, traits associated with dispersal are indeed generally related to invasiveness. Broad, comparative studies associate high fecundity (including production of many offspring, short juvenile periods, and/or long flowering seasons), small propagule size, and long-distance dispersal capability with high abundance or broad distribution of invasive plants (Richardson and Rejmánek 2004, Hamilton *et al.* 2005). For example, alien woody species with fleshy fruits (which can be carried great distances by their avian dispersers) expanded their ranges to a greater extent in New York City during the 20th century than species with other fruit types (Aronson *et al.* 2007). Having two modes of dispersal (e.g. sinking seeds and floating vegetative fragments in *Mimulus guttatus* on rivers of northern Europe; Truscott *et al.* 2006) or high plasticity in seed mass (e.g. *Ambrosia artemisiifolia* along European rivers; Fumanal *et al.* 2007) are other dispersal-related traits attributed to invasive plant species. For vertebrates, in contrast, body size and diet type are less important in predicting an invasive species' likelihood of reaching a new location than whether or not humans hunt it for sport – game species are brought across oceans and continents more frequently than are non-game species (Jeschke and Strayer 2006).

For our purposes, the question is whether the traits that increase the likelihood of a species reaching a new location are those that are likely to affect biodiversity, ecosystem functioning, and the strength of the relationship between and biodiversity ecosystem functioning. In the absence of human-mediated dispersal, the answer at this time appears to be 'no': in plants, studies specifically addressing this question have generally shown that no consistent relationship emerges between the traits of the seed stage, which is responsible for most dispersal, and the traits of the mature stage (Westoby 1998, Lavorel and Garnier 2002), in which effects on ecosystem functioning are greatest. Many more empirical tests of this relationship are needed to determine the robustness of this answer, however (Suding et al. 2008).

In contrast, species intentionally introduced and cultivated by humans are sometimes chosen because they are robust in the new location and are useful to humans in some way; these characteristics are related to successful establishment, spread, and effects on the new ecosystem (Alpert 2006) and therefore warrant special attention from management and policy communities in determining the balance of their risks and benefits (Section 16.4). Other intentionally introduced species are remarkably uninvasive. For example, maize (*Zea mays*) survives in the Old World only because of intense care through the use of fertilizer, irrigation, and a variety of biocides. The case of *un*intentional dispersal by humans is more ambiguous. For example, freshwater zooplankton disperse only slowly by natural means, but can drastically change ecosystem processes in new unintended locations (Havel and Medley 2006). Unintentionally introduced species are not constrained by a potential tradeoff between dispersal ability and competitive ability. They may simply be in the right place at the right time, and whether they affect resident diversity, functioning, or both in their new location may depend on traits that allow them to become established and to spread. It is these species that represent the greatest uncertainty regarding their potential threats to ecosystems. Consequently, they should be the focus of investigations for traits that predict the impact of colonizing species on biodiversity and ecosystem functioning.

16.3.2 The likelihood of establishment

Some species traits inherently confer a high likelihood of establishment success (e.g. Rejmanek and Richardson 1996, Daehler 1998), with high propagule output (Kolar and Lodge 2001), fast growth rate (Newsome and Noble 1986), and high adaptability (genetic variation in fitness traits or phenotypic plasticity; Poulin et al. 2007) appearing consistently across a wide range of species. Novel weapons (Abhilasha et al. 2008, Bais et al. 2003, Callaway and Ridenour 2004, Callaway et al. 2008) and a release from enemies (Keane and Crawley 2002, Mitchell and Power 2003) can also increase the chances of establishment by a novel species by conferring a competitive advantage over native species. Most species that become established become part of the resident community without noticeable effects on diversity or ecosystem functioning. Therefore, we ask again, which, if any, of these traits are most likely to cause a novel species to impact biodiversity, ecosystem functioning, and/or the shape or strength of the relationship between them?

High propagule output and other traits associated with juvenile stages are unlikely to have significant effects for the same reasons that we discussed for dispersal traits above. Fast growth rate and high adaptability may be associated with biodiversity and/or ecosystem functioning changes under certain circumstances, such as disturbed conditions, which may result in an immediate or latent impact of the invader on biodiversity or ecosystem functioning depending on when the disturbance happens during establishment. Competitive superiority, whether caused by enemy release, novel weapons, or some other mechanism, is more likely to have significant effects. In this case, only a one-for-one substitution within a functional group of a resident species by the novel species has no effect on biodiversity. A more likely outcome is a shift in species' abundances, which may include the complete elimination of one or more species or a change in abundances among functional groups. In any of these cases, ecosystem functioning may be impacted through a direct or indirect sampling effect. For example, Argentine ants (*Linepithema humile*) invading sub-tropical and temperate regions possess a different social structure that allows the formation of fast-growing, high-density 'super colonies' that deplete resources of an area faster than native ants can (Holway 1999). A direct effect of their resulting dominance may be changes in nutrient cycling due to differences in their nest construction from that of native ants (Holway *et al.* 2002), whereas an indirect effect could act through their increased predation on other invertebrates that affect pollination or decomposition. While sampling effects are common in BEF experiments, their significance in nature is not known (Cardinale *et al.* 2006a). Therefore, predicting novel species' effects on biodiversity, ecosystem functioning, or their relationship from traits that confer competitive advantage is tenuous.

On the other hand, some species become established not because of a specific trait that can be consistently traced across many communities, but because they use the environment in some novel way; i.e. they are initially complementary. For example, Clarke *et al.* (2005) showed that an invading grass can take advantage of both summer and winter rains, in contrast to natives, which used rain in only one of these seasons. Similarly, red brome (*Bromus madritensis* ssp. *rubens*) is an annual grass that can exploit water and other soil resources for 2–3 months before native perennials break dormancy in the Mojave Desert (DeFalco *et al.* 2007). We argue that traits that confer establishment success because of their complementarity (niche differences) will more consistently affect ecosystem functioning and the BEF relationship than will traits involved in a sampling effect by conferring competitive advantage. This is because complementarity implies a direct effect on resource use, which is a key part of ecosystem functioning (Vitousek 1990). Just how large these niche differences must be is a difficult question, however. The novel species needs to be similar enough to tolerate local environmental conditions, and disturbances can open space and release resources that allow species that do not differ from the residents to become established. The literature shows clearly that exotic diversity patterns mirror native diversity, suggesting that novel species are tracking similar conditions and resources (Stark *et al.* 2006). On the other hand, if the novel species is too similar to the resident species, its impact on biodiversity and ecosystem functioning will be negligible.

16.3.3 The impact of a colonizer on resident communities and ecosystem functioning

By definition (Box 16.1), an invader has a large impact on the native ecosystem. This usually involves a reduction of native species abundance or richness and/or a substantial change in ecosystem functioning. However, most species that colonize and become established in new environments have little or no impact (Williamson and Fitter 1996), and others even have a positive effect on the resident community (Hacker and Dethier 2006, Kondo and Tsuyuzaki 1999, Bruno *et al.* 2005). Determining the traits of the species and the characteristics of the corresponding receiving communities that lead to these different situations is a main goal of the science of invasion ecology, and it is equally important for determining how a new species will affect the relationship between biodiversity and ecosystem functioning.

A crucial, often overlooked, step in determining the traits that cause large impacts is a quantitative demonstration that a suspected invasive species actually does have an effect on the receiving community. Perceptions of species' impacts are often not substantiated by quantitative studies. For example, of 196 exotic species in the Chesapeake Bay region, 20 percent were thought to have a negative impact on a resident population, community, or process, but only 6 percent were actually documented to have a negative impact (Ruiz *et al.* 1999). In addition, there are reports of species widely considered to be invasive actually having no impact on individual native species of concern (e.g. Menke and Muir 2004) or on native richness or abundance (e.g. Treberg and Husband 1999). On the other hand, a novel species may ameliorate limiting conditions and positively affect native species. For example, a non-native larch species intensively planted on the lower slopes of a Japanese volcano after its eruption spread into unplanted areas on the volcano, making some consider it invasive (Kondo and Tsuyuzaki 1999). However, diversity and richness of native species during primary succession on the volcano was greater under this non-native tree than under a native tree (Titus and Tsuyuzaki 2003). In the US Pacific Northwest, *Spartina anglica*, a marine grass from the UK, facilitates the growth of native species by quickly accreting sediment and creating more hospitable habitats for growth in unvegetated estuarine habitats, but decreases native species diversity in other habitats (Hacker and Dethier 2006). These examples suggest that colonizers with positive impacts on residents often share traits with native species that play a similar role in succession.

Removal studies yield the strongest evidence for negative impacts of novel species on native abundance and richness (Levine et al. 2003). For example, growth rates and recruitment of two shrubs increased in response to the removal of three non-native grasses (D'Antonio et al. 1998). Hulme and Bremner (2006) observed a significant increase in α and γ diversity after the invasive riparian weed *Impatiens glandulifera* was removed. In these studies the role of competition is clear, but the traits that drive this competition vary among situations. Gould and Gorchov (2000) found that survival and fecundity of native annuals were greater when they were transplanted into forest plots without the invasive shrub *Lonicera mackii* than in plots where it was present. Although they did not investigate the specific traits responsible for interspecific competition, they suggested reduction of light availability early in the growing season due to the invader's longer phenology as one possibility. Dyer and Rice (1999) found that vegetative growth and reproductive output of a native bunch grass were greater when grown with conspecifics compared to when grown with exotic annual grasses at a variety of densities. In this case, the invasive species' early growth depleted shallow soil water resources and reduced light availability early in the growing season, thereby suppressing the root growth that the native perennial required for acquiring deep soil moisture later in the growing season. Other traits associated with competitive invasive species are those that confer high resource capture ability and utilization efficiency (Feng et al. 2007), the ability to forage on low-quality resources (Gido and Franssen 2007), and the ability to alter soil biotic communities (Callaway et al. 2004). A recurring theme in these traits is that they are somehow different from those of the species in the receiving community (D'Antonio and Hobbie 2005).

Once established, a species may also negatively impact resident species by disrupting their reproduction and dispersal. Animals may disrupt plant dispersal through displacement of more effective, native pollinators; predation of pollinators or dispersers; or destruction (eating or trampling) of flowers, pollen, or seeds (Traveset and Richardson 2006). A plant may impact dispersal of other plants by competing for pollinators or dispersers (Brown and Mitchell 2001), impeding dispersers (Traveset and Richardson 2006), or augmenting populations of generalist granivores (Ortega et al. 2004). For these interactions to cause substantial negative effects on the natives, however, these novel colonizers must somehow be more disruptive than the combined effect of all the other species already occurring in the community. Little information exists as to the traits that would yield this effect, but examples of such occurrences suggest that traits novel to the community, such as a carnivore in a previously carnivore-free system, are those that will have the largest impact.

Similarly, novel species that strongly affect ecosystem functioning also tend to do things differently than the residents (D'Antonio and Hobbie 2005). Vitousek (1990) described three ways that novel species alter ecosystem functioning: by altering resource supply rates, by changing trophic structure, and by modifying disturbance regimes. Plants that bring N_2-fixing bacteria into naturally nitrogen-poor systems and thereby increase nutrient availability and cycling rates (Vitousek and Walker 1989) are one of the classic examples of a trait that causes transformation by modifying resource supply rates. Novel species can also accumulate a resource (e.g. salt) to the point that concentrations are toxic to other species (Vivrette and Muller 1977). Ponds created by North American beavers in Chile increase retention of fine particulate organic matter in streams, leading to increased food availability for, and therefore productivity of, stream macroinvertebrates (Anderson and Rosemond 2007). Novel fish that affect feeding behaviour of herbivores alter the trophic structure of stream systems so that primary productivity increases and nitrogen dynamics are altered (Simon et al. 2004). The fine, quickly drying or highly flammable leaves of grasses, combined with their nearly continuous ground cover, allow novel grasses to carry fire through woody ecosystems that previously burned infrequently (D'Antonio and Vitousek 1992). This greater fire frequency, combined with the different structure of the novel and resident species, can shift the ecosystem from being a carbon sink to a carbon source (Bradley et al. 2006).

Currently, predicting the magnitude and direction of a species' effects on biodiversity and ecosystem functioning is tenuous at best. Novel species with traits similar to those of the resident species are less likely to have cascading effects on the ecosystem than novel species with different traits. Many uncertainties remain, however. *Myrica faya* is a novel functional type in Hawaii and has extensive effects on biodiversity and ecosystem functioning (Vitousek and Walker 1989). In contrast, *Acer platanoides* has many similar traits compared to the resident *Acer saccharum* (Kloeppel and Abrams 1995); however, the former can substantially reduce forest biodiversity in the northeastern United States whereas the latter does not (Webb *et al.* 2000). *Acer platanoides* has a fertilizing effect of forest soils, which increases the growth of tree seedlings of four different species (Gomez-Aparicio *et al.* 2008). Thus the invasion of *Acer platanoides* may reduce forest biodiversity but increase forest productivity. These uncertainties highlight the need to know whether and how current and future invaders will affect the relationship between biodiversity and ecosystem functioning.

16.3.4 Tying it together: invaders' effects on the BEF relationship

We know that the breakdown of dispersal barriers has allowed a degree of global biotic homogenization. In most cases, species novel to an ecosystem will not gain dominance and will simply blend into the saturating function of the complementarity effect. These species join the community because they have similar environmental tolerances to resident species; they displace few if any established individuals because they are complementary or equally competitive; and they increase functioning in a minor additive way, if at all. However, in some cases, novel species can have strong immediate or latent effects on biodiversity and ecosystem functioning. Here, the novel species may change the shape and trajectory of the BEF relationship by changing species richness, changing the point at which an ecosystem function saturates, and/or shifting the mechanism from complementarity to a sampling effect as the nature of species interactions shifts from resource partitioning to competition.

The traits of the species that substantially impact resident species or ecosystem processes vary considerably with the situation. Nonetheless, the overarching theme is that the traits must differ somewhat from those possessed by the species in the receiving community for novel species to impact both biodiversity and ecosystem functioning. Because functional diversity is a critical component of the biodiversity-ecosystem functioning relationship (see Chapter 4; Petchey and Gaston 2006; Wright *et al.* 2006), it is this difference between the colonizer and the resident species that determines whether a new species will impact biodiversity and ecosystem functioning. The challenge is learning how large of a difference between the colonizer and the resident species is required for the colonizer to have a significant effect on biodiversity and ecosystem functioning.

The type of effect that the novel species has on the BEF relationship will depend on the species interactions at play in the resident community and how the newcomer fits in. The diverse array of interactions witnessed in BEF studies (e.g. facilitation and competition) occur in experimental settings where extinction could occur but colonization of new species could not. Therefore it is to be expected that the arrival of a new species could have equally diverse ramifications. When a colonizer is common to the region ('native'), this process is referred to as succession. When the colonizing species is new to the region and has not coevolved with the resident species ('non-native' or 'exotic'), this process is generally called invasion if the newcomer achieves dominance (Box 16.1). The processes are fundamentally the same: a single species can have a small or large impact on the resident community or ecosystem functioning (Davis and Thompson 2000).

The lack of common evolutionary pressures between an exotic species and a receiving community makes it more likely that the new species will have a novel set of traits that allows it to be a 'super-competitor' that influences functioning through the sampling effect. The sampling effect is usually associated with a positive BEF relationship, but it can also create negative relationships when the competitive dominant does not have the greatest effect on functioning (the inverse sampling effect;

Loreau 2000, Engelhardt and Ritchie 2001, Weis et al. 2007, Jiang et al. 2008). We argue that the chance of an inverse sampling effect is higher when exotic species with novel sets of traits invade a system because the exotic species are less likely to follow the 'rules' that maintain biodiversity in the receiving community. Thus, an exotic invader may be a superior competitor but not be the most effective processor and transformer of nutrients and energy. Or, the exotic species might have a high growth rate but be a poor competitor. Indeed, in an experimental study that tested the effects of three native and one exotic species on wetland ecosystem functioning, Engelhardt and Ritchie (2001, 2002) found that the exotic species was the most productive species that retained the most nutrients, but it was a poor competitor. Hence ecosystem biomass production and nutrient retention in mixed communities was lower than in the monoculture of the exotic because interspecific competition led to the dominance of a poor biomass producer.

There is still much to be done in order to reliably predict the effects of individual species, no matter what their origin, on ecosystem functioning. It is clear, however, that understanding how the addition of a species impacts ecosystem functioning will require knowledge of species traits – those of the new species and those of the resident species. This focus on traits will occur in a BEF framework that so far has focused on the consequences of extinction on ecosystem functioning in closed systems. This framework is applicable to conservation biology and restoration science, which seek to understand how species extinctions or community assembly, respectively, may impact ecosystem processes (Figure 16.1). Opening BEF theory and experiments to unplanned species colonization poses five real challenges that must be overcome to better understand the effects of invasive species on biodiversity and ecosystem functioning. These are:

A. Biodiversity is both an independent and a dependent variable when species are added to the resident community through colonization, and species are potentially deleted through competitive interactions with the colonizer. We need to know how BEF studies, which typically manipulate diversity as the independent variable and only allow

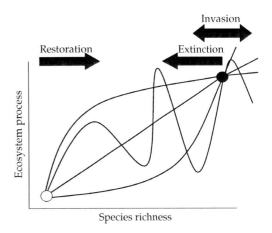

Figure 16.1 The effects of extinction, restoration, and invasion on species richness that translate into effects on ecosystem processes given linear, nonlinear, or idiosyncratic relationships between species richness on ecosystem processes. The filled circle depicts an intact system of resident species, the open circle a degraded system. Depending on the traits of the colonizing and resident species, invasion may increase, decrease, or not affect species richness, ecosystem processes, or the relationship between them.

extinction, can be redesigned to incorporate the inevitable and non-random process of new colonizations, in the form of succession or invasion. This would probably entail long-term experimental studies that would establish resident communities at a desired richness and then intentionally release one or more potentially invasive species. Changes in biodiversity and ecosystem functioning would need to be monitored through time using a design that accounts for changes in covarying environmental factors.

B. The establishment of a new species is influenced by biotic factors including the diversity of the resident community and the presence of enemies (Box 16.2). It is also influenced by abiotic factors such as disturbances, which affect the resident community even in the absence of the new species (Box 16.2). Thus, changes in a community's diversity in the face of a new addition may not be caused by the new species, but by abiotic conditions that drive changes in biodiversity and the successful establishment of the species. The challenge here is to test the relative impact of the biotic and abiotic factors that drive changes in biodiversity of the resident community. Again, a long-term experimental approach may be required as well as long-term monitoring of newly invaded systems.

C. Invaders can have a direct impact on ecosystem processes that can override or cancel the impacts of the resident community on ecosystem processes. However, invaders can also have an indirect impact on ecosystem processes by effecting change in the resident community. These direct and indirect effects of invaders need to be studied together to understand how the relationship between biodiversity and ecosystem functioning changes in communities that are open to colonization.

D. Many species that establish in a new location are not invasive and simply blend in to the saturating function of the BEF relationship. These species may be 'time bombs', however, that can have significant latent effects on biodiversity and ecosystem functioning when environmental conditions change. Similarly, some species may be invasive as soon as they establish at a new location, but their impact decreases through time. We need to know how frequent these latent effects are and whether they can be predicted, especially in the face of global climate change.

E. Finally, every species has a near endless number of traits. Thus it seems impossible to know whether a novel species will become invasive without knowing the traits of all species in a community and the traits of the colonizer. Little consensus exists about the functional traits of known invasive species and even less, if anything, is known about the traits of those yet to come, raising the question of whether the right traits have been measured. Instead of knowing all traits, which would be an unrealistic proposition, it might be useful to measure key functional traits that are reliable predictors of species' impacts on biodiversity and ecosystem functioning, and estimate the plasticity of these traits. Species can vary broadly in their traits between their native and introduced range (Siemann and Rogers 2001), suggesting that understanding key traits *and* their plasticity under different abiotic and biotic conditions will be important to assess which species pose the greatest risk of changing the BEF relationship.

16.4 Invasions and ecosystem services: assessing risk for better management

The globalization of international commerce presents a policy challenge: sales and movement of live organisms create wealth, but measures to prevent unintended movement of organisms have costs and there is potential for non-native species to cause considerable economic and environmental harm. Bioeconomic theory and modelling incorporate current understanding of species' impacts on biodiversity and ecosystem functioning to quantify the impact of non-native species on ecosystem services. This is an essential step in developing effective practices and policy for invasive species management. In this section, we take a brief look at how bioeconomic frameworks can be used to evaluate the economic costs and benefits of various actions regarding non-native species; then we illustrate how BEF principles are incorporated into a strategy for assessing the risk of economic impacts of non-native species.

16.4.1 The use of bioeconomic frameworks

Leung *et al.* (2002) presented a quantitative bioeconomic modeling framework to analyze risks from non-native species to economic activity and the environment. The model identifies the optimal allocation of resources to prevention versus control, acceptable invasion risks, and consequences of invasion to optimal investments (e.g. labour and capital). When applied to invasive zebra mussels (*Dreissena polymorpha*) in North America, the model indicated that society could benefit by spending up to US$324,000 per year to prevent invasions into a single lake. By contrast, the US Fish and Wildlife Service spent US$825,000 in 2001 alone to manage all aquatic invaders in all US lakes.

A bioeconomic approach was also used by Cook *et al.* (2007) to evaluate the economic benefit of biosecurity measures aimed at preventing the arrival and establishment of the parasitic bee mite, *Varroa destructor*, into Australia over the next 30 years. Specifically, this study evaluated the expected consequences of *Varroa* impact on feral bee populations and the flow-on effects in terms of loss of pollination services for the horticulture industry. The model estimated the benefits of exclusion to be between Aus$21.9 million and Aus$51.4 million per year, provided exclusion is maintained. The model further revealed that existing cost-sharing arrangements between government and industry do not accurately reflect the spread of potential benefits, such as the substantial benefits derived by the horticulture industry from 'free' pollination services of feral bees when they are not impacted by the mite. These studies are significant in demonstrating the potential for ex-ante evaluations of the economic impact of invasive species. Unfortunately, our knowledge of key economic variables, such as the value of biodiversity and the societal discount rates for environmental goods, is extremely limited and currently makes evaluation of non-market effects challenging. Identifying these gaps in our economic understanding highlights the need for interdisciplinary approaches in the development of improved policy frameworks for biosecurity and invasive species management (Thomas and Reid 2007; Wilson et al. 2007; and see Chapter 17).

16.4.2 Risk assessment

The bioeconomic examples above add weight to the recent conclusion of Keller et al. (2007) that risk assessment and screening protocols to limit the introduction of damaging species can deliver positive net economic benefits. Determining how to conduct an effective risk assessment and prioritize investment in biosecurity measures is complex. Risk assessments generally combine some measure of hazard with a measure of likelihood to score risk. With an invasive species, the threat or hazard is essentially determined by the magnitude of its impact and the rate at which this impact occurs. The magnitude of impact is nil to little when it is restricted to a local scale. Impact magnitude increases as species, communities, ecosystems and ecosystem functioning are adversely affected, and/or as the spatial and temporal extent of the impact increase. Similarly, some species will spread slowly and/or take a long time to have an impact and, at the other extreme, the impact of some invaders can be almost instantaneous.

In Fig. 16.2 we combine impact and rate to create a matrix to inform biosecurity strategy and investment priorities. In general, species populating the bottom left of the matrix (labelled class 1) might be considered lowest threat since the rate and magnitude of impact is small. As such, they likely represent low priorities for biosecurity investment, as the costs might be expected to outweigh the benefits.

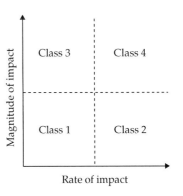

Figure 16.2 Overall impact of an invasive species is determined by the magnitude of its impact (magnitude increases as species, communities, ecosystems and ecosystem functioning are adversely affected, and/or as the spatial and temporal extent increases) and rate at which this occurs (some species will spread slowly and/or take a long time to have an impact while the impact of some invaders can be almost instantaneous). Species can be classed according to these properties and used to inform biosecurity strategy and control decisions.

Class 2' species have low impact but a more rapid rate of impact, as might result from higher rates of spread and/or lower impact–abundance thresholds. The higher a species is on the rate axis, the more important investment in biosecurity measures that prevent arrival and establishment (such as trade or movement barriers, quarantine and inspection) will be compared to investment in management measures aimed at mitigating the problem after arrival – if impact is rapid then prevention is better than cure. Nonetheless, with restricted impact magnitude class 2' species will still assume fairly low priority compared with species with greater, community- or ecosystem-level impacts (classes 3 and 4). Of these, species whose effects are rapid (class 4) represent the greatest priorities for preventive biosecurity measures, since the implications of an incursion are severe (e.g. a disease like foot and mouth, where a single case can impact a whole industry overnight). However, if the rate of impact is slow (class 3), this could create options for investment not just in preventive measures but also in mitigation measures aimed at longer-term control or eradication.

This broad framework rests on an understanding of the traits we discussed in the first section: those that determine dispersal and establishment influence the rate of impact, whereas those that determine

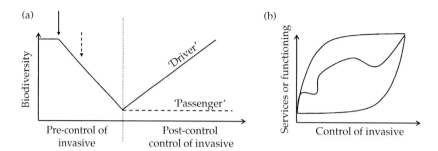

Figure 16.3 Benefit from control of an invasive species depends on the causal relationship between invasive species abundance and biodiversity. (a) If the invader is a 'driver' of biodiversity loss then decline in biodiversity will coincide with establishment of the species (indicated by the solid arrow) and subsequent control of the invasive will deliver a biodiversity benefit (example given for a linear impact function). If, on the other hand, the invader is a 'passenger' and biodiversity loss is due to some other environmental factor, then its arrival need not coincide with initial loss of diversity (indicated by the broken arrow which appears slightly after biodiversity has begun to fall) and even complete control may fail to deliver a biodiversity benefit. (b) Furthermore, even if there is a biodiversity benefit, the return of associated ecosystem services and function can be complex. Function can return at a higher rate than biodiversity (upper line), at a slower rate (lower line), or can exhibit an idiosyncratic pattern (middle line) depending on the order and rate at which functionally significant elements of biodiversity assemble.

how a species affects the receiving community and its ecosystem processes govern the magnitude of impact. Equally critical for determining the benefit of various actions, however, is the extent to which invasive species are a direct cause of biodiversity decline or whether they are simply responding to other forms of ecosystem change (Gurevitch and Padilla 2004). Whether an invasive species is a 'driver' of biodiversity change or a 'passenger' (MacDougall and Turkington 2005) has important implications for whether control of an invasive species is expected to increase biodiversity (Thomas and Reid 2007; Fig. 16.3(a)). Moreover, even if an invasive species is the cause of initial biodiversity loss, it need not necessarily follow that management of the invader will result in biodiversity recovery (Thomas and Reid 2007) because of dispersal limitation or local extirpation (Laughlin 2003) or because of lasting biotic (e.g. altered soil fauna and flora) or abiotic (e.g. altered stream flow) effects from the invader. Finally, even if structurally similar communities do re-establish, the order of species assembly can have a marked impact on the pattern and rate of functional recovery (Wilby and Thomas 2002; Kremen 2005; Hooper et al. 2005). If functionally significant species respond quickly to removal of the invader, then ecosystem services can recover at a more rapid rate than biodiversity overall (Thomas and Reid 2007; Fig. 16.3(b)). On the other hand, if functional species respond slowly, then even substantial recovery in biodiversity will not necessarily result in restoration of function (Thomas and Reid 2007; Fig. 16.3(b)). Such factors identify a clear need for understanding the impacts of non-native and potentially invasive species (and their control) from a BEF perspective.

16.5 Conclusions

Like BEF science, invasion biology has become its own branch of ecology. As we have shown here, the two branches are linked through their search for traits that determine the mechanisms by which species impact communities and ecosystems. Much still needs to be learned about how the traits can be used to forecast how novel species introduced to a new location will affect the relationship between biodiversity and ecosystem functioning. We offer the following conclusions as hypotheses that we hope will stimulate further research into the linkages between biodiversity and ecosystem functioning in *open* communities.

1) Traits that allow species to be good colonizers are poor predictors of species impacts on ecosystem functioning because traits at the juvenile stage are poorly related to traits at the mature stage when impacts on ecosystem processes are most likely to occur. Good colonizers are often poor competitors. Thus, good colonizers would be species with limited effects on biodiversity and ecosystem functioning and blend into the saturating function of the biodiversity-ecosystem functioning relationship.

2) Intentionally or unintentionally introduced species do not need to be good dispersers to get to a new location. Thus, exotic species do not necessarily play by the rules, and therefore have the greatest potential to affect biodiversity and ecosystem functioning. Two types of species with potentially large impacts are (a) those that directly impact ecosystem processes by occupying a new niche and changing environmental conditions to precipitate cascading effects on biodiversity, and (b) those that establish dominance by gaining a competitive advantage and affecting ecosystem processes through effect traits that differ from resident species.

3) The magnitude and direction of a biodiversity effect on ecosystem functioning differs between open and closed systems. The relationship in closed systems becomes stronger with time and switches from a sampling to a complementarity effect. The relationship in open systems is less strong and will be dominated by a sampling effect as an invasive species, by definition, drives community abundance and/or ecosystem processes.

To forecast the impact of invasive species on BEF and to evaluate risk of invasion, we need to know which species are most likely to cause cascading effects on biodiversity and ecosystem functioning. Much of this will rely on understanding similarities and dissimilarities in species traits. This is a major task, but an important one if we want to preserve biodiversity and ecosystem functioning within a changing world, where the introduction of novel species is an inevitable consequence of global trade and human travel.

Acknowledgments

We thank the NSF-supported BioMERGE Research Coordination Network and the DIVERSITAS eco-SERVICES Core Project for supporting the workshop 'The consequences of changing biodiversity – solutions and scenarios' (30 November–5 December 2006, Ascona, Switzerland) and the workshop organizers for giving us the opportunity to participate in this volume. We also thank Shahid Naeem, Diane Larson, Qinfeng Guo, and two anonymous reviewers for their thoughtful feedback that substantially improved this chapter.

CHAPTER 17

The economics of biodiversity and ecosystem services

Charles Perrings, Stefan Baumgärtner, William A. Brock, Kanchan Chopra, Marc Conte, Christopher Costello, Anantha Duraiappah, Ann P. Kinzig, Unai Pascual, Stephen Polasky, John Tschirhart, and Anastasios Xepapadeas

17.1 Introduction

The irreversible loss of genetic information (and the resulting loss of both evolutionary and technological options) caused by the extinction of species involves a global public good, the gene pool. Although important, it is not the only reason to be concerned about biodiversity change. As the Millennium Ecosystem Assessment (2005b) points out, another reason for concern is the role of biodiversity in the loss of ecosystem services. These also involve public goods, but unlike the public good associated with species extinction, they are almost always local or regional in extent. The conservation of species threatened with local extirpation protects a number of provisioning and cultural services, as well as the capacity of the local system to function over a range of environmental and market conditions. The latter may involve, for example, the regulation of specific biogeochemical cycles in different climatic conditions, or the protection of crop yields in the face of an array of pests and pathogens. In almost all cases, however, conservation of the functionality of particular ecosystems provides benefits to specific communities rather than to global society (Perrings and Gadgil 2003). Whether we focus on the gene pool or ecosystem services, however, biodiversity – the composition and relative abundance of species – is important because of its role in supporting the capacity of the system to deliver services over a range of environmental conditions. The economic problem of biodiversity, in this sense, differs from the economic problem of individual biological resources. The question is not at what rate to extract a particular resource, but how to balance the mix of species to assure a flow of benefits over a range of possible conditions.

Biodiversity conservation is frequently a public good. In many cases, nobody can be excluded from the benefits offered by the protection of assemblages, and if one person benefits it does not reduce the benefits to others. Because it is a public good, it will be 'undersupplied' if left to the market. The incentive that people have to free ride on the conservation activities of others means that people will collectively conserve too little biodiversity. At the same time the lack of markets for many of the biodiversity impacts of human activities mean that people are not confronted with the true cost of their decisions. Open access to scarce environmental resources is widely recognized to be a major cause of overexploitation.

Nowhere is this more clearly shown than in the world's fisheries. Worm *et al.* (2006) identified catches from 1950 to 2003 within all 64 large marine ecosystems worldwide: the source of 83 per cent of global catches over the past 50 years. They reported that the rate of fisheries collapses in these areas (catches less than 10 per cent of the recorded maximum) has been accelerating, and that 29 per cent of fished species were in a state of collapse in 2003. Cumulative collapses affected 65 per cent of all species fished. While property rights are generally better developed in terrestrial systems, many of the effects from anthropogenic land use change on biodiversity and ecosystem services are also not reflected by markets. Since habitat loss through land use change is the greatest single source of biodiversity loss, this is a major problem. Both the 'public good nature' of biodiversity conservation and the existence of biodiversity externalities mean that private decision-makers largely ignore the effect of their own behaviour on biodiversity, on ecosystem functioning, and, by and large, on the ecosystem services on which we all depend.

The economics of biodiversity and ecosystem services is largely about the failure of markets to signal the true cost of biodiversity change in terms of ecosystem services, the failure of governance systems to regulate access to the biodiversity embedded in 'common pool' environmental assets, and the failure of communities to invest in biodiversity conservation as an ecological 'public good'. This chapter reviews both the nature of the challenges posed by these failures and the options for addressing them. It requires that we are able to identify correctly both the private and social decision problems, and hence that we are able to value those non-marketed environmental effects that are ignored in many private decisions. Chapters 18 and 19 address the issues associated with the valuation and modelling, respectively, of the non-market effects of private decisions on biological resources. In addition, however, it requires that we are able to identify governance mechanisms, institutions, and instruments that will induce private decision-makers to behave in ways that are consistent with the social interest. This chapter focuses on the institutional and policy options for securing the socially optimal mix of species, given the role of biodiversity in assuring ecosystem services over a range of environmental conditions.

17.2 Biodiversity externalities in ecosystem services

Crocker and Tschirhart (1992) describe externalities of the kind described in the Millennium Ecosystem Assessment (2005b) as 'ecosystem externalities'. They define **ecosystem externalities** as: *market-driven actions that impact the wellbeing of either consumers or producers by altering the ecological functioning on which consumption or production depends, but where the welfare effects of those actions are ignored.*

Thus, an ecosystem externality refers to the case where an individual's economic activity generates a real change in ecosystem services that impacts the wellbeing of others, but which is ignored by that individual. This may be because the activity involves the use of a public good (often a common pool environmental resource) where there is scope for free-riding, because of the incompleteness of markets, or because of the system of governance. Where the change in ecosystem services is mediated by a change in biodiversity, we refer to biodiversity externalities in ecosystem services.

For example, nitrogen oxides emitted from coal-fired power plants and mobile sources are a serious air pollutant that can directly impact human health – a traditional externality. Urea, ammonia, or ammonium applied as fertilizer to agricultural lands contributes to nitrate pollution of ground and surface water. In bays and estuaries, nitrate pollution is a serious problem that causes algae blooms and reduces abundances of desirable food species. Nitrogen applied to land has a more direct effect on biodiversity. A twelve year study of Minnesota grasslands (Tilman *et al.* 2001) showed that added nitrogen decreased species diversity and dramatically changed community composition. Species richness declined by 50 per cent and native bunch grasses were replaced by weedy European grasses. In both cases there is an ecosystem externality. The pollutant causes adaptations in the ecosystems affected, which in turn reduce the flow of ecosystem services in the form of reduced fishing, grazing and recreational opportunities.

The effect of biodiversity change on ecological functioning and the provision of ecosystem services has been most convincingly demonstrated for grasslands. In another set of Minnesota grasslands experiments in which biodiversity was deliberately varied across replicate plots, the capacity of the system to function over a range of environmental conditions was lowest where species richness was lowest. These and other experiments suggest that the loss or removal of species that function effectively in specific conditions reduces the range of conditions over which the system as a whole can operate (Tilman and Downing 1994, McGrady-Steed *et al.* 1997, Naaem and Li 1997). Tilman *et al.* (1996) describe this result, using reasoning from economics, as a portfolio effect. Increasing the number of species that fluctuate independently will decrease system volatility, just as increasing the number of independent assets in a financial portfolio will decrease the volatility of returns. Greater species richness is equivalent to greater diversification, which leads to lower variance.

In a closely related line of reasoning, Perrings et al. (1995, p. 4) state that 'the importance of biodiversity is argued to lie in its role in preserving ecosystem resilience, by underwriting the provision of key ecosystem functions over a range of environmental conditions'. Conserving biodiversity maintains species that may look unimportant for ecosystem function under current conditions, but which may play a crucial role in a drought, pest infestation or other shock (Walker et al. 1999). As above, conserving diversity can increase the probability of maintaining the flow of desired services over a range of potential environmental conditions. Resilience has been defined in two ways in the ecological literature. One, due to Pimm (1984), is the speed with which an ecosystem returns to equilibrium after a shock. A second, due to Holling (1973), is the magnitude of the shock that can be absorbed by an ecosystem without losing functionality – effectively the maximum perturbation that be accommodated without flipping from one state (stability domain) to another. The Holling resilience of a system increases with both the resistance/robustness of that system (the extent to which a perturbation moves it from the equilibrium) and its flexibility/adaptability (its ability to accommodate perturbation without loss of functionality). The economics literature has tended to adopt the second of these two ideas. Under either definition, though, if the reference state is desirable, then greater resilience will increase welfare. If the reference state is undesirable, greater resilience will reduce welfare. The biodiversity problem, in this case, is to choose the mix of species that will maximize some index of human wellbeing, given both the expected range of environmental conditions (shocks) and people's aversion to risk. As Chapter 19 shows, the measure of biodiversity used will depend on the nature of the decision problem – on the aspects of biodiversity that matter for human wellbeing. If that involves maintaining the options open to society in an uncertain world, then the right way to think about biodiversity is as a portfolio of biological assets or a risk-pooling mechanism. That is, biodiversity limits the variability in the supply of provisioning and cultural services.

17.3 Biodiversity as insurance

To be more precise about this, consider a manager concerned to maintain the flow of some ecosystem service – say food crops in an agro-ecosystem – operating under uncertainty due to stochastic fluctuations in environmental conditions. We can analyze this problem using the approach developed by Baumgärtner (2007), Baumgärtner and Quaas (2006, 2008) and Quaas and Baumgärtner (2008). The manager chooses a level, v, of agrobiodiversity by selecting a portfolio of different crop varieties. Given this choice the manager realizes a crop yield at level s which is random. For simplicity we may assume that the agro-ecosystem service directly translates into monetary income and that the mean level, $Es = \mu$, of yields is independent of the level of biodiversity and is assumed to be constant. However, the variance of agro-ecosystem service depends on the level of agrobiodiversity:

$$\text{var } s = \sigma^2(v) \quad (17.1)$$

where $\sigma^{2'}(v) < 0$ and $\sigma^{2''}(v) \geq 0$. The farmer's private decision on the level of agrobiodiversity affects not only his private income risk, as expressed by the variance of on-farm agro-ecosystem service, but also causes external effects. Suppose that $B(v)$ defines the sum of external benefits of on-farm agrobiodiversity, and that this takes the form of a reduction in the variance of some public ecosystem service:

$$EB(v) = \Xi \quad (17.2)$$

$$\text{var } B(v) = \sum\nolimits^2(v) \quad (17.3)$$

where $\sum^{2'}(v) < 0$ and $\sum^{2''}(v) \geq 0$ To see the role of agrobiodiversity, suppose that the manager has the option of buying financial insurance by choosing some level $a \in [0,1]$ of insurance coverage, paying

$$a(Es - s) \quad (17.4)$$

to the insurance company as an actuarially fair premium if the farmer's realized income is below the mean income plus any transaction costs of insurance. The latter are measured by:

$$\delta a \operatorname{var} s, \quad (17.5)$$

where the parameter $\delta \geq 0$ describes the 'costs' of insurance. The higher the insurance coverage, a, the lower is the risk premium of the resulting income lottery. The farmer chooses the level of agrobiodiversity, v, and financial insurance coverage, a. A higher level of agrobiodiversity carries costs $c > 0$ per unit of agrobiodiversity. Hence the manager's (random) income is given by

$$y = (1-a)s - cv + aEs - \delta a \operatorname{var} s. \quad (17.6)$$

Increasing a to one allows the farmer to reduce the uncertain income component to zero. The mean and variance of the farmer's income are determined by the mean and variance of the agro-ecosystem service, which depends on the level of agro-biodiversity:

$$Ey = \mu - cv - \delta a \sigma^2(v) \quad (17.7)$$

$$\operatorname{var} y = (1-a)^2 \sigma^2(v). \quad (17.8)$$

Mean income is given by the mean level of agro-ecosystem service, μ, minus the costs of agrobiodiversity, cv, and the costs of financial insurance, $\delta a \sigma^2(v)$. For an actuarially fair financial insurance contract ($\delta = 0$), the mean income equals mean net income from agro-ecosystem use, $\mu - cv$. The variance of income vanishes for full financial insurance coverage, $a = 1$, and equals the full variance of agro-ecosystem service, $\sigma^2(v)$, without any financial insurance coverage, $a = 0$.

The farmer is assumed to be risk-averse with respect to his uncertain income y. Specifically, a general form of an expected utility function can be assumed, where $\rho > 0$ is a parameter describing the farmer's degree of risk aversion:

$$U = Ey - \frac{\rho}{2} \operatorname{var} y \quad (17.9)$$

Social welfare is assumed to be the expected welfare stemming from individual income and the public benefits of on-farm biodiversity. Furthermore, it is assumed that the private and the public risks associated with biodiversity are uncorrelated. Specifically, we assume an expected welfare function of the mean-variance type, where the parameter $\Omega > 0$ describes the degree of social risk aversion:

$$W = Ey + EB - \frac{\rho}{2} \operatorname{var} y - \frac{\Omega}{2} \operatorname{var} B. \quad (17.10)$$

In the private optimum, the farmer chooses the level of agrobiodiversity and financial insurance coverage so as to maximize his expected private utility (17.9) subject to constraints (17.7) and (17.8). The resulting allocation has the property that equilibrium levels of both agrobiodiversity and financial insurance coverage increase with the degree of risk-aversion:

$$\frac{dv^*}{d\rho} > 0 \text{ and } \frac{da^*}{d\rho} \geq 0, \quad (17.11)$$

with strict equality at the corner solution $a^* = 1$. The equilibrium level v^* of agrobiodiversity increases, and the equilibrium level a^* of financial insurance coverage decreases, with the costs of financial insurance:

$$\frac{dv^*}{d\delta} > 0 \text{ and } \frac{da^*}{d\delta} \leq 0, \quad (17.12)$$

with strict equality at the corner solution $a^* = 0$. The manager will choose the level of agrobiodiversity so as to equate its marginal benefits and marginal costs, where the marginal benefits comprise both the insurance value of agrobiodiversity and the reduction in payments for financial insurance that results from the reduced variance of agroecosystem service due to a marginal increase in agrobiodiversity. Where financial insurance is available, the manager will choose a level of agrobiodiversity that is below the one he would choose if financial insurance were not available.[1]

The socially optimal allocation $(\hat{v}, \hat{a})$ is derived by choosing the level of agrobiodiversity and financial insurance coverage so as to maximize social welfare (17.10), subject to constraints (17.2), (17.3), (17.7), and (17.8). The efficient allocation is such that both agrobiodiversity and financial insurance coverage increase with the degree of individual risk-aversion, i.e.:

$$\frac{d\hat{v}}{d\rho} > 0 \text{ and } \frac{d\hat{a}}{d\rho} \geq 0, \quad (17.13)$$

with strict equality in the corner solution $\hat{a} = 1$. The efficient level of agrobiodiversity increases with the degree of social risk-aversion, but the efficient level of financial insurance coverage is unaffected by the degree of social risk-aversion, i.e.:

$$\frac{d\hat{v}}{d\Omega} > 0 \text{ and } \frac{d\hat{a}}{d\Omega} = 0. \quad (17.14)$$

The efficient level $\hat{v}$ of agrobiodiversity increases with the costs of financial insurance, and the efficient level $\hat{a}$ of financial insurance coverage decreases with the costs of financial insurance:

$$\frac{d\hat{v}}{d\delta} > 0 \text{ and } \frac{d\hat{a}}{d\delta} \leq 0 \quad (17.15)$$

where equality may hold in the corner solution $\hat{a} = 0$.

The difference between the socially and privately optimal allocation is that the positive externality of a private farmer's effort is fully internalized in the socially optimal solution. By contrast, in the private optimum the manager chooses a level of agrobiodiversity that is too low. There are different ways that the social optimal solution can be reached by creating the right conditions for farmers. One possibility is by providing a subsidy, $\hat{\tau}$, on the conservation and utilization of agrobiodiversity. This

[1] This level can be determined from setting $a = 0$ in Eqn (17.11) and maximizing over v. It is strictly smaller than v^* for all $\delta < p$ and equals v^* for $\delta \geq p$, i.e. in cases where financial insurance is so expensive that an optimizing farmer would not buy it.

should align the private decisions with the social optimal agrobiodiversity level, i.e.:

$$\hat{\tau} = -\frac{\Omega}{2}\sum^{2'}(\hat{v}) > 0 \quad (17.16)$$

The size of the subsidy is increasing in the degree of social risk aversion, Ω, and decreasing in the degree of individual risk aversion, ρ, and the costs δ per unit of financial insurance:

$$\frac{d\hat{\tau}}{d\Omega} > 0, \quad \frac{d\hat{\tau}}{d\rho} < 0, \quad \frac{d\hat{\tau}}{d\delta} < 0. \quad (17.17)$$

The optimal subsidy, $\hat{\tau}$, can be interpreted as a measure of the financial flow needed to internalize the externality, i.e. to solve the public good problem. Thus it can also be interpreted as a measure of the size of the externality.

Although this problem is posed in the context of an agro-ecological problem, the same insights apply to the management of biodiversity in regulating the supply of the full range of provisioning and cultural services. Indeed, even though the Millennium Assessment (2005b) described the regulating services in terms of a very specific set of buffering functions, they actually summarize the role of the portfolio of biological assets in protecting us against the vagaries of both nature and society.

17.4 Biodiversity markets

In all cases, appropriate policy interventions depend on both a comparison between the privately and socially optimal outcomes, and the development of instruments that will induce private decision-makers to behave in ways that are consistent with the social interest.

In some cases, markets are already developing that allow biodiversity conservation to pay for itself (e.g. by establishing property rights in the effects of biodiversity change). As with the agro-ecological examples discussed in Section 17.3, these are cases where biodiversity supports the production of valuable goods and services that can, under the right circumstances, be sold in the market. Doing so may generate enough revenue to make conservation financially viable. This point is the core thesis behind several recent books (Heal 2000, Daily and

Ellison 2002, Pagiola 2002, OECD 2004). Market creation stems from a simple but powerful idea, i.e. markets can be devised to signal the opportunity cost to local land users of agricultural practices that affect biodiversity either positively or negatively. Ideally, such incentives need to address both 'forward' (or 'downstream') links from land users' decentralized decisions to biodiversity levels and 'backward' (or 'upstream') biodiversity linkages, i.e. from changing the stock of biodiversity level and its functional impacts on productivity to land users such as farmers and, thus, work at the landscape level (Pascual and Perrings 2007). But this implies that such incentives may affect the livelihoods of large numbers of land users. This adds a further layer of responsibility to public agencies to be aware of the distributional implications of alternative incentive measures.

One example in which conservation is currently being made financially attractive is ecotourism. The World Tourism Organization estimates that tourism generated revenues of $463 billion in 2001. One of the fastest growing segments of tourism may be nature-based or ecotourism. Some areas have had a long history of profiting from the richness of the local biodiversity, including Yellowstone National Park in the USA, Krueger National Park in South Africa, and a variety of National Parks in Kenya and Tanzania. Costa Rica has also done well promoting ecotourism, with approximately 1 million tourists spending $1 billion in 2000 (Daily and Ellison 2002, p. 178.). Several economic studies have found that ecotourism can generate significant revenues in a variety of developing country settings (e.g. Aylward *et al*. 1996, Lindberg 2001, Maille and Mendelsohn 1993, Wunder 2000).

A second example is bioprospecting for useful genetic material from plant or animal species that may lead to the development of valuable pharmaceuticals or other products. Pharmaceutical firms actively screen organisms in search of such active compounds as part of their intensive research and development programs. Bioprospecting is the term used to describe the process of testing natural organisms for these biochemically active compounds. If identified as active, a compound can result in the development of a new drug based on the natural compound itself (as in the case of vincristine and vinblastine found within the Rosy Periwinkle (*Vinca rosea*) or based on a synthetic compound developed from the blueprint provided by the natural compound. In either case, access to natural compounds is of great use in the research and development process. It has been estimated that 25 per cent of the drugs sold in developed countries and 75 per cent of those sold in developing countries were developed using natural compounds (Pearce and Puroshothamon 1995), suggesting that extant biodiversity is of value to pharmaceutical firms in their efforts to develop new drugs.

The CBD recommends a structure for bioprospecting agreements to accomplish three main goals: the conservation of biological diversity, the sustainable use of natural products, and the fair and equitable sharing of benefits derived from genetic resources (Article 1). They imply the existence of intellectual property rights in the face of current patent systems by which, for example, the pharmaceutical and seed industries can realize the monopoly benefits of new product development that are guaranteed by the ability to patent discoveries. Simpson, Sedjo, and Reid (1996) model the research and development process as a search through a list of research leads and conclude that the value of the marginal lead is generally insufficient for pharmaceutical firms to play a role in the conservation of biodiversity (see also Costello and Ward 2006). However, in the presence of competition, it is no longer the case that discovery of a single active compound is sufficient to guarantee a monopoly position within the market (Conte 2007). In a competitive search environment, the revenues associated with discovery will depend on the proportion of total successes controlled by the firm and a firm may have the incentive to preemptively exclude its competition from searching a portion of the research leads by signing bioprospecting agreements with host nations.

In recognition of the importance of IPRs to innovation, the Agreement on Trade Related Aspects of Intellectual Property Rights (1995) (TRIPS Agreement), mandates that all member nations of the World Trade Organization enact national legislation to provide minimum standards and scope of IPR protection (Strauss 1996). While all member nations have complied, there is still heterogeneity in the security of these property rights across nations, which might explain the pattern of existing agreements across countries. The importance of IPR security might also explain why some companies have made agreements with botanical gardens in developed countries for access to samples from tropical countries, as there is less uncertainty associated with the IPRs in developed nations (Sampath 2005).

There are unanswered questions about the optimal allocation of rents from a bioprospecting agreement. Consider, for example, the case of the rosy periwinkle mentioned earlier, a plant native to Madagascar that contains vincristine, a powerful cancer-fighting compound. No synthetic substitute for vincristine exists, and one ounce of vincristine requires 15 tons of periwinkle leaves. This has resulted in depletion of nearly the entire native periwinkle habitat in Madagascar (Koo and Wright 1999), though the plant has been extensively cultivated elsewhere. However, if drug companies do not keep a significant fraction of rents from developing new drugs they may not have sufficient incentive to develop new drugs via bioprospecting. Mendelsohn and Balick (1995) found a significant difference between likely social and private returns to development of new drugs. Koo and Wright (1999) also argue that biodiversity will be underprovided by the private sector via bioprospecting on the grounds that although the value of biodiversity is very large, market and social values are grossly misaligned.

It should also be noted that any added value of biological resources is created at each step of the innovation process – through the contributions of the local communities and research laboratories to industrial applications – and not only at the final stage of the innovation process. The existing IPR system only addresses the final stage of the innovation process, thus casting doubt as to whether IPRs are sufficient to induce the socially optimal level of conservation (Goeschl and Swanson, 2002; Dedeurwaerdere et al. 2007).

Indeed, both ecotourism and bioprospecting have been subject to criticism that revenues generated by conservation activities have not necessarily resulted in benefits to local communities. Local communities with no financial stake in conservation or that in fact suffer financial losses from conservation activities (e.g. wildlife damage to crops) might resent or actively oppose such activities, leading to a greater probability that conservation will fail. Trying to give local communities a stake in conservation has led to efforts to promote community-based conservation (Western and Wright 1994) and integrated conservation–development projects, or ICDPs (Wells and Brandon 1992). The goal of community-based conservation is to give local communities control over resources, thereby giving the community a stake in conservation. The most well-known community-based conservation program is the Communal Areas Management Program for Indigenous Resources (CAMPFIRE) in Zimbabwe (see Barbier 1992 for an early review and economic assessment). ICDPs try to 'link biodiversity conservation in protected areas with local socio-economic development' (Wells and Brandon 1992). Both approaches arose because of the failure of traditional protected areas conservation strategies that ignored the needs of local communities.

The extent to which conservation and local control over resources, or local economic development, are mutually consistent remains to be seen. Overall, community-based conservation and ICDPs have had mixed success to date. There is no guarantee that once they are given the choice, local communities will in fact choose to conserve. Cultural, social, or political factors may block conservation even when economic factors favour conservation. There is also no guarantee that conservation and local economic development are in fact consistent goals. Certainly in some communities with ecotourism potential, or where ecosystems provide valuable ecosystem services, conservation and development may go hand in hand. In other cases, the conservation of biodiversity and economic development may not be consistent. Because of the pervasive nature of external benefits created by biodiversity conservation, it may require more than just allowing local control and market forces to achieve an efficient level of conservation.

Recognition that the conservation of biodiversity may generate benefits that reach well beyond the local community provides a rationale for governments and non-governmental organizations to provide resources for conservation, and for the institution of national or international conservation policies. At present, though there are a number of policies to promote conservation, there are also a number of policies that have the opposite effect. Agricultural subsidies, subsidies to clearing land, resource extraction, and new development may all contribute to driving a further wedge between private and social returns from actions that conserve biodiversity. Perhaps the first rule for policy should be to 'do no harm'. Beyond doing no harm by eliminating perverse subsidies, however, positive external benefits from conservation require policies that create positive incentives to conserve.

Both governments and non-governmental organizations, such as the Nature Conservancy and World Wildlife Fund, are actively engaged in acquiring land for conservation and in other activities promoting conservation. Buying land is a direct and secure way to promote conservation, but it is often a costly instrument for protecting biodiversity. Boyd *et al.* (1999) find that acquisition is often 'conservation overkill'. Conservation easements that rule out certain incompatible land uses, but not all land uses, are often a far cheaper route to secure conservation objective than acquisition. Recently, interest has shifted away from land acquisition toward conservation easements and other ways of working with landowners to promote both conservation and landowner interests. For example, The Nature Conservancy's approach, once heavily weighted toward acquisition, now incorporates mechanisms such as community development projects to reduce the demand for fuelwood and the purchase of conservation easements to limit development (see http://www.nature.org/ for examples).

Acknowledging that donors from high-income nations invest billions of dollars toward ecosystem protection in low-income nations, a related literature debates the relative merits of direct conservation payments versus indirect mechanisms (e.g. payments to promote ecotourism which generates ecosystem protection as a joint product). Although indirect approaches are the predominant form of intervention in low-income countries, Ferraro and Simpson (2002) argue that

---

Box 17.1 The impact of market and non-market institutions on forest biodiversity and timber extraction: a study in northern India

The institutions that govern forests affect both the diversity of the forest stock and the mix of products and services that are extracted. A study of timber extraction from forests of the north Indian state of Uttar Pradesh from 1975 to 2000 shows how timber harvest is related to institutional conditions, species richness and the ecological characteristics of the forest, as well as to forest stocks, and harvest effort. Using a modified Gordon–Schaefer production function, and the assumption that forests are managed for 'sustainable timber extraction', the reduced form equations are derived and estimated (Chopra and Kumar 2004). The composition of products extracted is determined by their value, high value products being given priority. The modified model includes a bio-economic diversity index defined as $\sum_i (P_i Y_i / TR)^2$ where Y_i denotes harvest of the ith species, P_i denotes the price of the ith species, and $TR = \sum_i P_i Y_i$. The index is postulated to impact extraction as a shift factor in the extraction function. It is a weighted index of biodiversity in which prices are used as weights for the different Y_i. A loss of biodiversity is reflected in an increase in the value of the biodiversity index, which ranges between 0 and 1. The effect of the bio-economic diversity measure on timber productivity is captured by the introduction of an extra term, B, in the timber production function:

$$Y = qBEX \quad (17.1.1)$$

where E is effort and X the aggregate biomass of all species. Eqn. (17.1.1) implies that $Y/E = F(B, X)$, i.e. that the effort involved in timber extraction is inversely related to B. As B decreases (or as biodiversity increases), the extraction function shifts, resulting in a lower effort per unit effort. In other words, the model with the biodiversity index yields a lower level of Y for the same level of E, since $0 < B < 1$. If the forest manager is primarily interested in timber extraction, this results in a substitution of plantation forests for natural forests, so changing the ecological properties of the forest.

To capture this, Chopra and Kumar introduce W (the share of plantation forest in total area) in the growth function for timber biomass:

$$\dot{X} = rX(1 + eW - X/K) - qBEX \quad (17.1.2)$$

in which e is a coefficient for impact of W on growth of timber stock. They postulate that extraction increases as W increases. They further assume that $B = F(W)$. Since B increases as W increases, forests become less diverse. The estimated harvest equation is:

$$\log(U_{bt}/U_{bt-1}) = r + 0.0853\, E_t + 0.00169 U_b^*{}_t$$
$$+ 8.7454 W_t^{**} \quad (17.1.3)$$

in which U_{bt} is biodiversity adjusted extraction (per unit effort), * indicates significance of the coefficient at 5 per cent level and ** denotes significance at the 1 per cent level. Effort E_t is not, by itself, a significant determinant of trends in extraction in this formulation, but both U_{bt} and W_t are significant. Extraction increases over time as the plantation area increases. Since W is inversely related to the level of biodiversity in the forest, a decreasing biodiversity due to a larger ratio of plantation to natural forest leads to rising trends in extraction. Extraction per unit effort increases as U_{bt} increases. Further, assuming E to be constant, they show that U_{bt} (defined as Y/BE) may increase under the following conditions with respect to the biodiversity index B.

1. With a rising B (falling biodiversity), if Y rises faster than B (extraction rises faster than biodiversity falls) U_{bt} increases and extraction of timber over time decreases. A rising extraction with decreasing biodiversity of the forest pushes the system towards a state in which increases in extraction take place at an increasing rate.
2. With a falling B (rising biodiversity) U_{bt} could decrease provided Y is not rising faster than B is decreasing, leading to a decreasing trend in extraction in subsequent periods.
3. With a constant B (constant levels of biodiversity) increases in U_{bt} are determined by changes in Y.

direct payments can be far more cost-effective, often requiring no additional institutional infrastructure or donor sophistication.

17.5 Economic instruments

Where biodiversity markets do not exist, and where market-driven behaviour leads people to select a combination of species, ecosystems, and landscapes that is not socially optimal, economists have developed a range of market-like instruments for encouraging socially desirable behaviour. The application of these instruments has been widely endorsed, and a number of countries make use of one or more of them. The OECD (2004), for example, makes the following recommendation to member countries:

1) establish and apply a policy framework aimed at ensuring the efficient long-term conservation and sustainable use of biodiversity and its related resources. The overarching goal of such a framework should be to ensure maximum net benefits, both now and in the future, from the use and conservation of resources stemming from biodiversity – as well as an equitable sharing of these benefits that is consistent with national, and applicable international, legislation;
2) make greater and more consistent use of domestic economic instruments in the application of their biodiversity policy frameworks, while attempting to reach further agreement at the international level on the use of economic-based policy instruments with respect to biodiversity conservation and management;
3) integrate market and non-market (i.e. non-price) instruments – taking account of the respective advantages of each in lowering information and transactions costs, and in addressing the 'public' values of biodiversity – into an effective and efficient mix of policies; and
4) integrate biodiversity policy objectives in a cost-effective manner into government sectoral policies, in order to avoid undue adverse effects on biodiversity and its related resources.

The set of instruments proposed by the OECD is amongst those described in Table 17.1. In what follows we highlight those instruments that are currently attracting attention.

Payments for ecosystem services The most direct way to create positive incentives for conservation is to institute a system of payments for the provision of ecosystem services (ES). Payments (or Rewards)

Table 17.1 Instruments for encouraging the socially optimal use of biodiversity.

Category	Type of instrument	Example
Economic incentives	Fees, charges, and environmental taxes	Charges or non-compliance fees related to certain types of forestry activities
	Payments for ecosystem services	Liability fees for the maintenance or rehabilitation of ecologically sensitive lands
	Assignment of well-defined property rights	Fishing license fees or taxes (whose objective is resource management)
	Reform or removal of harmful subsidies	Levies for the abstraction of surface water or groundwater
		Liability payments for biodiversity damages (including interim losses)
		Charges for:
		Use of public lands for grazing in agriculture
		Use of sensitive lands
		Hunting or fishing of threatened species
		Tourism in natural parks
		Payments to farmers within a watershed for using farming techniques that maintain the quality of water resources
		Auctioned conservation contracts
Funds	Environmental funds and public financing	Global Environment Facility funding of local biodiversity conservation where there are global benefits
Framework incentives	Information provision, scientific and technical capacity building	Global Biodiversity Observation Systems
	Economic valuation	Development of natural resource accounts
	Market creation	Inclusion of biodiversity in Adjusted Net Savings Measures
	Institution-building and stakeholder involvement	Support for biodiversity-related labelling schemes
		Strengthening governance of local, regional and global common pool biological resources
		Transferable development rights

for ecosystems services, P(R)ES, are voluntary transactions, not necessarily of a financial nature, in the form of compensation flows for a well-defined ES, or land use likely to secure it. The notion of 'rewards' is used to acknowledge that transactions from beneficiaries to providers may not need to be based on a financial flow. It can also involve in-kind transactions that may include a myriad of valuable goods and services from the beneficiaries' point of view, which can take intangible forms in diverse situations, such as knowledge transfer. P(R)ES is paid/rewarded by the beneficiaries and shared by the providers of the ES after eventually securing such compensation. P(R)ES are often designed to address problems related to the decline in some environmental services, such as the provision of water, soil conservation, and carbon sequestration by upland farmers who manage forest-lands in upper watersheds. In essence, such compensations are intended to internalize the positive externalities generated by upland farmers who can maintain the flow of valuable services that benefit lowland farmers or urban dwellers. However, a key obstacle in the successful implementation of P(R)ES arises at the 'value demonstration' stage, especially due to the scientific uncertainties underpinning the linkages between alternative land uses and the provision of the targeted environmental services.

The country that has moved furthest in this direction is Costa Rica. The 1996 Forestry Law instituted payments for ecosystem services. The law recognizes four ecosystem services: mitigation of greenhouse gas emissions, watershed protection, biodiversity conservation, and scenic beauty. The National Forestry Financial Fund enters into contracts with landowners who agree to do forest preservation, reforestation, or sustainable timber management. Funds to pay landowners come from taxes on fuel use, sale of carbon credits, payments from industry, and the Global Environment Fund

(GEF). Many developed countries have adopted some form of 'green payments' in which agricultural support payments are targeted to farmers who adopt environmentally friendly management practices or land uses (OECD 2001).

P(R)ES cannot be properly designed or implemented without a clear understanding of the property right regimes. Property rights regimes in natural resource management comprise a structure of rights to resources, rules under which those rights are exercised, and duties binding both those who possess the right(s) and those who do not. As Bromley (1991, pp. 2) puts it, '[p]roperty *is not* an object but rather is a social relation that defines the property holder with respect to something of value...against all others'. In this context, Costa Rica is one of the few examples where an elaborate nationwide PES program is in place under a clear property rights regime. Under this program, only farmers with property rights to land can be paid for the environmental conservation they provide (Pagiola 2002).

A recent illustrative example of the potential effectiveness and flexibility of P(R)ES programs is that of RUPES: *Rewarding Upland Poor for Environmental Services*. The RUPES partnership comprises the International Fund for Agricultural Development (IFAD), the World Agroforestry Centre (ICRAF) and a group of local, national, and international partners.[2] RUPES aims to conserve environmental services at both global and local levels, while at the same time supporting the livelihoods of the upland poor in Asia. So far, the main focus has been on Nepal, the Philippines, and Indonesia, and the environmental services mostly include water flow and quality, biodiversity protection and carbon sequestration.

A variant of P(R)ES is the approach based on *direct compensation payments* (DCP) for taking private land out of production and into conservation (Swart 2003). Similar to other incentive mechanisms, the identification of the level of the efficient compensation payments to landowners requires the demonstration of an objective measure of its conservation value on both biological and economic grounds. In addition, the change in decentralized behaviour needs to be sustained into the future, which requires longer term political commitment. Asymmetric information between landowners and the compensating government agency is at least potentially problematic (Innes *et al.* 1998). If landowners expect compensation that is lower than the present value of the benefit stream arising from developing the land holding, they have a motive to develop quickly. Furthermore, even when exact compensation is foreseen by landowners, they may still have an incentive to intensify land use before compensation if this augments the market value of their property.

Transferable development rights Another approach to conservation is to institute a system of transferable development rights (TDR). TDR are virtually identical to cap-and-trade schemes to limit pollution emissions. In a TDR system, the conservation planner determines how much land can be developed in a given area. Development rights are then allocated and trades for the right to develop are allowed. Developers can increase density in a growth zone ('receiving area') only by purchasing development rights from the preservation area ('sending area'). The approach was developed and implemented extensively in the 1970s to direct development within urban areas (see Field and Conrad (1975) for what appears to be the first economic model of the supply and demand for development rights; see Mills (1980) for a model of TDR and a discussion of their appropriateness for use in protecting public goods).

Not until relatively recently have economists explicitly considered TDR as a mechanism to conserve biodiversity. Panayotou (1994) developed the TDR approach for conservation. He argued that 'biodiversity conservation is ultimately a development rather than a conservation issue' (Panayotou 1994, p. 91). Given that most biodiversity exists in the developing world, and that the public good nature of biodiversity requires a mechanism for paying developing countries to be stewards of this resource, Panayotou argues that TDR may also be an effective way to protect global (as well as local) biodiversity. Merrifield (1996) proposes use of a similar concept where 'habitat

---

[2] Some of the insights reflected here come from personal communication with Meine van Noordwijk, Tom Tomich and ICRAF personnel involved in RUPES program in Sumatra, Indonesia.

preservation credits' would be required for development. There is no guarantee that TDR schemes, like cap-and-trade schemes, will result in efficient outcomes unless the planner chooses the correct amount of rights/permits to allocate. An additional problem faced in TDR for conservation is deciding what are appropriate trades. Land units, unlike air emissions, have unique characteristics and may contribute to a number of conservation objectives. What constitutes an equal trade is not obvious. Similar problems over establishing the proper trading ratios exist in mitigation banking schemes for wetlands.

Having said this, TDR appears to be an innovative and cost-effective way to resolve the perverse incentives arising from DCPs. TDR extend the longstanding 'agro-ecological zoning' schemes, which aim to direct development to areas of high productivity potential and to restrict agricultural land use in ecologically significant and sensitive areas. However, such zoning programs do not allow for any substitutability between plots in meeting overall conservation goals. By providing a market-like alternative to the DCPs, flexibility in achieving conservation goals can be introduced. In this vein, the main advantage of TDR is that it can, in principle, encourage conservation on lands with low agricultural opportunity costs, while providing appropriate incentives to the affected landholders (Chomitz 1999).

In contrast to DCP, each landowner is issued tradable development permits by the government agency at an initial period. Subsequently, landowners hold the right to either develop or intensify their land holding. However, to develop that fraction of land a landowner needs to either use of one of the development permits (s)he holds or buy it from other landowners, who upon selling it can no longer develop their land fraction and instead must give it up for conservation. In this case, the government can share the cost of the 'takings', i.e. compulsory government land acquisition, with the landowners themselves.

Two main types of TDR programs exist at the landscape level: single and dual zoning. The former is similar to permit systems such as those used in transferable fishing quotas or pollution control. After the initial allocation of quotas, anyone within the program area may buy or sell the permits. The dual zone system instead explicitly designates both (permit) sending and receiving areas. This allows, for example, for new land use restrictions to be imposed on ecologically sensitive sending zone upon obtaining additional information about its higher conservation value and assigning TDRs to compensate for such additional restrictions. Usually, tight restrictions are also imposed on the receiving zone so as to increase the demand for TDRs (Chomitz 1999).

One of the forerunners of the TDR mechanism is in Brazil. While some initiatives have been proposed, the implementation is still under discussion. The basic idea is to give the opportunity for Brazilian agricultural land owners not complying with the National Forest Code (Law number 4771 approved on 15/09/1965) to buy forest reserves in other areas, normally in close proximity to their property. However, as pointed out by Pascual and Perrings (2007), a fully operational market for forest reserves is still to be implemented. Two examples are the National Provisionary Measure (Medida Provisória, Number 21666-67, approved on 24/08/2001), which amends the Forest Code and in the State of Sao Paulo (State Decree number 50889, approved on 16/06/2006).

Auction Contracts for Conservation (ACCs) One other way to induce private landowners to achieve desired level of supply of biodiversity conservation at the landscape level is by applying a competitive bidding or auction mechanism. An auction is a quasi-market institution with an interesting feature, i.e. it has a 'cost revealing' advantage compared to P(R)ES and DCP and can, in principle, be incorporated into a TDR system. In fact, the cost-revelation feature provides a way of generating important cost savings to governments. This is especially so when significant information asymmetry between farmers and conservation agencies exist regarding (i) the real opportunity cost of conservation and (ii) the ecological significance of the natural assets existing in farmlands. While the former is often better known by farmers themselves, the latter is normally better known by environmental experts. Such information asymmetries one reason for missing agrobiodiversity conservation markets. The idea is to use auctions to reveal the hidden information needed to recreate voluntary conservation contracts between landholders and the government.

In essence, landholders submit bids to win conservation contracts from the government. But while the latter prefers low bids, landowners need to submit bids that at least cover the opportunity cost of carrying out conservation activities. The problem is that information about such opportunity costs is often better known by resource users than by the government and the costs are also likely to be user-specific. Stoneham et al. (2007) provide a recent small-scale pilot case study of an auctioning system for biodiversity conservation contracts in Victoria, Australia, known as *BushTender*. The ACC involved 98 farmers, of whom 75 per cent obtained government contracts to conserve remnant vegetation in their farms after all farmers submitted sealed bids associated with their nominated conservation action plans. The selection of the farmers who won the contract was based on ranking the relative cost-effectiveness of each proposed contract. This involved weighting each private bid against the associated potential ecological impacts at the landscape level. Given a public budget of $400,000, contracts with bids that averaged about $4,600 were allocated and specified in management agreements over a three-year period. In total the contracts covered 3,160 ha of habitat on private land. Stoneham et al. (2007) have estimated that the *BushTender* mechanism has provided 75 per cent more biodiversity conservation compared to a fixed-price payment scheme (or DCP). In addition, they contend that given the relatively low enforcement costs in their pilot study, this ACC has interesting cost-effective properties. The pilot case study shows that it is possible to recreate the supply side of a market for agrobiodiversity conservation.

P(R)ES, DCP, TDP, and ACC all share an important characteristic for successful market creation for biodiversity conservation. For these mechanisms to be effective, accurate ecological and economic information at the demonstration, capture and sharing stages is needed. If it is not possible, or very costly, to convey clear and credible information about the nature of the services derived from biodiversity, the costs of supplying them, and the benefits derived for society, then the effect of implementing these economic mechanisms would be distorted and would lack precision. Moreover, it would be naïve to champion market creation for biodiversity conservation if other supporting institutions are also lacking, such as property rights to the resources in question (Pascual and Perrings, 2007). Furthermore, if markets for biodiversity are recreated without proper institutional and regulatory backup, then the social costs of such policies may well outweigh the benefits from conservation (Barrett and Lybbert 2000). In a second-best world where information is elusive, most policy initiatives pragmatically focus on ensuring that institutions are developed so as to keep future options open (Tomich et al. 2001). In fact, most conservation policies are aiming at developing flexible and open institutions that can mitigate the negative effects of intensification in agroecosystems, without foreclosing future land (de)intensification options.

An important qualification is that many market-like mechanisms have implications for the rights of the poor, particularly in low-income countries where people depend heavily on environmental resources (Dasgupta 2001). Pricing access to ecosystem services can cause the socially disadvantaged and vulnerable to be excluded from those services, and mechanisms need to be developed to address this. For example, in 1991 the Government of Uganda established a national park in the Bwindi forest to protect the mountain gorilla. This park was established with little consultation with the local populations who depended on the forests for their livelihood. As a result, poaching and encroachment were common. In 1995, the Mgahinga and Bwindi Impenetrable Forest Conservation Trust Fund was created, its proceeds being shared with the local communities to encourage sustainable development activities and conservation.

The general problem is that economic interventions that are efficient by the Pareto criterion (which states that an economic intervention is efficient if it benefits at least one person without leaving any other person worse off) may still leave people worse off in relative terms. One approach to this problem is to subject interventions to a second test: that the equity gap between individuals or groups after an economic intervention should be no larger than the gap before the intervention. In this way, if one individual has benefited from the economic instrument, then some transfer will need to take place to ensure that the gap between that individual and others will remain the same. In other words, some form of social redistribution mechanism will need to be institutionalized at the same time the

Box 17.2 The economics of the US Endangered Species Act (ESA)

While market-oriented policies have been of increasing importance in recent years, other important policies directed at the conservation of biodiversity, including the U.S. Endangered Species Act and the Convention on International Trade in Endangered Species, are at their core largely command and control regulatory regimes. The Endangered Species Act (ESA), enacted in 1973, changed conservation policy from a largely voluntary and toothless regime that existed prior to 1973 into a powerful environmental law capable of stopping large government projects and actions of private landowners (Brown and Shogren 1998). Section 7 of the ESA prohibits federal agencies from actions that cause 'jeopardy' (i.e. risk of extinction) to species listed as threatened or endangered. Section 9 prohibits public and private parties from 'taking' listed species. 'Taking' includes causing harm to species through adverse habitat modification from otherwise legal land uses, such as timber harvesting or building, as well as more obvious prohibitions against killing, injuring or capturing a listed species. The way the law is written, the ESA appears to have very limited scope for economic considerations. Sections 7 and 9 are absolute prohibitions. Biological criteria are the basis for listing species. In TVA v. Hill, the US Supreme Court wrote: 'The plain intent of Congress in enacting this statue was to halt or reverse the trend toward species extinction, whatever the cost' [437 U.S. 153, 184 (1978)]. When it looked like a small unremarkable fish (the snail darter) that was previously all but unknown would halt construction of a large dam backed by politically powerful members of Congress, Congress amended the ESA. They authorized the formation of the Endangered Species Committee ('The God Squad') to allow an exemption to the ESA if the benefits of doing so would clearly outweigh the costs. There are high hurdles to be met for convening this Committee and it has been used rarely.

Despite the fact that the law is written in a way that appears to marginalize economic considerations, it has proved impossible to administer the Act while totally ignoring economics. Several writers have noted that economic and political considerations influence agency actions at all stages of the ESA process including the listing stage, which is supposed to be done strictly on biological grounds (e.g. Bean 1991, Houck 1993). Endangered species whose protection threatens to impose large costs run into political opposition that translates into pressure on the Fish and Wildlife Service. This pressure appears to translate to lower probability of listing (Ando 1999). The benefits side of the equation also seems to affect listing and recovery spending even though the ESA does not base such decisions on the popularity of the species. Metrick and Weitzman (1996) found that more charismatic species were likely to be listed than uncharismatic species, and that once listed 'visceral characteristics play a highly significant role in explaining the observed spending patterns, while the more scientific characteristics appear to have little influence' (Metrick and Weitzman 1996, p. 3).

While much of the early regulatory activity under the ESA targeted government actions under Section 7, the 1990s saw an increase in the emphasis on conservation on private lands under Section 9. More than half of endangered species have over 80 per cent of their habitat on private land (USFWS 1997). Conservation on private lands presents a number of incentive issues (Innes et al. 1998). A landowner whose parcel contains an endangered species habitat may face restrictions on what activities may be undertaken. The landowner need not be compensated if restrictions are imposed and losses to the landowner result (though the law on regulatory takings is quite unsettled; see Polasky and Doremus 1998). The potential losses the ESA may impose on a landowner give rise to several perverse incentives. Innes (1997) shows that there can be a race to develop in order to beat the imposition of an ESA ruling. Similarly, there may be an incentive to 'shoot, shovel and shutup' in order to lower the likelihood of imposition of restrictions under the ESA (Stroup 1995). Further, because current law stipulates that acquiring specific information about species is a prerequisite to imposing restrictions on a landowner, there is no incentive for the landowner to cooperate in allowing biological information to be collected (Polasky and Doremus 1998).

There are several possible ways to reform the ESA to cure the worst of the perverse incentives. One method is to provide compensation. When eminent domain is used and there is a physical taking of property, the government is required to provide compensation equal to the market value of the property. The same approach could be taken when the government mandates conservation on private land. There are two potential problems with this approach. First, Blume et al. (1984) show that when landowners are fully compensated in the event of a taking, there is an incentive to over-invest. It is socially optimal to take account of the probability of future takings that render the investment worthless. The landowner is, however, fully reimbursed and so ignores this factor. Second, use of

Continues

Box 17.2 *(Continued)*
government funds to pay for compensation may be costly. On the other hand, others point out that there is an advantage to forcing regulators to understand the costs of imposing regulations by paying compensation (e.g. Stroup 1995). Rather than tying compensation to market value, paying compensation tied to the value of conservation along the lines of green payments discussed above, can generate efficient incentives to conserve (Hermalin 1995).

A different approach to reform is to allow landowners to avoid sanctions if they can prove that their proposed actions will not cause harm (Polasky and Doremus 1998). This type of approach is exemplified in the ESA by the provision to allow landowner actions that cause some minor and unintended harm to a listed species for landowners with approved Habitat Conservation Plans. The incentive for filing Habitat Conservation Plans was further sweetened by promises of 'no surprises' and 'safe harbors' that put the burden on the government for costs imposed by future regulatory actions.

---

economic instruments are being implemented. This however keeps the status quo of the existing equity gaps within society. A third test, which can be considered pro-poor, is that the net benefits accruing from the intervention are distributed according to some ratio whereby the increase in welfare of the worse-off individual is proportionately greater than the welfare increase of the best-off individual (Duraiappah 2006).

17.6 The international dimension

The problem of transboundary externalities resulting from the growth of international trade is the subject of a substantial literature. As with externalities in local markets, one focus has been the consequences of ill-defined property rights in environmental resources (Chichilnisky 1994, Brander and Taylor 1997, 1998, Rauscher 1997). The impact of trade on biodiversity as a specific environmental problem has been evaluated from two main perspectives. One focuses on the link between specialization under trade, habitat conversion, and species loss (Barbier and Schultz 1997, Polasky et al 2004). These studies calculate the impact of trade on biodiversity from the proportion of the land area that is converted to the production of primary commodities, the impact on existing species being taken from the species–area relationship (Macarthur and Wilson 1967). Polasky et al. (2004) extend the analysis to the two country case. The same mechanism operates in each country. They argue that if there are high levels of endemism in each country, and if consumers are concerned to protect local biodiversity, trade can reduce the level of welfare. But where species are common to both trading partners and consumers are interested in global rather than local levels of biodiversity, trade is unambiguously welfare-enhancing.

A second approach focuses on biological invasions as an externality of trade (Perrings et al. 2000, Perrings et al. 2002, Kohn and Capen 2002, Costello and McAusland 2003, McAusland and Costello 2004, Knowler and Barbier 2005). This literature considers both the problem of incentives to internalize biodiversity externalities of trade, and the problem of insufficient investment in biodiversity conservation as a public good. Costello and McAusland (2003) explore the use of tariffs on imports to reduce the damage costs from accidental introductions. While they show that import tariffs will always reduce import volumes of potentially invasive species, they find that tariffs can have adverse effects if they alter the composition of imports, or change land use in ways that make ecosystems more vulnerable to invasive species. McAusland and Costello (2004) consider the efficiency of port inspections combined with tariffs on imported goods, and find that the optimal tariff covers inspection costs plus the potential damage costs from outbreaks of pests undetected during inspections. The optimal level of tariffs in each case depends on the risk of biological invasions and the expected level of damage they cause. The public good problem in the case of invasive species involves the protection offered to all by measures to control the introduction of pests and pathogens. Since it is a 'weakest link' public good, the protection to all is frequently only as good as the protection offered by the weakest link in the chain. This has implications for the pattern of international investment in invasive species control (Perrings *et al.* 2002).

The persistence of international biodiversity externalities, like the persistence of other environmental externalities, has much to do with the ways in which international markets and the rules of international trade are structured. Unlike many other environmental resources, however, there does exist a treaty on the trade of biological species. The Convention on International Trade in Endangered Species (CITES) deals specifically with international markets for biological resources (as distinct from markets for the international benefits of local conservation effort). Its role is to reduce the impact of trade on the survival probability of rare and endangered species. It does this by imposing prohibitions. Examples include international bans on the trade in elephant ivory, and on timber obtained from certain endangered tree species (Barbier *et al.* 1994, Swanson 1995).

CITES has arguably been the international agreement that has had the greatest impact on conservation outcomes (WCMC 1992). CITES authorizes banning international trade in species listed under Appendix I, and regulating trade in species listed under Appendix II. In 1989, CITES initiated a ban on trade in ivory. In the 1970s and 1980s rampant poaching of elephants caused a drop in elephant populations of roughly 50 per cent (Barbier and Swanson 1990). Particularly threatened were elephant populations in east African countries. Elephant populations in southern African countries were less threatened. Imposing the ivory trade ban was controversial. Southern African countries with relatively healthy elephant populations (Botswana, Malawi, Namibia, South Africa, Zimbabwe) objected and did not sign on to the ban. Opponents of a ban argued that it would likely result in high ivory prices as supply was choked off, which would increase the rewards to poaching (Barbier and Swanson 1990). Opponents also argued that by denying rights to sell ivory legally there would be less financial reason to conserve elephant populations and less money available for enforcement efforts against poaching. Proponents of the ban, including east African countries and many developed countries, argued that without the ban elephant populations would continue to decline, as it was too easy to sell illegally harvested ivory and because anti-poaching efforts of impoverished governments were no match for well-organized poaching gangs. Van Kooten and Bulte (2000) summarize economic arguments about the ivory ban and present results from application of several dynamic models.

The argument for trade restrictions of the CITES type is that, in the absence of restrictions, there will be a 'race to the bottom'. Firms will seek to exploit the international advantages offered by relaxed labour and environmental conditions, and countries will use the lack of environmental protection to induce inward investment (Wheeler 2000). By this argument, biodiversity and other environmental externalities are not just an incidental product of market failures. They are the outcome of strategic decisions by governments and firms seeking a competitive advantage. The claim is that where trade agreements make it impossible either to induce inward investment or to protect domestic agriculture or industry through trade policy, countries may be encouraged to use environmental policies to the same effect. Specifically, they may be encouraged either to allow ecological dumping by relaxing environmental protection measures, or to use environmental regulation as trade protection measures.

The empirical evidence for a race to the bottom is mixed. The relocation of polluting industries from high-income to low-income countries is a part of the explanation for changes in environmental indicators observed in the Environmental Kuznets Curve literature (Barbier 1997, Cole *et al* 1997, Arrow *et al.* 1995). However, studies of the incentive effects of environmental regulation have concluded that the costs of compliance with environmental regulations are a sufficiently small proportion of total costs that they do not generally drive location decisions (Jaffe *et al.* 1995, Levinson 1996). Wheeler (2000) argues that the effects of income growth on environmental protection in low-income countries, along with the progressive empowerment of local communities affected by relocation, will be enough to avert a race to the bottom.

In fact, the environmental impacts of trade are one of the few acceptable justifications for imposing trade restrictions under the General Agreement on Tariffs and Trade (GATT). The exceptions allowable under

Article XX of the GATT, along with the Sanitary and Phytosanitary (SPS) Agreement, authorizes countries to impose restrictions on trade in order to protect human, animal and plant life. The evidence on the use of Article XX and the SPS Agreement provides some support for the notion that low-income countries do not, in general, use environmental measures to restrict trade. Both Article XX and the SPS Agreement have been successfully invoked in many circumstances. For example, in 1995–97 there were 724 measures notified under the SPS Agreement. Of these, 55 per cent were notified by high-income countries, 42 per cent by middle-income countries, and only 2 per cent by low-income countries (UNEP, 1999).

Despite Article XX and the SPS agreement, it is generally argued that the WTO is not the place to deal with the environmental effects of trade (Bhagwati 2000, Barrett 2000). This has led to growing pressure for the establishment of an environmental analogue to the WTO – either a World Environment Organization (WEO) (Whalley and Zissimos 2000) or a Global Environment Organization (GEO) (Runge 2001). It is interesting that the case for such an organization rests first and foremost on the fact that there is no other effective way of dealing with global environmental externalities. It is thought that a WEO/GEO could create new international markets for the global environmental benefits of local conservation effort. Existing examples of this, including joint implementation, bioprospecting contracts, debt-for-nature swaps, and transferable development rights, are first steps towards the creation of global markets for environmental benefits. The alternatives to date – multilateral environmental agreements – have been piecemeal, and have generally failed to address the important issues of compensation and penalties for non-compliance. While agreements involving small numbers of parties concerned with specific issues have been reasonably effective, the framework agreements involving much larger numbers of parties have been less effective (Barrett 1994, Barrett 2003).

17.7 Concluding remarks

The socially optimal use of biodiversity requires two problems to be solved. The first is the problem of local market failure, and is associated with the local public goods and biodiversity externalities. The second is the problem of international market failure, and is associated with the global public good protected by the international conservation effort and with the externalities of international trade. Both require the development of incentives to decision-makers to take the full costs of their actions into consideration, institutions for the regulation of access to biological resources, and an appropriate financial mechanism. The incentive problem has two elements. One is the generation of the correct incentives for the socially optimal use of biodiversity. The other is the discouragement of perverse incentives that work against this. The use of incentives to protect local public goods necessarily operates at the local level, where the millions of foresters, farmers, hunters, harvesters, herders, and fishers use environmental resources on a daily basis. It implies a package of direct incentives (taxes, subsidies, grants, compensation payments, user fees, and charges), indirect incentives (via fiscal, social, and environmental policies), and disincentives (prosecution leading to fines and other penalties).

Up to now the newer market-like mechanisms have emerged in areas where the capturable benefits are largest. The most direct attempts to do this involve the widening and deepening of markets for individual biological resources. Amongst the best-known examples concern the markets for forest-based pharmaceutical products. Bioprospecting contracts between individual pharmaceutical companies and developing countries, such as that between MERCK and INBIO in Costa Rica, have received a great deal of publicity. They seek to mobilize investment in biodiversity conservation by offering access to genetic resources, protected by the assignment of intellectual property in genetic 'discoveries' (Schulz and Barbier 1997). Although they are very well known, however, such contracts are not at all widespread, and have not generally yielded competitive rates of return (Barbier and Aylward 1996, Simpson et al. 1996, Pearce et al. 1999, Dedeurwaerdere et al. 2007).

A second set of markets offer biodiversity conservation benefits as a side-effect (an externality) of markets for unrelated effects. Joint implementation, or carbon offset arrangements, are promoted by the UN Framework Convention on Climate Change (FCCC). The arrangements allow one country – a high-income country – to meet its carbon emission targets under the Convention by investing in the reduction of carbon emissions or the sequestration of carbon in another country. The high-income country gains from the lower costs of reducing carbon emissions or sequestering carbon in the low-income country. The low-income country gains from the additional investment. Most joint implementation projects refer to improvements in energy efficiency along with investment in renewable energy or fuel switching. They do include some projects involving forest conservation and reforestation. However, the link between joint implementation and biodiversity conservation remains tenuous.

In conclusion, people's ability to maintain critical flows of ecosystem services are being lost because individual decision-makers have an incentive to destroy, or at least not sufficiently to protect, biodiversity. The remedy to this problem is to design governance mechanisms and incentives that encourage individuals to protect the common good, and to implement these incentives at all relevant scales through suitable policies, institutions, and financial mechanisms. A number of working examples of this approach already exist. They demonstrate its potential to improve the state of biodiversity and ecosystem services, and through that, to enhance human wellbeing.

Recent work in the field of ecological economics shows that stability adds additional economic value to ecosystem services in the form of insurance (Chapter 17), further underlining the importance of a thorough understanding of the effect of biodiversity on ecosystem functioning and associated services.

CHAPTER 18

The valuation of ecosystem services

Edward B. Barbier, Stefan Baumgärtner, Kanchan Chopra, Christopher Costello, Anantha Duraiappah, Rashid Hassan, Ann P. Kinzig, Mark Lehmann, Unai Pascual, Stephen Polasky, and Charles Perrings

18.1 Ecosystem services after the Millennium Ecosystem Assessment

The Millennium Ecosystem Assessment (2005b) has fundamentally changed the way in which scientists are thinking about the value of ecosystems. By harnessing recent results on the relationship between biodiversity and ecosystem functioning to an assessment of the valued services that people obtain from the natural environment, the Millennium Ecosystem Assessment has brought the analysis of ecosystems into the domain of economics. Ecosystem services are defined by the Millennium Ecosystem Assessment as the benefits that people obtain from ecosystems. Since the value of any asset is simply the discounted stream of benefits that are obtained from that asset, the benefit streams associated with ecosystem services may be used to estimate the value of the underlying ecological assets. Those assets are not the traditional stocks of resource economics – minerals, water, timber, and so on – but the systems that yield flows of such things.

The Millennium Ecosystem Assessment (2005b) distinguishes four broad benefit streams: provisioning services, cultural services, supporting services, and regulating services. The first of these is the most familiar. Provisioning services cover the renewable resources that have been the focus of much work in environmental and resource economics, including foods, fibres, fuels, water, biochemicals, medicines, pharmaceuticals, and genetic material. Many of these products are directly consumed, and are subject to reasonably well-defined property rights. They are priced in the market, and even though there may be important externalities in their production or consumption, those prices bear some relation to the scarcity of resources.

The second category, cultural services, is similarly quite familiar. These services include a range of largely non-consumptive uses of the environment, and reflect the fact that the diversity of ecosystems is mirrored in the diversity of human cultures. Cultural services include the spiritual, religious, aesthetic, and inspirational wellbeing that people derive from the 'natural' world around them. They include the sense of place that people have, as well as the totemic importance of particular landscapes and species. More importantly, they include (traditional and scientific) information, awareness, and understanding of ecosystems and their individual components offered by functioning ecosystems. Some cultural services, such as ecotourism, are offered through well-developed markets. Others are not. Many cultural services are still regulated by custom and usage, or by traditional taboos, rights, and obligations. Nevertheless, they are directly used by people, and so are amenable to valuation by methods designed to reveal people's preferences.

The category of support services captures the main ecosystem processes that underpin all other services. Examples offered by the Millennium Ecosystem Assessment include soil formation, photosynthesis, primary production, and nutrient, carbon, and water cycling. These services play out at quite different spatial and temporal scales. For example, nutrient cycling involves the maintenance of the roughly twenty nutrients essential for life, in different concentrations in different parts of the system. It is often localized, and is therefore at least partially captured by the price of the land on which it takes place. Carbon cycling, on the other hand, operates at a global scale, and is very poorly reflected in any set of prices, although carbon markets are now in development. Since many of these services are, in a sense, embedded in the other services, they are frequently captured in the valuation of those services.

The final set of ecosystem services identified by the Millennium Ecosystem Assessment is the set of regulating services. These include air quality regulation, climate regulation, hydrological regulation, erosion regulation or soil stabilization, water purification and waste treatment, disease regulation, pest regulation, and natural hazard regulation. There is a rich and rapidly growing literature on the services identified by the Millennium Ecosystem Assessment. Interested readers will find many ways into this literature through the references cited below. However, it is quite uneven. There are many more studies of the provisioning and cultural services than of the regulating services. As a result we pay special attention to this topic. The regulating services affect the impact of stresses and shocks to the system. Some – such as climate or disease regulation – are global public goods. Many are local public goods (Perrings and Gadgil 2003). That is, they offer non-exclusive and non-rival benefits to particular communities. The rest of the world may have minimal interest in such benefits. The fact that they are public goods means that, if left to the market, they will be undersupplied. But they will be supplied at some level. Indeed, the greater the local benefits to the provision of environmental public goods, the greater will be the local investment in their provision.

While the Millennium Ecosystem Assessment offers a useful way into the problem for economists, many others still have difficulty both with the notion that species have finite value, and with the fact that people routinely trade off ecosystems services. In this chapter we discuss the value of ecosystems and ecosystem services. We identify the main methods for valuing different types of ecosystem service, and the role of valuation in developing sustainability indicators. The sustainability of economic development requires that the value of the assets or capital stocks supporting development be maintained over time, and since capital includes produced, human, and natural capital, it is important to understand how the value of ecosystems may be changing relative to the value of other capital stocks.

18.2 The value of ecosystems

In what follows, value is understood to mean economic value. According to a classic definition, economics is the science which studies human behaviour with regard to the satisfaction of human needs and wants from scarce resources that have alternative uses (Robbins 1932, p. 15). That is, the economic value of an asset lies in its *instrumental* role in attaining human goals. The goal may be spiritual enlightenment, aesthetic pleasure, or the production of some marketed commodity. It does not matter. What matters is that there is a relation between the state of ecosystems and human wellbeing. We do not therefore discuss what some scientists refer to as *intrinsic* value, or value independent of human wellbeing. This concept is generally used very loosely as a way of trumping arguments based on anthropocentric values – or denying the trade-offs that people are willing to make between conservation and other objectives. Instead, we focus on the values generated by the preferences of the many *individuals* in the economy acting both independently and in concert.

Value is not an inherent property of an asset. Rather, it is attributed by economic agents through their willingness to pay for the services that flow from the asset. This depends partly on the objective (e.g. physical or ecological) properties of the asset, but also on the socio-economic *context* in which valuation takes place – on human preferences, institutions, culture, and so on. It accordingly reflects the current level of consumption of all goods, current preferences, the current distribution of income and wealth, the current state of the natural environment, current production technology, and current expectations about the future – irrespective of whether we would think of this state of the world as being good or bad. Economic value is also a *marginal* concept: it refers to the impact of small changes in the state of the world, and not the state of the world itself. In summary, the value of ecological assets, like the value of other assets, is instrumental, anthropocentric, individual-based and subjective, context dependent, marginal, and state-dependent (Goulder and Kennedy 1997, Nunes and van den Bergh 2001, Baumgärtner 2006, Baumgärtner *et al.* 2006).

The value of ecosystems derives from the services – the discounted stream of benefits – they produce. So, if B_t denotes the social benefits from the set of all services provided by an ecosystem at time *t*, then the present social value of that system is simply:

$$V_0 = \int_0^a e^{-\delta t} B_t dt$$

where δ is the social rate of discount. B_t is a measure of society's willingness to pay for that set of benefits. Resources only have value if they are scarce: i.e. if they have an opportunity cost – if obtaining an additional unit of the resource implies that something else must be given up. What makes ecological resources special is that whereas the services from most assets in an economy are marketed, the services yielded by ecosystems generally are not. If the aggregate willingness to pay for these benefits, is not revealed through market outcomes, then efficient management of such ecosystem services requires explicit methods to measure its social value (e.g. see Freeman 2003, Heal *et al.* 2005, Just *et al.* 2004). Indeed, the failure to take account of the value of non-marketed ecological services is one reason for excessive land conversion, habitat fragmentation, and pollution in aquatic and land-based food systems.

Although there are reasonable estimates of the value of many provisioning services – where markets are most well-developed – there are few reliable estimates of the value of most cultural services and all regulating services. A number of studies prior to the Millennium Ecosystem Assessment drew attention to the changes in ecosystem services and the importance of quantifying the value of these changes to human societies in terrestrial (e.g. Daily *et al.* 1997, Daily 1997), marine (e.g. Duarte 2000) and agroecosystems (Björklund *et al.* 1999). However, attempts to value these services at an aggregate level have generally been seen as problematic (Costanza *et al.* 1997, Bolund and Huhammar 1999, Norberg 1999, Limburg and Folke 1999, Woodward and Wui 2001). The Millennium Ecosystem Assessment (2005b) itself had great difficulty in attaching values to the observed changes in physical magnitudes.

One reason for this is that most ecosystem services are the result of a complex interaction between natural cycles operating over a wide range of space and time-scales (Daily 1997). Waste disposal, for example, depends both on highly localized life cycles of bacteria and the global cycles of carbon and nitrogen. Yet information on much of this is scarce. Many valuation studies depend on elicitation of the preferences of people for ecosystem stocks, when they have little information on the role of those stocks in the generation of ecosystem services, or of the link between ecosystem services and the production of commodities (Winkler 2006a). In fact, although the relationship between biodiversity and ecosystem functioning may be well understood, the current state of ecological knowledge is frequently insufficient to link biodiversity and ecosystem functioning to the provision of ecosystem services, or specific benefit streams. In other words, we may not understand ecosystem 'production functions' well enough to quantify how much service is produced, or how changes in ecosystem condition or function will translate into changes in amounts of ecosystem services produced (Daily *et al.* 1997).

The central issue here is that ecosystems and the services they provide are, for the most part, intermediate inputs into the goods and services that enter final demand – that satisfy people's various desires. Intermediate inputs are those used to produce the inputs that lead directly to economically valued goods or services. The value of intermediate inputs derives from the value of those goods and services. The directness with which ecosystem stocks are used is reflected in a literature that has sought to decompose *total economic value* into its constituent parts (Pearce 1993, Turner 1999). In this literature 'use value' normally comprises benefits deriving from consumptive or non-consumptive use by the individual, while 'non-use value' comprises benefits from consumptive or non-consumptive use by others (Krutilla 1967, Weisbrod 1964). This includes, for example, the benefits stemming from the ethical, spiritual, or religious desire to conserve biodiversity for future generations of humans, or for other species. *Direct use value* includes *consumptive use*, e.g. as food, fuelwood, or medicinal plants; *productive use*, e.g. as industrial resources, fuel, or construction material; and *non-consumptive use*, e.g. recreation, tourism, science and education. *Indirect use value* captures the derived demand for ecosystem stocks via ecosystem services (Barbier 1994, Hueting *et al.* 1998) that contribute to human wellbeing. Examples include the support of biological productivity in agro-ecosystems, climate regulation, maintenance of soil fertility, control of water runoff, and cleansing of water and air.

Even if an ecosystem (or component of it) has no current use, it may have an *option value*. For example, the future may bring human diseases or agricultural pests which are still unknown today. Today's biodiversity would then have an option value in so far as the variety of existing plants may already contain a cure against the yet unknown disease, or a biological control of the yet unknown pest (Heal *et al.* 2005, Polasky and Solow 1995, Polasky *et al.* 1993, Rausser and Small 2000, Simpson *et al.* 1996, Goeschl and Swanson 2003). In this sense, the option value of biodiversity conservation corresponds to an insurance premium (Perrings 1995, Weitzman 2000, Baumgärtner 2007, Quaas and Baumgärtner 2008) which one is willing to pay today in order to reduce the potential loss should an adverse event – such as a human disease or an agricultural pest – occur in the future.

Amongst non-use values, the same literature identifies *vicarious use value* (Watson *et al.* 1995) *bequest value* and *existence value* (Krutilla 1967). The first two derive from people's altruistic willingness to pay (or to forgo benefits) to ensure that other members of the present generation or future generations can enjoy the benefits from ecosystems. *Existence value*, originally defined as people's willingness to pay (or to forgo benefits) to ensure the continued existence of biodiversity irrespective of any actual or potential use by present or future generations of humans, may be seen as a form of altruism towards non-human species or nature in general, and, in most cases, rests on ethical or religious motives.

Note that final use includes conservation. For example, Balmford *et al.* (2002) used studies of specific ecosystems to ask whether conservation or development options generate greater value. Their study shows that conservation options are preferred, often by wide margins. Earlier work by Peters *et al.* (1989) reached a similar conclusion on conserving tropical forests. However, much of the work to date on ecosystem services requires making large leaps to overcome lack of data or understanding. Much greater understanding of ecosystem functioning, how these functions change with management actions, and how such changes impact on human values, is required before firm conclusions can be reached.

18.3 The valuation of provisioning and cultural services

As a recent report from the US National Academy of Science puts it, 'the fundamental challenge of valuing ecosystem services lies in providing an explicit description and adequate assessment of the links between the structure and functions of natural systems, the benefits (i.e. goods and services) derived by humanity, and their subsequent values' (Heal *et al.* 2005, p. 2). In recent years substantial progress has been made by economists working with ecologists and other natural scientists to improve environmental valuation methodologies. Table 18.1 indicates that there are now various methods that can be used for valuing the services derived from ecological regulatory and habitat functions. A discussion of these methods and their application to valuing ecosystem services can be found in Freeman (2003), Pagiola *et al.* (2004), and Heal *et al.* (2005).

For provisioning services (e.g. food, fuel, fibre, and fresh water production), functions of agrobiodiversity are better understood than for supporting (e.g. nutrient cycling and soil formation) and regulating (e.g. climate, flooding, disease regulation, or water purification) services that usually involve assemblages of species and guilds, each with a complex set of functions and interactions (Millennium Ecosystem Assessment, 2005b). Where there are prices for the outputs of activities, as is the case with many provisioning services, then derived demand methods are appropriate – and there are a growing number of studies that use such an approach (e.g. Barbier 2000, Matete and Hassan

Table 18.1 Valuation methods applied to ecosystem services.

Valuation method[a]	Types of value estimated[b]	Common types of applications	Ecosystem services valued
Travel cost	Direct use	Recreation	Maintenance of beneficial species, productive ecosystems and biodiversity
Averting behaviour	Direct use	Environmental impacts on human health	Pollution control and detoxification
Hedonic price	Direct and indirect use	Environmental impacts on residential property and human morbidity and mortality	Storm protection; flood mitigation; maintenance of air quality
Production function	Indirect use	Commercial and recreational fishing; agricultural systems; control of invasive species; watershed protection; damage costs avoided	Maintenance of beneficial species; maintenance of arable land and agricultural productivity; prevention of damage from erosion and siltation; groundwater recharge; drainage and natural irrigation; storm protection; flood mitigation
Replacement cost	Indirect use	Damage costs avoided; freshwater supply	Drainage and natural irrigation; storm protection; flood mitigation
Stated preference	Use and non-use	Recreation; environmental impacts on human health and residential property; damage costs avoided; existence and bequest values of preserving ecosystems	All of the above

Notes:
[a] See Freeman (2003), Heal et al. (2005), and Pagiola et al. (2004) for more discussion of these various valuation methods and their application to valuing ecosystem goods and services.
[b] Typically, use values involve some human 'interaction' with the environment whereas non-use values do not, as they represent an individual valuing the pure 'existence' of a natural habitat or ecosystem or wanting to 'bequest' it to future generations. Direct use values refer to both consumptive and non-consumptive uses that involve some form of direct physical interaction with environmental goods and services, such as recreational activities, resource harvesting, drinking clean water, breathing unpolluted air and so forth. Indirect use values refer to those ecosystem services whose values can only be measured indirectly, since they are derived from supporting and protecting activities that have directly measurable values.
Source: Barbier (2007), Table 2.

2006). Some of the most successful efforts to estimate the values of ecosystem services have focused on the production of specific tangible outputs, such as the production of fish and game species. Such outputs tend to be readily measurable and frequently have market prices (e.g. commercially harvested fish). The ecosystem (habitat) is a necessary input into the production of the output (species). Focusing on the input side, such work naturally fits into a classification of ecosystem services. One can just as easily focus on the output side, however, in which case such studies are best considered as studies of the values of individual species. Some research in economics has focused on how changes in ecosystem conditions translate into changes in the value of output (e.g. Brown and Hammack 1974, Lynne et al. 1981, Ellis and Fisher 1987, Swallow 1990).

One widely cited example of this approach relates to the provision of clean drinking water for New York City from watersheds in the Catskills. Increased housing development with septic systems, runoff from roads, and agriculture were causing water quality to decline. Continued declines in water quality would have forced the US Environmental Protection Agency to require New York City to build a water filtration plant. The total present value cost of building and operating the water filtration plant was estimated to be roughly $6–8 billion (Chilchilnisky and Heal 1998). Instead, New York City decided to invest $1 to 1.5 billion to conserve the Catskills watersheds to avoid building the water filtration plant. The value of the ecosystem service here is the saving provided by avoided cost – the cost of replacing the ecosystem service with some engineered alternative.

Market price-based 'dose–response' analysis is another example of this approach. While ecosystems may not be priced in the market, the commodities produced using ecosystems are. Hence, by directly relating the provision of the input to the level of the market output, it is possible to derive the financial returns provided by biodiversity. For example, Ricketts *et al.* (2004) found that the pollination contributes 7 per cent of total farm income to coffee producers in Costa Rica (in 2002–03 when coffee prices were depressed).

Travel cost. Where there are no prices for the outputs, as is the case with many cultural services, other methods are necessary. Many methods are specific to particular types of ecosystem services. So, for example, the travel cost method applies primarily to the value derived from recreation and tourism, averting behaviour models are best applied to the health effects arising from environmental pollution and hedonic wage and property models are used primarily for assessing work-related environmental hazards and environmental impacts on property values, respectively.

Contingent valuation. The most widely used class of methods for estimating willingness to pay for cultural services are 'stated preference methods'. These include contingent valuation, conjoint analysis and choice experiments. These valuation methods share a common approach with market research in that they survey individuals who benefit from an ecological service or range of services to elicit their willingness to pay for those services. They are flexible enough to include non-consumptive uses (Carson *et al.* 1996). Similarly, choice experiments and conjoint analysis, which ask respondents to rank, rate, or choose among various environmental outcomes or scenarios, have the potential to elicit the relative values that individuals place on different ecosystem services (see for example Carlsson *et al.* 2003).

However, implementation of stated preference methods requires that two key conditions be met: (1) the information must be available to describe the change in a natural ecosystem in terms of service that people care about, in order to place a value on those services; and (2) the change in the natural ecosystem must be explained in the survey instrument in a manner that people will understand (Heal *et al.* 2005). For many ecosystem services, one or both of these conditions may not hold. For instance, it has proven very difficult to describe accurately – through the hypothetical scenarios required by stated preference surveys – how changes in ecosystem processes and components affect ecosystem regulatory and habitat functions and thus the specific benefits arising from these functions that individuals value. Since there is considerable scientific uncertainty surrounding these linkages, is it difficult not only to construct such scenarios but also to elicit responses from individuals that accurately measure their willingness to pay for ecological services.

Production function methods. Whatever the method used to value willingness to pay for provisioning or cultural services, the value of the ecosystems that support those services can be obtained through the production function approaches. For many habitats where there is sufficient scientific knowledge of how these link to specific ecological services that support or protect economic activities, it is possible to employ the production function approach to value these services. The basic modelling approach underlying production function methods, also called 'valuing the environment as input', is similar to determining the additional value of a change in the supply of any factor input (Barbier 1994, 2000, 2007, Freeman 2003). If changes in the regulatory and habitat functions of ecosystems affect the marketed production activities of an economy, then the effects of these changes will be transmitted to individuals through the price system via changes in the costs and prices of final good and services. This means that any resulting 'improvements in the resource base or environmental quality' as a result of enhanced ecosystem services, 'lower costs and prices and increase the quantities of marketed goods, leading to increases in consumers' and perhaps producers' surpluses' (Freeman 2003, p. 259).

An adaptation of the production function methodology is required in the case where ecological regulatory and habitat functions have a protective value, through various ecological services such as storm protection, flood mitigation, prevention of erosion and siltation, pollution control, and maintenance of beneficial species (Table 18.1). In such cases, the environment may be thought of producing a non-marketed service, such as 'protection' of economic activity, property, and even human lives, which benefits individuals through limiting damages. Applying production function approaches requires modelling the 'production' of this

Box 18.1 Linking economic and ecological values of biodiversity using simulation modelling for Keoladeo National Park, India

Biodiversity in Keoladeo National Park is first valued using conventional travel cost methods. This provides a short-run valuation which assumes that the ecological health of the park is sustainable over the longer run. There is, however, uncertainty over the impact of various exogenous forcing factors on the state of the park. Indices of ecological health are constructed, along with scenarios of alternative states of ecological health of the park over a period of 25 years. This makes it possible to investigate the linkages between the economic value of the park in its current state, and its value in alternative states (Chopra and Adhikari 2004).

The literature on economic valuation typically focuses on use value in the short run, whether within or outside the market. Conservation biologists, on the other hand, are more concerned with the underlying long-run conservation value of the system being studied. Chopra and Adhikari (2004) investigate the link between these two aspects of value in the context of a wetland in Northern India. Keoladeo National Park has been designated as both a Ramsar site and a national park.

Assigning a value to an eco-system good or service implies an ability to define its role in increasing human welfare, whether or not it is marketed. This requires identification of (a) the physical or environmental linkages which result in the provision of the service, and (b) the economic linkages which help realize the value of the service.

Chopra and Adhikari (2004) use the travel cost method to estimate tourists' willingness to pay to visit the wetlands. It is assumed that household i derives utility from the consumption of a bundle of marketed goods, and from recreation or leisure activities:

$$U_i = f(x_1, x_2, ..., x_n, L_i)$$

where x_i is a marketed good and L_i is leisure. Demand for leisure is given by

$$D(L_i) = V_i = F(W_i, C_{ij}, H_p)$$

where W_i is travel cost, C_{ij} are other costs, and H_p is the ecological health of the park. The latter can be parameterized only on the basis of medium-run or long-run time series data and modelling of the wetland ecosystem. Application of the travel cost model indicates quite low elasticities with respect to cost incurred by visitors, W_i and C_{ij}.

By simulating ecology–economy interactions in the Keoladeo National Park it was possible to estimate the impact of exogenous changes, such as reductions in water inflow due to development of upstream agriculture, on both the ecological health and economic value of the park. Specifically, the simulations were used to estimate the effect of changes in ecological health on economic value (direct and indirect income derived from the park). The elasticities obtained varied from 0.6 to 1.7 in two scenarios. However, as pressured on the system increased and the wetland became less attractive as an ecosystem, the values of the elasticities fell, with a range from 0.04 to 0.124.

Given the present profile of visitors to KNP, visitation rates are not responsive to private cost. However, direct and indirect income obtained from the park are responsive to ecological health indices. Moreover, the elasticity with respect to ecological health is greater, the greater the value of the index, indicating that the impact of conservation on the attractiveness of the park is cumulative, and hence that the value of the park increases more than proportionately.

---

protection service and estimating its value as an environmental input in terms of the expected damages avoided by individuals (Barbier 2006).

An example of the use of production functions linked to stated preferences for the cultural services offered by totemic species is given by Allen and Loomis (2006). Following Goulder and Kennedy (1997), they use such an approach to derive the value of species at lower trophic levels from the results of surveys of willingness to pay for the conservation of species at higher trophic levels. Specifically, they derive the implicit willingness to pay for the conservation of prey species from direct estimates of willingness to pay for top predators. They refer to this as a form of quasi-benefit transfer.

They make the important point that it is not necessary for people to understand the trophic structure of an ecosystem, since their willingness to pay for top predators effectively captures their willingness to pay for the whole system. Economists first modelled this by assigning species the status of 'intermediate inputs' (Crocker and Tschirhart, 1992) due to their role in supporting more other species. The same idea can be generalized to say that species have value deriving from their role in the production of valuable goods and services, and that this is conditional on the state of the environment. So, for example, the derived value of members of a functional group of species, each of which performs differently in different environmental conditions, will vary with those conditions. Species that appear to be redundant in some conditions will still have value depending on the likelihood that the conditions in which they do have value will occur in the future (Loreau *et al.* 2002). It follows that the cost of species deletion is the cost of securing the same set of services in the conditions to which those species were adapted.

However, production function methods have their own measurement issues and limitations. For instance, applying the production function method raises questions about how changes in the ecological service should be measured, whether market distortions in the final goods market are significant, and whether current changes in ecological services may affect future productivity through biological 'stock effects'. A common approach in the literature is to assume that an estimate of ecosystem area may be included in the 'production function' of marketed output as a proxy for the ecological service input. For example, this is the standard approach adopted in coastal habitat–fishery production function models, as allowing wetland area to be a determinant of fish catch is thought by economists and ecologists to proxy some element of the productivity contribution of this important habitat function (Barbier 2000, 2007, Freeman 2003). A further measurement issue arises when stock effects need to be modelled explicitly. In the production function valuation literature, approaches that ignore stock effects are referred to as 'static models' of environmental change on a natural resource production system, whereas approaches that take into account the intertemporal stock effects of the environmental change are referred to as 'dynamic models' (Barbier 2000, 2007, Freeman 2003).

Benefit transfer. An important and growing trend in the valuation of all environmental impacts is the use of value (benefit) transfer techniques. Benefit transfer implies the use of value estimates for some ecosystem service derived from one location (or one set of locations) to infer the value of the same ecosystem service at another location, the 'policy site'. This is relatively straightforward in the case of, for example, carbon sequestration services, where the contribution of carbon sequestration to the general circulation system is independent of where it takes place (e.g. Songhen and Brown 2006). However, it raises more questions where the benefits of ecosystem services depend heavily on local conditions, or the characteristics of the communities that exploit those services are very different – in terms of age, income, asset holdings, education, culture, and so on (Vandenberg *et al.* 2001, Kirchoff *et al.* 1997). Given the cost of original valuation studies, the use of benefit transfer is growing rapidly, and economists have been investing considerable effort into refining procedures to extract as much information as possible from valuation studies of sites that overlap with the policy site in some dimension.

One popular approach is the use of meta-regression models – regression models based on a range of estimates of the value of some resource at different sites that are designed to estimate the value of resources at some policy site (Rosenberger and Loomis 2000, Shrestha and Loomis 2001, Johnston *et al.* 2005, Johnston *et al.* 2006). Although meta-analyses have been applied most frequently to stated preference studies where the cost of original research is highest, they have also been applied to studies based on other valuation methodologies, including hedonic valuation (Smith and Huang 1995, Smith and Pattanayak 2002, Mrozek and Taylor 2002). The adoption of Bayesian approaches in a meta-regression framework has significantly improved their capacity to estimate a relatively large set of parameters, even when the underlying meta-sample is small (Brunsdon and Willis 2002, Layton and Levine 2005, Moeltner *et al.* 2007).

In many cases, however, biodiversity impacts are highly context specific and thus not directly generalizable. Moreover, all such pathways can be interrelated or at least cannot be isolated. As a general rule, a *total valuation* of the potential value of ecological stocks should not be estimated by a piecewise independent valuation of each impact, and a subsequent aggregation of those impacts when these are interrelated (Randall, 2002). For instance if a dose–response approach to the value of agrobiodiversity were estimated and then added to the value derived from farmers themselves using a stated preference technique, it is likely that the double counting would occur. It is important to clearly delineate the different services and small number of different ecosystem functions. Then different valuation techniques can be used for each function.

An example of a study that attempts to decompose the value of non-marketed environmental resources in this way is Rodríguez *et al.*'s (2006) study of Opuntia scrubland in Ayacucho, Peru. Opuntia is important as the host plant for cochineal insects, the source of carminic acid, a natural dye used in the food, textile, and pharmaceutical industries. However, besides being an agricultural asset for the production of the cash crop of cochineal, the Opuntia scrubland protects slopes against erosion and flooding, as well as rehabilitating marginal lands by improving the levels of humidity and soil retention capabilities. Additionally, Opuntia scrubs are used for animal grazing all year round and can become an emergency feedstock in case of drought. In addition, its fruits and young cladodes have a considerable nutritional value and provide food for Andean farmers.

18.4 The valuation of regulating services

We have already noted that the Millennium Ecosystem Assessment drew attention – for the first time – to the value of services that regulate the Earth's capacity to respond to environmental shocks and stresses. Given current concern for the sustainability of environmental management strategies, this is of fundamental importance. While there are as yet few studies of the value of the regulating services, they are likely to be essential in the future. It is therefore also appropriate to identify the valuation options currently being considered for these services. The regulating services are closely connected to insurance role of functional diversity discussed in an economic context in Chapter 17 and in an ecological context in Chapters 4 and 6. Ecologists argue that over small scales an increase in species richness and the diversity of overlapping functional groups of species enhances the level of functional diversity, which, in turn, increases ecological stability (Tilman *et al.* 2005) and resilience (*sensu* Holling 1988). Following Carpenter *et al.* (2001), the resilience of an ecosystem is determined primarily by: (1) the amount of disturbance that the system can absorb and still remain within the same state or domain of attraction; (2) the degree to which the system is capable of self-organization versus the lack of organization, or organization forced by external factors, and (3) the degree to which the system can build and increase the capacity for learning and adaptation.

Maintaining a wider portfolio of options in ecosystems is likely to maintain or enhance the capacity to respond to short-run shocks and stresses in constructive and creative ways. In agro-ecological systems, for example, various recent studies have analyzed the contribution of crop diversity to the mean and variance of agricultural yields and farm income (Smale *et al.* 1998, Schläpfer *et al.* 2002, Widawsky and Rozelle 1998, Di Falco and Perrings, 2003, 2005, Birol *et al.* 2005). The '*option-insurance value*' of agrobiodiversity refers to the capacity of the agricultural system to adapt to economic or environmental external changes by either returning to its original state, thus making the agricultural system stable, or by evolving into a state preferable to the initial one. Extending genetic to species diversity, such as multi-crop and tree biodiversity in agro-ecosystems can enhance the ability of farmers to remain economically viable by diversifying their revenues to cope with volatile international market price changes of inputs, crops or tree products. Whereas the homogenization of agro-ecosystems can increase yields in the short run, it is often at the cost of increasing vulnerability to pests and pathogens. This in turn increases reliance on pesticides that are potentially damaging to both biodiversity and humans. So the cost of reducing the resilience of agricultural systems to cope with market (price fluctuations) and environmental (pests and diseases) vagaries can be seen as the loss in 'insurance value' of a portfolio of species.

Two things follow. First, the regulating services of ecosystems appear to be a function of biodiversity, and especially of functional diversity. Second, the value of the regulating services is a function of people's willingness to pay for a reduction in risk. Taking the first of these, results from experiments in grassland plots (Tilman *et al.* 1996) and in controlled environments (Naeem *et al.* 1995) found that increasing the number of species in the system tended to increase system productivity. Similar results were reported by Hector *et al.* (1999) in experiments across a number of European countries. In heterogeneous environments, having more species will generally allow the collection of species to better utilize all ecological niches and so be more productive (the niche differentiation effect). Tilman *et al.* (1997a) develop a model of niche differentiation that is formally similar to the Polasky *et al.* (1993) model of the probability that a set of species will contain a given trait. Both models show how coverage increases, either in niche space or genetic space, when more species are added.

In sum, there remain a number of unanswered questions about the production and value of ecosystem services. We currently lack understanding of the link between management actions and changes and ecosystem functioning. There remain questions about the link between ecosystem functioning and provision of ecosystem services. Finally, there remain questions about the link between the provision of ecosystem services and value of these services to humans. Provision of valuable ecosystem services may ultimately prove to be the most important reason to conserve biodiversity. At present, however, it is hard to know with any precision either the benefits of the provision of ecosystem services or the opportunity costs involved in ensuring their continued provision.

A promising alternative is the *expected damage function* (EDF) approach to estimating the total willingness to pay for the effect of a change in wetland area on expected damages from coastal storm events (Barbier 2007). This approach measures willingness to pay, by measuring the total expected damages, $E[D(S)]$, resulting from the change in ecosystem stocks. The approach assumes that the value of an asset that yields a benefit in terms of reducing the probability and severity of some economic damage is measured by the reduction in the expected damage. The essential step to implementing this approach, which is to estimate how changes in the asset affect the probability of the damaging event occurring, has been used routinely in risk analysis and health economics, as in the case of airline safety performance (Rose 1990), highway fatalities (Michener and Tighe 1992), drug safety (Olson 2004), and studies of the incidence of diseases and accident rates (Cameron and Trivedi 1998, Winkelmann 2003).

Barbier (2007) has shown that the EDF approach can also be applied, under certain circumstances, to value the protection service of coastal wetlands that reduce the probability and severity of economic damages from natural storm disasters. Two components are critical to implementing the EDF approach to estimating the changes in expected storm damages: the expected incidence of economically damaging natural disaster events as a function of wetland area, and some measure of the additional economic damage incurred per event. Provided that there are sufficient data on the incidence of past natural disasters and changes in wetland area in coastal regions, the marginal impact of wetland area on expected storm damage can be estimated using a *count data model* (see Box 18.2).

Box 18.2 The value of mangrove storm buffering

In the wake of the 26 December 2004 Asian tsunami, there have been widespread calls by governments, NGOs, and local communities to restore coastal mangrove forests and recover one of their putative 'ecosystem functions' – the protection of low-lying coastal communities from recurrent storm surges and unpredictable tsunamis. For example, Indonesia has promised to spend $22 million on mangrove restoration and has already planted 300,000 seedlings in Aceh Province. Malaysia is planning to spend $25 million on rehabilitating 4,000 hectares of mangrove area that has been deforested, and the state government of Kerala in India pledged $8 million for mangrove

Continues

Box 18.2 (*Continued*)

restoration (Check 2005). The governments of Sri Lanka and Thailand have also stated publicly their intentions to rehabilitate and replant large mangrove areas (Harakunarak and Aksornkoae 2005, UNEP 2005).

The rationale for these restoration schemes is the growing evidence that mangroves, marshlands and other natural barriers are critical components to the overall 'resilience' of coastal areas to threats posed by tsunamis, hurricanes, and other natural disasters (Adger *et al.* 2005). In the case of mangroves, the scientific evidence is now well documented. Mangroves are forested wetlands that grow along sheltered tropical and subtropical shores and estuaries. They attenuate wave amplitude and are thought to be particularly valuable in minimizing damage to property and loss of human life by acting as natural barriers against tropical cyclonic storms and tsunamis (Chong 2005, Massel *et al.* 1999, Mazda *et al.* 1997). Evidence from the aftermath of the 2004 Asian tsunami suggests that those coastal areas that had dense mangrove forests closer to pristine conditions suffered fewer losses and less damage to property than those areas in which mangroves had been degraded or converted to other land uses (Dahdouh-Guebas *et al.* 2005c, Danielsen *et al.* 2005, Kathiresan and Rajendran 2005, UNEP 2005, Dahdouh-Guebas and Koedam 2006).

Mangrove forests are among the most threatened global ecosystems. At least 35 per cent of global mangrove area has been lost in the past two decades, and in Asia, 36 per cent of mangrove area has been deforested, at the rate of ~1.5 per cent per year (Farnsworth and Ellison 1997, Valiela *et al.* 2001). Such widespread mangrove deforestation is believed to have increased the vulnerability of coastal areas and their inhabitants to the economic damages caused by natural disasters. Large-scale replanting and restoration of mangrove forest ecosystems as 'green shelterbelts' is therefore seen as an important solution to providing adequate protection to coastlines exposed by extensive mangrove loss.

All mangrove forests are not identical, however. As one post-tsunami assessment concluded, 'mangroves play a critical role in storm protection, but with the subtle point that this all depends on the quality of the mangrove forest' (Dahdouh-Guebas *et al.* 2005c, p. 446). The world's remaining mangrove forests have been fragmented, degraded, and mismanaged by over a century of small-scale conversion of forests to aquaculture (notably tiger-prawn farms); unsustainable harvesting of wood for lumber, pulp (for rayon), fuel, and charcoal production; construction within forests of housing and 'eco-resorts'; and outright neglect. Thus, while the decision by many Asian governments in the wake of the tsunami to invest in mangrove replanting and restoration is laudable, there are few intact, unfragmented mangrove forests available to use as reference states for successful restoration efforts.

Because of the lack of data on how this service arises from natural mangrove systems and how changes in these systems translate into increased risk of damage from storm events, the few valuation studies of coastal protection have used the 'replacement cost' method (Chong 2005). This involves using estimates of the cost of 'replacing' mangroves with human-built alternatives, such as storm walls and other artificial barriers, as approximations of the value of the coastal protection service. Such estimates not only make the methodological error of estimating a 'benefit' by a 'cost' (thus always assuming a unitary benefit–cost ratio) but also tend to yield unrealistically high estimates because removing all the mangroves and replacing them with constructed barriers is unlikely to be the least-cost alternative to providing storm prevention and flood mitigation services in any coastal area (Heal *et al.* 2005).

Barbier (2007) has demonstrated the use of the expected damage function (EDF) method as a preferred alternative to the replacement cost approach in valuing the storm buffer service of mangroves. Over the 1996 to 2004 period, one estimate suggests that in Thailand mangrove deforestation was 3.44 km^2 annually. Using the replacement cost method, which assumes that the equivalent cost of protecting the shoreline with a 75 m width stand of mangrove is approximately US$116,667 per ha (1996 prices), Barbier calculates an annual welfare loss of over $4.2 million in the storm protection service associated with that deforestation. This is equivalent to a net present value (10 per cent discount rate) of $24.3 million over the entire 1996–2004 period. In comparison, the EDF approach estimates an annual welfare loss in storm protection of just over $0.6 million, and a net present value (10 per cent discount rate) of $3.7 million over 1996–2004. The Thailand case study confirms that the replacement cost method tends to produce extremely high estimates – almost 8 times greater than the estimate provided by the EDF approach. This suggests that the replacement cost method should be used with caution, and when data are available, the expected damage function approach may provide more reliable values of the storm protection service of coastal wetlands such as mangroves.

18.5 Valuation of ecosystem assets and national sustainability indicators

The dominant economic approach to operationalizing sustainability is the capital approach. It requires that the value of a nation's inclusive wealth or aggregate capital stock is non-declining. This is also implicit in the basic definition of income: the maximum amount that can be spent on consumption in one period without reducing real consumption expenditure in future periods (Hicks 1939). Conventional macroeconomic income indicators like gross domestic product (GDP) are not suitable for assessing sustainability as they do not adequately consider human and natural capital. Attempts to find better sustainability indicators include the United Nations Development Program's Human Development Index (HDI), which adds proxies for human capital but is otherwise still based on GDP. Life expectancy at birth tells us something about the period over which we might expect people to be productive, while adult literacy tells us something about the average skill levels. But neither indicator tells us about the depreciation of human capital, nor do they tell us anything about the balance between depreciation and investment in human capital. More importantly, the HDI tells us nothing more about the rate of change in the value of natural capital than we get from GDP.

The same requirement is explicit in the investment rule identified by Hartwick (1977, 1978) and Solow (1986) to satisfy the sustainability criterion proposed by Solow (1974). The rule reflects the fact that a necessary condition for consumption based on the exploitation of natural resources to be sustainable is that investment be maintained at a level that at least compensates for the depletion of those resources. Current measures of *net* product, NDP and NNP, address the issue of depreciation of capital and so are to be preferred to measures of *gross* product. Similarly, measures of *national* product, GNP and NNP, address transboundary flows, and so are to be preferred to measures of *domestic* product. The most inclusive of all of these measures is therefore NNP. One starting point for the development of improved economic sustainability indicators is therefore the improvement of measures of NNP (Perrings and Vincent 2003).

Currently, most estimates of NNP still exclude depreciation of environmental capital. Moreover, although they include transboundary flows, they exclude transboundary environmental flows. Since the capital stock that supports the flow of services includes the ecosystems that are at the core of natural capital, and since environmental impacts involve transboundary effects, current NNP estimates are not adequate measures of sustainability. To make NNP estimates an adequate measure of sustainability requires that it include a measure of the change in the aggregate capital stock – sometimes called inclusive wealth.

The World Bank's adjusted net savings indicator is a direct attempt to measure net change in the value of a country's capital stocks, where that includes both natural and human capital (Hamilton and Clemens 1999). Dasgupta (2001) has developed the same idea, calling it genuine investment. It is a measure of the change in a country's wealth, where wealth includes not just produced capital, but also human capital (including the stock of knowledge) and natural capital. If wealth is defined as the value of the stock of all assets at some time, V_t, plus net investment at that time, dV_t/dt, it follows that a necessary and sufficient condition for wealth to be increasing over time is that $dV_t/dt > 0$. If the population is growing, then the appropriate measure of sustainability is the rate of change in per capita wealth. Using preliminary estimates of Adjusted Net Savings, Dasgupta (2001) showed that the average annual change in per capita wealth in Sub-Saharan Africa over the period 1970–93 was −3.4 per cent. This means that average wealth in Sub-Saharan Africa in 1993 was only half what it had been just twenty years earlier. This was partly a function of population growth, but it was also because of a systematic 'decumulation' of all forms of capital. The implication is that millions of people in Africa were dissaving to maintain consumption, often by running down environmental assets held in common property.

More narrowly focused attempts to develop resource accounts at the country level show that conditions are far from uniform in Sub-Saharan Africa. The value of at least some components of natural capital have been augmented in some countries (see Hassan 2003; Box 18.3). Nevertheless,

Box 18.3 Measuring net change in forest and woodland asset values in South Africa

The total contribution of forest and woodland resources to production in South Africa, measured as value added, amounted to R6,215 million in 1998 prices (Table 1). This included values already accounted for as well as values currently missing from the SNA. Of the total value of R6,215 million, 59 per cent came from natural forests and woodlands and 26 per cent from cultivated plantations, while 15 per cent was contributed by fynbos. Consumptive use value was the single largest component of total VAD, accounting for 73 per cent of the total economic contributions of forest and woodland resources, most of which is generated by natural forests and woodlands (59 per cent). Direct consumptive use activities in cultivated plantations consist of commercial logging as no other direct use benefits were reported for this forest resource type.

Carbon stocks: While net change in the physical stock of carbon stored in forest and woodland resources is not difficult to measure (usually proportional to timber volumes), establishing a value per unit of C released or sequestered is the most difficult task in the valuation exercise. The common practice in the literature is to use global value estimates such as those generated by climate change impact models (Nordhaus 1994). In countries where an environmental carbon charge is levied, such as Sweden, the used charge is applied (Hulkrantz, 1992). As no attempt has been made in South Africa to estimate carbon values, the present study could only borrow and adapt estimates from other parts of the world. Figures used in the literature to value carbon covered a very wide range between US$5 to US$130 per ton of C. If one uses Nordhaus's (1994) estimate of 1 per cent of GDP as the climate change damage cost, this will come to the equivalent of US$9 per ton of C for SA at 1995 prices. The fact that CO_2 contributes about 60 per cent of total greenhouse gas emissions in South Africa (Rowlands 1996, Scholes and van der Merwe 1995) leads to a carbon value of $5.4/ton for South Africa at 1995 prices. This is comparable to Nordhaus's (1994) global estimate of $5.2/ton of C in 1994, and is used estimate to value carbon stocks in both cultivated plantations and natural forests and woodlands.

Derived asset values were used to calculate estimates of asset depletion (or net accumulation). The change in assets' values indicates that the physical stocks and values of timber assets have been appreciating (net addition) in the case of cultivated plantations and natural forests and woodlands. The reverse however, is true for fynbos resources (net depletion). Net accumulation (depletion) of asset values in this case is derived simply as the difference

Table 1 Flow benefits of F&WLR resources in South Africa (Rm, 1998 prices).

Types of Values	Cultivated forest	Natural woodlands	Fynbos	Total	Percent of total
Direct consumptive use values[a]	1856	2613	79	4584	73.1%
Direct non-consumptive use values[b]	NAP[d]	NAV[e]	29	29	0.5%
Indirect use values	−225	1021	799	1595	25.7%
Water	−225	NAP	NAP	−225	−3.6%
Honey and pollination	NAP	NAV	786	786	12.7%
Livestock grazing	NAP	1021	13	1034	16.6%
Non-use values[c]	NAV	NAV	43	43	0.7%
Total value added (VAD)	**1631**	**3634**	**950**	**6215**	**100%**
Percent of total VAD	26.2%	58.5%	15.3%	100%	

[a.] Tangible timber and non-timber products for final consumption or intermediate use.
[b.] Intangible forest amenities, e.g. recreation.
[c.] Option and existence values
[d.] Not applicable (the sector's contribution to such values is insignificantly small).
[e.] Not available, indicating that the value of such services has not been estimated.

Continues

Box 18.3 *(Continued)*
between asset values estimated for each two consecutive periods.

The potential yield of pristine fynbos services is affected by aliens only through a reduction of the area in which the indigenous species can grow. Accordingly, a percentage decrease in fynbos yield potential proportional to the percentage cover of aliens was calculated for each category of alien infestation. More specifically, this meant that there is negligible change under conditions of light (occasional) infestation, but that only 87.5 per cent of the value remains in scattered infestations (12.5 per cent cover), 50 per cent in areas of medium infestation (50 per cent cover) and 12.5 per cent in densely infested areas (87.5 per cent cover) (Turpie et al. 2001). Net accumulation in timber values of cultivated plantations in 1998 (R1.09 billion) amounted to 0.15 per cent of the gross domestic product (GDP), 3.2 per cent of total value added in agriculture, forestry, and fisheries combined, and 63.4 per cent of value added generated in the cultivated plantations sector. Natural forests and woodlands, however, showed a total positive net accumulation (appreciation) in timber asset values of R7.09 billion, which amounted to more than four times the contribution of forestry to GDP and about 21 per cent of the value added in the combined agriculture, forestry, and fisheries in 1998. Net annual change in carbon stocks stored in cultivated plantations was equivalent to about 7 per cent of its officially recorded contribution to GDP in 1998. Natural forests and woodlands contributed about 21 per cent of the recorded value added of CPLNT in 1998 in net carbon stock accumulations. The net effect of alien invasion of fynbos resources led to a depreciation in its asset value of R0.62 billion in 1998 prices, equivalent to 36 per cent of the recorded contribution of cultivated plantations to GDP. In spite of the depreciation in fynbos resources, building up of timber stocks, especially in natural forests and woodlands, is the main source of net accumulation, contributing 94 per cent of the total annual change in the value of F & WLR assets.

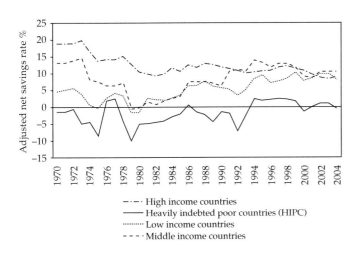

Figure 18.1 Adjusted Net Savings in high-, middle-, and low-income countries and in 'heavily indebted poor countries', 1970–2003. Negative Adjusted Net Savings indicate declining wealth. Calculated from World Bank data at http://web.worldbank.org/WBSITE/EXTERNAL/TOPICS/ENVIRONMENT/EXTDATASTA/0,,contentMDK:20502388~isCURL:Y~menuPK:2935543~pagePK:64168445~piPK:64168309~theSitePK:2875751,00.html

the most recent estimates of adjusted net savings show that many poorer countries continue to dissave (Figure 18.1).

18.6 Conclusions

The Millennium Ecosystem Assessment has refocused our attention on ecosystems as assets. They are important but often poorly understood components of wealth: the source of services that contribute to both material and spiritual wellbeing. The valuation of ecosystem services makes it possible to identify the opportunity cost of using ecological assets in particular ways. This in turn makes it possible to design instruments (payments for ecosystem services, prices, taxes, access charges, property rights, standards, and so on) for the efficient allocation of those assets. Beyond estimating the social opportunity cost of ecosystem services to support assessment of alternative uses of biodiversity, valuation serves another important purpose. It provides a means of testing the environmental sustainability of anthropogenic activity. More particularly, it provides a means of testing whether human activities are leading to a reduction in the value of ecosystems.

It is worth emphasizing that this is not equivalent to a reduction in the physical extent of those systems, or of aggregate productivity. The Millennium Ecosystem Assessment recorded trends in such physical measures in a number of systems, but was unable to record trends in the value of those systems. This is partly because the value of ecosystems depends on many factors other than their extent or productivity, so one may not be monotonically related to the other. But it is more because the work has not yet been done to support such an analysis. The adjusted net savings estimates produced by the World Bank are an important step in the right direction, but they are as yet limited to traditional natural resource stocks that correlate poorly to ecosystems. Nevertheless, since an understanding of changes in the value of ecosystem stocks is central to an understanding of the sustainability of economic development, this is work that needs to be done.

CHAPTER 19

Modelling biodiversity and ecosystem services in coupled ecological–economic systems

William A. Brock, David Finnoff, Ann P. Kinzig, Unai Pascual, Charles Perrings, John Tschirhart, and Anastasios Xepapadeas

19.1 Introduction

Many of the changes in biodiversity and ecosystem services described in this volume are the result of anthropogenic factors: the direct and indirect impacts of human decisions about the conservation and use of biodiversity, ecosystem functioning, and ecosystem services. In reality, since these impacts on biodiversity are seldom reflected in market prices, they are said to be external to the market and are generally ignored by decision-makers. Environmental, resource, and ecological economics are all concerned with the development of mechanisms to internalize such externalities, many of which are reviewed in Chapter 17. The way that feedbacks between economic and ecological systems are currently modelled by economists is reviewed by Wätzold et al. (2006). In this chapter, we consider how economists model biodiversity in coupled social ecological systems, taking two polar cases along with a more general problem.

All human decisions are taken to be purposive: that is, people have in mind some objective in deciding between alternative courses of action. They are assumed to optimize some objective function subject to some set of initial conditions, including the availability of information, to some set of resource constraints, and to the dynamics of the system being used. The two polar cases considered in this chapter involve quite different objectives, spanning the range of ecosystem services identified by the Millennium Ecosystem Assessment. They involve the preservation of ecosystems in 'close to natural' states, and the exploitation of ecosystems to produce foods, fuels, and fibres. The more general model involves the management of ecosystems to achieve a balance between a range of services. More particularly, the first case encompasses the preservation of remnant wilderness areas or protected parks and the restoration of altered ecosystems, and emphasizes non-consumptive cultural services. The second involves the transformation of ecosystems to yield provisioning services. The third encompasses mixed use, balancing non-consumptive cultural services with consumptive provisioning services.

Just as the objectives in each case are different, so are the concepts of biodiversity used. Indeed, these concepts are not independent of the decision-maker's objectives. In the first case, the measure of biodiversity is the difference in both the number and abundance of species from the 'natural' state. In the second case, we construct an index of biodiversity in agro-ecological systems that includes field margins, hedgerows, and other semi-natural habitats embedded in the cropping area. That is, the index reflects elements of landscape heterogeneity. In the more general model, we take the heterogeneity of the landscape – its patch structure – as given. The biodiversity in each patch is then measured by a vector, the components of which report the biomass of all the species on that patch. The optimal level of biodiversity is achieved through harvest (or equivalent controls), and the problem is modelled for both private users and society.

While the constrained optimization technique applied in all three cases may be unfamiliar to ecologists, we have tried to give the intuition behind it. We have also provided a verbal description of each of the three model structures developed. For those used to modelling aspects of the physical system, the chapter should show how the inclusion of decision-makers adds to our understanding of the dynamics of coupled systems. More importantly, it should show how biodiversity affects and is affected by the decisions that people make. In all cases the social and biogeophysical components of the coupled system are interdependent – connected through a series of feedback loops. Economists refer to such systems as 'general equilibrium systems'. That is, the dynamics of the system in some state are driven by a tendency towards the equilibrium corresponding to that state, and any perturbation has the potential to stimulate responses across the system.

19.2 Modelling the demand for 'naturalness'

The first case of interest is the classic concern of conservationists. This is the case where wellbeing depends on the preservation of the 'naturalness' of ecosystems. That is, the natural state of the system provides a higher level of non-consumptive (cultural) services than any other state. In this case people are assumed to value two ecosystem attributes above all others: (1) the diversity of species they contain – 'the appreciation for the variation or richness we observe in the ecosystems...based on the contemplation of the ecosystem as an ensemble of life forms...' (Goulder and Kennedy 1997, p. 34); and (2) the extent to which the ecosystem is not altered by humans (Eichner and Tschirhart 2007, Forsyth 2000).This preference for naturalness is illustrated by the Council Directive 92/43EEC on the conservation of natural habitats and of wild fauna and flora (Habitat Directive) of the European Union, and by the 1964 Wilderness Act of the United States. The latter defines wilderness as '...an area of undeveloped Federal land retaining its primeval character and influence, without permanent improvements or human habitation, which is protected and managed so as to preserve its natural conditions...'.

If naturalness is a proxy for various types of non-consumptive cultural services, then it is useful to measure naturalness in a concise and practical manner. Eichner and Tschirhart (2007) introduce a measure labelled the divergence from natural biodiversity (DNB). Consumers' preferences for biodiversity and naturalness are assumed to depend on DNB defined as:

$$s = S(h) = -\sum_{i=1}^{N}\left(\frac{n_i(h) - n_i(0)}{n_i(0)}\right)^2 \quad (19.1)$$

In Eqn. (19.1), s is a measure of deviation from 'natural' biodiversity (DNB), h denotes consumptive use such as harvesting, N is the total number of species, $n_i(h)$ is the population of species i as a function of consumptive use, and $n_i(0)$ is the natural, steady state population of species i. If there is no consumptive use, then $h = 0$ and $s = 0$. Consumptive use is a scalar here. However, h could be treated as a vector and additional consumptive use could include impacts on the ecosystem through pollution, habitat loss, introduction of invasive species, global climate change, and so on. As consumptive use increases, populations can be expected to diverge further from their natural levels and s decreases owing to the negative sign in Eqn. (19.1). As s approaches 0 from below, consumers are assumed to be better off because non-consumptive opportunities increase, other things equal. Of course, knowing what configuration of populations is feasible and how populations diverge with changes in h is central for using DNB, and this is taken up below.

DNB has several useful properties not shared by most biodiversity measures. First, the preferred value of DNB for ecosystems that are targeted for preservation or restoration is zero, and this is independent of the natural richness or evenness of species in the ecosystem. Second, consumptive use increases the value of DNB. Third, two species that diverge from their natural populations by the same percentages contribute the same to DNB. Thus, evenness, which has been shown to be negatively related to richness (Stirling and Wiley 2001), is not *per se* a positive property in DNB as it is with the Shannon index. Species that have stronger interactions with their neighbours than other species are often labelled 'keystone' species (Mills *et al.* 1993), and are expected to play a larger role in determining community structure. The third property of DNB does not imply that all species have equal impacts on the ecosystem and on $S(\mathbf{h})$. Population changes of keystone species relative to non keystone species will cause greater numbers of other species to deviate from their natural steady state populations. Therefore keystone species have a greater impact on $S(\mathbf{h})$. Fourth, with divergences of species from their natural populations DNB decreases ($\partial S/\partial n_i < 0$) at a decreasing rate ($\partial^2 S/\partial n_i^2 < 0$). Species whose populations are declining become increasingly difficult to recover (Beissinger and Perine 2001); therefore, DNB values reflect the difficulty of returning a community to its natural state.

If naturalness enters directly into social objectives, we can write the social welfare function as:

$$W\left(\mathbf{x}, \mathbf{h}, s(\mathbf{h})\right) \quad (19.2)$$

where $\mathbf{x}$ is a vector of manufactured goods, $\mathbf{h}$ a vector of harvest or other consumption of biological resources, and $s(\mathbf{h})$ the naturalness measure from Eqn. (19.1) and a proxy for non-consumptive ecosystem services. As described above, there is an inverse relationship between $\mathbf{h}$ and $\mathbf{s}$. Working with $W(\mathbf{x},h,s(\mathbf{h}))$ requires a bioeconomic model that accounts for aspects of both economic and ecological systems. The often used single-species harvesting models used in economics will not suffice for two reasons. First, they are in partial equilibrium on the economic side. Only one good, the harvested food, is present, so there is no account taken of how manufactured goods in the economy are affected by the harvests, and how the demand and supply of other ecosystem services are affected by the harvests. Second, they are in partial equilibrium on the ecological side. Only one species is present, so there is no account taken of how other species that provide ecosystem services are affected by the harvests.

At the same time, there are limits to how much detail about the economy or the ecosystem can be included in the bioeconomic model. Both systems are very complex and contain millions of agents making individual decisions in response to external incentives, and the individual decisions lead to aggregate outcomes that determine the external incentives. In other words, both systems are adaptive. In adaptive systems individuals process information about their environment and then alter their behaviour according to some objective (Hraber and Milne 1997). Ecosystems are clearly adaptive because individual plants and animals and even microorganisms act *as if* they have objectives and they alter their behaviour in response to environmental changes. Moreover, the individual behaviours lead to population dynamics that are often the cause of the environmental changes.

The most complete economic models are referred to as general equilibrium models that take into account all the market activity in an economy, whether that economy is a country, region, city, etc. General equilibrium (GE) theory has been referred to as the most important development in economics in the twentieth century (Sandler 2001). Including ecological variables in economic analyses without recognizing the general equilibrium (i.e. the interconnected and interdependent) nature of ecosystems can introduce errors for the same reasons that partial equilibrium compared to GE economic analyses can introduce errors in measuring welfare (Kokoski and Smith 1987). Recently, theoretical GE models have been developed for ecosystems (Eichner and Pethig 2005), and applied versions have been integrated with economic partial equilibrium models (Finnoff and Tschirhart 2003a, b) and an economic GE model (Finnoff and Tschirhart 2007). The authors refer to the applied GE ecosystem model as a General Equilibrium Ecosystem Model (GEEM).

GEEM involves a modified Computable General Equilibrium (CGE) economic model of predator–prey and competitive relationships. Predator–prey and competitive relationships are chosen because they contain key elements that drive ecosystem outcomes, and therefore can be building blocks for larger, ecosystem-wide CGE models. GEEM explains a fairly wide range of standard, and some not so standard, ecological outcomes. Applications include an Alaskan marine food web and the Alaskan economy (Finnoff and Tschirhart 2003ab), an early twentieth-century rodent invasion in California (Kim *et al.* 2007), invasions of sea lamprey in the Great Lakes (Finnoff *et al.* 2004), invasions of leafy spurge in the Western USA (Finnoff *et al.* 2007), and plant competition generally (Finnoff and Tschirhart 2005, 2006). One advantage of the approach is that ecosystem externalities that follow from the consumption of an ecosystem service can be taken into account. Thus, if economic decisions result in the consumptive use of an ecosystem service, then the model can track how other ecosystem services are altered, and then what these alterations mean for the economy. A model linking the Alaskan economy to a marine ecosystem comprising Alaska's Aleutian Islands and the Eastern Bering Sea is represented in Fig. 19.1.

The economy consists of Alaskan households and producing sectors, linked to one another and the rest of the world through commodity and factor markets. All species in the food web are linked together through predator–prey relationships and several species provide consumptive and non-consumptive ecosystem services. The prominent groundfish of the system, pollock, supports a very large fishery, and charismatic marine mammals – Steller sea lions (an endangered species), killer whales, and sea otter – provide non-consumptive ecosystem services to the state's recreation sector. Naturalness as defined in Eqn. (19.1) is not included below, but in Eichner and Tschirhart (2007) naturalness is included as a variable in a welfare function for the same Alaskan ecosystem examined here.

The three basic equations that comprise general equilibrium ecosystem model for this system are given by (19.3)–(19.5). The first equation is a general expression for the fitness net energy flow through a representative animal from species i.

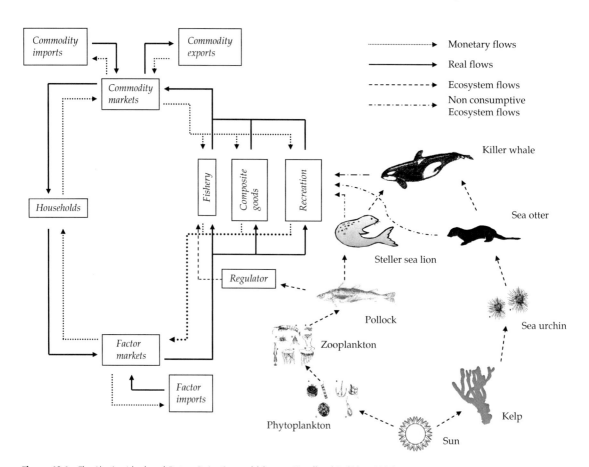

Figure 19.1 The Aleutian Islands and Eastern Bering Sea model (source: Finnoff and Tschirhart 2007).

$$R_i = \sum_{j=1}^{i-1}[e_j - e_{ij}]x_{ij} - \sum_{k=i+1}^{m} e_i[1 + t_i e_{ki}]y_{ik}$$
$$- f^i(\sum_{j=1}^{i-1} x_{ij}) - \beta_i \quad (19.3)$$

$$N_i x_{ij}(e_i) = N_j y_{ji}\left(x_j(e_j)\right) \quad (19.4)$$

$$N_i^{t+1} = N_i^t + N_i^t \left[\frac{2N_i^t}{s_i v_i^{ss} N_i^{ss}} \left(1 - \frac{N_i^t}{2N_i^{ss}}\right) (R_i(\cdot) + v_i) - \frac{1}{s_i} \right] \quad (19.5)$$

R_i is in power units (e.g. watts or kilocalories/time). According to Herendeen (1991) energy is the most frequently chosen maximand in ecological maximization models, and energy per time maximization, as adopted here, originates with Hannon (1973) and is expanded in Crocker and Tschirhart (1992) and to the individual level in Tschirhart (2000). Energy per time is also the individual's objective in the extensive optimum foraging literature (e.g. Stephens and Krebs 1986). The species in Eqn. (19.3) are arranged so that members of species i prey on organisms in lower-numbered species and are preyed on by members of higher-numbered species. The first term on the right-hand side is the inflow of energy from members of prey species (including plants) to the representative individual of species i. The choice variables or demands, x_{ij}, are the biomasses (in kilograms/time) transferred from the member of species j to the member of species i, e_j are the energies embodied in a unit of biomass (e.g. in kilocalories/kilogram) from a member of species j, and e_{ij} are the energies the member of species i must spend to locate, capture, and handle units of biomass of species j. These latter energies are the 'energy prices'.

The second term is the outflow of energy to animals of species k that prey on i. The e_i is the embodied energy in a unit of biomass from the representative individual of species i, and y_{ik} is the biomass supplied by i to k. The term in brackets is the energy the individual uses in attempts to avoid being preyed upon. It is assumed to be a linear function of the energy that its predators use in capture attempts: the more energy predators expend, the more energy the individual expends escaping. t_i is a tax on the individual because it loses energy above what it loses owing to being captured. The third and fourth terms in Eqn. (19.3) represent respiration energy lost to the atmosphere, which is divided into a variable component, $f^i(\cdot)$, that depends on energy intake and includes faeces, reproduction, defending territory, etc., and a fixed component, β_i, that is basal metabolism.

Time in the ecosystem is divided into reproductive periods, usually a year. In each period a general equilibrium is determined wherein the populations of all species are constant, each plant and animal is maximizing its net energy (using the derivatives of Eqn. (19.3) for first-order conditions), and aggregate demand equals aggregate supply between each predator and prey species. For each price that equates a demand and supply transfer there is an equilibrium equation given by Eqn. (19.4). Each plant and animal is assumed to be representative of individuals from its species; therefore the demand and supply sums are obtained by multiplying the representative individual's demands and supplies by the species populations given by the components of the vector **N**.

A representative plant or animal and its species may have positive, zero, or negative fitness net energy in equilibrium. Positive (zero, negative) net energy is associated with greater (constant, lesser) fitness and an increasing (constant, decreasing) population between periods. (The analogy in a competitive economy is the number of firms in an industry changing according to the sign of profits.) Fitness net energies, therefore, are the source of dynamic adjustments. If the period-by-period adjustments drive the net energies to zero, the system is moving to stable populations and a steady state. The predator–prey responses to changing energy prices tend to move the system to steady state.

The adjustment equation for the ith species (a top predator in this case) is given by Eqn. (19.5), where $R_i(\cdot) = R_i(x_{ij}; \mathbf{N}^t)$ is the optimum fitness net energy obtained by substituting the optimum demands and supplies as functions of energy prices into objective function Eqn. (19.3). $\mathbf{N}^t$ is a vector of all species' populations and it appears in $R_i(\cdot)$ to indicate that net energies in time period t depend on all populations in time period t. In the steady state,

$R_i(\cdot) = 0$. Also, s_i is the lifespan of the representative individual, v_i is the variable respiration, v_i^{ss} is the steady state variable respiration, and N_i^{ss} is the species steady state population. The first and second terms in brackets in Eqn. (19.4) are the birth and death rates. Expression (19.4) reduces to the steady state if $R_i(\cdot) = 0$ (in which case $v_i = v_i^{ss}$ and $N_i^t = N_i^{ss}$). Because the biomass demands depend on the period t populations of all species, the population adjustment for species i depends on the populations of all other species. In addition, out of steady state $R_i(\cdot)$ and v_i change across periods. These changes distinguish the GEEM approach from most all ecological dynamic population models, because the latter rely on fixed parameters in the adjustment equations that do not respond to changing ecosystem conditions.

In more detail, the simplified Alaskan economy is modelled as having three production sectors: the fishery F, recreation and tourism R, and composite goods C. The fishery is modelled as a single, vertically integrated industry consisting of catcher vessels, catcher processors, motherships, and inshore processors. Recreation and tourism represent the Census Bureau's classification of Wildlife Related Recreation (consumptive and non-consumptive), and composite goods is a catch-all for the residual private industries in Alaska. Profit-maximizing, price-taking firms employ harvests of pollock in the fishery, non-consumptive use of marine mammals (Steller sea lions, killer whales, and sea otter) in recreation, and capital and labour in all sectors, to produce their outputs in a continuous, non-reversible, and bounded process. Outputs from the fishery, recreation, and composite goods are sold in regional markets and exported out of the region, while regional production is differentiated from imports for fish and composite goods following Armington (1969). Capital, K, and labour, L, are homogeneous, perfectly mobile within the region, and defined in service units per period. Sector i factor employment levels are given by K_i and L_i ($i = F, R, C$). As a high proportion of fishery factors are owned and reside outside the region (specifically in Washington state), factors in the fishery are modelled as being interregionally mobile. This treatment is not extended to other sectors to allow a focus on the regional effects of policy changes.

Households consume fish, recreation, and composite goods and save for future consumption. Households first choose the proportion of income allocated between current and future consumption, where savings fund future consumption by adding to the capital stock in future periods. They then divide current disposable income between expenditures on recreation and an aggregate commodity. Finally, income allocated for the aggregate commodity is divided between purchases of fish and the composite good.

When the capital stock grows at the same rate as the effective labour force, the economy is on a balanced growth path; however, balanced growth is not a feature of the linked model, because species populations cannot grow indefinitely. Unlike the growth of the effective labour force and capital stock, ecosystem populations are limited by photosynthesis and may converge to steady states, with zero net growth. Thus the sectors reliant on ecosystem services will not be expected to grow at the rate of the human factor stocks.

Our welfare measures comprise annual equivalent variations EV_t for any single period across policy scenarios. Cumulative aggregate welfare measures (including a terminal term) are found using discounted summations of EV_t given by P_{EV}. These provide a comparative measure as they are based upon a common baseline price vector.

The recreation sector is reliant on marine mammals for a non-consumptive ecosystem service. The industry combines labour and capital with populations of marine mammals to deliver a 'recreational experience'. The recreation industry maximizes profits subject to a Cobb–Douglas production function (consistent with the fishery specification):

$$\max_{L_R, K_R} \pi_h = P_R^X X_R - WL_R - RK_R \text{ s.t.}$$

$$X_R = d^R L_R^{a^R} K_R^{b^R} N_{AT}^{c^R} \quad (19.6)$$

where a^R, c^R, and d^R are parameters, R subscripts for the recreation sector, and N_{AT} the contribution of Alaska's natural resources to the sector. This measure includes BSAI aggregate marine mammal populations taken as a linear combination of the three marine mammals. N_{AT} enters recreation production as a shift parameter, exogenous to the recreation firms but endogenous to the joint system. See Knowler (2002) for a discussion of models of these types, i.e. models that build on Ellis and Fisher (1987).

The linkage between the ecosystem and fishery requires two modifications to the standard fishery models. First, where most of the fishery literature employs effort as the single human factor of production, capital and labour must be included in CGE so that the fishery interacts with other sectors. Second, the non-fishery sectors hire capital and labour in service units per time period, in this case one year, but in the fishery factors are employed for considerably less than one year and may earn more per unit than in other sectors. This is due to restricted season lengths and potential unemployment during the off season.

Although ecosystems provide myriad services to economies, only one service is considered in most renewable resource models, usually harvesting. The bioeconomic model discussed here admits recreation as a second service, and more importantly accounts for how the two services are impacted by interactions within an eight species ecosystem. Here we illustrate some of the implications from creating the linkage between systems, based around shocks to the system from National Marine Fisheries Service (NMFS) policies. The NMFS in 2001 issued a Supplemental Environmental Impact Statement (SEIS) containing alternative management strategies that specify various pollock catch limits and no fishing zones to protect both the sea lions and the fishery. We consider rules for catch limits that are given by a regulator's choice of a quota for pollock. Here we consider quotas equal to the 1997 harvest levels and 30 per cent and 170 per cent of the 1997 harvest levels.

Steller sea lion recovery measures via alternative pollock quotas are shown to cause altered levels of all ecosystem populations, economic factor reallocation, changes in all regional prices, incomes, demands, outputs, imports, exports, and differential rates of factor accumulation. Of the eight species modelled, four are used directly in the economy either as consumption goods (fish) or non-consumption goods (marine mammals). While non-use values associated with the ecosystem are not considered, all species matter for the economy because the remaining species indirectly support the two modelled ecosystem services.

The results (reported in Finnoff and Tschirhart 2007) show that there are welfare gains from the reduced quota level of 30 per cent. A portion of the regional welfare gains from reduced quotas is due to the fact that the economy relies less on resource extraction and more on non-extraction – i.e. less on a consumptive ecosystem service and more on a non-consumptive ecosystem service. This result is consistent with a report from the Panel on Integrated Environmental and Economic Accounting which states: 'economic research indicates that many renewable resources, especially in the public domain, are today more valuable as sources of environmental service flows than as sources of marketed commodities' (Nordhaus and Kokkelenberg 1999, p. 177).

19.3 Modelling biodiversity and land use in agro-ecosystems

Our second case is at the opposite end of the conservation-use spectrum. It is the modelling of land use and biodiversity change in agro-ecosystems. The most important anthropogenic cause of agro-biodiversity loss is rapid land use and land cover change (LUCC) and the subsequent transformation of habitats (Millennium Ecosystem Assessment 2005b). In agricultural landscapes LUCC usually takes the form of land development. Most land development at the landscape level stems from decentralized economic decisions of economic agents, including small-scale farmers, agribusiness, and governments at different scales. The ecological causes and effects of such landscape transformations are increasingly well understood and documented, especially with regard to deforestation and desertification in developing regions (Perrings and Gadgil 2003). In agricultural landscapes, one impact of LUCC that is attracting increasing attention is the alteration of the flow of ecosystem services that are mediated by biodiversity (Millennium Ecosystem Assessment 2005b, Perrings *et al.* 2006). This has significant implications for biodiversity conservation strategies in agro-ecosystems.

Once again, we are interested in exploring the wedge between individual agents' *perceived* net benefits from LUCC actions and those realized by the community that is affected by those same actions (Swanson, 1995; Perrings, 2001; Millennium Ecosystem Assessment, 2005b). Part of the problem in understanding the social value of biodiversity change is that while some of the opportunity costs of conservation or foregone benefits from land development are easily identified, there remain important gaps in the understanding both the on- and off-farm benefits of agrobiodiversity conservation. The primary interest in agro-ecosystems is a set of provisioning services – the production of foods, fuels, and fibres – for which there are well-developed markets. There are, however, several effects of agricultural production that are not considered in market transactions for these services. One is the role of agricultural intensification as a means of reducing impacts on habitats for biodiversity at the landscape level (Green et al. 2005). A second is the contribution of crop diversity to productivity and variability of a main crop output and farm income (Smale et al. 1998, Widawsky and Rozelle 1998, Omer et al. 2007) and its variability under output uncertainty (Di Falco and Perrings 2003, 2005, Di Falco and Chavas 2007). The impacts we consider here relate to the application of fertilizers and pesticides. There are, of course, many distinguishing features of agro-ecosystems including modification of ecosystem processes (disturbance (tilling, burning), pest control, water input (irrigation), nutrient input (fertilizer), and the effects of herbivory (in the case of pastures or grain-fed livestock). All imply that nutrient cycling is spatially and temporally decoupled. Fertilizers and pesticides are just special cases of the general practice of regulating ecosystem processes to optimize local production.

Soule and Piper (1992) mention a number of adverse effects associated with industrial high-input agriculture, including spillovers from soil erosion into municipal water supplies, lakes, and reservoirs; erosion-induced suspended sediment carrying unwanted nutrients and pesticides; streams and wetlands filled with sediment causing less resilience to floods; soil pesticides adding to air pollution; salinization and other spillovers caused by irrigation; specialized breeding and failure to maintain the 'common pool' of genetic diversity because of optimization towards narrow productivity goals; biodiversity loss due to pursuit of agricultural monoculture in tropical forests and other common types of biodiversity reservoirs; concentration of disease-prone monocultures creating spillovers to the extent that the diseases are infectious across farms; and the development of resistance to pesticides.

Omer et al. (2007) explore the economic effects of biodiversity loss on marketable agricultural output for intensive agricultural systems, which require an increasing level of artificial capital inputs. They model the effect of the state of biodiversity on the optimal crop output both in the longer run and in the transitional path towards the steady state equilibrium. The hypothesized positive relationship between biodiversity stock and optimal levels of crop output is then empirically tested via a stochastic production frontier approach using panel data on UK cereal farms for the period 1989–2000. The results show that increases in associated agrobiodiversity can lead to a continual outward shift in the output frontier (although at a decreasing rate), controlling for the relevant set of labour and capital inputs. Agricultural transition towards biodiversity conservation may be consistent with an increase in crop output in already biodiversity-poor modern agricultural landscapes.

The model assumes that the allocation of agricultural inputs, for a given area of farmland, is driven by levels of crop output and by on-site biodiversity. It is assumed that decision-makers maximize the discounted present value of utility flows derived from both outputs. The stylized direct utility function is specified as $U = U(y_t, b_t)$, where y_t represents the flow of 'marketable' agricultural output at time t, and b_t is biodiversity loss, also a flow variable. This loss is attributable to intensive use of artificial inputs, x_t, which therefore negatively impacts on utility, i.e. $U_y > 0, U_{yy} < 0$, and $U_b < 0, U_{bb} < 0$, for a strictly concave and linearly separable utility function. This specification reflects a subset of economic decisions that would affect land use activities, and the welfare that these activities generate. The problem is to find the intertemporal optimal levels of utility yielding services (flows) based on (1) marketable agricultural supply and (2) physical depreciation of biodiversity. While b_t is an argument in the direct utility function, it depends on the level of application of artificial inputs and the 'current' level of biodiversity (z_t) through a 'biodiversity impact function'.

Following recent studies (e.g. Tscharntke et al. 2005), the crop production function is assumed to be affected by the stock of biodiversity, z_t, alongside the conventional agricultural input set x_t. In addition, the 'state of the art' of agricultural technology is captured by the parameter a_t, which acts as an exogenous shifter of the production possibility frontier, representing neutral technical progress. Normalizing the unit price of crop output, the value production function is represented by $f(x_t, z_t, a_t)$, which is assumed to exhibit well-behaved properties, i.e. $f_i > 0$, $f_{ii} < 0$ for $i = x_t, z_t$, and a_t, and to be linearly separable.

Given some managerial input, the stock z_t can be increased by conservation investment, c_t, which is given by the residual from the total proceeds from agricultural production, $f(x_t, z_t, a_t)$, less the proportion that is committed as 'marketable (value) output' y_t:

$$c_t = f(x_t, z_t, a_t) - y_t \quad (19.7)$$

If the focus is on the functional diversity of species, the effect of a change in z_t on the marginal product of x_t is likely to be different at each level or sublevel of z_t. For example, an increase in insect or microorganism diversity would increase the marginal product of fertilizer if it enhances soil productivity ($f_{xz} \geq 0$). On the other hand, an increase in natural vegetation diversity might decrease the marginal product of fertilizer if it increases the competition with cultivated crops ($f_{xz} \leq 0$). The biodiversity impact (or loss) function is expressed by $b_t = b(x_t, z_t)$. The ability of the agro-ecosystem to tolerate and overcome the potential adverse effects of agricultural land use activities depends on the current biodiversity stock, z_t such that $b_z < 0$, $b_{zz} > 0$. At the margin, biodiversity loss increases (decreases) at an increasing (decreasing) rate due to increases in input intensification, i.e. $b_x > 0$, $b_{xx} > 0$; for simplicity, $b_t = b(x_t, z_t)$ is also assumed to be linearly separable in x_t and z_t.

To maximize utility, the farmer chooses the level of the control variables y_t and x_t at each point in time, subject to the evolution of z_t. This evolution reflects agrobiodiversity stock, conservation investments, c_t, and artificial input use, x_t, that reflects the level of intensification.

$$\dot{z} = g(z_t, c_t, x_t) \quad (19.8)$$

The evolution of biodiversity is captured by Eqn. (19.8), which can be interpreted as a general form of the function:

$$\dot{z} = \alpha_1 z_t (1 - z_t/k) + \alpha_2 c_t - \alpha_3 x_t \quad (19.9)$$

The natural rate of growth of the biodiversity stock is given by $\alpha_1 > 0$. The parameter k reflects the maximum potential diversity that could be sustained in the ecological system. The parameter k, like the 'carrying capacity' in Lotka–Volterra population growth models, is the maximum potential level of diversity which could be sustained naturally. According to Eqn. (19.9), z_t is density dependent and increases with investment in conservation, α_2 being the rate of induced growth in z_t. The parameter α_3 may be interpreted as the marginal degradation in z_t caused by an increase in x_t. It is worth noting that while biodiversity is considered to be natural capital, it is assumed here that no depletion in biodiversity occurs as a result of its supporting role in the production process. In intensive agricultural systems, which are biodiverse poor relative to their potential maximum (e.g. their 'wild' state), the term z_t/k can be considered as negligible, and thus Eqn. (19.9) can be simplified through further approximation as:

$$\dot{z} = \alpha_1 z_t + \alpha_2 c_t - \alpha_3 x_t \quad (19.10)$$

The optimization problem is expressed, for a positive utility discount rate ($r > 0$) as:

$$\max_{y,x,c} W(y_t, b_t) = \int_{t=0}^{\infty} e^{-rt} U(y_t, b_t) dt \quad (19.11)$$

subject to (1) the environmental conservation investment function (Eqn. 19.7), (2) the evolution of z_t, (Eqn. 19.10), (3) the impact function $b(.)$, (4) the initial condition $z(0) = z_0$, and (5) the non-negativity constraints and $x \geq 0$ and $b \geq 0$. This yields the current-value Hamiltonian:

$$\tilde{H} = U(y_t, b_t) + \varphi(\alpha_1 z + \alpha_2 f(.) - \alpha_2 y_t - \alpha_3 x_t) \quad (19.12)$$

where φ is the current shadow value of biodiversity. Omer et al. (2007) show that where the stock of biodiversity at the steady state is greater than the initial stock, the optimal supply of marketable output can increase (albeit at a declining rate) along the transition path to the steady state.

In an application of the theoretical model summarized above, Omer et al. (2007) construct an aggregate biodiversity index that takes into account biodiversity of agricultural landscapes that include non-cropped areas such as field margins, hedgerows, and other semi-natural habitats embedded in the cropping area using UK farm and biodiversity data. The production data derive from a panel of 230 cereal producers from the East of England, for the period 1989–2000, yielding a total sample size of 2,778 observations in an unbalanced panel. The data are from the UK's annual Farm Business Survey (FBS) undertaken by the Department of Environment, Food and Rural Affairs of the UK. The UK Countryside Surveys undertaken in 1978, 1990, and 1998 were used to construct a farm-level biodiversity index (Haines-Young et al. 2000). The data set includes information on cereal output, level of input application, participation in and payments from agri-environmental schemes, and socioeconomic characteristics of the farm households. In addition, a variable measuring on-farm functional biodiversity is constructed. The per-hectare variables used in the econometric model are: crop enterprise output (marketed), hired and imputed family labour, use of machinery, fertilizers and pesticides, and the biodiversity index (BI). All the variables, except for BI, are derived from value measures deflated by the relevant agricultural price index (base year 1989), and are thus measures of volume.

This is consistent with a number of ecological studies (e.g. Altieri 1999, Tscharntke et al. 2005) that emphasize the role of landscape level biodiversity (associated and functional) affecting the ecological functioning of arable agro-ecosystems. The aggregation approach used to construct the biodiversity index is described by Wenum et al. (1999). The index (representing the variable z in the theoretical model) is given by:

$$z = \sum_j \sum_i a_j n_{ij} s_{ij} \quad (19.13)$$

where, s_{ij} is the mean plant species richness on a given plot located in aggregate vegetation class (AVC) i within broad habitat (BH) type j; n_{ij} stands for the measure of AVC-i dominance in BH type-j, i.e. the number of AVC-i plots in BH type-j relative to the total number of plots of all AVCs in BH-j; and lastly a_j is a scalar associated with BH-j dominance. In a second step, the evolution of the biodiversity index at the aggregate level is calibrated as a non-linear discrete-time aggregate version of Eqn. (19.10).

$$z_{t+1} - z_t = \alpha_1 \ln z_t + \alpha_2 c_t - \alpha_3 \ln x_t, \quad (19.14)$$

where c_t is proxied by a categorical dummy variable (1/0) showing whether the farmer is a beneficiary of the introduction of agri-environmental schemes following the EU's Common Agricultural Policy reform in 1992, and x_t is average per hectare pesticide use on the sample farms.

To test the proposition from the theoretical model, a reduced form dynamic parametric frontier model is used and fitted to data (Battese and Coelli 1995). A stochastic production frontier (SPF) approach then allows estimation of both the output production frontier that represents best practice among farmers (as assumed in the theoretical model) and the possibility of real deviations from the frontier attributed to the effects of variation in the sampled farmers' level of technical efficiency (TE). Controlling for TE, it is possible to qualify the key relationships derived from the theoretical model along the production frontier as it evolves over time. It should be noted that the frontier provides a closer approximation to the 'optimal path' than a more traditional econometric specification which (a) does not allow for technical inefficiency levels by farmers, (b) can be associated with their managerial ability, and (c) in turn may depend on factors such as information, experience, etc. Hence the data on marketed crop output is used to estimate the output optimal path via the function $y(x_t, z_t, a_t)$.

Omer et al. (2007) estimate a non-neutral inefficiency model fitted to the 12 years, $t = 1, 2, \ldots, T$, and using farm-level data, i.e. for farmer i, which takes the following general form:

$$y_{it} = \beta_0 + \sum_k \beta_k p_{kit} + v_{it} - u_{it} \quad (19.15)$$

where: y_{it} is the natural log of crop marketed output of farm i at time t; p_1 is the natural log of BI (biodiversity index); p_2 is the natural log of fertilizer input value; p_3 is the natural log of labour input value; p_4 is the natural log of machinery input value; p_5 is the natural log of pesticide input value; and p_6 is the year of observation, where $p_6 = 1, 2, \ldots, 12$. The v_{it}s are assumed to be independently and identically distributed random errors $N(0, \sigma_v^2)$, independent of the non-negative random error term, u_{it}, associated with technical inefficiency in production. β_k is the parameter vector to be estimated. Furthermore, it is also assumed that the inefficiency variable is also a function of farm-level characteristics:

$$u_{it} = \delta_0 + \sum_j \delta_j q_{jit} + \sum_j \sum_k \delta_{jk} p_{kit} q_{jit} + w_{it} \quad (19.16)$$

Allowing for interactions between farm-specific variables and the input variables in the stochastic frontier the elasticity of crop output with respect to all inputs, including agrobiodiversity, can be derived. The latter has two components: (1) the elasticity of frontier output with respect to the agrobiodiversity and (2) the elasticity of technical efficiency with respect to agrobiodiversity.

The results from the econometric model indicate that associated agrobiodiversity positively affects mean wheat output levels over the whole period, even though greater levels of agrobiodiversity appear to have negatively affected technical (managerial) efficiency in the intensive agricultural sector. As regards the former result, this is consistent with the hypothesis that there is a positive, although declining, effect. The data indicate that the frontier elasticities with respect to agrobiodiversity, while positive, have tended to decrease over the years. It also appears that the effect of the stock of biodiversity on technical efficiency has been different before and after 1996. While there is initially a negative technical efficiency elasticity, after 1996 the elasticity becomes positive. The net effect of biodiversity through the impacts on both frontier output and technical efficiency suggests that agrobiodiversity has negatively affected yields (average elasticity of –0.1) from 1989 to 1993, the year after broad environmental payments were introduced in the farming sector, as a strategic agri-environmetnal scheme. After the incorporation of the environmental payments for biodiversity conservation, the impact on mean output has reversed, with an elasticity in 2000 of 0.26. This implies that on average, during such agri-environmental scheme, a 1 per cent increase in agrobiodiversity has positively affected mean wheat output by 0.26 per cent. These results suggest that biodiversity conservation schemes have not undermined the productive performance of the cereal sector.

19.5 A generalized model of the consequences of biodiversity loss in exploited ecological–economic ecosystems

Although there is a high degree of uncertainty associated with species diversity and its loss, there is evidence that declining species diversity may adversely affect the performance of terrestrial ecosystems. Several empirical studies relate the number of species in ecosystems to plant productivity (Vandermeer 1989, Naeem et al. 1995, 1996, Tilman et al. 1996, Hooper and Vitousek 1997) which find that functional diversity is a principal factor explaining plant productivity. Given the relationship between human expansion, loss of biotic diversity and loss of ecosystem productivity, we need to be able to model the efficient management of the biotic diversity of ecosystems.

Equilibrium species diversity in resource-based models is governed by an exclusion principle that characterizes the way that undisturbed natural systems move towards a steady state equilibrium. This principle states that where there is multispecies competition for a limiting factor, in a patch free of disturbances, the species with the lowest resource requirement will competitively displace all other species in equilibrium. In this case the system is driven to a monoculture, and the equilibrium outcome of species competition is the survival of the competitive dominant species, i.e. the species with the lowest resource requirement. Of course there are a number of ways in which nature can produce equilibrium polycultures with one or more limited resources. As noted in Pacala and Tilman (1994), temperature-dependent growth and temperature variation in a habitat, spatial or temporal variations of resource ratios, differences in species palatabilities, and local abundance of herbivores, can all result in spatial or temporal variation of dominant competitors. In this way Nature's equilibrium is characterized by a Tilman set of species determined by limiting factors.

Given this, it is possible to develop a general model of species competition in a landscape consisting of potentially heterogeneous patches of land to explore the equilibrium state of an ecosystem that is managed for economic objectives. The growth of species is limited by resource availability, while species interact among themselves and compete for the limited resources. In this sense this model can be regarded as a generalization of Tilman's mechanistic resource-based model of competition, by including direct interactions among species. The multi-species Kolmogorov model and the Lotka–Volterra model can be also derived from this generalized model under appropriate assumptions. The benefits from species diversity are reflected in the value of a potentially large and diverse set of service flows generated by the ecosystem. Depending on whether these flows of services are taken into account in the design of management, the resulting management rules and the corresponding equilibrium state of the ecosystem can be analyzed. In this analysis Tilman's model with a single limited resource serves as a special case that provides some very interesting insights into the issue of specialization and biodiversity preservation (Brock and Xepapadeas 2002).

In particular, two basic management problems can be analyzed in this context. The first is the privately optimal management problem (POMP) where economic agents maximize solely profits from harvesting in their patch, subject to the evolution of the natural system, by regarding any interactions outside their patch as exogenous. In this context private agents do not account for the external costs and benefits of their management practices on ecosystem service flows. This means that the private agent ignores all interactions among patches, as well as any externalities that his or her actions might bring on other agents as these actions could affect the whole range of ecosystem service flows in the ecosystem. The second is the socially optimal management problem (SOMP), where a social planner maximizes utility from the whole ecosystem. Utility is derived both from harvesting and from the state of the ecosystem, with all interactions among species and resources both within and across patches taken into account. Thus, in the design of the socially optimal management rules values associated with flows of ecosystem services are internalized into the management problem.

By comparing the two management problems it is possible to explore how the decisions of private agents (who ignore biodiversity externalities) may deviate from the social optimum. In this sense the private and social optimal harvesting rules, and the state of the ecosystem as reflected in equilibrium biodiversity under private and social management can be derived and compared. Brock and Xepapadeas (2004) show this in the following manner. Let $i = 1, \ldots, n$ species exist in a given patch of land, for example, a hectare of forest stand, belonging to a broader landscape, and assume that the growth of the species in each patch is limited by resources $j = 1, \ldots, r$. They define a resource following Tilman (1982, p. 11) as: 'any substance or factor which can lead to increased growth rate [of an organism] as its availability in the environment is increased, and which is consumed by the organism'. For example, the growth rate of a plant measured as the rate of weight gain may be increased by the addition of nitrate which is consumed by the plant. Further, they denote by c a patch or a site; by $\mathbf{R}_c(t) = \left(R_{1c}(t), \ldots, R_{rc}(t)\right) \in \gamma \subset \mathfrak{R}_+^\rho$ a vector of available resources in that patch at time t; and by $\mathbf{B}_c(t) = \left(B_{1c}(t), \ldots, B_{rc}(t)\right) \in \beta \subset \mathfrak{R}_+^\rho$ a vector of the biomass of species in the patch at the same time. So $\mathbf{R}$ and $\mathbf{B}$ denote the vectors of resources and biomasses respectively in every patch, and $\mathbf{R}_{-c}$, $\mathbf{B}_{-c}$ the vectors of resources and biomasses and biomasses outside site c: i.e. $(\mathbf{R}, \mathbf{B}) = (\mathbf{R}_c, \mathbf{R}_{-c}, \mathbf{B}_c, \mathbf{B}_{-c})$. From this they are able to specify a general model of competition among species in each patch c by the system of differential equations:

$$\frac{\dot{B}_{ic}}{B_{ic}} = F_{ic}(\mathbf{B}_c, \mathbf{B}_{-c}) G_{ic}(\mathbf{R}_c, d_{ic}), B_{ic}(0)$$
$$= B_{ic}^0 > 0, \forall i, c \quad (19.17)$$

$$\dot{R}_{jc} = S_{jc}(\mathbf{R}_c, \mathbf{R}_{-c}) - D_{jc}(\mathbf{B}_c, \mathbf{B}_{-c}, \mathbf{R}_c, \mathbf{R}_{-c}), R_{jc}(0)$$
$$= R_{ic}^0 > 0, \forall j, c \quad (19.18)$$

Equation (19.17) describes the net rate of growth of the biomass of species i in patch c. The growth rates of a species in a habitat depend on a number of exceptionally complex processes involving soil building, nutrient uptake, and symbiotic relations among fungi, insects, mammals, and flora which are still not entirely understood (Scott 1998). Thus, the growth equation (19.17) attempts to capture the dependence of the growth rate of each species on resource availability in the patch, as well as its dependence on other species biomasses and resource availability in the whole lattice.

Equation (19.17) emphasizes the fact that the biomass vector $\mathbf{B}$ has $\dot{\mathbf{B}} = 0$ as a steady state. The function $G_{ic}(\mathbf{R}_c, d_{ic})$ captures the effects of resource availability in the patch on a species' rate of growth, with d_{ic} being a natural death rate parameter. By assumption, all resources in the patch are essential for species growth. The growth function is non-decreasing in $\mathbf{R}_c$ by the definition of a resource, and is assumed to be strictly concave in $\mathbf{R}_c$ (reflecting 'diminishing returns' in the resource use for all $R \geq 0$. The effect (complementary or competing) of one species on the growth rates of other species is captured by the function $F_{ic}(\mathbf{B}_c, \mathbf{B}_{-c})$.

Equation (19.18) describes the resource dynamics. $S_{jc}(\mathbf{R}_c, \mathbf{R}_{-c})$ is the amount of the resource supplied at time t in patch c and $-D_{jc}(\mathbf{B}_c, \mathbf{B}_{-c}, \mathbf{R}_c, \mathbf{R}_{-c})$ is consumption of the resource by all species. Brock and Xepapadeas assume that the availability of a resource can be affected by the availability of other resources both within and across sites, so that resources are themselves interacting entities with the interaction mechanism captured in the function $S_{jc}(\mathbf{R}_c, \mathbf{R}_{-c})$. It is further assumed that the resources cannot accumulate without bounds (even without consumption by species), and that Eqn. (19.18) is concave.

The model described by Eqns. (19.17) and (19.18) can be regarded as a generalization of a multispecies Kolmogorov model (Murray 2002) with the explicit introduction of the resource dynamics equation which makes it possible to directly analyze species competition through its influence on resource availability. It constitutes a resource-based model of competition between species in a given patch structure. A long-run equilibrium for every patch is defined as a situation where the net growths of the species and the resources are zero, or equivalently $\dot{\mathbf{B}} = \dot{\mathbf{R}} = 0$. If such an equilibrium exists, then the equilibrium biomass vector $\mathbf{B}_c^e$ describes the equilibrium biodiversity in patch c and $\mathbf{B}$ describes the equilibrium biodiversity of the whole system. Species for which $B_{kc}^e = 0$ are displaced by other species in patch c. Note that Tilman's mechanistic resource-based model (Tilman 1982, 1988, Pacala and Tilman 1994) is a special case of this generalized model. In this model each species affects all other species only through its effects on the availability of the limiting resource, that is, growth is density independent and no interactions among neighbouring patches take place. The single resource can be thought of, for example, as an index of soil quality that could include the health of microorganisms in the soil, the amount of useful humus, and so forth, and the resource supply can be regarded as the soil genesis function. The exclusion principle mentioned earlier is the main driving force in determining the long-run equilibrium for the ecosystem. If all species are ranked, by relabeling, according to their R_{ic}^e such that $R_{1c}^e < R_{2c}^e < \cdots < R_{nc}^e$, then according to the competitive displacement mechanism, species one will displace all other species in equilibrium.

Note that the competitive displacement of species according to R_{ic}^e occurs in the long-run equilibrium. Species with relatively high R_{ic}^e show an initial period of rapid growth relative to low species. However, as time goes by, the low R_{ic}^e species reduce the resource to levels insufficient for the survival of the high R_{ic}^e species, eventually displacing them. Thus high R_{ic}^e species tend to grow relatively faster than low R_{ic}^e species, but are eventually displaced in equilibrium. In an ecosystem with sites having different characteristics the exclusion principle will provide a c-specific monoculture, with a dominant c-competitor. Environmental heterogeneity across patches, associated with variations in temperature and the species thermal optimum, spatial variations of resource ratios, differences in species palatability, and local abundance of herbivores (Pacala and Tilman 1994) will lead to the coexistence of species (higher levels of biodiversity) equilibrium.

Similar results can be explored in a multi-resource context. In this case depending on the type of resources (e.g. substitutable, complementary, switching, essential), the species requirements and the resource supply (Tilman 1982, 1988) there might be species coexistence in equilibrium within the same c-patch. Allowing therefore for variation of characteristics across the lattice grid and/or multiple limiting resources, the mechanistic resource-based model predicts higher levels of biodiversity not only during transitional dynamics but also in equilibrium.

Given the parameters in Eqns. (19.17)–(19.18), undisturbed nature can produce a rich pattern of species in equilibrium. Therefore in an ecosystem where many species compete for limiting resources, undisturbed nature will provide a biodiversity pattern which depends mainly on the availability of resources and the variation of characteristics across the landscape. In this context, the purpose of management of the ecosystem can be seen as providing harvesting rules in order to optimize an appropriately defined objective function, subject to the constraints imposed by the dynamics of the resource-based model of species competition.

To model the economic management problem, Brock and Xepapadeas (2002) define an objective function that captures the benefits accruing from the landscape. A first type of benefit relates to the revenue yield of species that can be harvested in a given time period. Letting $\mathbf{R}_c(t) = \left(R_{1c}(t), ..., R_{rc}(t) \right) \in \gamma \subset \Re^p_+$ denote a vector of harvest of species in patch c at time t, the benefits of harvest are defined by a function $V(\mathbf{H}_c(t))$. They assume that this has the following specific form:

$$V\left(\mathbf{H}_c(t)\right) = \sum_{i=1}^{n} \left[P_i(t) H_{ic}(t) - C_i(H_{ic}(t)) \right] \quad (19.19)$$

where P_i is the market price of the harvested species and $C_i(H_{ic}(t))$ is a convex cost of harvesting function. The constraints on the above maximization problem are the growth rate of the biomasses adjusted to take into account the reduction in the growth rate due to harvesting.

Maximizing an objective function like (19.19) subject to Eqns. (19.17)–(19.18) implies that management focuses only on species that can provide commercially valuable biomass for harvesting, i.e. only on the provisioning services. It neglects many of the other services, especially the cultural and regulating services, identified by the Millennium Ecosystem Assessment. The private (optimal) management problem (POMP) involves the maximization of private profits, through an objective function defined solely in terms of harvesting and where the private agent is sufficiently small that it treats the resources and the biomasses in the rest of the ecosystem as fixed. On the other hand, the social (optimal) management problem (SOMP) is defined in terms of an objective function that includes not only consumptive benefits but also other sources of benefits related to the full range of services yielded by the ecosystem.

In terms of the Brock and Xepadadeas equations for species dynamics, if some part of the ecosystem has been converted into a monoculture, then terms of the type $\partial F_{ic}/\partial \mathbf{B}_{-c}$ could be positive and large in value when $\mathbf{B}_{-c} \to \mathbf{0}$, indicating that biodiversity loss ultimately reduces the growth of surviving species in the c-patch. Since each agent treats biodiversity outside his patch as fixed, these interactions are not taken into account in the private problem. Arguments have been put forward (Walker et al. 1999), suggesting that in a particular functional group, in an undisturbed ecosystem, one species serving a particular function tends to squeeze out other species in the same group. The functional group will be characterized by a few dominant abundant species and numerous species with almost zero abundance (which will nevertheless be functionally very close to the dominant species). Such minor species act as a kind of 'reserve species'. Walker et al. (1999) show evidence for this by looking at heavily grazed areas and lightly grazed areas for the same type of ranchland. In heavily grazed areas cattle eliminated the dominant grass and one of the 'minor' grasses took its place. If a shock destroys one or more of the abundant species, then there is room for the reserve species to take their place. When 'reserve' species exist, biodiversity provides resilience, especially for big shocks. Again these effects are not taken into account by private decision-makers.

While some of the services from existing species are internalized through the F functions, other effects, such as flood/erosion problems or contamination and salinization of water supplies, biodiversity loss, spillovers from energy use, are predominantly external. For example, erosion and flood control damages could be related to the fraction of patches covered with plants which are good at erosion and flood control like prairie. Other external effects associated with ecosystem state may be more direct. Suppose that in the exploitation of an ecosystem, the function $V(\mathbf{H})$ captures net revenues from harvesting species (an Millennium Ecosystem Assessment provisioning service), while $U(\mathbf{B})$ captures net benefits from the non-consumptive use of biodiversity (the sum of the Millennium Ecosystem Assessment cultural services – aesthetic, spiritual, scientific, and so on). Since these effects are external to the representative profit-maximizing agent in a c-patch, they can only be taken into account in the management problem for the ecosystem if the optimization is carried out by some social planner.

The social optimization management problem then supposes that the flow of benefits is determined both by consumptive (harvest) and non-consumptive utility or $V(\mathbf{H}(t)) + U(\mathbf{B}(t))$. Brock and Xepapadeas (2002) assume that $U(\mathbf{B})$ is non-decreasing and strictly concave with $\partial U/\partial B_i = m_i \geq 0$ as $B \to 0^+$. This defines a utility index from biodiversity as: $U(\mathbf{B}) = \left\{ \sum_{i=1}^{n} \left(u_i(B_i) \right)^\gamma \right\}, \gamma \in (0, 1],$ where the $u_i(B_i)$ are concave utility functions with $\gamma \left(u'_i(B_i) \right)^{\gamma-1} \to \infty$ for $B \to 0^+$ for all i. This specific utility index puts a premium on biodiversity since aggregate utility increases when one more species with positive biomass is added. Furthermore γ parametrizes the substitutability among species regarding tastes for them, since the elasticity of substitution between the utility derived from any two species is $1/(1 - \gamma) > 1$, if $\gamma = 1$ species are considered as perfect substitutes regarding individuals' tastes. If m_i is assumed to be very large (tends to infinity), then the assumption that the marginal utility of the biomass of a group of species becomes very large as the biomass becomes very small, reflects the fact that the particular species provides ecosystem services that are essential at the margin when extinction of the species is near. Alternatively, if species biomass is measured in terms of consumables (as the numeraire), then technological progress that makes consumables more plentiful relative to species implies that the marginal utility of a species near extinction becomes very high in terms of consumables. The assumption that $m_i \to \infty$ can be regarded as an Inada condition for species biomasses. On the other hand, if $m_i \to 0$ then the services of the ith species are not valued even if that species is close to extinction. A more general approach could be to think of some species as an invasive pathogen that can spread from patch to patch and destroy other species. In such a case if we partition the biodiversity vector into $(\mathbf{B}^u, \mathbf{B}^h)$, where the indices u, h correspond to useful and harmful species respectively, then marginal utilities will be defined as $\partial U^u/\partial B_i > 0$ and $\partial U^h/\partial B_i < 0$.

19.6 Conclusions

The modelling approaches described in this chapter are far from exhaustive, but they do illustrate the ways in which economists are beginning to approach the challenge of modelling biodiversity change in complex ecological–economic systems. Building on the simple single-species harvesting models that still dominate natural resource economics, economists are beginning to explore the implications of the fact that the ecosystem – like the economy – is a general equilibrium system. Its dynamics are given by the interactions between its component parts. While the models described here have been developed by economists, they build on insights from ecology into the implications of biodiversity change for the structure, productivity, and resilience of ecosystems. They also build on a growing body of collaborative research involving both disciplines.

The underlying economic problem remains – to identify allocations that best meet social goals in a world where the pursuit of private wealth does not yield the best outcome for all. However, economists now understand that the gap between private and social optima depends on a complex set of feedbacks between the ecological and economic components of the coupled system. Understanding and modelling those feedbacks is a first important step in developing strategies for the sustainable use of biodiversity, just as it is for the development of plans to manage and restore ecosystems.

PART 4
Summary and synthesis

CHAPTER 20

TraitNet: furthering biodiversity research through the curation, discovery, and sharing of species trait data

Shahid Naeem and Daniel E. Bunker

20.1 Introduction: traits and ecosystem functioning

Trait-based ecological and evolutionary research has undergone an extraordinary expansion in the last thirty years (Fig. 20.1), and species traits, broadly defined, have been key to advances in many fields of natural science (Table 20.1). For example, the evolutionary ecology of species' niches involves fundamental trade-offs in seed size (Moles et al. 2005), leaf economic traits (Ackerly 2004, Wright et al. 2004), and allometric constraints (West et al. 1997). Traits have been used to predict the risk of species invasions (Veltman et al. 1996, Kolar and Lodge 2002, Lloret et al. 2005, Ruesink et al. 2005), the risk of species extinction (Gittleman and Purvis 1998, Foufopoulos and Ives 1999, Purvis et al. 2000a), and crop responses to climate change (Lynch et al. 2004). Table 20.1 summarizes these and other examples of basic and applied research that are dependent on species trait data.

While early efforts to understand the effects of biodiversity on ecosystem functioning focused on species richness, recent efforts have recognized the role of functional diversity as the driver of these effects (Petchey et al. (Chapter 4), Diaz and Cabido 2001, Petchey et al. 2004b). The functional traits of species are the means by which species interact with their environment, and thus are directly responsible for the effects of species on ecosystem processes (e.g. Gordon 1998, Eviner and Chapin 2003, Kelso et al. 2003, Diaz et al. 2004, Eviner 2004), as well as the response of species to environmental change (Grime 1979, Tilman 1982, Huston and Smith 1987, Brown 1995, Enquist et al. 2003, Brown et al. 2004, Navas and Moreau-Richard 2005, McGill et al. 2006).

The BioMERGE Research Coordination Network strove to expand biodiversity–ecosystem functioning research to larger scales by developing models to predict functioning from species traits (Naeem et al. (Chapter 1), Duffy et al. (Chapter 5), Naeem et al. 2007, Naeem 2008). For example, Solan et al. (2004) used traits to estimate changes in estuarine sediment turnover in the face of biodiversity loss and Bunker et al. (2005) forecasted changes in forest carbon sequestration under different management practices. Another example is that of McIntyre et al. (2007) who estimated the influence of fish biodiversity on freshwater ecosystem functions. These early efforts primarily addressed selection effects (SE; see Chapter 7) by utilizing traits that predict *per capita* effects on functioning in combination with various traits associated with extinction risk. These efforts ignored complementarity effects (CE; see Chapter 7), in part because we still know relatively little about which traits may lead to complementarity, but also because species trait data are, at best, dispersed throughout the literature, and at worst lacking altogether. Indeed, even as the development of comprehensive vegetation databases (e.g. VegBank, Center for Tropical Forest Science, and SALVIAS) and phylogenetic databases

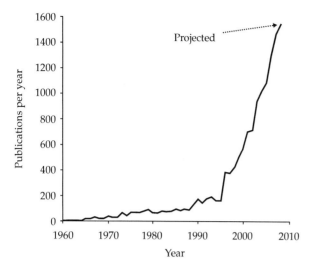

Figure 20.1 Exponential increase in ecological and evolutionary research using traits. Results from Scopus search for (("ecology" or "evolution") and "traits").

(e.g. Flora of North America, Tree of Life project) has proceeded rapidly, trait databases, where they exist at all, have remained specialized to particular regions, taxa, or sets of traits. This chapter describes TraitNet, a recently established research coordination network designed to facilitate trait-based research in general, and this new direction in biodiversity and ecosystem functioning research in particular.

20.2 TraitNet: enabling trait-based research

TraitNet aims to facilitate collaboration between ecologists and evolutionary biologists who utilize species trait data. Traits are used across a broad spectrum of disciplines, including ecology, evolution, and conservation biology. While each discipline has developed its own operational definitions, protocols, and databases, there is little coordination across disciplines. Because of these diverse uses and definitions of species traits, TraitNet takes an expansive view of what may constitute a 'trait', including most any character that can be quantified at the individual level. We take this approach because, while our efforts are motivated by the need to understand and quantify the traits that drive ecosystem functioning, we intend TraitNet to support the broad range of research that utilizes species trait data.

In order to facilitate this new, cross-disciplinary, collaborative, trait-based research, TraitNet will pursue five primary goals: (1) identify key questions and core hypotheses in trait-based research, (2) identify data gaps that hinder the advancement of intra- and inter-disciplinary trait-based research, (3) coordinate the standardization of collection and curation of trait data, (4) build a model database to address core hypotheses, and (5) facilitate the development of cross-disciplinary computational tools for merging, disseminating, and sharing trait data. There are several core activities, including workshops, online collaboratories, and database development (Fig. 20.2).

TraitNet aims to bring together species trait data from a variety of taxa across different trophic levels and from a variety of habitats and locations to address specific interdisciplinary hypotheses (see examples below). Our initial coverage will be greatest among terrestrial plants because several plant trait networks are well developed and will serve as useful starting points (Table 20.2). The interdisciplinary hypotheses chosen for study by the network of participants will likely require traits of herbivores, predators, detritivores, and other trophic groups in addition to traits of primary producers. Additionally, TraitNet will not limit itself to terrestrial systems. It will also examine aquatic ecosystems or transition habitats such as wetlands. To that end, we have assembled a group of core participants that is weighted towards terrestrial plant ecologists due to current advances in

Table 20.1 A broad sampling of trait-based research where trait is broadly defined. Note that additional research areas include trait-based taxonomy, trait-based phylogenetics, and morphometrics.

Subject area	Description	Example traits	References
Bioremediation	Using species to remediate pollution	Heavy metal resistance, specific root length, root surface area, root volume and average root diameter	Von Canstein et al. 2002, Pulford and Watson 2003, Merkl et al. 2005
Biodiversity and ecosystem functioning	Mechanisms by which changes in biodiversity change ecosystem functioning	Species response and effect traits	Lavorel and Garnier 2002, Solan et al. 2004, Bunker et al. 2005, Thompson et al. 2005
Comparative method	Using phylogenies and traits to understand evolutionary adaptation	Traits such as leaf mass per area, seed mass, genome size	Ackerly 2004, Moles et al. 2005
Community ecology	How trait filtering governs community composition and structure, including assembly rules, competition, facilitation and limiting similarity	Body size, height, leaf traits, trophic position, light requirements, clonality	Gaudet and Keddy 1988, Weiher and Keddy 1995a, Weiher and Keddy 1995b, Weiher and Keddy 1999, Ackerly et al. 2002, Suding et al. 2003, Cavender-Bares et al. 2004, Suding et al. 2005, Grime 2006, McGill et al. 2006
Conservation biology	Estimate threat levels for species or likelihood of extinction	Gestation period, range size, number of offspring, trophic position	Gittleman and Purvis 1998, Foufopoulos and Ives 1999, Purvis et al. 2000a
Ecosystem ecology	Trait-specific influences of organisms on ecosystem processes and biogeochemistry	Woody caudices, multi-branched rhizomes, N-fixing symbiotic associations, C3 or C4 pathway	Gordon 1998, Eviner and Chapin 2003, Kelso et al. 2003, Diaz et al. 2004, Eviner 2004
Gradient analysis	Mechanisms and patterns of biodiversity along ecological gradients	R^*, dispersal mode, reproductive structures, elements of leaf design	Tilman and Wedin 1991, Thuiller et al. 2004
Endemism	Determining what traits are associated with endemism	Stature, dispersal, pollen/ovule ratios, number of flowers	Lavergne et al. 2003, Lavergne et al. 2004
Fire ecology	Predicting fires based on plant traits	Resprouting capability, seed bank	Saha and Howe 2003, Pausas et al. 2004
Food webs	Structure and dynamics of communities governed by trophic interactions	Dietary or energy transfer linkages and trophic position, body size, morphological traits	Layman et al. 2005
Functional diversity	Identification and quantification of functional diversity	All functional traits	Petchey and Gaston 2002a, Mason et al. 2003, Botta-Dukat 2005, Mouillot et al. 2005a, Mouillot et al. 2005c
Guild analysis	Grouping species by environmental resource exploitation irrespective of taxonomy	C3, C4, annuals and biennial forbs, ephemeral spring forbs, spring forbs, summer/fall forbs, legumes, woody shrubs	Simberloff and Dayan 1991, Kindscher and Wells 1995, Blondel 2003
Heritability	Quantifying the heritability of species traits	Various traits	Iyengar et al. 2002, Caruso et al. 2005, Garant et al. 2005
Macroecology	Patterns of species adaptations at geographic scales	Body size, photosynthetic pathway, dispersal syndrome	Brandle et al. 2002, Burns 2004, Morin and Chuine 2006

(*Continues*)

Table 20.1 (*continued*)

Subject area	Description	Example traits	References
Metabolic theory of ecology	Metabolism as a basis for linking individual organisms to population, community, and ecosystem ecology.	Body size, physiological traits, and correlates such as growth, range	Brown 1995, West et al. 1997, Enquist et al. 2003, Brown et al. 2004
Natural selection	How species evolve in response to selective pressure	Body size, fledging weight, dispersal syndrome, palatability, host specificity	Boughman 2001, Etterson and Shaw 2001, Merila et al. 2001, Nosil et al. 2002, Beatty et al. 2004, Hoskin et al. 2005
Palaeobiology	Using leaf physiognomy to estimate past climate	Leaf size, leaf morphology	Royer et al. 2005, Royer and Wilf 2006
Plant ecological strategies	How plant species 'secure carbon profit during vegetative growth and ensure gene transmission into the future'	Leaf-span-per-area, leaf-lifespan	Hodgson et al. 1999, Westoby et al. 2002
Population ecology	Predicting properties of dynamics (e.g. probability of extinction)	Body size, age at first reproduction, or number of offspring	Fagan et al. 2001, McGill et al. 2006
Species invasions	Predicting invasiveness of species based on traits	Body size, endemism, reproductive rate	Veltman et al. 1996, Kolar and Lodge 2002, Lloret et al. 2004, Hamilton et al. 2005, Lloret et al. 2005, Ruesink et al. 2005
Succession	Temporal change in communities predicted by traits	Respiration rate, seed number, growth rate, maximum life span, induced dormancy, R^*, stress tolerance	Grime 1979, Tilman 1982, Huston and Smith 1987, Navas and Moreau-Richard 2005
Unified neutral theory of biodiversity and biogeography	Trait neutral core patterns in distribution and abundance forms contrasting hypotheses to trait-based patterns	Dispersal, growth rates	Hubbell 2001, Nee and Stone 2003, He and Hu 2005, Hubbell 2005, Ostling 2005, Wootton 2005

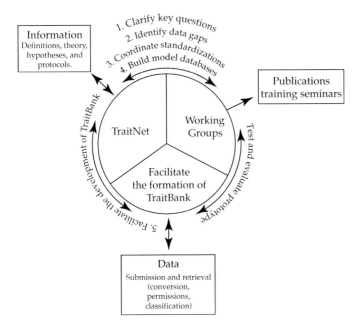

Figure 20.2 Architecture of TraitNet. Five objectives structure TraitNet, an NSF-funded Research Coordination Network that coordinates the development of cross-disciplinary trait-based research. Working groups consist of members across trait-based disciplines that will collectively clarify key questions within and across disciplines, identify data gaps, coordinate standardizations, and build model databases to test hypotheses that emerge from syntheses. Working groups will provide workshops and training sessions to broadly disseminate TraitNet results. Finally, TraitNet will coordinate the design of a prototype universal database entitled TraitBank.

the field, but which also includes researchers who specialize in insect, mammalian, microbial, aquatic, and disease ecology. These core participants are actively recruiting additional investigators within their respective areas of expertise. TraitNet participation is expected to grow substantially as we identify additional researchers who focus on other habitats, taxonomic groups, and trophic groups.

Hypotheses derived from cross-disciplinary research are often characterized by a scope that requires multiple traits collected uniformly from a diversity of species across widely dispersed localities. The TraitNet working groups will explore and identify such hypotheses, determine data needs and gaps, develop and test model datasets for addressing these hypotheses, and coordinate the establishment of trait databases, guidelines, and training for their use, and enable a variety of multidisciplinary research activities. As Table 20.1 illustrates, the potential number of cross-disciplinary approaches is very large. Here we provide three of the many possible hypotheses TraitNet will explore:

- *Dimensionality of life-history trade-offs.* While an endless number of traits can be measured on individuals and species, many traits are highly correlated with one another, and it has been suggested that a small number trade-off axes can explain the majority of variation in plant form and function (Grime 1979, Coley et al. 1985, Charnov 1997, Reich et al. 1999, Hubbell 2001, Westoby et al. 2002, Wright et al. 2004). How are species life histories constrained by these fundamental trade-offs, how many axes of differentiation exist, and how does the extent of these trade-offs vary across environmental gradients and among biomes? These key questions require data on multiple traits, collected from multiple species, from multiple sites, and standardized when different protocols were used.

- *Mechanisms of exotic species invasions.* The success of invasive species has often been attributed to an escape from natural enemies, whereby one would predict successful invaders to have 'better' traits than the native species they displace, such as greater height, lower R^*, or lower construction costs (Miller and Werner 1987, Gaudet and Keddy 1988, Nagel and Griffin 2001, Seabloom et al. 2003, Bunker 2004). Alternatively, the success of some invaders has been attributed to novel traits, such as nitrogen fixation (Vitousek and Walker 1989) or allelopathic effects (Bais et al. 2003), that allow them to dominate new habitats. While both mechanisms certainly play a strong role, the relative importance of each in driving species invasions is not clear. An effective

Table 20.2 Examples of databases that include trait data. Current collaborators in bold.

Database name	Description
BiolFlor	Focuses on the German flora and includes > 60 traits and > 3600 plant species
BioPop	Database of plant traits of the Mid-European flora including 60 traits and > 4,700 species
Center for Tropical Forest Science Trait Database	A newly initiated effort to collect functional trait data for 6,200 tree species found in 18 large forest dynamics plots located in 14 tropical countries
Ecological flora of the British Isles	Database of plant traits of the flora of the British Isles including > 130 traits and > 1700 plant species
Ecological flora of California	A database of ecological characteristics, including life history, phenology, morphology and other traits for the California flora. Under development by David Ackerly
FishBase	Worldwide fish species database with more than 29,000 species
Glopnet	Global compilation of leaf economic traits from > 2500 plant species. Initiated by Peter Reich, Ian Wright, and Mark Westoby
Hawaii Plant Trait Database	A database of systematic, biogeographic, functional, physiological, and ecological data for Hawaii's native and alien flora. Initiated by Rebecca Montgomery, Lawren Sack, Becky Ostertag, Susan Cordell, and Jon Price
LEDA Traitbase	Focuses on the Northwest European flora and plant traits that describe three key features of plant dynamics: persistence, regeneration, and dispersability
NatureServe Explorer	Conservation data on more than 50,000 plants, animals, and ecological communities of the USA and Canada
NatureServe InfoNatura	Conservation information on the more than 5,500 birds, mammals, and amphibians of Latin America and the Caribbean
Seed Information Database, Kew Botanic Garden	Database of seed characteristics, including storage behavior, weight, dispersal, germination, oil content, protein content, morphology, for several thousand plant species, with plans to include > 24,000 species
TRY	Global plant trait database to develop new plant functional classifications for earth system modelling
USDA PLANTS	The PLANTS Database provides standardized information about the vascular plants, mosses, liverworts, hornworts, and lichens of the USA and its territories. It includes names, distributional data, characteristics, images, and crop information

test would require species trait data on plant invader species, on the native species they may displace, on palatability to native herbivores, and data on traits of potential natural enemies such as body size, diet, and growth rate.

• *Predicting species, community and ecosystem responses to global change.* Predicting the response of species to climate change, pollution, and land use change is a key challenge to ecologists. These predictions could be developed by correlating species traits with either observed responses to global drivers or across natural environmental gradients. In either case trait data from a wide variety of species, across multiple trophic levels, from a variety of habitats would be required. Similarly, predicting the effects of these global drivers on ecosystem function will require additional trait data that mechanistically link species with their *per capita* effects on ecosystem functioning (Etterson and Shaw 2001, Solan *et al.* 2004, Bunker *et al.* 2005).

20.3 The challenges of data integration

Much of what TraitNet aims to accomplish will rely on integrating data from disparate sources. Integrating disparate data is a complex process with many challenges. These challenges are not trivial. Here we outline the main issues and describe our approach to addressing them. In order to meet these challenges, we have included several informatics experts in our group of core participants, representing several organizations including the Science Environment for Ecological Knowledge project (SEEK), the Pacific Ecoinformatics and Computational Ecology Lab, the National Center for Ecological Analysis and Synthesis (NCEAS), the National Evolutionary Synthesis Center, and the Microsoft European Science Initiative. To address our Core Hypotheses, we will build a model database that also will serve as a training ground for building a fully accessible and open source trait data archive termed TraitBank.

20.3.1 Intellectual property rights

Intellectual property rights are a critical issue for any research network and even more so when data are aggregated from multiple sources. Trait-based research progresses best when data sharing is maximal, but currently the sharing of raw data is not common except within groups. Workshops, collaboratories, training sessions, and the TraitNet website will provide a forum for discussion of the many issues surrounding intellectual property rights and how they would affect database tools, resources, and the design and implementation of TraitBank in the future.

To that end, we propose that data owners will retain full rights and full control over their data. TraitNet will facilitate collaborations between participants that would otherwise be less likely to occur. Our model database will be fully searchable, whereby one could search for all available traits for a particular species, or all species with a particular trait, or for a set of traits among a set of species. If participants so choose, we can set up the search system to return only whether the data exist, and who owns the data. It would then be up to the participant to contact the data owner and propose collaboration.

20.3.2 Taxonomic standardization

Definitions of biological taxa change with taxonomic revisions over time. For instance, a single species may be split by one revision into several species, and then lumped back into a single species in subsequent revisions. A trait value measured on the lumped species cannot be assigned to any one of the split species, and a trait value measured on one of the split species cannot be assumed to represent the entirety of the lumped species. In addition, species are often cited with only the name authority, but not the underlying taxon concept reference. For these reasons, taxonomic names by themselves cannot be considered a unique index for TraitNet datasets. This obstacle applies to all data that are specific to individual species, such as GenBank and VegBank.

Fortunately, efforts are under way to address this complex issue. The SEEK Taxon project has created an internationally accepted standard for taxonomic data, the Taxonomic Concept Schema (TCS), and work is under way to implement the Taxonomic Object Service (TOS), a repository and web service allowing for translation between taxonomic concept authorities (Graham and Kennedy 2007). TraitNet will collaborate with SEEK Taxon to ensure that our database schema will function with the TOS when it is fully functional and populated with taxon concepts.

To ensure that TraitNet data will be compatible with TOS, TraitNet will require that participants include name authorities and taxon concept references, as well as subspecies when appropriate, in their data submissions. Only in this way can we specify, for each record, the taxonomic concept upon which the measurement was taken. For instance, a full taxon concept reference might be: "*Aus beus* Sarg. 1893 Smith 1989," where Smith 1989 is a link to the reference in which the concept is described or defined (Fig. 20.3).

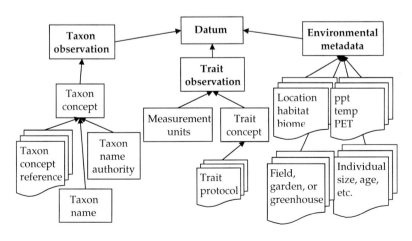

Figure 20.3 TraitNet trait observation schema. The ideal trait observation will include critical metadata, such as trait protocols, taxon concepts and environmental data. These metadata will make trait observations robust to various future uses of the trait data and thus ensure the longevity and usefulness of trait data collections.

20.3.3 Trait protocol standardization

Variations in collection protocols introduce challenges that are similar to those introduced by taxonomic revisions. A trait concept may remain fixed, but the protocol used to quantify the trait may change as new protocols are introduced. A trait database must be able to incorporate revised trait collection protocols as they are developed to ensure that data produced though all protocols for a given trait concept are quantitatively comparable. For example, wood density is the trait concept of mass per unit volume. However the protocol to collect wood density varies. Wood mass may be measured on oven-dry wood samples or on air-dried samples with 12–15 per cent moisture content. Both metrics quantify wood density, but data from air-dried samples must be corrected to account for the moisture content. TraitNet will define trait concepts and associated trait collection protocols. Each trait protocol for a given trait concept must be quantitatively comparable (Fig. 20.3).

20.4 Integrating TraitNet into ongoing ecoinformatics frameworks

TraitNet will build on current ecoinformatic efforts to address these issues. Ecological Metadata Language (EML) has been developed by NCEAS' Knowledge Network for Biocomplexity project and is widely considered the standard for documenting metadata for ecological datasets. The SEEK project has extended and formalized critical aspects of EML in the Observation Ontology (OBOE), a formal model of scientific observations that includes trait measurements (Madin *et al.* 2007, Madin *et al.* 2008). Thus TraitNet will use and extend EML to specify the Taxon Concepts, Trait Concepts, and associated environmental data outlined in Fig. 20.3. Eventually, these trait concepts will be included in SEEK's formal ontologies such as OBOE.

Because species trait data are used for such diverse research ends, the data must be collected and archived with sufficient metadata to ensure wide applicability to potentially unforeseen research questions. For instance, an investigator may collect wood density data with the intention of calculating above-ground biomass at a given site. However, future investigators may ask how wood density varies within species at a given site, within species across environmental gradients, among species, among size classes, or even throughout the year. To ensure that a given trait observation contains maximum scientific value, the collector will want to document explicitly the conditions under which it is collected, including latitude and longitude, habitat, individual age or body size, time and date of collection, etc. Many trait observations, such as wood density, require substantial effort and/or expense to collect. Only minimal additional effort is required to collect substantial metadata that will ensure that a trait observation has lasting scientific value (Fig. 20.3).

Finally, an ideal network environment would also: (1) allow for automated integration of trait data contributions; (2) include a web-enabled search engine that would allow user-friendly access to the general public, including students, educators, and policymakers; and (3) enable seamless access to related community, phylogenetic, and environmental databases.

As the understanding and appreciation of functional biodiversity grows among the public, TraitBank has the potential to be much more than a resource for natural scientists. For example, a birder may wish to learn more about the functional traits of a group of species found at a particular site, or a farmer or natural resource manager might want to gain additional insights into the types of weeds or invasive species found on site. A teacher on a field trip may wish to design an exercise based on local species, or a student may require trait data for their paper or science project. In this way, TraitBank will offer the public detailed trait data for individual species, and thus augment existing resources such as the Global Biodiversity Information Facility (http://www.gbif.org/).

20.5 Species traits, functional diversity, and the future of biodiversity research

TraitNet is designed to coordinate a wide array of scientific disciplines that are centred on a specific research theme but would benefit enormously from cross-disciplinary coordination. TraitNet will facilitate cross-disciplinary research among ecological and evolutionary fields centered on trait-based research. All of these diverse activities rely strongly on understanding the functional characteristics, or traits, of species. It is only through these traits that scientists can understand and predict the responses of species to their environment as well as species effects on ecosystem functioning and the ecosystem services that humans increasingly demand from both natural and managed ecosystems (Naeem *et al.*, Chapter 5; Jackson *et al.*, Chapter 13). Indeed, this shift towards trait-based ecology and functional diversity will bring a new perspective on biological diversity to a wide array of fundamental and applied scientists, students, and educators in ecology, evolution, and environmental biology.

CHAPTER 21

Can we predict the effects of global change on biodiversity loss and ecosystem functioning?

Shahid Naeem, Daniel E. Bunker, Andy Hector, Michel Loreau, and Charles Perrings

21.1 Efficacy, practicability, and social will

The efficacy and practicability of an idea, and the will of individuals or society to explore it, determine whether it catalyzes change or merely enters the vast store of quiescent ideas that make up the bulk of humanity's collective wisdom. As we noted in the *Introduction*, the idea that biodiversity influences ecosystem functioning is not new. There are, for example, similarities between the *Hortus Gramineus Woburnensis* experiment of 1817 (Hector and Hooper 2002) and BIODEPTH experiments (Hector *et al.* 1999) almost two centuries apart. The *Hortus Gramineus*, however, was nearly forgotten. Perhaps it was forgotten because the idea of improving yield by manipulating vegetation gave way in the 1840s to Justus von Liebig's idea that yield was controlled by the availability of limiting mineral nutrients. In contrast, in the 1990s, individual and social concerns over the environmental consequences of worldwide changes in biodiversity raised questions about ecosystem functions in general, not just yield (Loreau *et al.* 2002). Because primary production is a convenient measure of ecosystem functioning, it has been emphasized in biodiversity and ecosystem functioning work, which creates the uncanny resemblance between modern experiments and the *Hortus Gramineus*. Although biodiversity and ecosystem functioning research shows no sign of abating some 15 years later, we might nevertheless ask whether it too will be forgotten like the *Hortus Gramineus*?

As in all science, there remain differences among researchers on the interpretation of biodiversity and ecosystem functioning research, but the efficacy of the idea that the diversity of life, not just its mass, influences both the biogeochemical and biotic properties of ecosystems, is well established. Even in 1997, although they guessed a stronger, less complex role for diversity than meta-analyses would eventually support (Chapter 2), researchers had the right sense of things with just a small number of studies to hand. The rapid rise in numbers of studies (Chapter 2), their influence on the literature (Chapter 3), incorporation of the idea into the Millennium Assessment framework (Millennium Ecosystem Assessment 2003), and the achievement of scientific consensus (Loreau *et al.* 2001, Hooper *et al.* 2005), all suggest that today the efficacy of the idea is no longer in doubt. Many questions concerning mechanisms, generality, and the relative strength of biodiversity effects compared to other factors that influence ecosystem functioning, such as temperature, precipitation, ocean depth, and physical substrate, remain, but few question that changes in biodiversity influence ecosystem functioning.

Although efficacy may be less of an issue than it was in the 1990s, practicability and societal will remain significant challenges. By *practicability* we mean the ability to test the idea empirically and translate it into real-world applications, and by *societal will* we mean the willingness of individuals and society to adopt the idea as a foundation for decision-making. In this chapter, we look across the many contributions in this volume and consider a few messages the current field of biodiversity and ecosystem functioning research give us concerning efficacy, practicability and societal will.

From a rich set of cross-cutting ideas embodied in this book we focus on just three that are shaping the trends in biodiversity and ecosystem functioning research. First, concerning efficacy, there is a struggle in the discipline to make the research more realistic. Unfortunately, what constitutes realism in ecology can sometimes be subjective, thus if biodiversity and ecosystem functioning research is to avoid another round of debate, further clarity on the issue of realism is needed. Second, concerning practicability, we revisit the Millennium Assessment's framework and restructure it based on current empirical and theoretical findings. Our hope is that this modified framework points the way to practicable, coupled, natural–social research and policy. Finally, in order to facilitate individual and societal will, we provide a graphical device that may better communicate the core idea of the importance of biodiversity to ecosystem functioning and link it more directly to sustainable development. We suggest that the preservation, management, and intelligent use of biodiversity may be our only hope for achieving environmental sustainability which, in turn, is our only hope for achieving overall sustainable development and its many goals (e.g. the United Nations' Millennium Development Goals) of social and economic equity across the globe.

21.2 Efficacy and realism in biodiversity research

Since its inception, biodiversity and ecosystem functioning research has sought to encapsulate the key elements of biodiversity and ecosystem functioning in its theory and experiments, but every study has been hounded by the question of realism. The full complexity of biodiversity, whose ecological and evolutionary processes scale from microscopic to planetary, can never be entirely captured in any one experiment, nor does it have to be. Rather, researchers ask focused questions and make decisions about what is and is not necessary to test a particular idea. Even focused questions about biodiversity and ecosystem functioning, however, require fairly elaborate experiments (see Chapters 2 and 7). Research in biodiversity and ecosystem functioning has pushed the envelope of empirical ecology, establishing some of the largest (e.g. Roscher *et al.* 2005, Spehn *et al.* 2005, Tilman *et al.* 2001) and most elaborate micro- and mesocosm studies ever conducted (e.g. Naeem and Li 1998, McGrady-Steed *et al.* 1997, Downing and Liebold 2002, Fukami and Morin 2003, Fox 2004a, Bell *et al.* 2005b, Cadotte and Fukami 2005). The trend of increasing the size or the number of replicates and the complexity of experimental design reflects attempts to continuously improve experimental realism. Increasing plot size, for example, is based on the notion that, in nature, some ecological processes operate at larger scales. Likewise, the use of microbial communities whose small spatial scales can be readily accommodated using bottles and Petri dishes, allows for multiple generations. The presumption here is that multiple generations better approximate the temporal scale at which ecological processes operate in nature (Petchey *et al.* 2002, Raffaelli *et al.* 2002). Microcosms also allow for much more community and trophic complexity, again presuming that greater complexity better approximates nature (Petchey *et al.* 2002, Raffaelli *et al.* 2002). Exploration of different types of systems, such as wetlands (e.g. Engelhardt and Ritchie 2001), estuarine (e.g. Duffy *et al.* 2005) and marine ecosystems (e.g, Emmerson *et al.* 2001, Stachowicz *et al.* 2002, Bracken *et al.* 2008), and organisms other than plants, such as fungi (e.g. Tiunov and Scheu 2005, Van der Heijden *et al.* 1998), soil fauna (e.g. Mikola and Setälä 1998), and zooplankton (e.g. Norberg 2000), also reflects attempts to test the generality of findings. With every year, the cumulative range of spatial and temporal scales, community complexity, and the scope of taxonomic and ecological diversity explored by biodiversity and ecosystem functioning research has grown.

The question of realism, however, continues to dog biodiversity and ecosystem functioning research (Raffaelli 2004). Clearly one should put more stock in the findings of a more realistic experiment, but how does one evaluate realism in ecological experiments? There are two features of biodiversity and ecosystem functioning studies that determine how comparable they are, both of which are determined by a large number of decisions that researchers make when conducting their studies. First, in any biodiversity and ecosystem functioning study, researchers must decide what *biodiversity gradient* is appropriate for the question they wish to address. Decisions concerning the biodiversity gradient include:

1) choosing the appropriate measure of diversity,
2) determining the size of the species pool to be used in the experiment,
3) establishing the low-diversity endpoint (most often monocultures) and
4) establishing the high-diversity endpoint of the gradient (most often all species in the pool).

Biodiversity gradients in experiments range from high-diversity end points that are simply convenient (e.g. the 16 plant species out of over 800 at Cedar Creek, Minnesota, were selected in part because they were known to do well in experimental settings and seeds were commercially available; S. Naeem, *pers. comm.*) to high-diversity endpoints that contain as many species as possible in abundances commonly observed in the field (e.g. when the high-diversity endpoint is an unmanipulated plot). Low-diversity endpoints are often simply monocultures, but here too one may elect to set the low-diversity endpoint at a higher richness level (and use monocultures only to calculate null predictions, as reviewed in Chapter 7).

Second, researchers must also make decisions concerning *species selection*, or which species should be observed or manipulated, since it is generally not possible to study every species, especially microorganisms, in ecosystems. Decisions concerning species selection include:

1) whether or not one should include exotics (naturalized, domestic, invasive, or other non-resident species);
2) what range of biotic interactions should be included among the selected species (i.e. should predators, diseases, mutualists, and other interacting species be used or just competitors for the same limiting resources); and
3) whether the subset of species should be selected at random or based on some other criteria, such as commonness, cultivability, or traits related to the likelihood of extinction.

Species selections in experiments range from investigators using whatever is convenient (e.g. whatever can be cultivated or manipulated), to selecting only biogeographically coherent sets of species (i.e. only sets of species observed to co-occur in nature), to using all species in an ecosystem.

In silico studies represent a recent, promising trend in biodiversity and ecosystem functioning experiments, but their gains in realism made possible through enormous numbers of replicates come at the cost to realism in estimating ecosystem functioning. *In silico* studies, such as those by Solan *et al.* (2004), Bunker *et al.* (2005), McIntyre *et al.* (2007), and Bracken *et al.* (2008) can generate thousands to millions of species combinations, thereby eliminating the practical constraints of field research that is often limited to hundreds of species combinations. *In silico* experiments require researchers to make the same decisions concerning species selection and the biodiversity gradient, but with fewer constraints on the size of the species pool, the number of levels of biodiversity, and the number of replicates. Ecosystem function, however, must be estimated, which is usually done by algorithms that translate individual species abundances and functional traits into likely ecosystem function, and it is here where uncertainty lies. Current *in silico* experiments ignore the multiple interactions and modes of functional complementarity between species, and hence have other limitations regarding realism. *In silico* experiments will become more realistic and accurate as we develop a greater understanding of the mechanisms of complementarity and better data on the species traits that lead to them (Chapters 5 and 20).

Considering the biodiversity gradient and species selections of multiple studies provides a basis for comparing studies. Figure 21.1 graphically illustrates how studies relate to one another. The endpoints of the biodiversity gradient axis reflect extremes in decision making by researchers. 'Fully combinatorial' refers to an experiment that uses every possible combination of species possible, irrespective of what is found in nature. 'Trait-based extinctions' refers to experiments in which combinations are constrained to those in which the presence or absence of species is determined by the particular traits of species. For example, for an *in silico* study of mammalian bush meat production, where body size determines probability of extirpation by hunting, the gradient will range from species-rich communities with both small and large

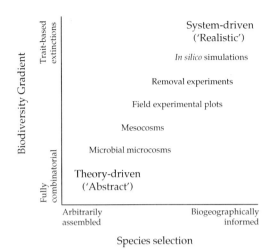

Figure 21.1 The biodiversity gradient and species selection of different kinds of studies in biodiversity and ecosystem functioning research. Each study makes decisions concerning how it establishes the gradient in biodiversity it explores as well as how it selects species. See text for discussion.

sized animals to communities composed only of small sized animals, as large mammal populations will decline first in response to hunting pressure (Cardillo et al. 2005).

The endpoints of the species selection axis in Fig. 21.1 consist of studies in which species were selected by the researcher because they were convenient (e.g. common and cultivable) or through other arbitrary decisions to selecting combinations of species currently found co-occurring in nature or which are likely to be found in the future after extinction takes its toll. 'Theory-driven' studies are often typified by biodiversity gradients and selection well suited to testing the theory, but are perhaps not reflective of what is observed in nature. 'System-driven' studies tend to be closely modelled on the ecosystem under investigation. Obviously, all experiments have their virtues, but whether their findings refer broadly to theory or more specifically to particular ecosystems depends on the biodiversity gradient and species selection of the study. We observe that there is a tendency to consider 'realism' in studies that are more system-driven.

The biodiversity gradient and species selection properties of experiments make clear that realism in biodiversity and ecosystem functioning research is a complex issue. Our deliberations here illustrate that the value of realism is its ability to provide a reference for our findings, not to pass judgement on the efficacy of a study. The day may come when several million field plots, each fifty hectares in size, have had their diversity manipulated across every trophic level according to trait-based extinction scenarios and at densities observed in nature, and both microbial to macrobial species are manipulated, and run for a century or more. Such an experiment might be hailed as the ultimate realistic study, but it might not be the most efficient and economical way to do science, and so laboratory microcosm and in silico experiments would not lose their value.

21.3 Biodiversity, ecosystem functioning, and human wellbeing

The biodiversity → ecosystem functioning → ecosystem services → human wellbeing framework of the Millennium Assessment was a brilliant synthesis that united the natural and social environmental sciences by linking biodiversity, ecosystem processes, ecosystem functioning, and the services of ecosystems (see Introduction, Chapter 1). For the first time, it made it possible to see ecosystems as social assets whose value lies in the flow of social benefits (services) they yield. Although invaluable as a conceptual framework, however, it is too simplistic to serve as a guide for the development of practical biodiversity-based solutions to environmental problems. The biodiversity and ecosystem functioning research reviewed in this volume suggest two areas that need to be refined. First, the biodiversity → ecosystem functioning part of the Millennium Assessment conceptual framework needs to recognize the interdependency between the biotic and abiotic (the biological and geochemical) components of the system and its functioning. The value of ecosystems lies in their capacity to deliver services. Since the supporting services identified in the Millennium Assessment are just the processes that underpin ecosystem functioning, they are an integral part of the ecosystem as an asset – a functional unit. The supporting services accordingly need to be considered separately from regulating, provisioning, and cultural services (see Chapter 18 for a detailed treatment of ecosystem services). Second, while the conceptual framework provides a nice link between ecosystem services and human wellbeing, it does not reflect the critical importance of globalization – the closer integration of human society through trade and interactions among human populations. The interconnections between ecosystem services at different spatial and temporal scales turn out to be highly sensitive to the degree of globalization. A variant of the Millennium Assessment framework that reflects these concerns is illustrated in Fig. 21.2 and we explain these modifications below.

The ecosystem as a functional unit. The assets from which humans extract provisioning, cultural, and regulating services are functioning ecosystems. These comprise both abiotic and biotic components, and biogeochemical processes that underpin ecosystem functioning. In the absence of biological processes on Earth, geochemistry governs surface conditions as on any planet. The inclusion of biological organisms alters geochemical processes. In this modified framework, stocks are functioning ecosystems and flows are the services those systems yield. The elements within the system comprise the biotic and abiotic (atmospheric, lithospheric, and hydrospheric) pools of carbon, nutrients, and water, together with the plants, animals, and microorganisms that move carbon, nutrients, and water into and out of these biotic and abiotic pools.

If one eliminates the biota in the ecosystem, as we did in the thought exercise in the *Introduction*, the only fluxes in the pools of carbon, nutrients, and water would be those induced by geochemical processes. In a system with both biotic and abiotic elements, these fluxes are modified by biological processes. The resulting ecosystem processes are the basis for the flows of interest to human societies: the provisioning and cultural services, and their variability (the regulating services). We note that flows, in this sense, are not generally the same as fluxes in pools of carbon,

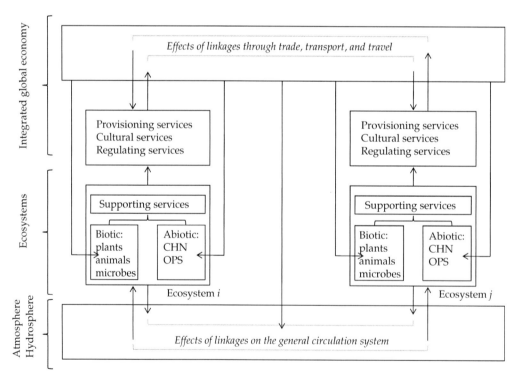

Figure 21.2 A coupled social–natural biodiversity and ecosystem functioning framework. Individual ecosystems worldwide, such as ecosystems *i* and *j* in this figure, are inextricably linked, both by market forces in the global economy and by biogeochemical fluxes through the atmosphere and hydrosphere. While more complex than the Millennium Assessment's framework, it eliminates ambiguities and facilitates integration of research, analyses, and policy development. See text for further explanation.

nutrient, or water. They are the benefit streams yielded by durable assets – functioning ecosystems in this case. It follows that the Millennium Assessment's supporting services are distinct from the other ecosystem services in that they are elements of the functioning ecosystems that yield the other ecosystem services as flows. By analogy, an automobile is a system whose components include a large number of parts organized so as to yield a flow of personal transport services. The value of a functional automobile is greater than the sum of its parts. The transport services it yields depend, *inter alia*, on the combustion processes occurring within the engine. Like the Millennium Assessment's supporting services, combustion processes help make the vehicle functional. In most managed ecosystems, the supporting services may be tailored to the production of specific services, either through direct modification of biogeochemical processes or indirectly through modifications of biodiversity, which influence biogeochemical processes. In Fig. 21.2 we therefore depict these Millennium Assessment 'services' as structuring elements of ecosystems.

Globalization and the closer integration of the biosphere. Humans impact their local biota through direct exploitation of local ecosystems, but they also impact biodiversity outside their geographic boundaries through the indirect impacts of trade, transport and travel (Chapter 17, but see also Kohn and Capen 2002, Perrings *et al.* 2002). For example, China's increasing demand for natural resources affects its own biodiversity both directly, through the exploitation of resources in China, and indirectly, through the effect of species introduced along with resources imported from other parts of the world. At the same time, imports to China affect biodiversity in the exporting countries through the indirect impact they have on, for example, rates of land conversion (and biodiversity loss) in those countries (Aide and Grau 2004), while exports from China to bioclimatically similar trading partners increase the risks that accompanying species will establish and spread in those countries (Costello *et al.* 2007). The same mechanisms operate for all trading countries.

Closer integration of world markets has another important impact on local biota. By increasing the number and accessibility of substitutes for particular ecosystem services, it reduces the cost of running down the associated assets – the ecosystems themselves. One manifestation of this is the 'roving bandit' phenomenon in the exploitation of open ocean fisheries, which has led to the sequential depletion of one fish stock after another (Berkes 2005, Worm *et al.* 2006). Another manifestation is the substitution between, for example, food sources. So reductions in West African fish supplies due to overharvesting have increased demand for bush meat as an alternative source of protein (Brashares *et al.* 2004). The role of the integrated economy in affecting local ecosystems is captured in the trade-mediated feedbacks between those systems in Fig. 21.2.

As the chapters in Part 3 all indicate, and especially Chapters 17–19, a more synthetic framework is needed if we are to move forward on finding biodiversity-based solutions to environmental problems. Brock *et al.* (Chapter 19) note: 'economists now understand that the gap between private and social optima depends on a complex set of feedbacks between the ecological and economic components of the coupled system'. Understanding the pathways and feedbacks that link biogeochemical (i.e. ecological) and social (i.e. economic) systems is critical to the development of practicable solutions to environmental problems that involve biodiversity change of one kind or another.

21.4 Implications for sustainable development

What is the main message of biodiversity and ecosystem functioning research? Can it be effectively and clearly communicated to the public and policy-makers? Is it likely to resonate with their perceptions of the environmental dimensions of a sustainable development strategy?

The main message from this volume, but particularly from Part 3, is that biodiversity conservation is an essential element in any strategy for sustainable development. In 1987, *Our Common Future*, also known as the Brundtland Report (World Commission on Environment and Development 1987), laid down a convincing argument that the benefits to humanity of the last century's economic development were tremendous, but that they were experienced largely by rich developed nations and came at the cost of severe depletion of the world's natural capital.

The goal of sustainable development currently enjoys an enormous subscription among policy-makers (e.g. World Commission on Environment and Development 1987, Reid 1989, Annan 2000, National Research Council 2000, Kates *et al.* 2001, Folke *et al.* 2002, Raven 2002, Sachs 2004). The largest international summits in human history, the UN Conference on Environment and Development, also known as the Earth Summit, held in Rio de Janeiro in 1992, and the 2002 World Summit on Sustainable Development in Johannesburg, were centred on the ideas of sustainable development. The UN Convention on Biological Diversity and the Millennium Assessment are also founded on the idea that biodiversity conservation is essential for achieving sustainable development. This has stimulated an intense effort to understand the scientific implications of the concept (Clark 1987, Kates 2001).

Communicating a complex message simply. The Brundtland Commission's call to abandon development by unsustainable spending of natural capital was essentially a call for biodiversity conservation in some measure to secure ecosystem services. It has been interpreted as a call to protect the value of natural capital, where natural capital comprises both the biotic and the abiotic components of the natural environment. In the wake of the Millennium Assessment we can think of the world's ecosystems as being amongst the most important elements of this. It thus includes biodiversity – the mix of plants, animals, microorganisms – and the ecosystem processes it supports. Most ecosystems have been shaped or at least impacted by human actions, but all still rely on a set of processes that are independent of human action. Man may have structured the system to promote or reduce particular processes, but the processes – along with the organisms and abiotic components they interact with – are properties of the natural world.

Alongside natural capital, it is conventional to identify at least two other forms of capital: produced or manmade capital and human capital (Fig. 21.3). Manmade capital, by contrast, involves assets that are produced, and that do not replicate nature – factories, roads, bridges, power plants, financial institutions and assets, etc. Human capital is the store of knowledge, culture, and social structure of people. Humans benefit from natural capital directly (e.g. natural resource extraction, such as lumber or fish harvesting) or indirectly (e.g. processing lumber in sawmills or transforming landscapes into agricultural systems). Fossil resources, such as petroleum, natural gas, and aquifers of fossil water, can supplement natural resource inputs, but renewal of these resources is so slow (tens of thousands to tens of millions of years) that they are best considered as non-renewable. Essentially, humanity controls the levers that open or close the flow of energy, nutrients, and water to either manmade or natural capital.

Sustainable development requires that the value of all three sets of assets is not declining over time. It allows for substitution between the different forms of capital, but respects the fact that there are not manmade substitutes for all forms of natural capital. It also respects the fact that ecosystems, like human technology and preferences, evolve over time. Hence sustainable development involves a strategy that builds the aggregate wealth of countries whilst allowing for their evolution. Biodiversity is critical in this for three main reasons.

First, a mix of species enhances the functioning of ecosystems and hence the value of those systems, regardless of the state of nature. That is, in any given state of nature, any positive diversity–functioning relationship that does not rely on sampling effects (i.e. that enhances the efficiency of resource exploitation through niche partitioning) implies complementarity between species. So too does any obligate or symbiotic relationship between species. The complementarity between species in this sense, like the complementarity between factors of production in economic systems, enhances the productivity and hence value of ecosystems.

Second, the redundancy of some species in functional groups provides insurance against changes in conditions that compromise the ability of other species in the same groups. In this sense, biodiversity is like a portfolio of assets. The value of the portfolio depends on both the range of conditions that is expected to occur, and the covariance in the performance of all assets in the portfolio over that range of conditions.

Third, and related to this last point, the evolutionary potential of the system is an increasing

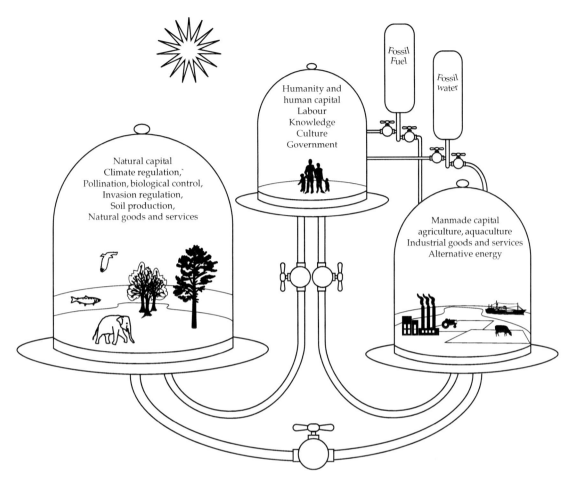

Figure 21.3 Human domination of the biosphere: a graphical device for communicating the importance of biodiversity and ecosystem services in environmental sustainability and sustainable development. This figure illustrates humanity and human capital supported by inputs from natural and manmade capital as well as current inputs from fossil fuel and water supplies. The series of pipes and valves illustrate how humanity's wellbeing and fate are controlled by the ways in which we balance the flows of nutrients, energy, and water among the different compartments. For example, if we allocate greater flows to manmade capital, natural capital shrinks, as do the services we derive from natural capital. See text for further explanation.

functioning of the gene combinations that enable species to exploit novel conditions. In economic terms, gene combinations have an option value – more particularly a quasi option value (the potential value of the yet to be uncovered information they offer).

Note that this does not mean that more biodiversity is always more valuable than less biodiversity. Indeed, simplification of ecosystems to enhance the productivity of one or more services has been the cornerstone of development in the past. The elimination of pests, predators, and pathogens has substantially enhanced human wellbeing in many cases. The problem identified in Chapters 17 and 18 is that neglect of the external effects of decisions to convert habitat to 'productive' uses, or to eliminating pests, competitors, predators, and pathogens to enhance productivity, may have undesirable and potentially unexpected consequences, for instance on other ecosystem services whose costs and benefits are externalized. Only by understanding the relationship between biodiversity, ecosystem functioning, and the production of ecosystem services is it possible to identify the degree to which biodiversity in any given system should be conserved.

21.5 Concluding comments

Hooper *et al.* (2005) listed five areas in need of expansion or resolution in biodiversity and ecosystem functioning research. Although published in 2005, the first outline of the consensus was produced in 2000, but even with this earlier date it remains remarkable that all five areas have been explored. These five areas were: (1) the relationship between taxonomic and functional diversity, (2) the importance of multiple trophic levels, (3) effects on temporal stability, (4) the relative influence of extrinsic factors versus biodiversity effects, and (5) the exploration of a wider array of ecosystems. Much, much more needs to be done, but expansion into these areas has strengthened the central message that biodiversity influences ecosystem functioning.

Current challenges to the field are multifold. Researchers must strive to: (1) incorporate greater realism into experimental approaches, (2) unify natural and social science methodology to address the full scope of the effects of diversity on human wellbeing, and (3) convey our findings to the non-scientific community, where environmental decisions are made and policy developed. These issues will dominate the field for the next phase of research into the effects of biodiversity on human wellbeing. The loss of biodiversity beyond levels that contribute to human wellbeing will decelerate only once the interactions between biodiversity, ecosystem functioning, ecosystem services, and human wellbeing are properly understood, since it is only then that the consequences of excessive rates of biodiversity loss will become apparent.

References

Aarssen, L. W. (1997) High productivity in grassland ecosystems: effected by species diversity or productive species? *Oikos*, **80**, 183–4.

Abhilasha, D., Quintana, N., Vivanco, J., and Joshi, J. (2008) Do allelopathic compounds in invasive *Solidago Canadensis S. l.* restrain the native European flora? *Journal of Ecology*, **96**, 993–1001.

Achtman, M. and Wagner, M. (2008) Microbial diversity and the genetic nature of microbial species. *Nature Reviews Microbiology*, **6**, 431–40.

Ackerly, D. D. (2004) Adaptation, niche conservatism, and convergence: comparative studies of leaf evolution in the California chaparral. *American Naturalist*, **163**, 654–71.

Ackerly, D. D. and Cornwell, W. K. (2007) A trait-based approach to community assembly: partitioning of species trait values into within- and among-community components. *Ecology Letters*, **10**, 135–45.

Ackerly, D. D., Knight, C. A., Weiss, S. B., Barton, K., and Starmer, K. P. (2002) Leaf size, specific leaf area and microhabitat distribution of chaparral woody plants: contrasting patterns in species level and community level analyses. *Oecologia*, **130**, 449–57.

Adger, W. N., Hughes, T. P., Folke, C., Carpenter, S. R., and Rockström, J. (2005) Social-ecological resilience to coastal disasters. *Science* **309**, 1036–9.

Aerts, R. (1995) The advantages of being evergreen. *Trends in Ecology & Evolution*, **10**, 402–7.

Aerts, R. (1997) Climate, leaf litter chemistry and leaf litter decomposition in terrestrial ecosystems: a triangular relationship. *Oikos*, **79**, 439–49.

Aerts, R. and Chapin, F. S. (2000) The mineral nutrition of wild plants revisited: a re-evaluation of processes and patterns. *Advances in Ecological Research*, **30**, 1–67.

Aguilar, R., Ashworth, L., Galetto, L., and Aizen, M. A. (2006) Plant reproductive susceptibility to habitat fragmentation: review and synthesis through a meta-analysis. *Ecology Letters*, **9**, 968–80.

Aide, T. M. and Grau, H. R. (2004) ECOLOGY: enhanced: globalization, migration, and Latin American ecosystems. *Science*, **305**, 1915–16.

Aigner, P. A. (2004) Ecological and genetic effects on demographic processes: pollination, clonality and seed production in *Dithyrea maritima*. *Biological Conservation*, **116**, 27–34.

Aizen, M. A., Garibaldi, L. A., Cunningham, S. A., and Klein, A. M. (2008a) Long-term global trends in crop yield and production reveal no current pollination shortage but increasing pollinator dependency. *Current Biology*, **18**, 1–4.

Aizen, M. A., Morales, C. L., and Morales, J. M. (2008b) Invasive mutualists erode native pollination webs. *PLoS Biol*, **6**(2), e31.

Albert, R. and Barabasi, A. L. (2002) Statistical mechanics of complex networks. *Reviews of Modern Physics*, **74**, 47–97.

Albrecht, M., Duelli, P., Müller, C. B., Kleijn, D., and Schmid, B. (2006) The Swiss agri-environment scheme enhances pollinator diversity and plant reproductive success in nearby intensively managed farmland. *Journal of Applied Ecology*, **76**, 1015–25.

Allen, B. P. and Loomis, J. B. (2006) Deriving values for the ecological support function of wildlife: an indirect valuation approach. *Ecological Economics*, **56**, 49–57.

Allen, M. F., Swenson, W., Querejeta, J. I., Egerton-Warburton, L. M., and Treseder, K. K. (2003) Ecology of mycorrhizae: a conceptual framework for complex interactions among plants and fungi. *Annual Review of Phytopathology*, **41**, 271–303.

Allison, G. (2004) The influence of species diversity and stress intensity on community resistance and resilience. *Ecological Monographs*, **74**, 117–34.

Allison, G. W. (1999) The implications of experimental design for biodiversity manipulations. *American Naturalist*, **153**, 26–45.

Alpert, P. (2006) The advantages and disadvantages of being introduced. *Biological Invasions*, **8**, 1523–34.

Altieri, M. (1999) The ecological role of biodiversity in agroecosystems. *Agriculture, Ecosystems, and Environment*, **74**, 19–31.

Altieri, M. (2004) Linking ecologists and traditional farmers in the search for sustainable agriculture. *Frontiers in Ecology and the Environment*, **2**, 35–42.

Amano, T. and Yamaura, Y. (2007) Ecological and life-history traits related to range contractions among breeding birds in Japan. *Biological Conservation*, **137**, 271–82.

Amarasekare, P. (2003) Competitive coexistence in spatially structured environments: a synthesis. *Ecology Letters*, **6**, 1109–22.

Anderson, C. B. and Rosemond, A. D. (2007) Ecosystem engineering by invasive exotic beavers reduces in-stream diversity and enhances ecosystem function in Cape Horn, Chile. *Oecologia*, **154**, 141–53.

Anderson, J. M. (1991) The effects of climate change on decomposition processes in grassland and coniferous forests. *Ecological Applications*, **1**, 326–47.

Anderson, J. E., Kriedemann, P. E., Austin, M. P., and Farquhar, G. D. (2000) Eucalypts forming a canopy functional type in dry sclerophyll forests respond differentially to environment. *Australian Journal of Botany*, **48**, 759–75.

Anderson, P. K., Cunningham, A. A., Patel, N. G., Morales, F. J., Epstein, P. R., and Daszak, P. (2004) Emerging infectious diseases of plants: pathogen pollution, climate change, and agrotechnology drivers. *Trends in Ecology and Evolution*, **119**, 535–44.

Ando, A. (1999) Waiting to be protected under the Endangered Species Act: The political economy of regulatory delay. *Journal of Law and Economics*, **42**, 29–60.

Andow, D. A. (1991) Vegetational diversity and arthropod population response. *Annual Review of Entomology*, **36**, 561–86.

Andrén, H. (1994) Effects of habitat fragmentation on birds and mammals in landscapes with different proportions of suitable habitat: a review. *Oikos*, **71**, 355–66.

Angermeier, P. L. (1995) Ecological attributes of extinction-prone species: loss of freshwater fishes of Virginia. *Conservation Biology*, **9**, 143–58.

Annan, K. A. (2000) *We, the Peoples: The Role of the United Nations in the 21st Century*. United Nations, New York.

Appanah, S. and Weinland, G. (1992) Will the management systems for hill Dipterocarp forests stand up? *Journal of Tropical Forest Science*, **3**, 140–58.

Armbrecht, I. and Perfecto, I. (2003) Litter-twig dwelling ant species richness and predation potential within a forest fragment and neighboring coffee plantations of contrasting habitat quality in Mexico. *Agriculture Ecosystems & Environment*, **97**, 107–15.

Armbrecht, I., Rivera, L., and Perfecto, I. (2005) Reduced diversity and complexity in the leaf-litter ant assemblage of Colombian coffee plantations. *Conservation Biology*, **19**, 897–907.

Armington, P. (1969) A theory of demand for products distinguished by place of production. *IMF Staff Papers*, **16**, 158–76.

Armstrong, R. A. (1976) Fugitive species – experiments with fungi and some theoretical considerations. *Ecology*, **57**, 953–63.

Armsworth, P. R. and Roughgarden, J. (2003) The economic value of ecological stability. *Proceedings of the National Academy of Sciences of the USA*, **100**, 7147–51.

Aronson, J. and Van Andel, J. (2005) Challenges for ecological theory. In Van Andel, J. and Aronson, J. (eds.) *Restoration Ecology: The New Frontier*. Blackwell Publishing, Malden, MA.

Aronson, M. F. J., Handel, S. N., and Clemants, S. E. (2007) Fruit type, life form and origin determine the success of woody plant invaders in an urban landscape. *Biological Invasions*, **9**, 465–75.

Arrow, K., Bolin, B., Costanza, R., et al. (1995) Economic growth, carrying capacity and the environment. *Science*, **268**, 520–1.

Ashman, T. L., Knight, T. M., Steets, J. A., Amarasekare, P., Burd, M., Campbell, D. R., et al. (2004) Pollen limitation of plant reproduction: ecological and evolutionary causes and consequences. *Ecology*, **85**, 2408–21.

Austin, A. T. (2002) Differential effects of precipitation on production and decomposition along a rainfall gradient in Hawaii. *Ecology*, **83**, 328–38.

Austin, M. P. (1999) The potential contribution of vegetation ecology to biodiversity research. *Ecography*, **22**, 465–84.

Aylward, B. Allen, K., Echeverria, J., and Tosi, J. (1996) Sustainable ecotourism in Costa Rica: The Monte Verde Cloud Forest Preserve. *Biodiversity and Conservation*, **5**(3), 315–43.

Backhed, F., Ding, H., Wang, T., Hooper, L. V., L. V. Koh, L. V., Nagy, A., Semenkovich, C. F., and Gordon, J. I. (2004) The gut microbiota as an environmental factor that regulates fat storage. *Proceedings of the National Academy of Sciences of the USA*, **101**, 15718–23.

Bacompte, J., Jordano, P., and Olesen, J. M. (2006) Asymmetric coevolutionary networks facilitate biodiversity maintenance. *Science*, **312**, 431–3.

Bady, P., Doledec, S., Fesl, C., Gayraud, S., Bacchi, M., and Scholl, F. (2005) Use of invertebrate traits for the biomonitoring of european large rivers: the effects of sampling effort on genus richness and functional diversity. *Freshwater Biology*, **50**, 159–73.

Bai, Y., Xingguo, H., Jianguo, W., Zuozhong, C., and Linghao, L. (2004) Ecosystem stability and compensatory effects inthe Inner Mongolia grassland. *Nature*, **431**, 181–4.

Bailey, P. (2007) On the trail of a killer. *UC Davis Magazine*, **24**.

Bais, H. P., Vepachedu, R., Gilroy, S., Callaway, R. M., and Vivanco, J. M. (2003) Allelopathy and exotic plant invasion: from molecules and genes to species interactions. *Science*, **301**, 1377–80.

Balmford, A., Bruner, A., Cooper, P., *et al.* (2002) Economic reasons for conserving wild nature. *Science*, **297**, 950–3.

Balvanera, P., Kremen, C., and Martinez-Ramos, M. (2005) Applying community structure analysis to ecosystem function: examples from pollination and carbon storage. *Ecological Applications*, **15**, 360–75.

Balvanera, P., Pfisterer, A. B., Buchmann, N., *et al.* (2006) Quantifying the evidence for biodiversity effects on ecosystem functioning and services. *Ecology Letters*, **9**, 1146–56.

Barbarika, A. (2005) *FY 2005 Annual Crop Contract Summary*. Natural Resources Analysis Group, Economic and Policy Analysis Staff, Farm Service Agency. USDA.

Barber, C. B., Dobkin, D. P., and Huhdanpaa, H. (1996) The quickhull algorithm for convex hulls. *ACM Transactions on Mathematical Software*, **22**, 469–83.

Barbier, E. B. (1992) Community-based development in Africa. In T. M. Swanson and E. Barbier (eds.) *Economics for the Wilds: Wildlife, Wildlands, Diversity, and Development*, pp. 103–35. Earthscan, London.

Barbier, E. B. (1994) Valuing environmental functions, tropical wetlands. *Land Economics*, **70**, 155–73.

Barbier, E. B. (1997) Introduction to the environmental Kuznets curve special issue. *Environment and Development Economics*, **2**(4), 369–82.

Barbier, E. B. (2000) Valuing the environment as input: review of applications to mangrove–fishery linkages. *Ecological Economics*, **35**, 47–61.

Barbier, E. B. (2007) Valuing ecosystem services as productive inputs. *Economic Policy*, **22**(49), 177–229.

Barbier, E. B. and Aylward, B. A. (1996) Capturing the pharmaceutical value of biodiversity in a developing country. *Environmental and Resource Economics*, **8**(2), 157–91.

Barbier, E. B. and Schulz, C. (1997) Wildlife, biodiversity and trade. *Environment and Development Economics*, **2**, 145–72.

Barbier, E. B. and Swanson, T. (1990) Ivory: the case against the ban. *New Scientist*, **1743**, 52–4.

Barbier, E. B., Burgess, J. C., and Folke, C. (1994) *Paradise Lost? The Ecological Economics of Biodiversity*. Earthscan, London.

Bardgett, R. D. and Shine, A. (1999) Linkages between plant litter diversity, soil microbial biomass and ecosystem function in temperate grasslands. *Soil Biology and Biochemistry*, **31**, 317–21.

Bardgett, R. D., Freeman, C., and Ostle, N. J. (2008) Microbial contributions to climate change through carbon cycle feedbacks. *The ISME Journal*, **2**(8), 805–14.

Bardgett, R. D., Smith, R. S., Shiel, R. S., *et al.* (2006) Parasitic plants indirectly regulate below-ground properties in grassland ecosystems. *Nature*, **439**, 969–72.

Bärlocher, F. and Corkum, M. (2003) Nutrient enrichment overwhelms diversity effects in leaf decomposition by stream fungi. *Oikos*, **101**, 247–52.

Bärlocher, F. and Graca, M. A. S. (2002) Exotic riparian vegetation lowers fungal diversity but not leaf decomposition in Portuguese streams. *Freshwater Biology*, **47**, 1123–35.

Barnett, A. and Beisner, B. E. (2007) Zooplankton biodiversity and lake trophic state: explanations invoking resource abundance and distribution. *Ecology*, **88**, 1675–86.

Barnett, A. J., Finlay, K., and Beisner, B. E. (2007) Functional diversity of crustacean zooplankton communities: towards a trait-based classification. *Freshwater Biology*, **52**, 796–813.

Barrett, C. B. and Lybbert, T. J. (2000) Is bioprospecting a viable strategy for conserving tropical ecosystems? *Ecological Economics*, **34**, 293–300.

Barrett, S. (1994) The biodiversity supergame. *Environmental and Resource Economics*, **4**(1), 111–22.

Barrett, S. (2000) Trade and the environment: local versus multilateral reforms. *Environment and Development Economics*, **5**(4), 349–60.

Barrett, S. (2003) *Environment and Statecraft*. Oxford University Press, Oxford.

Bascompte, J., Jordano, P., Meliàn, C. J., and Olesen, J. M. (2003) The nested assembly of plant–animal mutualistic networks. *Proceedings of the National Academy of Science USA*, **100**, 9383–7.

Battese, G. and Coelli, T. (1995) A model for technical inefficiency effects in a stochastic frontier production function for panel data. *Empirical Economics*, **20**, 325–32.

Baumgärtner, S. (2006) *Natural Science Constraints in Environmental and Resource Economics*. University of Heidelberg Publications Online, Heidelberg (http://www.ub.uni-heidelberg.de/archiv/6593).

Baumgärtner, S. (2007) The insurance value of biodiversity in the provision of ecosystem services. *Natural Resource Modeling*, **20**, 87–127.

Baumgärtner, S. and Quaas, M. F. (2006) The private and public insurance value of conservative biodiversity management. Manuscript (available at http://ssrn.com/abstract=892101).

Baumgärtner, S. and Quaas, M. F. (2008), Agro-biodiversity as natural insurance and the development of financial insurance markets. In A. Kontoleon, U. Pascual and M. Smale (eds.), *Agrobiodiversity and Economic Development*, Routledge.

Baumgärtner, S., Becker, C., Manstetten, R., and Faber, M. (2006) Relative and absolute scarcity of nature: assessing the roles of economics and ecology for biodiversity conservation. *Ecological Economics*, **59**, 487–98.

Baur, B., Cremene, C., Groza, G., et al. (2006) Effects of abandonment of subalpine hay meadows on plant and invertebrate diversity in Transylvania, Romania. *Biological Conservation*, **132**, 261–73.

Bawa, K. S. (1990) Plant–pollinator interactions in tropical rain forests. *Annual Review of Ecology and Systematics*, **21**, 299–422.

Bawa, K., Joseph, G., and Setty, S. (2007) Poverty, biodiversity, and institutions in forest–agriculture ecotones in the Western Ghats and Eastern Himalaya ranges of India. *Agriculture, Ecosystems, and Environment*, **121**, 287–95.

Bean, J. M. (1991) Looking back over the first fifteen years. In K. A. Kohm (ed.) *Balancing on the Brink of Extinction: Endangered Species Act and Lessons for the Future*, pp. 37–42. Island Press. Washington, DC.

Beatty, C. D., Beirinckx, K., and Sherratt, T. N. (2004) The evolution of Mullerian mimicry in multispecies communities. *Nature*, **431**, 63–7.

Beckage, B. and Gross, L. J. (2006) Overyielding and species diversity: what should we expect? *New Phytologist*, **172**, 140–8.

Begon, M. (2008) Effects of host diversity on disease dynamics. In R. S. Ostfeld, F. Keesing, and V. T. Eviner (eds.) *Infectious Disease Ecology: Effects of Ecosystems on Disease and of Disease on Ecosystems*, pp. 12–29. Princeton University Press, Princeton, NJ.

Begon, M., Townsend, C. A., and Harper, J. L. (2006) *Ecology: From Individuals to Ecosystems*, 4th edn, Wiley-Blackwell, Oxford.

Beissinger, S. R. and Perrine, J. D. (2001) Extinction, recovery, and the Endangered Species Act. In J. Shogren and J. Tschirhart (eds.) *Protecting Endangered Species in the United States*, pp. 51–71. Cambridge University Press, New York.

Belgrano, A., Scharler, U. M., Dunne, J., and Ulanowicz, R. E. (2005) *Aquatic Food Webs: an Ecosystem Approach*. Oxford University Press, New York.

Bell, G. (1990) The ecology and genetics of fitness in Chlamydomonas, II. The properties of mixtures of strains. *Proceedings of the Royal Society B: Biological Sciences*, **240**, 323–50.

Bell, G. (1991) The ecology and genetics of fitness in Chlamydomonas, IV. The properties of mixtures of genotypes of the same species. *Evolution*, **45**, 1036–46.

Bell, T., Newman, J. A., Lilley, A. K., and van der Gast, C. (2005a) Bacteria and island biogeography. *Science*, **309**, 1997–8.

Bell, T., Newman, J. A., Silverman, B. W., Turner, S. L., and Lilley, A. K. (2005b) The contribution of species richness and composition to bacterial services. *Nature*, **436**, 1157–60.

Bellwood, D. R., Hoey, A. S., and Choat, J. H. (2003) Limited functional redundancy in high diversity systems: resilience and ecosystem function on coral reefs. *Ecology Letters*, **6**, 281–5.

Bellwood, D. R., Hughes, T. P., Folke, C., and Nystrom, M. (2004) Confronting the coral reef crisis. *Nature*, **429**, 827–33.

Bellwood, D. R., Hughes, T. P., and Hoey, A. S. (2006) Sleeping functional group drives coral-reef recovery. *Current Biology*, **16**, 2434–9.

Benedetti-Cecchi, L. (2005) Unanticipated impacts of spatial variance of biodiversity on plant productivity. *Ecology Letters*, **8**, 791–9.

Benton, T. G., Solan, M., Travis, J. M. J. et al. (2007) Microcosm experiments can inform global ecological problems. *Trends in Ecology and Evolution*, **22**, 516–21.

Berg, M. P., Stoffer, M., and van den Heuvel, H. H. (2004) Feeding guilds in *collembola* based on digestive enzymes. *Pedobiologia*, **48**, 589–601.

Berish, C. W. and Ewel, J. J. (1988) Root development in simple and complex tropical successional ecosystems. *Plant and Soil*, **106**, 73–84.

Berkes, F., Hughes, T. P., Steneck, R. S., Wilson, J. A., Bellwood, D. R., Crona, B., Folke, C., Gunderson, L. H., Leslie, H. M., Norberg, J., Nyström, M, Olsson, P., Österblom, H., Scheffer, M., and Worm, B. (2005) Globalization, roving bandits, and marine resources. *Science*, **311**, 1557–8.

Berlow, E. L., Navarrete, S. A., Briggs, C. J., Power, M. E. and Menge, B. A. (1999) Quantifying variation in the strengths of species interactions. *Ecology*, **80**, 2206–24.

Berlow, E. L., Neutel, A. M., Cohen, J. E., de Ruiter, P. C., Ebenman, B., Emmerson, M., Fox, J. W., Jansen, V. A. A., Jones, J. I., Kokkoris, G. D., Logofet, D. O., McKane, A. J., Montoya, J. M., and Petchey, O. (2004) Interaction strengths in food webs: issues and opportunities. *Journal of Animal Ecology*, **73**, 585–98.

Bernhardt, E. S., Palmer, M. A., Allan, J. D., et al. (2005) Ecology – synthesizing US river restoration efforts. *Science*, **308**, 636–7.

Bernhardt, E. S., Sudduth, E. B., Palmer, M. A., et al. (2007) Restoring rivers one reach at a time: Results from a survey of U. S. River restoration practitioners. *Restoration Ecology*, **15**, 482–93.

Betts, R. A., Malhi, Y., and Roberts, J. T. (2008) Review. The future of the Amazon: new perspectives from climate, ecosystem and social sciences. *Philosophical Transactions of the Royal Society B: Biological Sciences*, **363**, 1729–35.

Bezemer, T. M., De Deyn, G. B., Bossinga, T. M., van Dam, N. M., Harvey, J. A., and Van der Putten, W. H. (2005) Soil community composition drives aboveground plant–herbivore–parasitoid interactions. *Ecology Letters*, **8**, 652–61.

Bezemer, T. M., Harvey, J. A., Kowalchuk, G. A., Korpershoek, H., and van der Putten, W. H. (2006). Interplay between *Senecio jacobaea* and plant, soil, and aboveground insect community composition. *Ecology*, **87**, 2002–13.

Bhagwati, J. (2000) On thinking clearly about the linkage between trade and the environment. *Environment and Development Economics*, **5**, 483–529.

Bianchi, F., Booij, C. J. H., and Tscharntke, T. (2006) Sustainable pest regulation in agricultural landscapes: a review on landscape composition, biodiversity and natural pest control. *Proceedings of the Royal Society B: Biological Sciences*, **273**, 1715–27.

Biesmeijer, J. C., Roberts, S. P. M., Reemer, M., *et al.* (2006) Parallel declines in pollinators and insect-pollinated plants in Britain and the Netherlands. *Science*, **313**, 351–4.

Biles, C. L., Solan, M., Isaksson, I., *et al.* (2003) Flow modifies the effect of biodiversity on ecosystem functioning: an *in situ* study of estuarine sediments. *Journal of Experimental Marine Biology and Ecology*, **285**, 165–77.

Birol, E., Kontoleon, A., and Smale, M. (2005) Farmer demand for agricultural biodiversity in Hungary's transition economy: A choice experiment approach. In Smale, M. (ed.) *Valuing Crop Genetic Biodiversity on Farms during Economic Change*, pp. 32–47. CAB International, Wallingford, UK.

Bishop, J. A. and Armbruster, W. S. (1999) Thermoregulatory abilities of Alaskan bees: effects of size, phylogeny and ecology. *Functional Ecology*, **13**, 711–24.

Bjelke, U. and Herrmann, J. (2005) Processing of two detritus types by lake-dwelling shredders: species-specific impacts and effects of species richness. *Journal of Animal Ecology*, **74**, 92–8.

Björklund, J., Limburg, K. E., and Rydberg, T. (1999) Impact of production intensity on the ability of the agricultural landscape to generate ecosystem services: an example from Sweden. *Ecological Economics*, **29**, 269–91.

Blackburn, T. M., Cassey, P., Duncan, R. P., Evans, K. L., and Gaston, K. J. (2004) Avian extinction and mammalian introductions on oceanic islands. *Science*, **305**, 1955–8.

Blackburn, T. M., Petchey, O. L., Cassey, P., and Gaston, K. J. (2005) Functional diversity of mammalian predators and extinction in island birds. *Ecology*, **86**, 2916–23.

Blanford, S., Thomas, M. B., Pugh, C., and Pell, J. K. (2003) Temperature checks the Red Queen? Resistance and virulence in a fluctuating environment. *Ecology Letters*, **6**, 2–5.

Blondel, J. (2003) Guilds or functional groups: does it matter? *Oikos*, **100**, 223–31.

Blüthgen, N., Menzel, F., Hovestadt, T., Fiala, B., and Blüthgen, N. (2007) Specialization, constraints, and conflicting interests in mutualistic networks. *Current Biology*, **17**, 1–6.

Bodmer, R. E., Eisenberg, J. F., and Redford, K. H. (1997) Hunting and the likelihood of extinction of Amazonian mammals. *Conservation Biology*, **11**, 460–6.

Bohart, G. E. (1947) Wild bees in relation to alfalfa pollination. *Farm and Home Science*, **8**, 13–14.

Bolker, B. M., Pacala, S. W., Bazzaz, F. A., Canham, C. D., and Levin, S. A. (1995) Species-diversity and ecosystem response to carbon-dioxide fertilization – conclusions from a temperate forest model. *Global Change Biology*, **1**, 373–81.

Bolund, P. and Hunhammar, S. (1999) Ecosystem services in urban areas. *Ecological Economics*, **29**, 293–301.

Bond, E. M. and Chase, J. M. (2002) Biodiversity and ecosystem functioning at local and regional spatial scales. *Ecology Letters*, **5**, 467–70.

Bond, W. J. and Midgley, G. F. (2000) A proposed CO_2-controlled mechanism of woody plant invasion in grasslands and savannas. *Global Change Biology*, **6**, 865–9.

Bonkowski, M. and Roy, J. (2005) Soil microbial diversity and soil functioning affect competition among grasses in experimental microcosms. *Oecologia*, **143**, 232–40.

Bonkowski, M., Geoghegan, I. E., Birch, A. N. E., and Griffiths, B. S. (2001) Effects of soil decomposer invertebrates (protozoa and earthworms) on an aboveground phytophagous insect (cereal aphid) mediated through changes in the host plant. *Oikos*, **95**, 441.

Bonnie, R., Carey, M., and Petsonk, A. (2002) Protecting terrestrial ecosystems and the climate through a global carbon market. *Philosophical Transactions of the Royal Society of London Series A: Mathematical Physical and Engineering Sciences*, **360**, 1853–73.

Bonsall, M. B. and Hassell, M. P. (2007) Predator–prey interactions. In R. M. May and A. McLean (eds.) *Theoretical Ecology: Principles and Applications*, 3rd edn, pp. 46–61. Oxford University Press, Oxford.

Booth, R. E. and Grime, J. P. (2003) Effects of genetic impoverishment on plant community diversity. *Journal of Ecology*, **91**, 721–30.

Borer, E. T., Seabloom, E. W., Shurin, J. B., Anderson, K. E., Blanchette, C. A., Broitman, B., Cooper, S. D., and Halpern, B. S. (2005) What determines the strength of a trophic cascade? *Ecology*, **86**, 528–37.

Borrvall, C. and Ebenman, B. (2006) Early onset of secondary extinctions in ecological communities following the loss of top predators. *Ecology Letters*, **9**, 435–42.

Borrvall, C., Ebenman, B., and Jonsson, T. (2000) Biodiversity lessens the risk of cascading extinction in model food webs. *Ecology Letters*, **3**, 131–6.

Bos, M. M., Veddeler, D., Bogdanski, A. K., *et al.* (2007) Caveats to quantifying ecosystem services: fruit abortion blurs benefits from crop pollination. *Ecological Applications*, **17**, 1841–9.

Botta-Dukat, Z. (2005) Rao's quadratic entropy as a measure of functional diversity based on multiple traits. *Journal of Vegetation Science*, **16**, 533–40.

Boughman, J. W. (2001) Divergent sexual selection enhances reproductive isolation in sticklebacks. *Nature*, **411**, 944–8.

Boyd, J. and Simpson, R. D. (1999) Economics and biodiversity conservation options: an argument for continued experimentation and measured expectations. *The Science of the Total Environment*, **240**, 91–105.

Boyer, A. G. (2008) Extinction patterns in the avifauna of the Hawaiian islands. *Diversity and Distributions*, **14**, 509–17.

Bracken, M. E. S. and Stachowitz, J. J. (2006) Seaweed diversity enhances nitrogen uptake via complementary use of nitrate and ammonium. *Ecology*, **87**, 2397–404.

Bracken, M. E., Friberg, S. E., Gonzales-Dorantes, C. A., and Williams, S. L. (2008) Functional consequences of realistic biodiversity changes in a marine ecosystem. *Proceedings of the National Academy of Sciences of the USA*, **105**, 924–8.

Bradford, M. A., Tordoff, G. M., Eggers, T., Jones, T. H., and Newington, J. E. (2002) Microbiota, fauna, and mesh size interactions in litter decomposition. *Oikos*, **99**, 317–23.

Bradley, B. A., Houghton, R. A., Mustard, J. F. and Hamburg, S. P. (2006) Invasive grass reduces aboveground carbon stocks in shrublands of the western US. *Global Change Biology*, **12**, 1815–22.

Bradshaw, A. D. (1987) Restoration: an acid test for ecology. In W. R. Jordan, M. E. Gilpin, and J. D. Aber (eds.) *Restoration Ecology*. Cambridge University Press, Cambridge.

Bradshaw, A. D. (1997) The importance of soil ecology in restoration science. In K. M. Urbanska, N. R. Webb, and P. J. Edwards (eds.) *Restoration Ecology and Sustainable Development*. Cambridge University Press, New York.

Bradshaw, A. D. (2000) The use of natural processes in reclamation – advantages and difficulties. *Landscape and Urban Planning*, **51**, 89–100.

Bradshaw, A. D. and Chadwick, M. J. (1980) *The Restoration of Land: the Ecology and Reclamation of Derelict Land and Degraded Land*. Blackwell Scientific Publications, Oxford.

Brander, J. A. and Taylor, M. S. (1997) International trade and open-access renewable resources: the small open economy case. *Canadian Journal of Economics*, **30**, 526–52.

Brander, J. A. and Taylor, M. S. (1998) Open access renewable resources: trade and trade policy in a two-country model. *Journal of International Economics*, **44**, 181–209.

Brandle, J. R., Johnson, B. B., and Akeson, T. (1992) Field windbreaks: are they economical? *Journal of Production Agriculture*, **5**, 393–8.

Brandle, M., Ohlschlager, S., and Brandl, R. (2002) Range sizes in butterflies: correlation across scales. *Evolutionary Ecology Research*, **4**, 993–1004.

Brashares, J. S. (2005) Ecological, behavioral, and life-history correlates of mammal extinctions in West Africa. *Conservation Biology*, **17**, 733–43.

Brashares, J. S., Arcese, P., Sam, M. K., Coppolillo, P. B., Sinclair, A. R. E., and Balmford, A. (2004) Bushmeat hunting, wildlife declines, and fish supply in West Africa. *Science*, **306**, 1180–3.

Bret-Harte, M. S., Garcia, E. A., Sacre, V. M., et al. (2004) Plant and soil responses to neighbour removal and fertilization in Alaskan tussock tundra. *Journal of Ecology*, **92**, 635–47.

Briggs, C. J. and Borer, E. T. (2005) Why short-term experiments may not allow long-term predictions about intraguild predation. *Ecological Applications*, **15**, 1111–17.

Briske, D. D., Fuhlendorf, S. D., and Smeins, F. E. (2006) A unified framework for assessment and application of ecological thresholds. *Rangeland Ecology and Management*, **59**, 225–36.

Brisson, D., Dykhuizen, D. E. and Ostfeld, R. S. (2008) Conspicuous impacts of inconspicuous hosts on the human Lyme disease epidemic. *Proceedings of the Royal Society of London B: Biological Sciences*, **275**, 227–35.

British Columbia Ministry of Agriculture and Food (1998) *Planning for profit: alfalfa seed, Peace River, Spring 1998*, published online at http://www.agf.gov.bc.ca/busmgmt/budgets/budget_pdf/grain_oilseed/alfalf98.pdf

Brock, W. A. and Xepapadeas, A. (2002) Biodiversity management under uncertainty: species selection and harvesting rules. In B. Kristrom, P. Dasgupta, and K. Lofgren (eds.) *Economic Theory for the Environment: Essays in Honour of Karl-Goran Maler*, pp. 62–97. Edward Elgar, Cheltenham.

Brock, W. A. and Xepapadeas, A. (2004) Optimal management when species compete for limited resources. *Journal of Environmental Economics and Management*, **44**, 189–220.

Brockhurst, M. A., Buckling, A., and Gardner, A. (2007) Cooperation peaks at intermediate disturbance. *Current Biology*, **17**, 761–5.

Brockhurst, M. A., Hochberg, M. E., Bell, T., and Buckling, A. (2006) Character displacement promotes cooperation in bacterial biofilms. *Current Biology*, **16**, 2030–4.

Brook, B. W., Sodhi, N. S., and Ng, P. K. L. (2003) Catastrophic extinctions follow deforestation in Singapore. *Nature*, **424**, 420–3.

Brook, B. W., Sodhi, N. S., and Bradshaw, C. J. A. (2008) Synergies among extinction drivers under global change. *Trends in Ecology & Evolution*, **23**, 453–60.

Brooks, T. M., Pimm, S. L., and Oyugi, J. O. (1999) Time lag between deforestation and bird extinction in tropical forest fragments. *Conservation Biology*, **13**, 1140–50.

Brose, U., Jonsson, T., Berlow, E. L., Warren, P., Banasek-Richter, C., Bersier, L. F., Blanchard, J. L., Brey, T., Carpenter, S. R., Blandenier, M. F. C., Cushing, L., Dawah, H. A., Dell, T., Edwards, F., Harper-Smith, S., Jacob, U., Ledger, M. E., Martinez, N. D., Memmott, J., Mintenbeck, K., Pinnegar, J. K., Rall, B. C., Rayner, T. S., Reuman, D. C., Ruess, L., Ulrich, W., Williams, R. J., Woodward, G., and Cohen, J. E. (2006) Consumer–resource body-size relationships in natural food webs. *Ecology*, **87**, 2411–17.

Brown, A. W. A. (1977) Yellow fever, dengue and dengue haemorrhagic fever. In G. M. Howe (ed.) *A World Geography of Human Diseases*, pp. 271–317. Academic Press, London.

Brown, B. J. and Mitchell, R. J. (2001) Competition for pollination: effects of pollen of an invasive plant on seed set of a native congener. *Oecologia*, **129**, 43–9.

Brown, G. M. and Hammack, J. (1974) *Waterfowl and Wetlands, Toward Bioeconomic Analysis*. Johns Hopkins Press, Baltimore, MD.

Brown, J. H. (1995) *Macroecology*. University of Chicago Press, Chicago.

Brown, J. H. and Kodricbrown, A. (1977) Turnover rates in insular biogeography – effect of immigration on extinction. *Ecology*, **58**, 445–9.

Brown, R. L. and Peet, R. K. (2003) Diversity and invasibility of southern Appalachian plant communities. *Ecology*, **84**, 32–9.

Brown, J. H., Ernest, S. K. M., Parody, J. M., and Haskell, J. P. (2001) Regulation of diversity: maintenance of species richness in changing environments. *Oecologia*, **126**, 321–2.

Brown, J. H., Gillooly, J. F., Allen, A. P., Savage, V. M., and West, G. B. (2004) Toward a metabolic theory of ecology. *Ecology*, **85**, 1771–89.

Bruno, J. F. and Cardinale, B. J. (2008) Cascading effects of predator richness. *Frontiers in Ecology and the Environment*, **6**(10), 539–46.

Bruno, J. F. and O'Connor, M. I. (2005) Cascading effects of predator diversity and omnivory in a marine food web. *Ecology Letters*, **8**, 1048–56.

Bruno, J. F., Boyer, K. E., Duffy, J. E., *et al.* (2005a) Effects of macroalgal species identity and richness on primary production in benthic marine communities. *Ecology Letters*, **8**, 1165–74.

Bruno, J. F., Fridley, J. D., Bromberg, K. D., and Bertness, M. D. (2005b) Insights into biotic interactions from studies of species invasions. In D. F. Sax, J. J. Stachowicz, and S. D. Gaines (eds.) *Species Invasions: Insights into Ecology, Evolution, and Biogeography*, pp. 13–40. Sinauer Associates, Inc., Sunderland, MA.

Bruno, J. F., Lee, S. C., Kertesz, J. S., *et al.* (2006) Partitioning the effects of algal species identity and richness on benthic marine primary production. *Oikos*, **115**, 170–8.

Brunsdon, C. and Willis, K. G. (2002) Meta-analysis: a Bayesian perspective. In R. J. G. M. Florax, P. Nijkamp, and K. G. Willis (eds.) *Comparative Environmental Economic Assessment*. Edward Elgar, Cheltenham.

Brush, S. B. (2004) *Farmers' Bounty: Locating Crop Diversity in the Contemporary World*. Yale University Press, New Haven, CT.

Buchmann, N. and Schulze, E. (1999) Net CO_2 and H_2O fluxes of terrestrial ecosystems. *Global Biochemical Cycles*, **13**, 751–60.

Buckling, A., Kassen, R., Bell, G., and Rainey, P. B. (2000) Disturbance and diversity in experimental microcosms. *Nature*, **408**, 961–4.

Bulling, M. T., White, P. C. L., Raffaelli, D., *et al.* (2006) Using model systems to address the biodiversity–ecosystem functioning process. *Marine Ecology Progress Series*, **311**, 295–309.

Bullock, J. M., Pywell, R. F., Burke, M. J. W., *et al.* (2001) Restoration of biodiversity enhances agricultural production. *Ecology Letters*, **4**, 185–9.

Bullock, J. M., Pywell, R. F., Coulson, S. J., Nolan, A. M., and Caswell, H. (2002) Plant dispersal and colonisation processes at local and landscape scales. In J. M. Bullock, R. E. Kenward, and R. Hails (eds.) *Dispersal Ecology*. Blackwell Science, Oxford.

Bullock, J. M., Pywell, R. F., and Walker, K. J. (2007) Long-term enhancement of agricultural production by restoration of biodiversity. *Journal of Applied Ecology*, **44**, 6–12.

Bunker, D. E. (2004) *The Application of Competition Theory to Invaders and Biological Control: a Test Case with Purple Loosestrife (Lythrum salicaria), Broad-Leaved Cattail (Typha latifolia), and a Leaf-Feeding Beetle (Galerucella calmariensis)*. University of Pittsburgh, Pittsburgh.

Bunker, D. E., DeClerck, F., Bradford, J. C., *et al.* (2005) Species loss and above-ground carbon storage in a tropical forest. *Science*, **310**, 1029–31.

Burke, M. J. W. and Grime, J. P. (1996) An experimental study of plant community invasibility. *Ecology*, **77**, 776–90.

Burns, K. C. (2004) Scale and macroecological patterns in seed dispersal mutualisms. *Global Ecology and Biogeography*, **13**, 289–93.

Butler, J. L., Parker, M. S., and Murphy, J. T. (2006) Efficacy of flea beetle control of leafy spurge in Montana and South Dakota. *Rangeland Ecology & Management*, **59**, 453–61.

Butler, S. J., Vickery, J. A., and Norris, K. (2007) Farmland biodiversity and the footprint of agriculture. *Science*, **315**, 381–4.

Byrnes, J., Stachowicz, J. J., Hultgren, K. M., Hughes, A. R., Olyarnik, S. V., and Thornber, C. S. (2006) Predator diversity strengthens trophic cascades in kelp forests by modifying herbivore behaviour. *Ecology Letters*, **9**, 61–71.

Cadotte, M. W. and Fukami, T. (2005) Dispersal, spatial scale, and species diversity in a hierarchically structured experimental landscape. *Ecology Letters*, **8**, 548–57.

Caldeira, M. C., Hector, A., Loreau, M., and Pereira, J. S. (2005) Species richness, temporal variability and resistance of biomass production in a Mediterranean grassland. *Oikos*, **110**, 115–23.

Callaway, J. C., Sullivan, G., and Zedler, J. B. (2003) Species-rich plantings increase biomass and nitrogen accumulation in a wetland restoration experiment. *Ecological Applications*, **13**, 1626–39.

Callaway, R. M. and Ridenour, W. M. (2004) Novel weapons: invasive success and the evolution of increased competitive ability. *Frontiers in Ecology and the Environment*, **2**, 436–43.

Callaway, R. M., Thelen, G. C., Rodriguez, A., and Holben, W. E. (2004) Soil biota and exotic plant invasion. *Nature*, **427**, 731–3.

Callaway, R. M., Cipollini, D., Barto, K., Thelen, G. C., Hallett, S. G., Prati, D., Stinson, K., and Klironomos, J. (2008) Novel weapons: invasive plant suppresses fungal mutualists in America but not in its native Europe. *Ecology*, **89**, 1043–55.

Callenbach, E. (2000) *Bring Back the Buffalo! A Sustainable Future for America's Great Plains*. Island Press, Berkeley, CA.

Cameron, C. A. and Trivedi, P. (1998) *Regression Analysis of Count Data*. Cambridge University Press, Cambridge.

Cameron, T. (2002) 2002: the year of the diversity–ecosystem function debate. *Trends in Ecology and Evolution*, **17**, 495–6.

Camill, P., Mckone, M. J., Sturges, S. T., *et al.* (2004) Community- and ecosystem-level changes in a species-rich tallgrass prairie restoration. *Ecological Applications*, **14**, 1680–94.

Canadell, J. G. and Raupach, M. R. (2008) Managing forests for climate change mitigation. *Science*, **320**, 1456–7.

Cardillo, M., Mace, G. M., Jones, K. E., *et al.* (2005) Multiple causes of high extinction risk in large mammal species. *Science*, **309**, 1239–41.

Cardillo, M., Purvis, A., Sechrest, W., Gittleman, J. L., Bielby, J., and Mace, G. M. (2004) Human population density and extinction risk in the world's carnivores. *PLoS Biology*, **2**, 909–14.

Cardinale, B. J. and Palmer, M. A. (2002) Disturbance moderates biodiversity–ecosystem function relationships: experimental evidence from caddisflies in stream mesocosms. *Ecology*, **83**, 1915–27.

Cardinale, B. J., Nelson, K., Palmer, M. A. (2000) Linking species diversity to the functioning of ecosystems: on the importance of environmental context. *Oikos*, **91**, 175–83.

Cardinale, B. J., Palmer, M. A. and Collins, S. L. (2002) Species diversity enhances ecosystem functioning through interspecific facilitation. *Nature*, **415**, 426–9.

Cardinale, B. J., Harvey, C. T., Gross, K., and Ives, A. R. (2003) Biodiversity and biocontrol: emergent impacts of a multi-enemy assemblage on pest suppression and crop yield in an agroecosystem. *Ecology Letters*, **6**, 857–65.

Cardinale, B. J., Ives, A. R., and Inchausti, P. (2004) Effects of species diversity on the primary productivity of ecosystems: extending our spatial and temporal scale of inference. *Oikos*, **104**, 437–450.

Cardinale, B. J., Palmer, M. A., Ives, A. R., *et al.* (2005) Diversity–productivity relationships in streams vary as a function of the natural disturbance regime. *Ecology*, **86**, 716–26.

Cardinale, B. J., Srivastava, D. S., Duffy, J. E., *et al.* (2006a) Effects of biodiversity on the functioning of trophic groups and ecosystems. *Nature*, **443**, 989–92.

Cardinale, B. J., Weis, J. J., Forbes, A. E., Tilmon, K. J., and Ives, A. R. (2006b). Biodiversity as both a cause and consequence of resource availability: a study of reciprocal causality in a predator–prey system. *Journal of Animal Ecology*, **75**, 497–505.

Cardinale, B. J., Wright, J. P., Cadotte, M. W., *et al.* (2007) Impacts of plant diversity on biomass production increase through time because of species complementarity. *Proceedings of the National Academy of Sciences of the USA*, **104**, 18123–8.

Cardinale, B. J., Srivastava, D. S., Duffy, J. E., *et al.*, (2009) Effects of biodiversity on the functioning of ecosystem: a summary of 164 experimental manipulations of species richness. *Ecology*, **90**, 854.

Carlander, K. D. (1952) Farm Fish Pond Research in Iowa. *The Journal of Wildlife Management*, **16**, 258–61.

Carnevale, N. J. and Montagnini, F. (2002) Facilitating regeneration of secondary forests with the use of mixed and pure plantations of indigenous tree species. *Forest Ecology and Management*, **163**, 217–27.

Carney, K. M., Matson, P. A., and Bohannan, B. (2004) Diversity and composition of tropical soil nitrifiers across a plant diversity gradient and among land-use types. *Ecology Letters*, **7**, 684–94.

Carpenter, S. R. (1996) Microcosm experiments have limited relevance for community and ecosystem ecology. *Ecology*, **77**, 677–80.

Carpenter, S. R., Kitchell, J. F., Hodgson, J. R., Cochran, P. A., Elser, J. J., Elser, M. M., Lodge, D. M., Kretchmer, D., He, X., and Vonende, C. N. (1987) Regulation of lake primary productivity by food web structure. *Ecology*, **68**, 1863–76.

Carson, R. (2008) *Contingent Valuation: a Comprehensive Bibliography and History*. Edward Elgar, Cheltenham.

Caruso, C. M., Maherali, H., Mikulyuk, A., Carlson, K., and Jackson, R. B. (2005) Genetic variance and covariance for physiological traits in *Lobelia*: are there constraints on adaptive evolution? *Evolution*, **59**, 826–37.

Caspersen, J. P. and Pacala, S. W. (2001) Successional diversity and forest ecosystem function. *Ecological Research*, **16**, 895–903.

Catovsky, S., Bradford, M. A., and Hector, A. (2002) Biodiversity and ecosystem productivity: implications for carbon storage. *Oikos*, **97**, 443–8.

Cavender-Bares, J., Ackerly, D. D., Baum, D. A., and Bazzaz, F. A. (2004) Phylogenetic overdispersion in Floridian oak communities. *American Naturalist*, **163**, 823–43.

Cavigelli, M. A., and Robertson, G. P. (2000) The functional significance of denitrifier community composition in a terrestrial ecosystem. *Ecology*, **81**, 1402–14.

Cecen, S., Gurel, F., and Karaca, A. (2008) Impact of honeybee and bumblebee pollination on alfalfa seed yield. *Acta Agriculturae Scandinavica Section B – Soil and Plant Science*, **58**, 77–81.

Chagnon, M., Gingras, J., and De Oliveira, D. (1993) Complementary aspects of strawberry pollination by honey and indigenous bees (Hymenoptera). *Journal of Economic Entomology*, **86**, 416–20.

Chalcraft, D. R. and Resetarits, W. J. (2003) Predator identity and ecological impacts: functional redundancy or functional diversity? *Ecology*, **84**, 2407–18.

Chapin, F. S. (2003) Effects of plant traits on ecosystem and regional processes: a conceptual framework for predicting the consequences of global change. *Annals of Botany*, **91**, 455–63.

Chapin, F. S., Schulze, E., and Mooney H. (1992) Biodiversity and ecosystem processes. *Trends in Ecology and Evolution*, **7**, 107–8.

Chapin, F. S., Bret-Harte, M. S., Hobbie, S. E., and Hailan, Z. (1996) Plant functional types as predictors of transient responses of Arctic vegetation to global change. *Journal of Vegetation Science*, **7**, 347–58.

Chapin, F. S., Walker, B. H., Hobbs, R. J., Hooper, D. U., Lawton, J. H., Sala, E. O., and Tilman, D. (1997) Biotic control over the functioning of ecosystems. *Science*, **277**, 500–4.

Chapin, F. S., Eugster, W., McFadden, J., Lynch, A., and Walker, D. (2000a) Summer differences among Arctic ecosystems in regional climate forcing. *Journal of Climate*, **13**, 2002–10.

Chapin, F. S., McGuire, A., Randerson, J., et al. (2000b) Arctic and boreal ecosystems of western North America as components of the climate system. *Global Change Biology*, **6**, 211–23.

Chapin, F. S., Zavaleta, E. S., Eviner, E. T., et al. (2000c) Consequences of changing biodiversity. *Nature*, **405**, 234–42.

Chapin, F. S., Sturm, M., Serreze, M. C., et al. (2005) Role of land-surface changes in Arctic summer warming. *Science*, **310**, 657–60.

Chapin, F. S., Trainor, S. F., Huntington, O., et al. (2008) Increasing wildfire in Alaska's boreal forest: pathways to potential solutions of a wicked problem. *Bioscience*, **58**, 531–40.

Chapin, S. I., Oe, S., Burke, I., et al. (1998) Ecosystem consequences of changing biodiversity. *BioScience*, **48**, 45–52.

Chapman, K., Whittaker, J. B. and Heal, O. W. (1988) Metabolic and faunal activity in litters of tree mixtures compared with pure stands. *Agriculture Ecosystems and Environment*, **24**, 33–40.

Charnov, E. L. (1997) Trade-off-invariant rules for evolutionarily stable life histories. *Nature*, **387**, 393–4.

Charrette, N. A., Cleary, D. F. R., and Mooers, A. Ø. (2006) Range-restricted, specialist Bornean butterflies are less likely to recover from ENSO-induced disturbance. *Ecology*, **87**, 2330–7.

Chave, J. (2004) Neutral theory and community ecology. *Ecology Letters*, **7**, 241–53.

Chazdon, R. L. (2008) Beyond deforestation: restoring forests and ecosystem services on degraded lands. *Science*, **320**, 1458–60.

Check, E. (2005) Roots of recovery. *Nature*, **438**, 910–11.

Chesson, P. (2000) Mechanisms of maintenance of species diversity. *Annual Review of Ecology and Systematics*, **31**, 343–66.

Chesson, P., Pacala, S., and Neuhauser, C. (2002) Environmental niches and ecosystem functioning. In A. P. Kinzig, S. W. Pacala, and D. Tilman (eds.) *The Functional Consequences of Biodiversity*, pp. 213–45. Princeton University Press, Princeton, NJ.

Chiari, W. C., de Alencar Arnaut de Toledo, V., Ruvolo-Takasusuki, M. C. C., et al. (2005) Pollination of soybean (*Glycine max* L. Merril) by honeybees (*Apis mellifera* L.). *Brazilian Archives of Biology and Technology*, **48**, 31–6.

Chichilnisky, G. (1994) North–south trade and the global environment. *American Economic Review*, **84**, 851–74.

Chichilnisky, G. and Heal, G. (1998) Economic returns from the biosphere. *Nature*, **391**, 629–30.

Chittka, L. and Schürkens, S. (2001) Successful invasion of a floral market. *Nature*, **411**, 653.

Chomitz, K. E. (1999) *Transferable Development Rights and Forest Protection: an Exploratory Analysis*. Paper prepared for the Workshop on Market-Based Instruments for Environmental Protection, July 1999. John F. Kennedy School of Government, Harvard University.

Chong, J. (2005) *Protective Values of Mangrove and Coral Ecosystems: a Review of Methods and Evidence*. IUCN, Gland, Switzerland.

Chopra, K. and Adhikar, S. (2004) Environment- development linkages: a dynamic modeling for a wetland eco-system. *Environment and Development Economics*, **9**(1), 19–45.

Chopra, K. and Kumar, P. (2004) Forest biodiversity and timber extraction: an analysis of the interaction of market and non-market mechanisms. *Ecological Economics*, **49**, 135–48.

Christian, J. M. and Wilson, S. D. (1999) Long-term ecosystem impacts of an introduced grass in the northern great plains. *Ecology*, **80**, 2397–407.

Christianou, M. and Ebenman, B. (2005) Keystone species and vulnerable species in ecological communities: strong or weak interactors? *Journal of Theoretical Biology*, **235**, 95–103.

Chu, Y., He, W. M., Liu, H. D., Liu, J., Zhu, X. W., and Dong, M. (2006) Phytomass and plant functional diversity in early restoration of the degraded, semi-arid grasslands in northern china. *Journal of Arid Environments*, **67**, 678–87.

Cianciaruso, M. V., Batalha, M. A., Gaston, K. J., and Petchey O. L. (2009) Including intraspecific variability in functional diversity, **90**: 81–89.

Clark, W. C. and Munn, R. E. (1987) *Sustainable development of the biosphere*. Cambridge University Press, New York.

Clarke, P. J., Latz, P. K., and Albrecht, D. E. (2005) Long-term changes in semi-arid vegetation: invasion of an exotic perennial grass has larger effects than rainfall variability. *Journal of Vegetation Science*, **16**, 237–48.

Clay, J. W. (2004) *World Agriculture and the Environment: a Commodity-by-Commodity Guide to Impacts and Practices*. Island Press, Washington, DC.

Cleland, E. E., Smith, M. D., Andelman, S. J., Bowles, C., Carney, K. M., Claire Horner-Devine, M., Drake, J. M., Emery, S. M., Gramling, J. M., and Vandermast, D. B. (2004) Invasion in space and time: non-native species richness and relative abundance respond to interannual variation in productivity and diversity. *Ecology Letters*, **7**, 947–57.

Cole, M. A., Rayner, A. J., and Bates, J. M. (1997) The environmental Kuznets curve: an empirical analysis. *Environment and Development Economics*, **2**(4), 401–16.

Coley, P. (1983) Herbivory and defensive characteristics of tree species in a lowland tropical forest. *Ecological Monographs*, **53**, 209–33.

Coley, P. D., Bryant, J. P., and Chapin, F. S. (1985) Resource availability and plant antiherbivore defense. *Science*, **230**, 895–9.

Collen, B., Bykova, E., Ling, S., Milner-Gulland, E. J., and Purvis, A. (2006) Extinction risk: a comparative analysis of central Asian vertebrates. *Biodiversity and Conservation*, **15**, 1859–71.

Collins, W. W. and Qualset, C. O. (1999) *Biodiversity in Agroecosystems*. CRC Press, Boca Raton, FL.

Connell, J. H. (1978) Diversity in tropical rain forests and tropical reefs. *Science*, **199**, 1302–10.

Conte, M. N. (2007) *Competitive Search and Preemptive Exclusion*. University of California, Santa Barbara, working paper.

Cook, D. C., Thomas, M. B., Cunningham, S. A., Anderson, D. L., and De Barro, P. J. (2007) Predicting the economic impact of an invasive species on an ecosystem service. *Ecological Applications*, **17**, 1832–40.

Cooper, N., Bielby, J., Thomas, G. H., and Purvis, A. (2008) Macroecology and extinction risk correlates of frogs. *Global Ecology and Biogeography*, **17**, 211–21.

Cornelissen, J. H. C. (1996) An experimental comparison of leaf decomposition rates in a wide range of temperate plant species and types. *Journal of Ecology*, **84**, 573–82.

Cornelissen, J., Perez-Harguindeguy, N., Díaz, S., et al. (1999) Leaf structure and defence control litter decomposition rate across species and life forms in regional floras on two continents. *New Phytologist*, **143**, 191–200.

Cornelissen, J. H. C. and Thompson, K. (1997) Functional leaf attributes predict litter decomposition rate in herbaceous plants. *New Phytologist*, **135**, 109–14.

Cornelissen, J. H. C., Lavorel, S., Garnier, E., et al. (2003) A handbook of protocols for standardised and easy measurement of plant functional traits worldwide. *Australian Journal of Botany*, **51**, 335–80.

Cornwell, W. K., Schwilk, D. W., and Ackerly, D. D. (2006) A trait-based test for habitat filtering: convex hull volume. *Ecology*, **87**, 1465–71.

Cornwell, W. K., Cornelissen, J. H. C., Amatangelo, K., et al. (2008) Plant species traits are the predominant control on litter decomposition rates within biomes worldwide. *Ecology Letters*, **11**, 1065–71.

Cortez, J., Garnier, E., Perez-Harguindeguy, N., Debussche, M., and Gillon, D. (2007) Plant traits, litter quality and decomposition in a Mediterranean old-field succession. *Plant and Soil*, **296**, 19–34.

Costanza, R. and Folke, C. (1997) Valuing ecosystem services with efficiency, fairness and sustainability as goals.

In G. C. Daily (ed.) *Nature's services*. Island Press, Washington, DC.

Costanza, R., d'Arge, R., de Groot, R., et al. (1997) The value of the world's ecosystem services and natural capital. *Nature*, **387**, 253–9.

Costello, C. and McAusland, C. (2003) Protectionism, trade and measures of damage from exotic species introduction. *American Journal of Agricultural Economics*, **85**(4), 964–75.

Costello, C. and Ward, M. (2006) Search, bioprospecting and biodiversity conservation. *Journal of Environmental Economics and Management*, **52**, 615–26.

Costello, C., Springborn, M., McAusland, C., and Solow, A. (2007) Unintended biological invasions: Does risk vary by trading partner? *Journal of Environmental Economics and Management*, **54**, 262–76.

Cottingham, K. L., Brown, B. L., and Lennon, J. T. (2001) Biodiversity may regulate the temporal variability of ecological systems. *Ecology Letters*, **4**, 72–85.

Covich, A. P., Austen, M., Bärlocher, F., et al. (2004) The role of biodiversity in the functioning of freshwater and marine benthic ecosystems. *Bioscience*, **54**, 767–75.

Cox, F. E. G. (2001) Concomitant infections, parasites and immune responses. *Parasitology*, **122**, S23–S38.

Cox-Foster, D. L., Conlan, S., Holmes, E. C., et al. (2007) A metagenomic survey of microbes in honey bee colony collapse disorder. *Science*, **318**, 283–7.

Craft, C., Reader, J., Sacco, J. N., and Broome, S. W. (1999) Twenty-five years of ecosystem development of constructed *Spartina alterniflora* (loisel) marshes. *Ecological Applications*, **9**, 1405–19.

Cragg, R. G. and R. D. Bardgett (2001) How changes in soil faunal diversity and composition within a trophic group influence decomposition processes. *Soil Biology & Biochemistry*, **33**, 2073–81.

Craine, J. M., Tilman, D., Wedin, D., Reich, P., Tjoelker, M., and Knops, J. (2002) Functional traits, productivity and effects on nitrogen cycling of 33 grassland species. *Functional Ecology*, **16**, 563–74.

Cramer, J. M., Mesquita, R. C. G., and Williamson, G. B. (2007) Forest fragmentation differentially affects seed dispersal of large and small-seeded tropical trees. *Biological Conservation*, **137**, 415–23.

Crawford, J. W., Harris, J. A., Ritz, K., and Young, I. M. (2005) Towards an evolutionary ecology of life in soil. *Trends in Ecology & Evolution*, **20**, 81–7.

Crawley, M. J. and Harral, J. E. (2001) Scale dependence in plant biodiversity. *Science*, **291**, 864–8.

Crews, T., Kitayama, K., Fownes, J., et al. (1995) Changes in soil-phosphorus fractions and ecosystem dynamics across a long chronosequence in Hawaii. *Ecology*, **76**, 1407–24.

Critchley, C. N. R., Burke, M. J. W., and Stevens, D. P. (2004) Conservation of lowland semi-natural grasslands in the UK: a review of botanical monitoring results from agri-environment schemes. *Biological Conservation*, **115**, 263–78.

Crocker, T. D. and Tschirhart, J. (1992) Ecosystems, externalities, and economics. *Environmental and Resource Economics*, **2**, 551–67.

Cross, M. S. and Harte, J. (2007) Compensatory responses to loss of warming-sensitive plant species. *Ecology*, **88**, 740–8.

Crutsinger, G. M., Collins, M. D., Fordyce, J. A., Gompert, Z., Nice, C. C., and Sanders, N. J. (2006) Plant genotypic diversity predicts community structure and governs an ecosystem process. *Science*, **313**, 966–8.

Curtis, T. P. and Sloan, W. T. (2004) Prokaryotic diversity and its limits: microbial community structure in nature and implications for microbial ecology. *Current Opinion in Microbiology*, **7**, 221–6.

Daehler, C. C. (1998) Variation in self-fertility and the reproductive advantage of self-fertility for an invading plant (*Spartina alterniflora*). *Evolutionary Ecology*, **12**, 553–68.

Dahdouh-Guebas, F. and Koedam, N. (2006) Coastal vegetation and the Asian tsunami. *Science*, **311**(5757), 37–8.

Dahdouh-Guebas, F., Jayatissa, L. P., Di Nitto, D., Bosire, J. O., Lo Seen, D., and Koedam, N. (2005) How effective were mangroves as a defence against the recent tsunami? *Current Biology*, **15**(12), 443–7.

Daily, G. (ed.) (1997) *Nature's Services, Societal Dependence on Natural Ecosystems*. Island Press, Washington, DC.

Daily, G. and Ellison, K. (2002) *The New Economy of Nature*. Island Press, Washington, DC.

Daily, G. C., Alexander, S., Ehrlich, P. R., et al. (1997) Ecosystem services: benefits supplied to human societies by natural ecosystems. *Issues in Ecology*, **1**(2), 1–18.

Dang, C. K., Chauvet, E., and Gessner, M. O. (2005) Magnitude and variability of process rates in fungal diversity–litter decomposition relationships. *Ecology Letters*, **8**, 1129–37.

Danielsen, F., Sørensen, M. K., Olwig, M. F., et al. (2005) The Asian tsunami: A protective role for coastal vegetation. *Science*, **310**(5748), 643.

D'Antonio, C. M. and Vitousek, P. M. (1992) Biological invasions by exotic grasses, the grass fire cycle, and global change. *Annual Review of Ecology and Systematics*, **23**, 63–87.

D'Antonio, C. M., Hughes, R. F., Mack, M., Hitchcock, D., and Vitousek, P. M. (1998) The response of native species to removal of invasive exotic grasses in a seasonally dry Hawaiian woodland. *Journal of Vegetation Science*, **9**, 699–712.

D'Antonio, C. M. and Hobbie, S. E. (2005) Plant species effects on ecosystem processes: insights from invasive species. In D. F. Sax, J. J. Stachowicz, and S. D. Gaines (eds.) *Species Invasions: Insights into Ecology, Evolution, and Biogeography*, pp. 65–84. Sinauer Associates, Inc. Publishers, Sunderland, MA.

Darwin, C. (1859) *On the Origin of Species by Means of Natural Selection, or the Preservation of Favoured Races in the Struggle for Life*. John Murray, London.

Dasgupta, P. (2001) *Human Well-Being and the Environment*. Oxford University Press, Oxford.

Davic, R. D. (2003) Linking keystone species and functional groups: a new operational definition of the keystone species concept – response. *Conservation Ecology*, **7**.

Davies, K. F., Margules, C. R., and Lawrence, J. F. (2000) Which traits of species predict population declines in experimental forest fragments? *Ecology*, **81**, 1450–61.

Davies, K. F., Margules, C. R., and Lawrence, J. F. (2004) A synergistic effect puts rare, specialized species at greater risk of extinction. *Ecology*, **85**, 265–71.

Davies, R. G., Orme, C. D. L., Storch, D., Olson, V. A., Thomas, G. H., Ross, S. G., Ding, T. S., Rasmussen, P. C., Bennett, P. M., Owens, I. P. F., Blackburn, T. M., and Gaston, K. J. (2007) Topography, energy and the global distribution of bird species richness. *Proceedings of the Royal Society B: Biological Sciences*, **274**, 1189–97.

Davis, M. A. and Thompson, K. (2000) Eight ways to be a colonizer; two ways to be an invader: a proposed nomenclature scheme for invasion ecology. *Bulletin of the Ecological Society of America*, **81**, 226–30 (Abstract)

Davis, M. A., Wrage, K. J., and Reich, P. B. (1998) Competition between tree seedlings and herbaceous vegetation: support for a theory of resource supply and demand. *Journal of Ecology*, **86**, 652–61.

Davis, M. A., Wrage, K. J., Reich, P. B., Tjoelker, M. G., Schaeffer, T., and Muermann, C. (1999) Survival, growth, and photosynthesis of tree seedlings competing with herbaceous vegetation along a water–light–nitrogen gradient. *Plant Ecology*, **145**, 341–50.

Deacon, L. J., Pryce-Miller, E. J., Frankland, J. C., Bainbridge, B. W., Moore, P. D., and Robinson, C. H. (2006) Diversity and function of decomposer fungi from a grassland soil. *Soil Biology & Biochemistry*, **38**, 7–20.

De Angelis, D. L. (1992) *Dynamics of Nutrient Cycling and Food Webs*. Chapman & Hall, London.

de Bach, P. (1974) *Biological Control by Natural Enemies*. Cambridge University Press, London.

de Bello, F., Leps, J., and Sebastia, M. T. (2006) Variations in species and functional plant diversity along climatic and grazing gradients. *Ecography*, **29**, 801–10.

de Bello, F., Leps, J., Lavorel, S., and Moretti, M. (2007) Importance of species abundance for assessment of trait composition: an example based on pollinator communities. *Community Ecology*, **8**, 163–70.

Debras, J. F., Torre, F., Rieux, R., *et al.* (2006) Discrimination between agricultural management and the hedge effect in pear orchards (south-eastern France). *Annals of Applied Biology*, **149**, 347–55.

DeClerck, F. A. J., Barbour, M. G., and Sawyer, J. O. (2006a) Species richness and stand stability in conifer forests of the Sierra Nevada. *Ecology*, **87**, 2787–99.

DeClerck, F., Ingram, J. C., and del Rio, C. M. R. (2006b) The role of ecological theory and practice in poverty alleviation and environmental conservation. *Frontiers in Ecology and the Environment*, **10**, 533–40.

Decocq, G., and Hermy, M. (2003) Are there herbaceous dryads in temperate deciduous forests? *Acta Botanica Gallica*, **150**, 373–82.

De Deyn, G. B. and Van der Putten, W. H. (2005) Linking aboveground and belowground diversity. *Trends in Ecology & Evolution*, **20**, 625–33.

De Deyn, G. B., Raaijmakers, C. E., van Ruijven, J., Berendse, F., and van der Putten, W. H. (2004) Plant species identity and diversity effects on different trophic levels of nematodes in the soil food web. *Oikos*, **106**, 576–86.

DeFalco, L. A., Fernandez, G. C. J., and Nowak, R. S. (2007) Variation in the establishment of a non-native annual grass influences competitive interactions with Mojave Desert perennials. *Biological Invasions*, **9**, 293–307.

Degens, B. P. (1998) Microbial functional diversity can be influenced by the addition simple organic substrates to soil. *Soil Biology & Biochemistry*, **30**, 1981–8.

Delaplane, K. S. and Mayer, D. F. (2000) *Crop Pollination by Bees*. CABI Publishing, New York.

De Mazancourt, C., Loreau, M., and Abbadie, L. (1998) Grazing optimization and nutrient cycling: when do herbivores enhance plant production? *Ecology*, **79**, 2242–52.

De Mesel, I., Derycke, S., Swings, J., Vincx, M., and Moens, T. (2006) Role of nematodes in decomposition processes: does within-trophic group diversity matter? *Marine Ecology – Progress Series*, **321**, 157–66.

Demott, W. R. (1998) Utilization of a cyanobacterium and a phosphorus-deficient green alga as complementary resources by daphnids. *Ecology*, **79**, 2463–81.

Dennehy, J. J., Friedenberg, N. A., Yang, Y. W., and Turner, P. E. (2007) Virus population extinction via ecological traps. *Ecology Letters*, **10**, 230–40.

Denslow, J. (1987) Tropical rain-forest gaps and tree species-diversity. *Annual Review of Ecology and Systematics*, **18**, 431–51.

de Ruiter, P. C., Neutel, A. M., and Moore, J. C. (1995) Energetics, patterns of interaction strengths, and stability in real ecosystems. *Science*, **269**, 1257–60.

de Ruiter, P. C., Wolters, V., Moore, J. C., and Winemiller, K. O. (2005) ECOLOGY: food web ecology: playing Jenga and beyond. *Science*, **309**, 68–71.

Descamps-Julien, B. and Gonzalez, A. (2005) Stable coexistence in a fluctuating environment: an experimental demonstration. *Ecology*, **86**, 2815–24.

Deurwaerdere, T., Krishna, V., and Pascual, U. (2007) An evolutionary institutional economics approach to the economics of bioprospecting. In A. Kontoleon, U. Pascual, and T. Swanson (eds.) *Biodiversity Economics: Principles, Methods and Applications*, pp. 417–45. Cambridge University Press, Cambridge.

De Wit, C. T. and Van den Bergh, J. P. (1965) Competition between herbage plants. *Netherlands Journal of Agricultural Science*, **13**, 212–21.

Dhôte, J.-F. (2005) Implication of forest diversity in resistance to strong winds. In M. Scherer-Lorenzen, C. Körner, and E.-D. Schulze (eds.) *The Functional Significance of Forest Diversity*. Springer-Verlag, Berlin.

Di Falco, S. and Chavas, J. P. (2007) On the role of crop biodiversity in the management of environmental risk. In A. Kontoleon, U. Pascual, and T. Swanson, eds. *Biodiversity Economics: Principles, Methods and Applications*, pp. 581–93. Cambridge University Press, Cambridge.

Di Falco, S. and Perrings, C. (2003) Crop genetic diversity, productivity and stability of agroecosystems: a theoretical and empirical investigation. *Scottish Journal of Political Economy*, **50**, 207–16.

Di Falco, S. and Perrings, C. (2005) Crop biodiversity, risk management and the implications of agricultural assistance. *Ecological Economics*, **55**(4), 459–66.

Diamond, J. M. (1972) Biogeographic kinetics: estimation of relaxation times for avifaunas of Southwest Pacific Islands. *Proceedings of the National Academy of Sciences of the USA*, **69**, 3199–203.

Díaz, S. and Cabido, M. (1997) Plant functional types and ecosystem function in relation to global change. *Journal of Vegetation Science*, **8**, 463–74.

Díaz, S. and Cabido, M. (2001) Vive la différence: plant functional diversity matters to ecosystem processes. *Trends in Ecology and Evolution*, **16**, 646–55.

Díaz, S. and Cáceres, D. (2000) Ecological approaches to rural development projects. *Cadernos de Saúde Publica*, **16**, 7–14.

Díaz, S., Symstad, A., Chapin, F., Wardle, D., and Huenneke, L. (2003) Functional diversity revealed by removal experiments. *Trends in Ecology & Evolution*, **18**, 140–6.

Díaz, S., Hodgson, J. G., Thompson, K., et al. (2004) The plant traits that drive ecosystems: evidence from three continents. *Journal of Vegetation Science*, **15**, 295–304.

Díaz, S., Tilman, D., Fargione, J., et al. (2005) Biodiversity regulation of ecosystem services. In Hassan, R., Scholes, R., and Ash, N. (eds.) *Ecosystems and Human Well-Being. Current State and Trends – Findings of the Condition and Trends Working Group of the Millennium Ecosystem Assessment*. Island Press, Washington, DC.

Díaz, S., Fargione, J., Chapin, F., and Tilman, D. (2006) Biodiversity loss threatens human well-being. *PLoS Biology*, **4**(8), e277.

Díaz, S., Lavorel, S., McIntyre, S., et al. (2007) Plant trait responses to grazing – a global synthesis. *Global Change Biology*, **13**, 313–41.

Didham, R. K., Hammond, P. M., Lawton, J. H., Eggleton, P., and Stork, N. E. (1998a). Beetle species responses to tropical forest fragmentation. *Ecological Monographs*, **68**, 295–323.

Didham, R. K., Lawton, J. H., Hammond, P. M., and Eggleton, P. (1998b). Trophic structure stability and extinction dynamics of beetles (Coleoptera). in tropical forest fragments. *Philosophical Transactions of the Royal Society of London Series B: Biological Sciences*, **353**, 437–51.

Dierssen, K. (2006) Indicating botanical diversity – structural and functional aspects based on case studies from northern Germany. *Ecological Indicators*, **6**, 94–103.

Dimitrakopoulos, P. G. and Schmid, B. (2004) Biodiversity effects increase linearly with biotope space. *Ecology Letters*, **7**, 574–83.

Dimitrakopoulos, P. G., Siamantziouras, A. S. D., Galanidis, A., Mprezetou, I., and Troumbis, A. Y. (2006) The interactive effects of fire and diversity on short-term responses of ecosystem processes in experimental Mediterranean grasslands. *Environmental Management*, **37**, 826–39.

Dirzo, R. and Raven, P. H. (2003) Global state of biodiversity and loss. *Annual Review of Environment and Resources*, **28**, 137–67.

Dizney, L. J., and L. A. Ruedas. In press. Increased species diversity decreases prevalence of a directly transmitted zoonosis. *Emerging Infectious Diseases*.

Doak, D. F., Bigger, D., Harding-Smith, E., Marvier, M. A., O'Malley, R. and Thomson, D. (1998) The statistical inevitability of stability–diversity relationships in community ecology. *American Naturalist*, **151**, 264–76.

Dobson, A. P. (2004) Population dynamics of pathogens with multiple host species. *American Naturalist*, **164**, S64–S78.

Dobson, A. P. and Foufopoulos, J. (2001) Emerging infectious pathogens in wildlife. *Philosophical Transactions of the Royal Society of London B*, **356**, 1001–12.

Dobson, A., Cattadori, I., Holt, R. D., et al. (2006) Sacred cows and sympathetic squirrels: the importance of biological diversity to human health. *PLoS Medicine*, **3**, 714–18.

Dobson, A., Lodge, D., Alder, J. et al. (2006) Habitat loss, trophic collapse, and the decline of ecosystem services. *Ecology*, **87**, 1915–24.

Dodd, M. E., Silvertown, J., McConway, K., Potts, J., and Crawley, M. (1994) Stability in the plant communities of the park grass experiment: the relationships between species richness, soil pH and biomass variability. *Philosophical Transactions: Biological Sciences*, **346**, 185–93.

Donald, P. F. (2004) Biodiversity impacts of some agricultural commodity production systems. *Conservation Biology*, **18**, 17–37.

Downey, D. L. and Winston, M. L. (2001) Honey bee colony mortality and productivity with single and dual infestations of parasitic mite species. *Apidologie*, **32**, 567–75.

Downing, A. L. (2005) Relative effects of species composition and richness on ecosystem properties in ponds. *Ecology*, **86**, 701–15.

Downing, A. L. and Liebold, M. A. (2002) Ecosystem consequences of species richness and composition in pond food webs. *Nature*, **416**, 837–41.

Drenovsky, R. E., Martin, C. E., Falasco, M. R., and James, J. J. (2008) Variation in resource acquisition and utilization traits between native and invasive perennial forbs. *American Journal of Botany*, **95**, 681–7.

Drever, C. R., Peterson, G., Messier, C., Bergeron, Y., and Flannigan, M. (2006) Can forest management based on natural disturbances maintain ecological resilience? *Annual Review of Ecology and Systematics*, **36**, 2285–95.

Duarte, C. M. (2000) Marine biodiversity and ecosystem services: an elusive link. *Journal of Experimental Marine Biology and Ecology*, **250**(1–2), 117–31.

Duarte, C. M., Marba, N., and Holmer, M. (2007) Rapid domestication of marine species. *Science*, **316**, 382–3.

Duarte, S., Pascoal, C., Cassio, F., and Bärlocher, F. (2006) Aquatic hyphomycete diversity and identity affect leaf litter decomposition in microcosms. *Oecologia*, **147**, 658–66.

Duffy, E. (2003) Biodiversity loss, trophic skew and ecosystem functioning. *Ecology Letters*, **6**, 680–7.

Duffy, J. E. (2008) Why biodiversity is important to functioning of real-world ecosystems. *Frontiers in Ecology and the Environment* (in press).

Duffy, J. E. (2002) Biodiversity and ecosystem function: the consumer connection. *Oikos*, **99**, 201–19.

Duffy, J. E., Macdonald, K. S., Rhode, J. M., and Parker, J. D. (2001) Grazer diversity, functional redundancy, and productivity in seagrass beds: an experimental test. *Ecology*, **82**, 2417–34.

Duffy, J. E., Richardson, J. P., and France, K. E. (2005) Ecosystem consequences of diversity depend on food chain length in estuarine vegetation. *Ecology Letters*, **8**, 301–9.

Duffy, J. E., Cardinale, B. J., France, K. E., McIntyre, P. B., Thebault, E., and Loreau, M. (2007) The functional role of biodiversity in ecosystems: incorporating trophic complexity. *Ecology Letters*, **10**, 522–38.

Dukes, J. S. (2001) Biodiversity and invasibility in grassland microcosms. *Oecologia*, **126**, 563–8.

Dulvy, N. K., Ellis, J. R., Goodwin, N. B., Grant, A., Reynolds, J. D., and Jennings, S. (2004) Methods of assessing extinction risk in marine fishes. *Fish and Fisheries*, **5**, 255–76.

Dulvy, N. K., Jennings, S., Goodwin, N. B., Grant, A., and Reynolds, J. D. (2005) Comparison of threat and exploitation status in North-East Atlantic marine populations. *Journal of Applied Ecology*, **42**, 883–91.

Dumay, O., Tari, P. S., Tomasini, J. A., and Mouillot, D. (2004) Functional groups of lagoon fish species in Languedoc Roussillon, southern France. *Journal of Fish Biology*, **64**, 970–83.

Duncan, R. P. and Young, J. R. (2000) Determinants of plant extinction and rarity 145 years after European settlement of Auckland, New Zealand. *Ecology*, **81**, 3048–61.

Dunne, J. A., Williams, R. J., and Martinez, N. D. (2002a). Food-web structure and network theory: the role of connectance and size. *Proceedings of the National Academy of Sciences of the USA*, **99**, 12917–22.

Dunne, J. A., Williams, R. J., and Martinez, N. D. (2002b) Network structure and biodiversity loss in food webs: robustness increases with connectance. *Ecology Letters*, **5**, 558–67.

Dunne, J. A., Williams, R. J., and Martinez, N. D. (2004) Network structure and robustness of marine food webs. *Marine Ecology–Progress Series*, **273**, 291–302.

Dunne, J. A., Williams, R. J., Martinez, N. D., Wood, R. A., and Erwin, D. H. (2008) Compilation and network analyses of Cambrian food webs. *PLoS Biology*, **6**, 693–708.

Duraiappah, A. K. (2006) *Markets for Ecosystem Services*. International Institute for Sustainable Development, Winnipeg, Canada.

Duraiappah, A. K. and Naeem, S. (2005) Synthesis report on biodiversity. In *Millennium Ecosystem Assessment. Ecosystems and Human Well-Being: Synthesis*. Island Press, Washington, DC.

Dyer, A. R. and Rice, K. J. (1999) Effects of competition on resource availability and growth of a California bunchgrass. *Ecology*, **80**, 2697–710.

Dyson, K. E., Bulling, M. T., Solan, M., *et al.* (2007) Influence of macrofaunal assemblages and environmental heterogeneity on microphytobenthic production in experimental systems. *Proceedings of the Royal Society B: Biological Sciences*, **274**, 2547–54.

Ebeling, A., Klein, A. M., Schumacher, J., Weisser, W. W., and Tscharntke, T. (2008) How does plant richness affect pollinator richness and temporal stability of flower visits? *Oikos*, **117**, 1808–15.

Ebenman, B. and Jonsson, T. (2005) Using community viability analysis to identify fragile systems and keystone species. *Trends in Ecology & Evolution*, **20**, 568–75.

Ebenman, B., Law, R., and Borrvall, C. (2004) Community viability analysis: the response of ecological communities to species loss. *Ecology*, **85**, 2591–600.

Edwards, E. J., Still, C. J., and Donoghue, M. J. (2007) The relevance of phylogeny to studies of global change. *Trends in Ecology & Evolution*, **22**, 243–9.

Ehrenfeld, J. G. and Toth, L. A. (1997) Restoration ecology and the ecosystem perspective. *Restoration Ecology*, **5**, 307–17.

Eichner, T. and Pethig, R. (2005) Ecosystem and economy: an integrated dynamic general equilibrium approach. *Journal of Economics*, **85**, 213–49.

Eichner, T. and Tschirhart, J. (2007) Efficient ecosystem services and naturalness in an ecological/economic model. *Environmental and Resource Economics*, **37**, 733–55.

Eklöf, A. and Ebenman, B. (2006) Species loss and secondary extinctions in simple and complex model communities. *Journal of Animal Ecology*, **75**, 239–46.

Ellis, A. R., Hubbell, S. P., and Potvin, C. (2000) *In situ* field measurements of photosynthetic rates of tropical tree species: a test of the functional group hypothesis. *Canadian Journal of Botany*, **78**, 1336–47.

Ellis, G. M. and Fisher, A. C. (1987) Valuing the environment as input. *Journal of Environmental Management*, **25**, 149–56.

Elmqvist, T., Folke, C., Nystrom, M., *et al.* (2003) Response diversity, ecosystem change, and resilience. *Frontiers in Ecology and the Environment*, **1**, 488–94.

Elser, J. J. and Sterner, R. (2002) *Ecological Stoichiometry: The Biology of Elements from Molecules to the Biosphere*. Princeton University Press, Princeton.

Elser, J. J., Elser, M. M., Mackay, N. A., and Carpenter, S. R. (1988) Zooplankton-mediated transitions between N-limited and P-limited algal growth. *Limnology and Oceanography*, **33**, 1–14.

El Serafy, S. (1989) The proper calculation of income from depletable natural resources. In Y. J. Ahmad, S. El Serafy and E. Lutz (eds.) *Environmental Accounting for Sustainable Development*. The World Bank, Washington, DC.

Elton, C. S. (1958) *The Ecology of Invasions by Animals and Plants*. Methuen, London.

Emmerson, M. and Huxham, M. (2002) Population and ecosystem level processes in marine habitats. In M. Loreau, S. Naeem, and P. Inchausti (eds.) *Biodiversity and Ecosystem Functioning. Synthesis and Perspectives*, pp. 139–46. Oxford University Press, Oxford.

Emmerson, M. C. and Raffaelli, D. G. (2000) Detecting the effects of diversity on measures of ecosystem function: experimental design, null models and empirical observations. *Oikos*, **91**, 195–203.

Emmerson, M. C., Solan, M., Emes, C., Paterson, D. M., and Raffaelli, D. G. (2001) Consistent patterns and the idiosyncratic effects of biodiversity in marine ecosystems. *Nature*, **411**, 73–7.

Engelhardt, K. A. M. (2006) Relating effect and response traits in submersed aquatic macrophytes. *Ecological Applications*, **16**, 1808–20.

Engelhardt, K. A. M. and Kadlec, J. A. (2001) Species traits, species richness and the resilience of wetlands after disturbance. *Journal of Aquatic Plant Management*, **39**, 36–9.

Engelhardt, K. A. M. and Ritchie, M. E. (2001) Effects of macrophyte species richness on wetland ecosystem functioning and services. *Nature*, **411**, 687–9.

Engelhardt, K. A. M. and Ritchie, M. E. (2002) The effect of aquatic plant species richness on wetland ecosystem processes. *Ecology*, **83**, 2911–24.

Englund, G. and Hamback, P. A. (2007) Scale dependence of immigration rates: models, metrics and data. *Journal of Animal Ecology*, **76**, 30–5.

Enquist, B. J. and Niklas, K. J. (2001) Invariant scaling relations across tree-dominated communities. *Nature*, **410**, 655–60.

Enquist, B. J., West, G. B., Charnov, E. L., and Brown, J. H. (1999) Allometric scaling of production and life-history variation in vascular plants. *Nature*, **401**, 907–11.

Enquist, B. J., Economo, E. P., Huxman, T. E., Allen, A. P., Ignace, D. D., and Gillooly, J. F. (2003) Scaling metabolism from organisms to ecosystems. *Nature*, **423**, 639–42.

Enquist, B. J., Kerkhoff, A. J., Stark, S. C., Swenson, N. G., McCarthy, M. C., and Price, C. A. (2007) A general integrative model for scaling plant growth, carbon flux, and functional trait spectra. *Nature*, **449**, 218–22.

Epps, K. Y., Comerford, N. B., Reeves, J. B., Cropper, W. P., and Araujo, Q. R. (2007) Chemical diversity – highlighting a species richness and ecosystem function disconnect. *Oikos*, **116**, 1831–40.

Ernst, R., Linsenmair, K. E., and Rodel, M. O. (2006) Diversity erosion beyond the species level: dramatic loss of functional diversity after selective logging in two tropical amphibian communities. *Biological Conservation*, **133**, 143–55.

Espinosa-García, F. J., Villaseñor, J. L., and Vibrans, H. (2004) The rich generally get richer, but there are exceptions: correlations between species richness and native plant species and alien weeds in Mexico. *Diversity and Distributions*, **10**, 399–407.

Etterson, J. R. and Shaw, R. G. (2001) Constraint to adaptive evolution in response to global warming. *Science*, **294**, 151–4.

Evans, J. and Turnbull, J. (2004) *Plantation Forestry in the Tropics*. Oxford University Press, Oxford.

Eviner, V. T. (2004) Plant traits that influence ecosystem processes vary independently among species. *Ecology*, **85**, 2215–29.

Eviner, V. T. and Chapin, F. S. (2003) Functional matrix: a conceptual framework for predicting multiple plant effects on ecosystem processes. *Annual Review in Ecology, Evolution, and Systematics*, **34**, 455–85.

Eviner, V. T., Chapin, F. S., and Vaughn, C. E. (2006) Seasonal variations in plant species effects on soil n and p dynamics. *Ecology*, **87**, 974–86.

Ewel, J. (1986) Designing agricultural ecosystems for the humid tropics. *Annual Rewiew of Ecology and Systematics*, **17**, 245–71.

Ewers, R. M., and Didham, R. K. (2006) Confounding factors in the detection of species responses to habitat fragmentation. *Biological Reviews*, **81**, 117–42.

Ezenwa, V. O., Godsey, M. S., King, R. J., and Guptill, S. C. (2006) Avian diversity and West Nile virus: Testing associations between biodiversity and infectious disease risk. *Proceedings of the Royal Society B: Biological Sciences*, **273**, 109–17.

Fabos, J. G. (2004) Greenway planning in the United States: its origins and recent case studies. *Landscape and Urban Planning*, **68**, 321–42.

Fagan, W. F., Meir, E., Prendergast, J., Folarin, A., and Karieva, P. (2001) Characterizing population vulnerability for 758 species. *Ecology Letters*, **4**, 132–8.

Fahnestock, J. T. and Detling, J. K. (2002) Bison–prairie dog–plant interactions in a North American mixed-grass prairie. *Oecologia*, **132**, 86–95.

Fahrig, L. (2003) Effects of habitat fragmentation on biodiversity. *Annual Review of Ecology Evolution and Systematics*, **34**, 487–515.

Falk, D. A., Palmer, M. A., and Zedler, J. B. (eds.) (2006) *Foundations of Restoration Ecology*. Island Press, Washington, DC.

Falkowski, P. G., Fenchel, T., and Delong, E. F. (2008) The microbial engines that drive Earth's biogeochemical cycles. *Science*, **320**, 1034–9.

FAO (2003) *World Agriculture: Towards 2015/2030. An FAO Perspective*. Rome.

FAO (1998) Food and Agriculture Organization of the United Nations (Sustainable Development Department). *Soil and Microbial Biodiversity*. http://www.fao.org/sd/EPdirect/EPre0045.htm.

FAOstat data (2006) http://faostat.fao.org/.

Fargione, J. E. and Tilman, D. (2005) Diversity decreases invasion via both sampling and complementarity effects. *Ecology Letters*, **8**, 604–11.

Fargione, J. Brown, C. S., and Tilman, D. (2003) Community assembly and invasion: an experimental test of neutral versus niche processes. *Proceedings of the National Academy of Sciences of the USA*, **100**, 8916–20.

Fargione, J., Tilman, D., Dybzinski, R., *et al.* (2007) From selection to complementarity: shifts in the causes of biodiversity–productivity relationships in a long-term biodiversity experiment. *Proceedings of the Royal Society B: Biological Sciences*, **274**, 871–6.

Farnsworth, E. J. and Ellison, A. M. (1997) The global conservation status of mangroves. *Ambio*, **26**(6), 328–34.

Farrelly, V., Rainey, F. A., and Stackebrandt, E. (1995) Effect of genome size and Rrn Gene Copy Number on PCR amplification of 16s ribosomal-RNA genes from a mixture of bacterial species. *Applied and Environmental Microbiology*, **61**, 2798–801.

Fearnside, P. M. (2006a). Mitigation of climatic change in the Amazon. In Laurance, W. F. and Peres, C. A. (eds.) *Emerging Threats to Tropical Forests*. University of Chicago Press, Chicago.

Fearnside, P. M. (2006b). Tropical deforestation and global warming. *Science*, **312**, 1137.

Fearnside, P. M. and Barbosa, K. I. (2004) Accelerating deforestation in Brazilian Amazonia: towards answering open questions. *Environmental Conservation*, **31**, 7–10.

Fédoroff, E., Ponge, J. F., Dubs, F., Fernandez-Gonzalez, F., and Lavelle, P. (2005) Small-scale response of plant species to land-use intensification. *Agriculture Ecosystems & Environment*, **105**, 283–90.

Feehan, J., Gillmor, D. A., and Culleton, N. (2005) Effects of an agri-environment scheme on farmland biodiversity in Ireland. *Agriculture Ecosystems & Environment*, **107**, 275–86.

Feeley, K. J., Gillespie, T. W., Lebbin, D. J., and Walter, H. S. (2007) Species characteristics associated with extinction vulnerability and nestedness rankings of birds in tropical forest fragments. *Animal Conservation*, **10**, 493–501.

Feng, Y. L., Auge, H., and Ebeling, S. K. (2007) Invasive *Buddleja davidii* allocates more nitrogen to its photosynthetic machinery than five native woody species. *Oecologia*, **153**, 501–10.

Fenner, M. and Thompson, K. (2005) *The Ecology of Seeds*. Cambridge University Press, Cambridge.

Ferrari, J., Muller, C. B., Kraaijeveld, A. R., and Godfray, H. C. J. (2001) Clonal variation and covariation in aphid resistance to parasitoids and a pathogen. *Evolution*, **55**, 1805–14.

Ferraro, P. J. and Simpson, R. D. (2002) The cost-effectiveness of conservation payments. *Land Economics*, **78**, 339–53.

Field, B. C. and Conra, J. M. (1975) Economic issues in programs of transferable development rights. *Land Economics*, **4**, 331–40.

Field, C. B., Campbell, J. E., and Lobell, D. B. (2008) Biomass energy: the scale of the potential resource. *Trends in Ecology & Evolution*, **23**, 65–72.

Fierer, N., Bradford, M. A., and Jackson, R. B. (2007) Toward an ecological classification of soil bacteria. *Ecology*, **88**, 1354–64.

Finke, D. L. and Denno, R. F. (2004) Predator diversity dampens trophic cascades. *Nature*, **429**, 407–10.

Finke, D. L. and Denno, R. F. (2005) Predator diversity and the functioning of ecosystems: the role of intraguild predation in dampening trophic cascades. *Ecology Letters*, **8**, 1299–306.

Finke, D. L. and Denno, R. F. (2006) Spatial refuge from intraguild predation: Implications for prey suppression and trophic cascades. *Oecologia*, **149**, 265–75.

Finnoff, D. and Tschirhart, J. (2003a). Protecting an endangered species while harvesting its prey in a general equilibrium ecosystem model. *Land Economics*, **79**, 160–80.

Finnoff, D. and Tschirhart, J. (2003b). Harvesting in an eight species ecosystem. *Journal of Environmental Economics and Management*, **45**, 589–611.

Finnoff, D., and Tschirhart, J. (2005) Identifying, preventing and controlling successful invasive plant species using their physiological traits. *Ecological Economics*, **52**, 397–416.

Finnoff, D. and Tschirhart, J. (2006) Using oligopoly theory to examine individual plant versus community optimization and evolutionary stable objectives. *Natural Resource Modeling*, **20**, 61–86.

Finnoff, D. and Tschirhart, J. (2007) *Linking Dynamic Economic and Ecological General Equilibrium Models*. Working paper, Dept. of Economics, University of Wyoming.

Finnoff, D., Strong, A., and Tschirhart, J. (2007) Stocking regulations and the spread of invasive plants. In J. F. Shogren *et al.* (eds.) *Integrating Economics and Biology for Bioeconomic Risk Assessment/Management of Invasive Species in Agriculture*. Report to Economic Research Service, USDA, Washington DC.

Fischer, J., Lindenmayer, D. B., Blomberg, S. P., Montague-Drake, R., Felton, A., and Stein, J. A. (2007) Functional richness and relative resilience of bird communities in regions with different land use intensities. *Ecosystems*, **10**, 964–74.

Fischlin, A., Midgley, G., Price, J., *et al.* (2007) Ecosystems, their properties, goods, and services. In IPCC (ed.) *Climate Change 2007: Impacts, Adaptation and Vulnerability. Working Group II Contribution to the Intergovernmental Panel on Climate Change Fourth Assessment Report*. Cambridge University Press, Cambridge.

Fisher, D. O., Blomberg, S. P., and Owens, I. P. F. (2003) Extrinsic versus intrinsic factors in the decline and extinction of Australian marsupials. *Proceedings of the Royal Society of London Series B: Biological Sciences*, **270**, 1801–8.

Fisher, S. G. and Likens, G. E. (1973) Energy flow in Bear Brook, New Hampshire: an integrative approach to stream ecosystem metabolism. *Ecological Monographs*, **43**, 421–39.

Flint, R. W. and Kalke, R. D. (2005) Reinventing the wheel in ecology research? *Science*, **307**, 1875–6.

Flynn, D. F. B., He, J.-S., Wolf-Bellin, K. S., Schmid, B., and Bazzaz, F. A. (2008) Hierarchical reliability in experimental plant communities. *Journal of Plant Ecology*, **1**, 59–65.

Foley, J. A., DeFries, R., Asner, G. P., *et al.* (2005) Global consequences of land use. *Science*, **309**, 570–4.

Folke, C., Carpenter, S., Elmqvist, T., Gunderson, L., Holling, C. S., and Walker, B. (2002) Resilience and sustainable development: building adaptive capacity in a world of transformations. *AMBIO: A Journal of the Human Environment*, **31**, 437–40.

Folke, C., Holling, C. S., and Perrings, C. (1996) Biological diversity, ecosystems and the human scale. *Ecological Applications*, **6**, 1018–24.

Fonseca, C. R. and Ganade, G. (2001) Species functional redundancy, random extinctions and the stability of ecosystems. *Journal of Ecology*, **89**, 118–25.

Fontaine, S., Bardoux, G., Abbadie, L., and Mariotti, A. (2004) Carbon input to soil may decrease soil carbon content. *Ecology Letters*, **7**, 314–20.

Forsyth, M. (2000) On estimating the option value of preserving a wilderness area. *Canadian Journal of Economics*, **33**, 413–34.

Forup, M. L., Henson, K. S. E., Craze, P. G., and Memmott, J. (2008) The restoration of ecological interactions: plant–pollinator networks on ancient and restored heathlands. *Journal of Applied Ecology*, **45**, 742–52.

Foufopoulos, J. and Ives, A. R. (1999) Reptile extinction on land-bridge islands: life-history attributes and vulnerability to extinction. *American Naturalist*, **153**, 1–25.

Foufopoulos, J. and Mayer, G. C. (2007) Turnover of passerine birds on islands in the Aegean Sea (Greece). *Journal of Biogeography*, **34**, 1113–23.

Fox, J. W. (2004a). Effects of algal and herbivore diversity on the partitioning of—biomass within and among trophic levels. *Ecology*, **85**, 549–59.

Fox, J. W. (2004b). Modelling the joint effects of predator and prey diversity on total prey biomass. *Journal of Animal Ecology*, **73**, 88–96.

Fox, J. W. (2005a). Biodiversity, food web structure, and the partitioning of biomass within and among trophic levels. In P. C. de Ruiter, V. Wolters, and J. C. Moore (eds.) *Dynamic Food Webs: Multispecies Assemblages, Ecosystem Development, and Environmental Change*, pp. 283–94. Elsevier, Amsterdam.

Fox, J. W. (2005b). Interpreting the 'selection effect' of biodiversity on ecosystem function. *Ecology Letters*, **8**, 846–56.

Fox, J. W. (2006) Using the Price Equation to partition the effects of biodiversity loss on ecosystem function. *Ecology*, **87**, 2687–96.

Fox, J. W. and Harpole, W. S. (2008) Revealing how species loss affects ecosystem function: the trait-based price equation partition. *Ecology*, **89**, 269–79.

France, K. E. and Duffy, J. E. (2006a). Consumer diversity mediates invasion dynamics at multiple trophic levels. *Oikos*, **113**, 515–29.

France, K. E. and Duffy, J. E. (2006b). Diversity and dispersal interactively affect predictability of ecosystem function. *Nature*, **441**, 1139–43.

Frank, S. A. (1995) George Price's contributions to evolutionary genetics. *Journal of Theoretical Biology*, **175**, 373–88.

Frank, S. A. (1997) The Price Equation, Fisher's fundamental theorem, kin selection and causal analysis. *Evolution*, **51**, 1712–29.

Franklin, R. B., Garland, J. L., Bolster, C. H., and Mills, A. L. (2001) Impact of dilution on microbial community structure and functional potential: comparison of numerical simulations and batch culture experiments. *Applied and Environmental Microbiology*, **67**, 702–12.

Franzén, D. (2004) Plant species coexistence and dispersion of seed traits in a grassland. *Ecography*, **27**, 218–24.

Free, J. B. (1993) *Insect Pollination of Crops*. Academic Press, London.

Freeman, A. M. (2003) *The Measurement of Environmental and Resource Values: Theory and Methods*. Resources For the Future, Washington, DC.

Freville, H., McConway, K., Dodd, M., and Silvertown, J. (2007) Prediction of extinction in plants: interaction of extrinsic threats and life history traits. *Ecology*, **88**, 2662–72.

Fridley, J. D., Stachowicz, J. J., Naeem, S., *et al.* (2007) The invasion paradox: reconciling pattern and process in species invasions. *Ecology*, **88**, 3–17.

Frivold, L. and Frank, J. (2002) Growth of mixed birch–coniferous stands in relation to pure coniferous stands at similar sites in south-eastern Norway. *Scandinavian Journal of Forest Research*, **17**, 139–49.

Fuhrman, J. A., Hewson, I., Schwalbach, M. S., Steele, J. A., Brown, M. V., and Naeem, S. (2006) Annually reoccurring bacterial communities are predictable from ocean conditions. *Proceedings of the National Academy of Sciences of the USA*, **103**, 13104.

Fukami, T. and Morin, P. J. (2003) Productivity–biodiversity relationships depend on the history of community assembly. *Nature*, **424**, 423–6.

Fukami, T. and Wardle, D. A. (2005) Long-term ecological dynamics: reciprocal insights from natural and anthropogenic gradients. *Proceedings of the Royal Society B: Biological Sciences*, **272**, 2105–15.

Fukami, T. Naeem, S., and Wardle, D. (2001) On similarity among local communities in biodiversity experiments. *Oikos*, **95**, 340–8.

Fukami, T., Bezemer, T. M., Mortimer, S. R., and Van der Putten, W. H. (2005) Species divergence and trait convergence in experimental plant community assembly. *Ecology Letters*, **8**, 1283–90.

Fukami, T., Beaumont, H. J., Zhang, X. X., and Rainey, P. B. (2007) Immigration history controls diversification in experimental adaptive radiation. *Nature*, **446**, 436–9.

Fumanal, B., Chauvel, B., Sabatier, A., and Bretagnolle, F. (2007) Variability and cryptic heteromorphism of *Ambrosia artemisiifolia* seeds: what consequences for its invasion in France? *Annals of Botany*, **100**, 305–13.

Gabriel, D., Roschewitz, I., Tscharntke, T., and Thies, C. (2006) Beta diversity at different spatial scales: plant communities in organic and conventional agriculture. *Ecological Applications*, **16**, 2011–21.

Gamfeldt, L., Hillebrand, H., and Jonsson, P. R. (2005) Species richness changes across two trophic levels simultaneously affect prey and consumer biomass. *Ecology Letters*, **8**, 696–703.

Gamfeldt, L., Hillebrand, H., and Jonsson, P. R. (2008) Multiple functions increase the importance of biodiversity for overall ecosystem functioning. *Ecology*, **89**, 1223–31.

Garant, D., Kruuk, L. E. B., Wilkin, T. A., McCleery, R. H., and Sheldon, B. C. (2005) Evolution driven by differential dispersal within a wild bird population. *Nature*, **433**, 60–5.

Garcia-Romero, A., Oropeza-Orozco, O., and Galicia-Sarmiento, L. (2004) Land-use systems and resilience of tropical rain forests in the Tehuantepec Isthmus, Mexico. *Environmental Management*, **34**, 768–85.

Garnier, E., Cortez, J., Billes, G., Navas, M. L., Roumet, C., Debussche, M., Laurent, G., Blanchard, A., Aubry, D., Bellmann, A., Neill, C., and Toussaint, J. P. (2004) Plant functional markers capture ecosystem properties during secondary succession. *Ecology*, **85**, 2630–7.

Garnier, E., Navas, M. L., Austin, M. P., Lilley, J. M., and Gifford, R. M. (1997) A problem for biodiversity–productivity studies: how to compare the productivity of multispecific plant mixtures to that of monocultures? *Acta Oecologica*, **18**, 657–70.

Gartner, T. B. and Cardon, Z. G. (2004) Decomposition dynamics in mixed-species leaf litter. *Oikos*, **104**, 230–46.

Gascon, C. and Lovejoy, T. E. (1998) Ecological impacts of forest fragmentation in central Amazonia. *Zoology – Analysis of Complex Systems*, **101**, 273–80.

Gastine, A., Scherer-Lorenzen, M., Leadley, P. W. (2003) No consistent effects of plant diversity on root biomass, soil biota and soil abiotic conditions in temperate grassland communities. *Applied Soil Ecology*, **24**, 101–11.

Gaston, K. J. and Blackburn, T. M. (1995) Birds, body-size and the threat of extinction. *Philosophical Transactions of the Royal Society of London Series B: Biological Sciences*, **347**, 205–12.

Gaston, K. J., Blackburn, T. M., and Goldewijk, K. K. (2003) Habitat conversion and global avian biodiversity loss. *Proceedings of the Royal Society Series B: Biological Sciences*, **270**, 1293–300.

Gaudet, C. L. and Keddy, P. A. (1988) A comparative approach to predicting competitive ability from plant traits. *Nature*, **334**, 242–3.

Geist, C. and Galatowitsch, S. M. (1999) Reciprocal model for meeting ecological and human needs in restoration projects. *Conservation Biology*, **13**, 970–9.

Geist, H. J. and Lambin, E. F. (2002) Proximate causes and underlying driving forces of tropical deforestation. *Bioscience*, **52**, 143–50.

Genghini, M., Gellini, S., and Gustin, M. (2006) Organic and integrated agriculture: The effects on bird communities in orchard farms in northern Italy. *Biodiversity and Conservation*, **15**, 3077–94.

Ghazoul, J. (2002) Flowers at the front line of invasion? *Ecological Entomology*, **27**, 638–40.

Ghazoul, J. (2006) Floral diversity and the facilitation of pollination. *Journal of Ecology*, **94**, 295–304.

Ghazoul, J. (2007) Challenges to the uptake of the ecosystem service rationale for conservation. *Conservation Biology*, **21**, 1651–2.

Gibson, R. H., Nelson, I. L., Hopkins, G. W., Hamplett, B. J., and Memmott, J. (2006) Pollinator webs, plant communities and the conservation of rare plants: arable weeds as a case study. *Journal of Applied Ecology*, **43**, 246–57.

Gido, K. B. and Franssen, N. R. (2007) Invasion of stream fishes into low trophic positions. *Ecology of Freshwater Fish*, **16**, 457–64.

Gilbert, F., Gonzalez, A., and Evans-Freke, I. (1998) Corridors maintain species richness in the fragmented landscapes of a microecosystem. *Proceedings of the Royal Society of London B*, **265**, 577–82.

Giller, P. S., Hillebrand, H., Berninger, U. G., et al. (2004) Biodiversity effects on ecosystem functioning: emerging issues and their experimental test in aquatic environments. *Oikos*, **104**, 423–36.

Gillespie, T. W. (2001) Application of extinction and conservation theories for forest birds in Nicaragua. *Conservation Biology*, **15**, 699–709.

Gillison, A. N., Liswanti, N., Budidarsono, S., van Noordwijk, M., and Tomich, T. P. (2004) Impact of cropping methods on biodiversity in coffee agroecosystems in Sumatra, Indonesia. *Ecology and Society*, **9**, 7.

Girvan, M. S., Campbell, C. D., Killham, K., Prosser, J. I., and Glover, L. A. (2005) Bacterial diversity promotes community stability and functional resilience after perturbation. *Environmental Microbiology*, **7**, 301–13.

Gitay, H., Suárez, A., Watson, R., and Dokken, D. (2002) *Climate Change and Biodiversity*. IPCC Technical Paper V – April 2002. IPCC, Geneva, Switzerland.

Gittleman, J. L. and Purvis, A. (1998) Body size and species-richness in carnivores and primates. *Proceedings of the Royal Society of London B*, **265**, 113–19.

Givnish, T. J. (1994) Does diversity beget stability? *Nature*, **371**, 113–14.

Glor, R. E., Flecker, A. S., Benard, M. F., and Power, A. G. (2001) Lizard diversity and agricultural disturbance in a Caribbean forest landscape. *Biodiversity and Conservation*, **10**, 711–23.

Godbold, J. A. (2008) Marine benthic biodiversity–ecosystem function relations in complex systems. *Unpublished Ph.D. Thesis*, University of Aberdeen, UK.

Goeschl, T. and Swanson, T. (2002), On the economic limits of technological potential: will industry resolve the resistance problem? In T. Swanson (ed.) *The Economics of Managing Biotechnologies*, pp. 99–128. Kluwer Academic Publishers, Dordrecht/London/Boston.

Goeschl, T. and Swanson, T. (2003) Pests, plagues, and patents. *Journal of the European Economic Association*, **1**(2–3), 561–75.

Gomez-Aparicio, L., Canham, C. D., and Martin, P. H. (2008) Neighbourhood models of the effects of the invasive *Acer platanoides* on tree seedling dynamics:

linking impacts on communities and ecosystems. *Journal of Ecology*, **96**, 78–90.

Gonzalez-Megias, A., Menendez, R., Roy, D., Brereton, T., and Thomas, C. D. (2008) Changes in the composition of British butterfly assemblages over two decades. *Global Change Biology*, **14**, 1464–74.

Gonzalez, A. (2000) Community relaxation in fragmented landscapes: the relation between species, area and age. *Ecology Letters*, **3**, 441–6.

Gonzalez, A. and Chaneton, E. (2002) Heterotroph species extinction, abundance and biomass dynamics in an experimentally fragmented microecosystem. *Journal of Animal Ecology*, **71**, 594–602.

Gonzalez, A. and Descamps-Julien, B. (2004) Population and community variability in randomly fluctuating environments. *Oikos*, **106**, 105–16.

Gonzalez, A., Mouquet, N., and Loreau, M. (2009) Biodiversity as spatial insurance: the effects of fragmentation and dispersal on ecosystem functioning. In S. Naeem, D. Bunker, A. Hector, M. Loreau, and C. Perrings (eds.) *Biodiversity and Human Impacts*. Oxford University Press, Oxford.

Gordon, C., Manson, R., Sundberg, J., and Cruz-Angon, A. (2007) Biodiversity, profitability, and vegetation structure in a Mexican coffee agroecosystem. *Agriculture Ecosystems & Environment*, **118**, 256–66.

Gordon, D. R. (1998) Effects of invasive, non-indigenous plant species on ecosystem processes: lessons from Florida. *Ecological Applications*, **8**, 975–89.

Gottschalk, T. K., Diekotter, T., Ekschmitt, K., et al. (2007) Impact of agricultural subsidies on biodiversity at the landscape level. *Landscape Ecology*, **22**, 643–56.

Gough, L., Shrestha, K., Johnson, D. R., and Moon, B. (2008) Long-term mammalian herbivory and nutrient addition alter lichen community structure in Alaskan dry heath tundra. *Arctic, Antarctic, and Alpine Research*, **40**, 65–73.

Gould, A. M. A. and Gorchov, D. L. (2000) Effects of the exotic invasive shrub *Lonicera maackii* on the survival and fecundity of three species of native annuals. *American Midland Naturalist*, **144**, 36–50.

Goulder, L. H. and Kennedy, D. (1997) Valuing ecosystem services: philosophical bases and empirical methods. In G. C. Daily (ed.) *Nature's Services*, pp. 23–48. Island Press, Washington, DC.

Gower, S. T. (2003) Patterns and mechanisms of the forest carbon cycle. *Annual Review of Environment and Resources*, **28**, 169–204.

Grace, J. B., Michael Anderson, T., Smith, M. D., et al. (2007) Does species diversity limit productivity in natural grassland communities? *Ecology Letters*, **10**, 680–9.

Graham, M. and Kennedy, J. (2007) Visual exploration of alternative taxonomies through concepts. *Ecological Informatics*, **2**, 248–61.

Grasman, R. and Gramacy, R. B. (2008) Geometry: mesh generation and surface tesselation. R package version 0.1-1.

Grau, H. R., Aide, T. M., Zimmerman, J. K., Thomlinson, J. R., Helmer, E., and Zou, X. M. (2003) The ecological consequences of socioeconomic and land-use changes in postagriculture Puerto Rico. *Bioscience*, **53**, 1159–68.

Green, J. L., Holmes, A. J., Westoby, M., Oliver, I., Briscoe, D., Dangerfield, M., Gillings, M., and Beattie, A. J. (2004) Spatial scaling of microbial eukaryotic diversity. *Nature*, **432**, 747–50.

Green, R. E., Cornell, S. J., Scharlemann, J. P. W., and Balmford, A. (2005) Farming and the fate of wild nature. *Science*, **307**, 550–5.

Greenleaf, S. A. and Kremen, C. (2006a). Wild bees enhance honeybees' pollination of hybrid sunflower. *Proceedings of the National Academy of Sciences of the USA*, **103**, 13890–5.

Greenleaf, S. A. and Kremen, C. (2006b). Wild bee species increase tomato production and respond differently to surrounding land use in Northern California. *Biological Conservation*, **133**, 128–35.

Griffin, J., O'Gorman, E., Emmerson, M., et al. (2009) Biodiversity and the stability of ecosystem functioning. In S. Naeem, D. Bunker, A. Hector, M. Loreau, and C. Perrings (eds.) *Biodiversity and Human Impacts*. Oxford University Press, Oxford.

Griffiths, B. S., Ritz, K., Bardgett, R. D., et al. (2000) Ecosystem response of pasture soil communities to fumigation-induced microbial diversity reductions: an examination of the biodiversity–ecosystem function relationship. *Oikos*, **90**, 279–94.

Griffiths, B. S., Bonkowski, M., Roy, J., and Ritz, K. (2001) Functional stability, substrate utilisation and biological indicators of soils following environmental impacts. *Applied Soil Ecology*, **16**, 49–61.

Griffiths, B. S., Kuan, H. L., Ritz, K., Glover, L. A., McCaig, A. E., and Fenwick, C. (2004) The relationship between microbial community structure and functional stability, tested experimentally in an upland pasture soil. *Microbial Ecology*, **47**, 104–13.

Griffiths, G. J. K., Holland, J. M., Bailey, A., and Thomas, M. B. (2008) Efficacy and economics of shelter habitats for conservation biological control. *Biological Control*, **45**, 200–9.

Grime, J. P. (1979) *Plant Strategies and Vegetation Processes*. Wiley, New York.

Grime, J. P. (1998) Benefits of plant diversity to ecosystems: immediate, filter and founder effects. *Journal of Ecology*, **86**, 902–10.

Grime, J. P. (2006) Trait convergence and trait divergence in herbaceous plant communities: mechanisms and consequences. *Journal of Vegetation Science*, **17**, 255–60.

Gross, K., and Cardinale, B. J. (2005) The functional consequences of random vs. ordered species extinctions. *Ecology Letters*, **8**, 409–18.

Gross, M. (2008) Algal biofuel hopes. *Current Biology*, **18**, 46–7.

Grover, J. P. and Loreau, M. (1996) Linking communities and ecosystems: trophic interactions as nutrient cycling pathways. In M. E. Hochberg, J. Clobert, and R. Barbault (eds.) *Aspects of the Genesis and Maintenance of Biological Diversity*. Oxford University Press, Oxford.

Guariguata, M. R., Rheingans, R., and Montagnini, F. (1995) Early woody invasion under tree plantations in Costa Rica: Implications for forest restoration. *Restoration Ecology*, **3**, 252–60.

Gunderson, L. H. (2000) Ecological resilience – in theory and application. *Annual Review of Ecology and Systematics*, **31**, 425–39.

Guo, L. B. and Gifford, R. M. (2002) Soil carbon stocks and land use change: a meta analysis. *Global Change Biology*, **8**, 345–60.

Gurevitch, J. and Padilla, D. K. (2004) Are invasive species a major cause of extinctions? *Trends in Ecology & Evolution*, **19**, 470–4.

Guterman, L. (2000) Have ecologists oversold biodiversity? Some scientists question experiments on how numerous species help ecosystems. *The Chronicle of Higher Education*, **47**, A24–A26.

Haberl, H., Erb, K. H., Krausmann, F., et al. (2007) Quantifying and mapping the human appropriation of net primary production in earth's terrestrial ecosystems. *Proceedings of the National Academy of Sciences of the USA*, **104**, 12942–7.

Hacker, S. D. and Dethier, M. N. (2006) Community modification by a grass invader has differing impacts for marine habitats. *Oikos*, **113**, 279–86.

Haggar, J. P. and Ewel, J. J. (1997) Primary productivity and resource partitioning in model tropical ecosystems. *Ecology*, **78**, 1211–21.

Haines-Young, R. H., Barr, C. J., Black, H. I. J., et al. (2000) *Accounting for Nature: Assessing Habitats in the UK Countryside*. Department of the Environment, Transport and the Regions, London.

Hairston, N. G., Smith, F. E., and Slobodkin, L. B. (1960) Community structure, population control, and competition. *American Naturalist*, **94**, 421–5.

Halpern, B. S., Walbridge, S., Selkoe, K. A., et al. (2008) A global map of human impact on marine ecosystems. *Science*, **319**, 948–52.

Hambäck, P. A. (2001) Direct and indirect effects of herbivory: feeding by spittlebugs affects pollinator visitation rates and seedset of *Rudbeckia hirta*. *Ecoscience*, **8**, 45–50.

Hamilton, K. and Clemens, M. (1999) Genuine savings rates in developing countries. *World Bank Economic Review*, **13**, 333–56.

Hamilton, M. A., Murray, B. R., Cadotte, M. W., et al. (2005) Life-history correlates of plant invasiveness at regional and continental scales. *Ecology Letters*, **8**, 1066–74.

Hannon, B. (1973) The structure of ecosystems. *Journal of Theoretical Biology*, **41**, 535–46.

Hanski, I. and Ovaskainen, O. (2002) Extinction debt at extinction threshold. *Conservation Biology*, **16**, 666–73.

Harakunarak, A. and Aksornkoae, S. (2005) Life-saving belts: Post-tsunami reassessment of mangrove ecosystem values and management in Thailand. *Tropical Coasts*, July, 48–55.

Harcourt, A. H., Coppeto, S. A., and Parks, S. A. (2002) Rarity, specialization and extinction in primates. *Journal of Biogeography*, **29**, 445–56.

Harper, J. L. (1977) *Population Biology of Plants*. Academic Press, London.

Harpole, W. S. and Tilman, D. (2007) Grassland species loss resulting from reduced niche dimension. *Nature*, **446**, 791–3.

Harte, J. and Shaw, R. (1995) Shifting dominance within a montane vegetation community – results of a climate-warming experiment. *Science*, **267**, 876–80.

Hartwick, J. M. (1977) Intergenerational equity and the investing of rents from exhaustible resources. *American Economic Review*, **66**, 972–4.

Hartwick, J. M. (1978) Substitution among exhaustible resources and intergenerational equity. *Review of Economic Studies*, **45**(2), 347–54.

Harvey, C. A., Medina, A., Sanchez, D. M., et al. (2006) Patterns of animal diversity in different forms of tree cover in agricultural landscapes. *Ecological Applications*, **16**, 1986–99.

Hassan, R. (2003) Measuring asset values and flow benefits of non-traded products and ecosystems services of forest and woodland resources in South Africa. *Environment, Development and Sustainability*, **5**, 403–18.

Hassell, M. P., Comins, H. N., and May, R. M. (1994) Species coexistence and self-organizing spatial dynamics. *Nature*, **379**, 290–2.

Hasselwandter, K. (1997) Soil micro-organisms, mycorrhiza, and restoration ecology. In K. M. Urbanska, N. R. Webb, and P. J. Edwards (eds.) *Restoration Ecology and Sustainable Development*. Cambridge University Press, New York.

Hättenschwiler, S. (2005) Effect of tree species diversity on litter quality and decomposition. In M. Scherer-Lorenzen, C. Körner, and E.-D. Schulze (eds.) *The Functional Significance of Forest Diversity*, Springer-Verlag, Berlin, Heidelberg.

Hättenschwiler, S. and Gasser, P. (2005) Soil animals alter plant litter diversity effects on decomposition. *Proceedings of the National Academy of Sciences of the USA*, **102**, 1519–24.

Hättenschwiler, S., Tiunov, A. V., and Scheu, S. (2005) Biodiversity and litter decomposition in terrestrial ecosystems. *Annual Review of Ecology, Evolution, and Systematics*, **36**, 191–218.

Havel, J. E. and Medley, K. A. (2006) Biological invasions across spatial scales: intercontinental, regional, and local dispersal of Cladoceran zooplankton. *Biological Invasions*, **8**, 459–73.

Hawes, C., Begg, G. S., Squire, G. R., and Iannetta, P. P. M. (2005) Individuals as the basic accounting unit in studies of ecosystem function: functional diversity in shepherd's purse, *Capsella*. *Oikos*, **109**, 521–34.

He, F. L. and Hu, X. S. (2005) Hubbell's fundamental biodiversity parameter and the Simpson diversity index. *Ecology Letters*, **8**, 386–90.

Heal, G. (2000) Biodiversity as a commodity. In S. A. Levin (ed.) *Encyclopedia of Biodiversity*, Vol. 1, pp. 359–76. Academic Press, New York.

Heal, G. M., Barbier, E. B., Boyle, K. J., *et al.* (2005) *Valuing Ecosystem Services: Toward Better Environmental Decision Making*. The National Academies Press, Washington, DC.

Hector, A. (1998) The effect of diversity on productivity: detecting the role of species complementarity. *Oikos*, **82**, 597–9.

Hector, A. and Bagchi, R. (2007) Biodiversity and ecosystem multifunctionality. *Nature*, **448**, 188–90.

Hector, A. and Hooper, R. (2002) Darwin and the first ecological experiment. *Science*, **295**, 639–40.

Hector, A., Schmid, B., Beierkuhnlein, C., *et al.* (1999) Plant diversity and productivity experiments in european grasslands. *Science*, **286**, 1123–7.

Hector, A., Beale, A. J., Minns, A., Otway, S. J., and Lawton, J. H. (2000) Consequences of the reduction of plant diversity for litter decomposition: effects through litter quality and microenvironment. *Oikos*, **90**, 357–71.

Hector, A., Dobson, K., Minns, A., Bazeley-White, E., and Lawton, J. H. (2001) Community diversity and invasion resistance: an experimental test in a grassland ecosystem and a review of comparable studies. *Ecological Research*, **16**, 819–31.

Hector, A., Bazeley-White, E., Loreau, M., Otway, S., and Schmid, B. (2002a) Overyielding in plant communities: Testing the sampling effect hypothesis with replicated biodiversity experiments. *Ecology Letters*, **5**, 502–11.

Hector, A., Loreau, M., Schmid, B. and The BIODEPTH Project (2002b) Biodiversity manipulation experiments: studies replicated at multiple sites. In M. Loreau, S. Naeem, and P. Inchausti (eds.) *Biodiversity and Ecosystem Functioning: Synthesis and Perspectives*. Oxford University Press, Oxford.

Hector, A., Joshi, J., Scherer-Lorenzen, M., *et al.* (2007) Biodiversity and ecosystem functioning: reconciling the results of experimental and observational studies. *Functional Ecology*, **21**, 998–1002.

Hedlund, K., Regina, I. S., Van der Putten, W. H., Leps, J., Diaz, T., Korthals, G. W., Lavorel, S., Brown, V. K., Gormsen, D., Mortimer, S. R., Barrueco, C. R., Roy, J., Smilauer, P., Smilauerova, M., and Van Dijk, C. (2003) Plant species diversity, plant biomass and responses of the soil community on abandoned land across Europe: idiosyncrasy or above–belowground time lags. *Oikos*, **103**, 45–58.

Heemsbergen, D. A., Berg, M. P., Loreau, M., van Hal, J. R., Faber, J. H., and Verhoef, H. A. (2004) Biodiversity effects on soil processes explained by interspecific functional dissimilarity. *Science*, **306**, 1019.

Heino, J. (2005) Functional biodiversity of macroinvertebrate assemblages along major ecological gradients of boreal headwater streams. *Freshwater Biology*, **50**, 1578–87.

Heisse, K., Roscher, C., Schumacher, J., and Schulze, E. D. (2007) Establishment of grassland species in monocultures: different strategies lead to success. *Oecologia*, **152**, 435–47.

Henle, K., Davies, K. F., Kleyer, M., Margules, C. and Settele, J. (2004) Predictors of species sensitivity to fragmentation. *Biodiversity and Conservation*, **13**, 207–51.

Henle, K., Dziock, F., Foeckler, F., *et al.* (2006) Study design for assessing species environment relationships and developing indicator systems for ecological changes in floodplains – the approach of the Riva project. *International Review of Hydrobiology*, **91**, 292–313.

Herendeen, R. (1991) Do economic-like principles predict ecosystem behavior under changing resource constraints? In M. Higashi and T. Burns (eds.) *Theoretical Studies of Ecosystems: the Network Perspective*. Cambridge University Press, New York.

Hero, J. M., Williams, S. E., and Magnusson, W. E. (2005) Ecological traits of declining amphibians in upland areas of eastern Australia. *Journal of Zoology*, **267**, 221–32.

Herrera, C. M. (1988) Variation in mutualisms: the spatiotemporal mosaic of a pollinator community. *Biological Journal of the Linnean Society*, **35**, 95–125.

Herrmann, F., Westphal, C., Moritz, R. F. A., and Steffan-Dewenter, I. (2007) Genetic diversity and mass resources promote colony size and forager densities of a social bee (*Bombus pascuorum*) in agricultural landscapes. *Molecular Ecology*, **16**, 1167–78.

Hicks, J. R. (1939) *Value and Capital*. Clarendon Press, Oxford.

Hildén, M., Furman, E., Varjopuro, R., *et al.* (2006) Views on biodiversity research in Europe. *Reports of Finnish Environment Institute*, No. 16. Finland.

Hilderbrand, R. H., Watts, A. C., and Randle, A. M. (2005) The myths of restoration ecology. *Ecology and Society*, **10**, Article 19.

Hill, A. R. (1996) Nitrate removal in stream riparian zones. *Journal of Environmental Quality*, **25**, 743–55.

Hillebrand, H. and Cardinale, B. J. (2004) Consumer effects decline with prey diversity. *Ecology Letters*, **7**, 192–201.

Hiremath, A. J. and Ewel, J. J. (2001) Ecosystem nutrient use efficiency, productivity, and nutrient accrual in model tropical communities. *Ecosystems*, **4**, 669–82.

Hirsch, J. E. (2005) An index to quantify an individual's scientific research output. *Proceedings of the National Academy of Sciences of the USA*, **102**, 16569–72.

Hobbie, S. E. (1992) Effects of plant species on nutrient cycling. *Trends in Ecology and Evolution*, **7**, 336–9.

Hobbs, R. J. (2006) Foreword. In D. A. Falk, M. A. Palmer and J. B. Zedler (eds.) *Foundations of Restoration Ecology*. Island Press, Washington, DC.

Hobbs, R. J. and Huenneke, L. F. (1992) Disturbance, diversity, and invasion – implications for conservations. *Conservation Biology*, **6**, 324–37.

Hodgson, D. J., Rainey, P. B., and Buckling, A. (2002) Mechanisms linking diversity, productivity and invasibility in experimental bacterial communities. *Proceedings of the Royal Society of London Series B: Biological Sciences*, **269**, 2277–83.

Hodgson, J. G., Wilson, P. J., Hunt, R., Grime, J. P., and Thompson, K. (1999) Allocating C-S-R plant functional types: a soft approach to a hard problem. *Oikos*, **85**, 282–94.

Hodgson, J. R. and Illius, A. W. (eds.) (1996) *The Ecology and Management of Grazing Systems*. CAB International, New York.

Hoehn, P., Tscharntke, T., Tylianakis, J. M., and Steffan-Dewenter, I. (2008) Functional group diversity of bee pollinators increases crop yield. *Proceedings of the Royal Society of London, Series B*, **275**, 2283–91.

Holdren, J. P. (2008) Presidential Address: Science and Technology for Sustainable Well-Being. *Science*, **319**, 424–34.

Hole, D. G., Perkins, A. J., Wilson, J. D., Alexander, I. H., Grice, F., and Evans, A. D. (2005) Does organic farming benefit biodiversity? *Biological Conservation*, **122**, 113–30.

Holling, C. S. (1973) Resilience and the stability of ecological systems. *Annual Review of Ecology and Systematics*, **4**, 1–23.

Holling, C. S. (1988) Temperate forest insect outbreaks, tropical deforestation and migratory birds. *Memoirs of the Entomological Society of Canada*, **146**, 21–32.

Holling, C. S. (1996) Engineering resilience versus ecological resilience. In P. Schulze (ed.) *Engineering Within Ecological Constraints*, pp. 31–44. National Academy Press, Washington, DC.

Hollowell, V. C. (ed.) (2001) *Managing Human Dominated Ecosystems*. Missouri Botanical Garden Press, St Louis.

Holmes, R. T., Bonney, R. E. J., and Pacala, S. W. (1979) Guild structure of the Hubbard Brook bird community: a multivariate approach. *Ecology*, **60**, 512–20.

Holt, R. D. (1990) The microevolutionary consequences of climate change. *Trends in Ecology and Evolution*, **5**, 311–15.

Holt, R. D. (1993) Ecology at the mesoscale: the influence of regional processes on local communities. In R. E. Ricklefs and D. Schluter (eds.) *Species Diversity in Ecological Communities: Historical and Geographical Perspectives*, pp. 77–88. University of Chicago Press, Chicago.

Holt, R. D. (2008) The community context of disease emergence: could changes in predation be a key driver? In R. S. Ostfeld, F. Keesing, and V. T. Eviner (eds.) *Infectious Disease Ecology: Effects of Ecosystems on Disease and of Disease on Ecosystems*, pp. 324–46. Princeton University Press, Princeton, NJ.

Holt, R. D. and Loreau, M. (2002) Biodiversity and ecosystem functioning: the role of trophic interactions and the importance of system openness. In A. P. Kinzig, S. W. Pacala, and D. Tilman (eds.) *The Functional Consequences of Biodiversity*, pp. 213–45. Princeton University Press, Princeton, NJ.

Holt, R. D. and Polis, G. A. (1997) A theoretical framework for intraguild predation. *American Naturalist*, **149**, 745–64.

Holt, R. D., Grover, J., and Tilman, D. (1994) Simple rules for interspecific dominance in systems with exploitative and apparent competition. *American Naturalist*, **144**, 741–71.

Holway, D. A. (1999) Competitive mechanisms underlying the displacement of native ants by the invasive Argentine ant. *Ecology*, **80**, 238–51.

Holway, D. A., Lach, L., Suarez, A. V., Tsutsui, N. D., and Case, T. J. (2002) The causes and consequences of ant invasions. *Annual Review of Ecology and Systematics*, **33**, 181–233.

Holyoak, M. (2000) Habitat subdivision causes changes in food web structure. *Ecology Letters*, **3**, 509–15.

Holyoak, M., Leibold, M. A., and Holt, R. D. (2005) *Metacommunities: Spatial Dynamics and Ecological Communities*. Chicago Press.

Holzschuh, A., Steffan-Dewenter, I., Leijn, D., and Tscharntke, T. (2007) Diversity of flower-visiting bees in cereal fields: effects of farming system, landscape composition and regional context. *Journal of Applied Ecology*, **44**, 41–9.

Hooper, D. U. and Dukes, J. S. (2004) Overyielding among plant functional groups in a long-term experiment. *Ecology Letters*, **7**, 95–105.

Hooper, D. U. and Vitousek, P. M. (1997) The effects of plant composition and diversity on ecosystem processes. *Science*, **277**, 1302–5.

Hooper, D. and Vitousek, P. (1998) Effects of plant composition and diversity on nutrient cycling. *Ecological Monographs*, **68**, 121–49.

Hooper, D., Bignell, D., Brown, V., et al. (2000) Interactions between aboveground and belowground biodiversity in terrestrial ecosystems: patterns, mechanisms, and feedbacks. *Bioscience*, **50**, 1049–61.

Hooper, D. U., Solan, M., Symstad, A., et al. (2002) Species diversity, functional diversity, and ecosystem functioning. In M. Loreau, S. Naeem, and P. Inchausti (eds.) *Biodiversity and Ecosystem Functioning: Synthesis and Perspectives*. Oxford University Press, New York.

Hooper, D. U., Chapin, F. S., and Ewel, J. J. et al. (2005) Effects of biodiversity on ecosystem functioning: a consensus of current knowledge. *Ecological Monographs*, **75**, 3–35.

Horner-Devine, M. C., Carney, K. M., and Bohannan, B. J. M. (2004) An ecological perspective on bacterial biodiversity. *Proceedings of the Royal Society of London Series B: Biological Sciences*, **271**, 113–22.

Hoskin, C. J., Higgie, M., McDonald, K. R., and Moritz, C. (2005) Reinforcement drives rapid allopatric speciation. *Nature*, **437**, 1353–6.

Howarth, R. B. and Farber, S. (2002) Accounting for the value of ecosystem services. *Ecological Economics*, **41**, 421–9

Hoyle, M. (2004) Causes of the species–area relationship by trophic level in a field-based microecosystem. *Proceedings of the Royal Society of London Series B: Biological Sciences*, **271**, 1159–64.

Hraber, P. T. and Milne, B. T. (1997) Community assembly in a model ecosystem. *Ecological Modelling*, **103**, 267–85.

Hubbell, S. P. (2001) *The Unified Neutral Theory of Biodiversity and Biogeography*. Princeton University Press, Princeton.

Hughes, A. R. and Stachowicz, J. J. (2004) Genetic diversity enhances the resistance of a seagrass ecosystem to disturbance. *Proceedings of the National Academy of Sciences of the USA*, **101**, 8998–9002.

Hughes, A. R., Daily, G. C., and Ehrlich, P. R. (1997) Population diversity: its extent and extinction. *Science*, **278**, 689–92.

Hughes, A. R., Inouye, B. D., Johnson, M. T. C., Underwood, N., and Vellend, M. (2008) Ecological consequences of genetic diversity. *Ecology Letters*, **11**, 609–23.

Hughes, J. B. and Petchey, O. L. (2001) Merging perspectives on biodiversity and ecosystem functioning. *Trends in Ecology and Evolution*, **16**, 222–3.

Hughes, J. B., Hellmann, J. J., Ricketts, T. H., and Bohannan, B. J. M. (2001) Counting the uncountable: statistical approaches to estimating microbial diversity. *Applied Environmental Microbiology*, **67**, 4399–406.

Hughes, T. P. (1994) Catastrophes, phase-shifts, and large-scale degradation of a Caribbean coral-reef. *Science*, **265**, 1547–51.

Hulkrantz, L. (1992) National accounts of timber and forest environmental resources in Sweden. *Environmental and Resource Economics*, **2**, 283–305.

Hulme, P. E. and Bremner, E. T. (2006) Assessing the impact of *Impatiens glandulifera* on riparian habitats: partitioning diversity components following species removal. *Journal of Applied Ecology*, **43**, 43–50.

Hulot, F. D., Lacroix, G., Lescher-Moutoué, F. and Loreau, M. (2000) Functional diversity governs ecosystem response to nutrient enrichment. *Nature*, **405**, 340–4.

Hunter, T., Peacock, L., Turner, H., and Brain, P. (2002) Effect of plantation design on stem-infecting form of rust in willow biomass coppice. *Forest Pathology*, **32**, 87–97.

Hurd, L. E. and Wolf, L. L. (1974) Stability in relation to nutrient enrichment in arthropod consumers of old-field successional ecosystems. *Ecological Monographs*, **44**, 465–82.

Huston, M. A. and McBride, A. (2002) Evaluating the relative strengths of biotic versus abiotic controls of ecosystem processes. In M. Loreau, S. Naeem, and P. Inchausti (eds.) *Biodiversity and Ecosystem Functioning*. Oxford University Press, Oxford.

Huston, M. A. (1997) Hidden treatments in ecological experiments: re-evaluating the ecosystem function of biodiversity. *Oecologia*, **110**, 449–60.

Huston, M. A. and Smith, T. (1987) Plant succession: life hisory and competition. *American Naturalist*, **130**, 168–9.

Hutchings, J. A. and Baum, J. K. (2005) Measuring marine fish biodiversity: temporal changes in abundance, life history and demography. *Philosophical Transactions of the Royal Society B: Biological Sciences*, **360**, 315–38.

Hutchinson, G. E. (1961) The paradox of the plankton. *American Naturalist*, **95**, 137.

Hutton, S. A. and Giller, P. S. (2003) The effects of the intensification of agriculture on northern temperate dung beetle communities. *Journal of Applied Ecology*, **40**, 994–1007.

Huxel, G. R. and McCann, K. (1998) Food web stability: the influence of trophic flows across habitats. *American Naturalist*, **152**, 460–9.

Ieno, E. N., Solan, M., Batty, P., et al. (2006) How biodiversity affects ecosystem functioning: roles of infaunal species richness, identity and density in the marine benthos. *Marine Ecology Progress Series*, **311**, 263–71.

IGBP (1998) The terrestrial carbon cycle: Implications for the Kyoto Protocol. *Science*, **280**, 1393–4.

Imhoff, M. L., Bounoua, L., Ricketts, T., Loucks, C., Harriss, R. and Lawrence, W. T. (2004) Global patterns in human consumption of net primary production. *Nature*, **429**, 870–3.

Innes, R., Polasky, S., and Tschirhart, J. (1998) Takings, compensation and endangered species protection on private lands. *Journal of Economic Perspectives*, **12**, 35–52.

IPCC (2007) *Climate Change 2007: Impacts, Adaptation and Vulnerability. Contribution of Working Group II to the Fourth Assessment Report of the Intergovernmental Panel on Climate Change.* M. L. Parry, O. F. Canziani, J. P. Palutikof, P. J. v. d. Linden and C. E. Hanson (eds.) Cambridge University Press, Cambridge.

Ives, A. R. and Cardinale, B. J. (2004) Food-web interactions govern the resistance of communities after nonrandom extinctions. *Nature*, **429**, 174–7.

Ives, A. R. and Carpenter, S. R. (2007) Stability and diversity of ecosystems. *Science*, **317**, 58–62.

Ives, A. R. and Hughes, J. B. (2002) General relationships between species diversity and stability in competitive systems. *American Naturalist*, **159**, 388–95.

Ives, A. R., Cardinale, B. J., and Snyder, W. E. (2005) A synthesis of subdisciplines: predator–prey interactions, and biodiversity and ecosystem functioning. *Ecology Letters*, **8**, 102–16.

Ives, A. R., Gross, K., and Klug, J. L. (1999) Stability and variability in competitive communities. *Science*, **286**, 542–4.

Ives, A. R., Klug, J. L., and Gross, K. (2000) Stability and species richness in complex communities. *Ecology Letters*, **3**, 399–411.

Ives, A. R., Woody, S. T., Nordheim, E. V., Nelson, C., and Andrews, J. H. (2004) The synergistic effects of stochasticity and dispersal on population densities. *American Naturalist*, **163**, 375–87.

Iyengar, V. K., Reeve, H. K., and Eisner, T. (2002) Paternal inheritance of a female moth's mating preference. *Nature*, **419**, 830–2.

Jackson, L. E., Mayberry, K., Laemmlen, Koike, S., Schulbach, K., and Chaney, W. (1996) *Iceberg Lettuce Production in California*. University of California Division of Agriculture and Natural Resources.

Jackson, L. E., Pascual, U., and Hodgkin, T. (2007) Utilizing and conserving agrobiodiversity in agricultural landscapes. *Agriculture, Ecosystems, and Environment*, **121**, 196–210.

Jackson, R. B., Fierer, N., and Schimel, J. P. (2007) New directions in microbial ecology. *Ecology*, **88**, 1343–4.

Jackson, R. B., Jobbagy, E. G., Avissar, R., *et al.* (2005) Trading water for carbon with biological carbon sequestration. *Science*, **310**, 1944–7.

Jactel, H., Brockerhoff, E., and Duelli, P. (2005) A test of the biodiversity–stability theory: meta-analysis of tree species diversity effects on insect pest infestations, and re-examination of responsible factors. In M. Scherer-Lorenzen, C. Koerner, and E.-D. Schulze (eds.) *Forest Diversity and Function*. Springer-Verlag, Berlin.

Jaffe, A. B., Peterson, S. R., Portney, P. R., and Stavins, R. N. (1995) Environmental regulation and the competitiveness of US manufacturing: what does the evidence tell us? *Journal of Economic Literature*, **33**, 132–63.

Jandl, R., Lindner, M., Vesterdal, L., *et al.* (2007) How strongly can forest management influence soil carbon sequestration? *Geoderma*, **137**, 253–68.

Janzen, D. (1970) Herbivores and the number of tree species in tropical forests. *American Naturalist*, **104**, 501.

Jax, K. (2005) Function and "functioning" in ecology: what does it mean? *Oikos*, **111**, 641–8.

Jenkins, D. G., Brescacin, C. R., Duxbury, C. V., Elliott, J. A., Evans, J. A., Grablow, K. R., Hillegass, M., Lyono, B. N., Metzger, G. A., Olandese, M. L., Pepe, D., Silvers, G. A., Suresch, H. N., Thompson, T. N., Trexler, C. M., Williams, G. E., Williams, N. C., and Williams, S. E. (2007) Does size matter for dispersal distance? *Global Ecology and Biogeography*, **16**, 415–25.

Jenkinson, D. (1971) Studies on the decomposition of ^{14}C-labelled organic matter in soil. *Soil Science*, **111**, 64–70.

Jennings, S., Greenstreet, S. P. R., and Reynolds, J. D. (1999) Structural change in an exploited fish community: a consequence of differential fishing effects on species with contrasting life histories. *Journal of Animal Ecology*, **68**, 617–27.

Jennings, S., Melin, F., Blanchard, J. L., Forster, R. M., Dulvy, N. K., and Wilson, R. W. (2008) Global-scale predictions of community and ecosystem properties from simple ecological theory. *Proceedings of the Royal Society B: Biological Sciences*, **275**, 1375–83.

Jennings, S., Reynolds, J. D., and Mills, S. C. (1998) Life history correlates of responses to fisheries exploitation. *Proceedings of the Royal Society of London Series B: Biological Sciences*, **265**, 333–9.

Jensen, J. L. (1906) Sur les functions convexes et les inegaliatiés entre les valeurs moyennes. *Acta Mathematica*, **30**, 175–93.

Jeschke, J. M. and Strayer, D. L. (2006) Determinants of vertebrate invasion success in Europe and North America. *Global Change Biology*, **12**, 1608–19.

Jessup, C. M., Kassen, R., Forde, S. E., Kerr, B., Buckling, A., Rainey, P. B., and Bohannan, B. J. M. (2004) Big questions, small worlds: microbial model systems in ecology. *Trends in Ecology & Evolution*, **19**, 189–97.

Jessup, C. M., Forde, S. E., and Bohannan, B. J. M. (2005) Microbial experimental systems in ecology. *Advances in Ecological Research*, **37**, 273–307.

Jiang, L. (2007) Negative selection effects suppress relationships between bacterial diversity and ecosystem functioning. *Ecology*, **88**, 1075–85.

Jiang, L. and Morin, P. J. (2005) Predator diet breadth influences the relative importance of bottom-up and top-down control of prey biomass and diversity. *American Naturalist*, **165**, 350–63.

Jiang, X. L., Zhang, W. G., and Wang, G. (2007) Effects of different components of diversity on productivity in artificial plant communities. *Ecological Research*, **22**, 629–34.

Jiang, L., Pu, Z., and Nemergut, D. R. (2008) On the importance of the negative selection effect for the relationship between biodiversity and ecosystem functioning. *Oikos*, **117**, 488–93.

Jiguet, F., Gadot, A. S., Julliard, R., Newson, S. E., and Couvet, D. (2007) Climate envelope, life history traits and the resilience of birds facing global change. *Global Change Biology*, **13**, 1672–84.

Jobbagy, E. G. and Jackson, R. B. (2000) The vertical distribution of soil organic carbon and its relation to climate and vegetation. *Ecological Applications*, **10**, 423–36.

Johnson, C. N., Delean, S., and Balmford, A. (2002) Phylogeny and the selectivity of extinction in Australian marsupials. *Animal Conservation*, **5**, 135–42.

Johnson, S. D., Peter, C. I., Nilsson, L. A., and Agren, J. (2003) Pollination success in a deceptive orchid is enhanced by co-occurring rewarding magnet plants. *Ecology*, **84**, 2919–17.

Johnston, R. J., Besedin, E. Y., Iovanna, R., Miller, C. J., Wardwell, R. F., and Ranson, M. H. (2005) Systematic variation in willingness to pay for aquatic resource improvements and implications for benefit transfer, a meta-analysis. *Canadian Journal of Agricultural Economics*, **53**, 221–48.

Johnston, R. J., Ranson, M. H., Besedin, E. Y., and Helm, E. C. (2006) What determines willingness to pay per fish? A meta-analysis of recreational fishing values. *Marine Resource Economics*, **21**, 1–32.

Jones, C. G., Lawton, J. H., and Shachak, M. (1994) Organisms as ecosystem engineers. *Oikos*, **69**, 373–86.

Jones, H. E., McNamara, N., and Mason, W. L. (2005) Functioning of mixed-species stands: evidence from a long-term forest experiment. In M. Scherer-Lorenzen, C. Körner, and E.-D. Schulze (eds.) *The Functional Significance of Forest Diversity*, Springer-Verlag, Berlin.

Jones, K. E., Purvis, A., and Gittleman, J. L. (2003) Biological correlates of extinction risk in bats. *American Naturalist*, **161**, 601–14.

Jones, K. E., Patel, N. G., Levy, M. A., Storeygard, A., Balk, D., Gittleman, J. L., and Daszak, P. (2008) Global trends in emerging infectious diseases. *Nature*, **451**, 990–4.

Jonsson, L. M., Nilsson, M. C., Wardle, D. A., and Zackrisson, O. (2001) Context dependent effects of ectomycorrhizal species richness on tree seedling productivity. *Oikos*, **93**, 353.

Jonsson, M. (2006) Species richness effects on ecosystem functioning increase with time in an ephemeral resource system. *Acta Oecologica – International Journal of Ecology*, **29**, 72–7.

Jonsson, M. and Malmqvist, B. (2000) Ecosystem process rate increases with animal species richness: evidence from leaf-eating, aquatic insects. *Oikos*, **89**, 519–23.

Jonsson, M., Dangles, O., Malmqvist, B., and Guerold, F. (2002) Simulating species loss following perturbation: assessing the effects on process rates. *Proceedings of the Royal Society of London Series B: Biological Sciences*, **269**, 1047–52.

Jordan, N., Boody, G., Broussard, W., et al. (2007) Environment – sustainable development of the agricultural bio-economy. *Science*, **316**, 1570–1.

Jordano, P., Bascompte, J., and Olesen, J. M. (2003) Invariant properties in coevolutionary networks of plant–animal interactions. *Ecology Letters*, **6**, 69–81.

Joshi, J., Matthies, D., and Schmid, B. (2000) Root hemi-parasites and plant diversity in experimental grassland communities. *Journal of Ecology*, **88**, 634–44.

Joshi, J., Schmid, B., Caldeira, M., et al. (2001) Local adaptation enhances performance of common plant species. *Ecology Letters*, **4**, 536–44.

Joyce, C. (2001) The sensitivity of a species-rich flood-meadow plant community to fertilizer nitrogen: the Luznice river floodplain, Czech Republic. *Plant Ecology*, **155**, 47–60.

Just, R. E., Hueth, D. L., and Schmitz, A. (2004) *The Welfare Economics of Public Policy: A Practical Approach to Project and Policy Evaluation*. Edward Elgar, Cheltenham.

Kahmen, A., Perner, J., and Buchmann, N. (2005) Diversity-dependent productivity in semi-natural grasslands following climate perturbations. *Functional Ecology*, **19**, 594–601.

Kaiser, J. K. (2000) Rift over biodiversity divides ecologists. *Science*, **289**, 1282–3.

Kareiva, P. and Wennergren, U. (1995) Connecting landscape patterns to ecosystem and population processes. *Nature*, **373**, 299–302.

Kareiva, P., Watts, S., McDonald, R., and Boucher, T. (2007) Domesticated nature: shaping landscapes and ecosystems for human welfare. *Science*, **316**, 1866–9.

Kark, S., Mukerji, T., Safriel, U. N., Noy-Meir, I., Nissani, R., and Darvasi, A. (2002) Peak morphological diversity in an ecotone unveiled in the chukar partridge by a novel estimator in a dependent sample (eds). *Journal of Animal Ecology*, **71**, 1015–29.

Kates, R. W. and Parris, T. M. (2003) Long-term trends and a sustainability transition. *Proceedings of the National Academy of Sciences of the USA*, **100**, 8062–7.

Kates, R. W., Clark, W. C., Corell, R., et al. (2001) Environment and development: sustainability science. *Science*, **292**, 641–2.

Kathiresan, K. and Rajendran, N. (2005) Coastal mangrove forests mitigated tsunami. *Estuarine Coastal and Shelf Science*, **65**, 601–6.

Keane, R. M. and Crawley, M. J. (2002) Exotic plant invasions and the enemy release hypothesis. *Trends in Ecology & Evolution*, **17**, 164–70.

Kearns, C., Inouye, D., and Waser, N. (1998) Endangered mutualisms: the conservation of plant–pollinator interactions. *Annual Review of Ecology and Systematics*, **29**, 83–112.

Keer, G. H. and Zedler, J. B. (2002) Salt marsh canopy architecture differs with the number and composition of species. *Ecological Applications*, **12**, 456–73.

Keesing, F., Holt, R. D., and Ostfeld, R. S. (2006) Effects of species diversity on disease risk. *Ecology Letters*, **9**, 485–98.

Keitt, T. H., Urban, D. L., and Milne, B. T. (1997) Detecting critical scales in fragmented landscapes. *Conservation Ecology*, **1**, 4; URL: http://www.consecol.org/vol1/iss1/art4/.

Keller, R. P., Lodge, D. M., and Finnoff, D. C. (2007) Risk assessment for invasive species produces net bioeconomic benefits. *Proceedings of the National Academy of Sciences of the USA*, **104**, 203–7.

Kelso, S., Bower, N. W., Heckmann, K. E., Beardsley, P. M., and Greve, D. G. (2003) Geobotany of the Niobrara chalk barrens in Colorado: a study of edaphic endemism. *Western North American Naturalist*, **63**, 299–313.

Kemp, D. R., King, W. M., Gilmore, A. R., Lodge, G. M., Murphy, S. R., Quigley, P. E., Sandford, P., and Andrew, M. H. (2003) SGS biodiversity theme: impact of plant diversity on the productivity and stability of grazing systems across southern Australia. *Australian Journal of Experimental Agriculture*, **43**, 961–75.

Kennedy, A. D., Biggs, H., and Zambatis, N. (2003) Relationship between grass species richness and ecosystem stability in Kruger National Park, South Africa. *African Journal of Ecology*, **41**, 131–40.

Kennedy, T. A., Naeem, S., Howe, K. M., Knops, J. M. H, Tilman, D., and Reich, P. (2002) Biodiversity as a barrier to ecological invasion. *Nature*, **417**, 636–8.

Kiessling, W. and Aberhan, M. (2007) Geographical distribution and extinction risk: lessons from Triassic–Jurassic marine benthic organisms. *Journal of Biogeography*, **34**, 1473–89.

Kim, J., Williams, N., and Kremen, C. (2006) Effects of cultivation and proximity to natural habitat on ground-nesting native bees in California sunflower fields. *Journal of the Kansas Entomological Society*, **79**, 309–20.

Kim, S., Tschirhart, J., and Buskirk, S. (2007) Reconstructing past population processes with general equilibrium models: house mice in Kern County, California, 1926–27. *Ecological Modelling*, **209**, 235–48.

Kindscher, K. and Tieszen, L. L. (1998) Floristic and soil organic matter changes after five and thirty-five years of native tallgrass prairie restoration. *Restoration Ecology*, **6**, 181–96.

Kindscher, K. and Wells, P. V. (1995) Prairie plant guilds: a multivariate analysis of prairie species based on ecological and morphological traits. *Vegetatio*, **117**, 29–50.

King, E. G. and Hobbs, R. J. (2006) Identifying linkages among conceptual models of ecosystem degradation and restoration: towards an integrative framework. *Restoration Ecology*, **14**, 369–78.

Kinzig, A. P., Pacala, S. W., and Tilman, D. (eds.) (2001) *The Functional Consequences of Biodiversity: Empirical Progress and Theoretical Extensions*. Princeton University Press, Princeton, NJ.

Kirchhoff, S., Colby, B. G., and LaFrance, J. T. (1997) Evaluating the performance of benefit transfer, an empirical inquiry. *Journal of Environmental Economics and Management*, **33**, 75–93.

Kirwan, L., Luescher, A., Sebastia, M. T., et al. (2007) Evenness drives consistent diversity effects in intensive grassland systems across 28 European sites. *Journal of Ecology*, **95**, 530–9.

Kleijn, D. and Sutherland, W. J. (2003) How effective are European agri-environment schemes in conserving and promoting biodiversity? *Journal of Applied Ecology*, **40**, 947–69.

Kleijn, D., Baquero, R. A., Clough, Y., et al. (2006) Mixed biodiversity benefits of agri-environment schemes in five European countries. *Ecology Letters*, **9**, 243–54.

Klein, A. M., Tscharntke, T., Steffan-Dewenter, I., and Buchori, D. (2002) Effects of land-use intensity in tropical agroforestry systems on coffee flower-visiting and trap-nesting bees and wasps. *Conservation Biology*, **16**, 1003–14.

Klein, A. M., Steffan-Dewenter, I., and Tscharntke, T. (2003a). Fruit set of highland coffee increases with the diversity of pollinating bees. *Proceedings of the Royal Society, Series B*, **270**, 955–61.

Klein, A. M., Steffan-Dewenter, I., and Tscharntke, T. (2003b) Pollination of *Coffea canephora* in relation to local and regional agroforestry management. *Journal of Applied Ecology*, **40**, 837–45.

Klein, A. M., Steffan-Dewenter, I., and Tscharntke, T. (2006) Rain forest promotes trophic interactions and diversity of trap-nesting hymenoptera in adjacent agroforestry. *Journal of Animal Ecology*, **75**, 315–23.

Klein, A. M., Vaissière, B. E., Cane, J. H., et al. (2007) Importance of pollinators in changing landscapes for world crops. *Proceedings of the Royal Society, Series B*, **274**, 303–13.

Klein, A. M., Cunningham, S., Bos, M., and Steffan-Dewenter, I. (2008) Advances in pollination ecology from tropical plantation crops. *Ecology*, **89**, 935–43.

Kloeppel, B. D. and Abrams, M. D. (1995) Ecophysiological attributes of the native Acer-Saccharum and the exotic Acer-Platanoides in urban oak forests in Pennsylvania, USA. *Tree Physiology*, **15**, 739–46.

Knohl, A., Kolle, O., Minayeva, T. Y., et al. (2002) Carbon dioxide exchange of a Russian boreal forest after disturbance by wind throw. *Global Change Biology*, **8**, 231–46.

Knops, J. M. H., Tilman, D., Haddad, N. M., et al. (1999) Effects of plant species richness on invasions dynamics, disease outbreaks, insect abundances, and diversity. *Ecology Letters*, **2**, 286–93.

Knowler, D. (2002) A review of selected bioeconomic models with environmental influences in fisheries. *Journal of Bioeconomics*, **4**, 163–81.

Knowler, D. and Barbier, E. (2005) Importing exotic plants and the risk of invasion: are market-based instruments adequate? *Ecological Economics*, **52**, 341–54.

Koenig W. D. (1999) Spatial autocorrelation of ecological phenomena. *Trends in Ecology & Evolution*, **14**, 22–6.

Kohn, R. E. and Capen, D. (2002) Optimal volume of environmentally damaging trade. *Scottish Journal of Political Economy*, **49**, 22–38.

Kokkoris, G. D., Troumbis, A. Y., and Lawton, J. L. (1999) Patterns of species interaction strength in assembled theoretical competition communities. *Ecology Letters*, **2**, 70–4.

Kolar, C. S. and Lodge, D. M. (2001) Progress in invasion biology: predicting invaders. *Trends in Ecology & Evolution*, **16**, 199–204.

Kolar, C. S. and Lodge, D. M. (2002) Ecological predictions and risk assessment for alien fishes in North America. *Science*, **298**, 1233–6.

Kolasa, J. and Li, B. (2003) Removing the confounding effect of habitat specialization reveals the stabilizing contribution of diversity to species variability. *Proceedings of the Royal Society B: Biological Sciences*, **270**, S198–S201.

Kondo, T. and Tsuyuzaki, S. (1999) Natural regeneration patterns of the introduced larch, *Larix kaempferi* (Pinaceae), on the volcano Mount Koma, northern Japan. *Diversity and Distributions*, **5**, 223–33.

Kondoh, M. (2003) Foraging adaptation and the relationship between food-web complexity and stability. *Science*, **299**, 1388–91.

Kondoh, M. (2006) Does foraging adaptation create the positive complexity–stability relationship in realistic food-web structure? *Journal of Theoretical Biology*, **238**, 646–51.

Koo, B. and Wright, B. D. (1999) The role of biodiversity products as incentives for conserving biological diversity: some instructive examples. *The Science of the Total Environment*, **240**, 21–30.

Kopp, K. and Jokela, J. (2007) Resistant invaders can convey benefits to native species. *Oikos*, **116**, 295–301.

Korhonen, K. P. C., Karjalainen, R., and Stenlid, J. (1998) Distribution of *Heterobasidion annosum* intersterility groups in Europe. In R. Karjalainen, R. and A. Hüttermann (eds.) *Heterobasidion annosum. Biology, Ecology, Impact and Control*. CABI, Willinford.

Koricheva, J., Mulder, C. P. H., Schmid, B., Joshi, J., and Huss-Danell, K. (2000) Numerical responses of different trophic groups of invertebrates to manipulations of plant diversity in grasslands. *Oecologia*, **125**, 271–82.

Koricheva, J., Vehviläinen, H., Riihimaki, J., Ruohomaki, K., Kaitaniemi, P., and Ranta, H. (2006) Diversification of tree stands as a means to manage pests and diseases in boreal forests: Myth or reality? *Canadian Journal of Forest Research*, **36**, 324–36.

Körner, C. (2003) Slow in, rapid out – carbon flux studies and Kyoto targets. *Science*, **300**, 1242–3.

Körner, C. (2004) Through enhanced tree dynamics carbon dioxide enrichment may cause tropical forests to lose carbon. *Proceedings of the Royal Society London B*, **359**, 493–8.

Körner, C. (2005) An introduction to the functional diversity of temperate forest trees. In M. Scherer-Lorenzen, C. Körner, and E.-D. Schulze (eds.) *The Functional Significance of Forest Diversity*. Springer-Verlag, Berlin.

Korthals, G. W., Smilauer, P., Van Dijk, C., and Van der Putten, W. H. (2001) Linking above- and below-ground biodiversity: abundance and trophic complexity in soil as a response to experimental plant communities on abandoned arable land. *Functional Ecology*, **15**, 506–14.

Kotanen, P. M. (1997) Effects of experimental soil disturbance on revegetation by natives and exotics in coastal

Californian meadows. *Journal of Applied Ecology*, **34**, 631–44.

Kotiaho, J. S., Kaitala, V., Komonen, A., and Paivinen, J. (2005) Predicting the risk of extinction from shared ecological characteristics. *Proceedings of the National Academy of Sciences of the USA*, **102**, 1963–7.

Kowalchuk, G. A., Buma, D. S., de Boer, W., Klinkhamer, P. G. L., and van Veen, J. A. (2002) Effects of above-ground plant species composition and diversity on the diversity of soil-borne microorganisms. *Antonie Van Leeuwenhoek International Journal of General and Molecular Microbiology*, **81**, 509–20.

Kraaij, T. and Ward, D. (2006) Effects of rain, nitrogen, fire and grazing on tree recruitment and early survival in bush-encroached savanna, South Africa. *Plant Ecology*, **186**, 235–46.

Krab, E. J., Cornelissen, J. H. C., Lang, S. I., and van Logtestijn, R. S. P. (2008) Amino acid uptake among wide-ranging moss species may contribute to their strong position in higher-latitude ecosystems. *Plant and Soil*, **304**, 199–208.

Kraus, B. and Page, R. E., Jr. (1995) Effect of *Varroa jacobsoni* (Mesostigmata: Varroidae) on feral *Apis mellifera* (Hymenoptera: Apidae) in California. *Environmental Entomology*, **24**, 1473–80.

Krebs, C. J. (1972) *Ecology: the Experimental Analysis of Distribution and Abundance*. Harper & Row, New York.

Krebs, C. J. (2001) *Ecology: the Experimental Analysis of Distribution and Abundance*, 5th edn. Benjamin Cummings, San Francisco.

Kremen, C. (2005) Managing ecosystem services: what do we need to know about their ecology? *Ecology Letters*, **8**, 468–79.

Kremen, C. and Chaplin-Kramer, B. (2007) Insects as providers of ecosystem services: crop pollination and pest control. In A. J. A. Stewart, T. R. New, and O. T. Lewis (eds.) *Insect Conservation Biology*, pp. 349–83. 23rd Symposium of the Proceedings of the Royal Entomological Society.

Kremen, C., Niles, J., Dalton, M., *et al.* (2000) Economic incentives for rain forest conservation across scales. *Science*, **288**, 1828–32.

Kremen, C., Williams, N. M., and Thorp, R. W. (2002) Crop pollination from native bees at risk from agricultural intensification. *Proceedings of the National Academy of Sciences of the USA*, **99**, 16812–16.

Kremen, C., Williams, N. M., Bugg, R. L., Fay, J. P., and Thorp, R. W. (2004) The area requirements of an ecosystem service: crop pollination by native bee communities in California. *Ecology Letters*, **7**, 1109–19.

Kremen, C., Williams, N. M., Aizen, M. A., *et al.* (2007) Pollination and other ecosystem services produced by mobile organisms: a conceptual framework for the effects of land-use change. *Ecology Letters*, **10**, 299–314.

Kruess, A. and Tscharntke, T. (1994) Habitat fragmentation, species loss, and biological control. *Science*, **264**, 1581–4.

Kruger, O. and Radford, A. N. (2008) Doomed to die? Predicting extinction risk in the true hawks Accipitridae. *Animal Conservation*, **11**, 83–91.

Krutilla, J. V. (1967) Conservation reconsidered. *The American Economic Review* **57**(4), 777–86.

Laakso, J. and Setälä, H. (1999) Sensitivity of primary production to changes in the architecture of belowground food webs. *Oikos*, **87**, 57–64.

Lafferty, K. D., Dobson, A. P., and Kuris, A. M. (2006) Parasites dominate food web links. *Proceedings of the National Academy of Sciences of the USA*, **103**, 11211–16.

Laidre, K. L., Stirling, I., Lowry, L. F., Wiig, O., Heide-Jorgensen, M. P., and Ferguson, S. H. (2008) Quantifying the sensitivity of Arctic marine mammals to climate-induced habitat change. *Ecological Applications*, **18**, S97–S125.

Lal, R. (2004) Soil carbon sequestration impacts on global climate change and food security. *Science*, **304**, 1623–7.

Lal, R. (2005) Soil carbon sequestration in natural and managed tropical forest ecosystems. *Journal of Sustainable Forestry*, **21**, 1–30.

Laland, K. N. and Sterelny, K. (2006) Perspective: seven reasons not to neglect niche construction. *Evolution*, **60**, 1751–62.

Lambin, E. F., Geist, H. J., and Lepers, E. (2003) Dynamics of land-use and land-cover change in tropical regions. *Annual Review of Environment and Resources*, **28**, 205–41.

Landsberg, J. (1999) Response and effect – different reasons for classifying plant functional types under grazing. In Eldridge, D. and Freudenberger, D. (eds.) *People and Rangelands: Building the Future*. Proceedings of the VIth International Rangeland Congress, Aitkenvale, Queensland.

Landsberg, J., Lavorel, S., and Stol, J. (1999) Grazing response groups among understorey plants in arid rangelands. *Journal of Vegetation Science*, **10**, 683–96.

Larsen, T. H., Williams, N. M., and Kremen, C. (2005) Extinction order and altered community structure rapidly disrupt ecosystem functioning. *Ecology Letters*, **8**, 538–47.

Larson, D. L., Royer, R. A., and Royer, M. R. (2006) Insect visitation and pollen deposition in an invaded prairie plant community. *Biological Conservation*, **130**, 148–59.

Laughlin, D. C. (2003) Lack of native propagules in a Pennsylvania, USA, limestone prairie seed bank: futile hopes for a role in ecological restoration. *Natural Areas Journal*, **23**, 158–64.

Laurance, W. F. (1991) Ecological correlates of extinction proneness in Australian tropical rain-forest mammals. *Conservation Biology*, **5**, 79–89.

Laurance, W. F. (2000) Edge effects and ecological processes: are they on the same scale? Reply. *Trends in Ecology & Evolution*, **15**, 373.

Laurance, W. F., Laurance, S. G., Ferreira, L. V., Rankin-de Merona, J. M., Gascon, C., and Lovejoy, T. E. (1997) Biomass collapse in Amazonian forest fragments. *Science*, **278**, 1117–18.

Laurance, W. F., Nascimento, H. E. M., Laurance, S. G., et al. (2006) Rain forest fragmentation and the proliferation of successional trees. *Ecology*, **87**, 469–82.

Lavergne, S., Garnier, E., and Debussche, M. (2003) Do rock endemic and widespread plant species differ under the Leaf–Height–Seed plant ecology strategy scheme? *Ecology Letters*, **6**, 398–404.

Lavergne, S., Thompson, J. D., Garnier, E., and Debussche, M. (2004) The biology and ecology of narrow endemic and widespread plants: a comparative study of trait variation in 20 congeneric pairs. *Oikos*, **107**, 505–18.

Lavorel, S. and Garnier, E. (2001) Aardvarck to zyzyxia – functional groups across kingdoms. *New Phytologist*, **149**, 360–3.

Lavorel, S. and Garnier, E. (2002) Predicting changes in community composition and ecosystem functioning from plant traits: revisiting the Holy Grail. *Functional Ecology*, **16**, 545–56.

Lavorel, S., Grigulis, K., McIntyre, S. et al. (2008) Assessing functional diversity in the field – methodology matters! *Functional Ecology*, **22**, 134–47.

Lawlor, S. P., Armesto, J. J., and Kareiva, P. (2002) *How relevant to conservation are studies linking biodiversity and ecosystem functioning?* In A. P. Kinzig, S. W. Pacala, and D. Tilman (eds.) *The Functional Consequences of Biodiversity*, pp. 213–45. Princeton University Press, Princeton, NJ.

Lawton, J. H. (1995) Ecological experiments with model systems. *Science*, **269**, 328–31.

Lawton, J. H. (1996) The Ecotron facility at Silwood Park: The value of "big bottle" experiments. *Ecology*, **77**, 665–9.

Lawton, J. H. and May, R. M. (1995) *Extinction Rates*. Oxford University Press, Oxford.

Lawton, J. H., Naeem, S., Woodfin, R. M., et al. (1993) The Ecotron: a controlled environmental facility for the investigation of populations and ecosystem processes. *Philosophical Transactions of the Royal Society of London B*, **341**, 181–94.

Layman, C. A., Langerhans, R. B., and Winemiller, K. O. (2005) Body size, not other morphological traits, characterizes cascading effects in fish assemblage composition following commercial netting. *Canadian Journal of Fisheries and Aquatic Sciences*, **62**, 2802–10.

Layton, D. F. and Levine, R. A. (2005) Bayesian approaches to modeling stated preference data. In R. Scarpa and A. Alberini (eds.), *Applications of Simulation Methods in Environmental and Resource Economics*, pp. 187–208. Kluwer Academic Press, Dordrecht.

Leach, M. K. and Givnish, T. J. (1996) Ecological determinants of species loss in remnant prairies. *Science*, **273**, 1555–8.

Lecerf, A., Dobson, M., Dang, C. K., and Chauvet, E. (2005) Riparian plant species loss alters trophic dynamics in detritus-based stream ecosystems. *Oecologia*, **146**, 432–42.

Lecerf, A., Risnoveanu, G., Popescu, C., Gessner, M. O., and Chauvet, E. (2007) Decomposition of diverse litter mixtures in streams. *Ecology*, **88**, 219–27.

Lee, M. T., Peet, R. K., Roberts, S. D., and Wentworth, T. R. (2007) *CVS-EEP Protocol for Recording Vegetation*. NCEEP, Raleigh, NC.

Lehman, C. L. and Tilman, D. (2000) Biodiversity, stability, and productivity in competitive communities. *American Naturalist*, **156**, 534–52.

Leibold, M. A. (1989) Resource edibility and the effects of predators and productivity on the outcome of trophic interactions. *American Naturalist*, **134**, 922–49.

Leibold, M. A. and Norberg, J. (2004) Biodiversity in metacommunities: plankton as complex adaptive systems? *Limonology and Oceanography*, **49**, 1278–89.

Leibold, M. A., Holyoak, M., Mouquet, N., Amarasekare, P., Chase, J. M., Hoopes, M. F., Holt, R. D., Shurin, J. B., Law, R., Tilman, D., Loreau, M., and Gonzalez, A. (2004) The metacommunity concept: a framework for multi-scale community ecology. *Ecology Letters*, **7**, 601–13.

Lenoir, J., Gegout, J. C., Marquet, P. A., de Ruffray, P., and Brisse, H. (2008) A significant upward shift in plant species optimum elevation during the 20th century. *Science*, **320**, 1768–71.

Lepš, J. (2004) What do the biodiversity experiments tell us about consequences of plant species loss in the real world? *Basic and Applied Ecology*, **5**, 529–34.

Lepš, J., Osbornovakosinova, J., and Rejmanek, M. (1982) Community stability, complexity and species life-history strategies. *Vegetatio*, **50**, 53–63.

Lepš, J., Brown, V. K., Diaz Len, T. A., et al. (2001) Separating the chance effect from other diversity effects in the functioning of plant communities. *Oikos*, **92**, 123–34.

Lepš, J., de Bello, F., Lavorel, S. and Berman, S. (2006) Quantifying and interpreting functional diversity of natural communities: practical considerations matter. *Preslia*, **78**, 481–501.

Leung, B., Lodge, D. M., Finnoff, D., Shogren, J. F., Lewis, M. A., and Lamberti, G. (2002) An ounce of prevention or a pound of cure: bioeconomic risk analysis of invasive species. *Proceedings of the Royal Society of London Series B: Biological Sciences*, **269**, 2407–13.

Levine, J. M. (2000) Species diversity and biological invasions: relating local process to community pattern. *Science*, **288**, 852–4.

Levine, J. M. and D'Antonio, C. M. (1999) Elton revisited: a review of evidence linking diversity and invisibility. *Oikos*, **87**, 15–26.

Levine, J. M., Vila, M., D'Antonio, C. M., Dukes, J. S., Grigulis, K., and Lavorel, S. (2003) Mechanisms underlying the impacts of exotic plant invasions. *Proceedings of the Royal Society of London Series B: Biological Sciences*, **270**, 775–81.

Levinson, A. (1996) Environmental regulations and industry location: international and domestic evidence. In J. Bhagwati and R. Hudec (eds.) *Fair Trade and Harmonization: Prerequisites for Free Trade?*, Vol. 1, pp. 429–57. MIT Press, Cambridge, MA.

Liebman, M. and Staver, C. P. (2001) Crop diversification for weed management. In M. Liebman, C. L. Mohler, and C. P. Staver (eds.) *Ecological Management of Agricultural Weeds*. Cambridge University Press, Cambridge.

Limburg, K. and Folke, C. (1999) The ecology of ecosystem services: introduction to the special issue. *Ecological Economics*, **29**, 179–82.

Lindberg, K. A. (2001) Economic impacts. In D. Weaver (ed.) *The Encyclopedia of Ecotourism*, pp. 363–77. CABI, Wallingford.

Lindberg, N., Engtsson, J. B., and Persson, T. (2002) Effects of experimental irrigation and drought on the composition and diversity of soil fauna in a coniferous stand. *Journal of Applied Ecology*, **39**, 924–36.

Lips, K. R., Reeve, J. D., and Witters, L. R. (2003) Ecological traits predicting amphibian population declines in Central America. *Conservation Biology*, **17**, 1078–88.

Lloret, F. and Vila, M. (2003) Diversity patterns of plant functional types in relation to fire regime and previous land use in mediterranean woodlands. *Journal of Vegetation Science*, **14**, 387–98.

Lloret, F., Medail, F., Brundu, G., and Hulme, P. E. (2004) Local and regional abundance of exotic plant species on Mediterranean islands: are species traits important? *Global Ecology and Biogeography*, **13**, 37–45.

Lloret, F., Penuelas, J., and Estiarte, M. (2005) Effects of vegetation canopy and climate on seedling establishment in Mediterranean shrubland. *Journal of Vegetation Science*, **16**, 67–76.

Lodge, D. M. (1993) Biological invasions – lessons for ecology. *Trends in Ecology & Evolution*, **8**, 133–7.

LoGiudice, K., Ostfeld, R. S., Schmidt, K. A., and Keesing, F. (2003) The ecology of infectious disease: effects of host diversity and community composition on Lyme disease risk. *Proceedings of the National Academy of Sciences of the USA*, **100**, 567–71.

LoGiudice, K., Duerr, S., Newhouse, M., Schmidt, K. A., Killilea, M., and Ostfeld R. S. (2008). Impact of community composition on Lyme disease risk. *Ecology* **89**, 2841–2849

Lomolino, M. V. and Heaney, L. R. (eds.) (2004) *Frontiers of Biogeography*, Sinauer Associates, Sunderland.

Long, Z. T., Steiner, C. F., Krumins, J. A., and Morin, P. J. (2006) Species richness and allometric scaling jointly determine biomass in model aquatic food webs. *Journal of Animal Ecology*, **75**, 1014–23.

Lopezaraiza-Mikel, M. E., Hayes, R. B., Whalley, M. R., and Memmott, J. (2007) The impact of an alien plant on a native plant–pollinator network: an experimental approach. *Ecology Letters*, **10**, 539–50.

Loranger, G., Ponge, J. F., Blanchart, E., and Lavelle, P. (1998) Influence of agricultural practices on arthropod communities in a vertisol (Martinique). *European Journal of Soil Biology*, **34**, 157–65.

Loreau, M. (1994) Material cycling and stability in ecosystems. *American Naturalist*, **143**, 508–13.

Loreau, M. (1995) Consumers as maximizers of matter and energy flow in ecosystems. *American Naturalist*, **145**, 22–42.

Loreau, M. (1998a) Biodiversity and ecosystem functioning: a mechanistic model. *Proceedings of the National Academy of Sciences of the USA*, **95**, 5632–6.

Loreau, M. (1998b) Separating sampling and other effects in biodiversity experiments. *Oikos*, **82**, 600–2.

Loreau, M. (2000) Biodiversity and ecosystem functioning: recent theoretical advances. *Oikos*, **91**, 3–17.

Loreau, M. (2001) Microbial diversity, producer–decomposer interactions and ecosystem processes: a theoretical model. *Proceedings of the Royal Society London B*, **268**, 303–9.

Loreau, M. (2004) Does functional redundancy exist? *Oikos*, **104**, 606–11.

Loreau, M. and Behera, N. (1999) Phenotypic diversity and stability of ecosystem processes. *Theoretical Population Biology*, **56**, 29–47.

Loreau, M. and Hector, A. (2001) Partitioning selection and complementarity in biodiversity experiments. *Nature*, **412**, 72–6.

Loreau, M. and Mouquet, N. (1999) Immigration and the maintenance of local species diversity. *American Naturalist*, **154**, 427–40.

Loreau, M., Naeem, S., Inchausti, P., *et al.* (2001) Biodiversity and ecosystem functioning: current knowledge and future challenges. *Science*, **294**, 804–8.

Loreau, M., Downing, A., Emmerson, M., Gonzalez, A., Hughes, J., Inchausti, P., Joshi, J., Norberg, J., and Sala, O.

(2002) A new look at the relationship between diversity and stability. In M. Loreau, S. Naeem, and P. Inchausti (eds.) *Biodiversity and Ecosystem Functioning: Synthesis and Perspectives*, pp. 79–91. Oxford University Press, Oxford.

Loreau, M., Naeem, S., and Inchausti, P. (eds.) (2002c) *Biodiversity and Ecosystem Functioning: Synthesis and Perspectives*. Oxford University Press, Oxford.

Loreau, M., Mouquet, N., and Holt, R. D. (2003a) Metaecosystems: a theoretical framework for spatial ecosystem ecology. *Ecology Letters*, **6**, 673–9.

Loreau, M., Mouquet, N., and Gonzalez, A. (2003b) Biodiversity as spatial insurance in heterogeneous landscapes. *Proceedings of the National Academy of Sciences of the USA*, **100**, 12765–70.

Loreau, M., Oteng-Yeboah, A., Arroyo, M. T. K., et al. (2006) Diversity without representation. *Nature*, **442**, 245–6.

Lortie, C. J. and Aarsen, L. W. (1999) The advantage of being tall: higher flowers receive more pollen in *Verbascum thapsus* (Scrophulariaceae). *Ecoscience*, **6**, 68–71.

Losey, J. E. and Denno, R. F. (1998) Positive predator–predator interactions: enhanced predation rates and synergistic suppression of aphid populations. *Ecology*, **79**, 2143–52.

Losey, J. E. and Vaughan, M. (2006) The economic value of ecological services provided by insects. *Bioscience*, **56**, 311–23.

Lovell, S. T. and Sullivan, W. C. (2006) Environmental benefits of conservation buffers in the United States: Evidence, promise, and open questions. *Agriculture Ecosystems & Environment*, **112**, 249–60.

Lubchenco, J., Olson, A. M., Brubaker, L. B., et al. (1991) The sustainable biosphere initiative: an ecological research agenda. *Ecology*, **72**, 371–412.

Luyssaert, S., Schulze, E.-D., Borner, A., et al. (2008) Old-growth forests as global carbon sinks. *Nature*, **455**, 213–15.

Lynam, T., De Jong, W., Sheil, D., Kusumanto, T., and Evans, K. (2007) A review of tools for incorporating community knowledge, preferences, and values into decision making in natural resources management. *Ecology and Society*, **12**, 5.

Lynch, J. M., Benedetti, A., Insam, H., et al. (2004) Microbial diversity in soil: ecological theories, the contribution of molecular techniques and the impact of transgenic plants and transgenic microorganisms. *Biology and Fertility of Soils*, **40**, 363–85.

Lynne, G. D., Conroy, P., and Prochaska, F. J. (1981) Economic valuation of marsh areas for marine production processes. *Journal of Environmental Economics and Management*, **8**(2), 175–86.

Lyons, K. G. and Schwartz, M. W. (2001) Rare species loss alters ecosystem function – invasion resistance. *Ecology Letters*, **4**, 358–65.

Macarthur, R. (1955) Fluctuations of animal populations, and a measure of community stability. *Ecology*, **36**, 533–6.

MacArthur, R. H. (1972) *Geographical Ecology*. Princeton University Press, Princeton, NJ.

MacArthur, R. H. and Wilson, E. O. (1967) *The Theory of Island Biogeography*. Princeton University Press, Princeton, NJ.

Macdougall, A. S. and Turkington, R. (2005) Are invasive species the drivers or passengers of change in degraded ecosystems? *Ecology*, **86**, 42–55.

Madin, J., Bowers, S., Schildhauer, M., Krivov, S., Pennington, D., and Villa, F. (2007) An ontology for describing and synthesizing ecological observation data. *Ecological Informatics*, **2**, 279–96.

Madin, J. S., Bowers, S., Schildhauer, M. P., and Jones, M. B. (2008) Advancing ecological research with ontologies. *Trends in Ecology & Evolution*, **23**, 159–68.

Madritch, M. D. and Hunter, M. D. (2002) Phenotypic diversity influences ecosystem functioning in an oak sandhills community. *Ecology*, **83**, 2084–90.

Madritch, M. D. and Hunter, M. D. (2004) Phenotypic diversity and litter chemistry affect nutrient dynamics during litter decomposition in a two species mix. *Oikos*, **105**, 125–31.

Madritch, M. D. and Hunter, M. D. (2005) Phenotypic variation in oak litter influences short- and long-term nutrient cycling through litter chemistry. *Soil Biology and Biochemistry*, **37**, 319–27.

Madritch, M., Donaldson, J., and Lindroth, R. (2006) Genetic identity of populus tremuloides litter influences decomposition and nutrient release in a mixed forest stand. *Ecosystems*, **9**, 528–37.

Maherali, H. and Klironomos, J. N. (2007) Influence of phylogeny on fungal community assembly and ecosystem functioning. *Science*, **316**, 1746.

Maille, P. and Mendelsohn, R. (1993) Valuing ecotourism in Madagascar. *Journal of Environmental Management*, **38**, 213–18.

Mangel, M. and Hilborn, R. (1997) *The Ecological Detective*. Princeton University Press, Princeton.

Marba, N., Duarte, C. M., and Agusti, S. (2007) Allometric scaling of plant life history. *Proceedings of the National Academy of Sciences of the USA*, **104**, 15777–80.

Marquez, C. O., Cambardella, C. A., Isenhart, T. M., and Schultz, R. C. (1999) Assessing soil quality in a riparian buffer by testing organic matter fractions in central Iowa, USA. *Agroforestry Systems*, **44**, 133–40.

Martinez, N. D. (1992) Constant connectance in community food webs. *American Naturalist*, **139**, 1208–18.

Martiny, J. B. H., Bohannan, B. J. M., Brown, J. H., Colwell, R. K., Fuhrman, J. A., Green, J. L., Horner-Devine, M. C., Kane, M., Krumins, J. A., Kuske, C. R., Morin, P. J., Naeem, S., Ovreas, L., Reysenbach, A. L., Smith, V. H., and Staley, J. T. (2006) Microbial biogeography: putting microorganisms on the map. *Nature Reviews Microbiology*, **4**, 102–12.

Mason, N. W. H., MacGillivray, K., Steel, J. B., and Wilson, J. B. (2003) An index of functional diversity. *Journal of Vegetation Science*, **14**, 571–8.

Mason, N. W. H., Irz, P., Lanoiselee, C., Mouillot, D., and Argillier, C. (2008) Evidence that niche specialization explains species–energy relationships in lake fish communities. *Journal of Animal Ecology*, **77**, 285–96.

Mason, N. W. H., Lanoiselee, C., Mouillot, D., Irz, P., and Argillier, C. (2007) Functional characters combined with null models reveal inconsistency in mechanisms of species turnover in lacustrine fish communities. *Oecologia*, **153**, 441–52.

Mason, N. W. H., Mouillot, D., Lee, W. G., and Wilson, J. B. (2005) Functional richness, functional evenness and functional divergence: the primary components of functional diversity. *Oikos*, **111**, 112–18.

Mason, W. K., Lamb, K., and Russell, B. (2003) The sustainable grazing systems program: new solutions for livestock producers. *Australian Journal of Experimental Agriculture*, **43**, 663–72.

Massel, S. R., Furukawa, K., and Brinkman, R. M. (1999) Surface wave propagation data in mangrove forests. *Fluid Dynamics Research*, **24**, 219–49.

Matamala, R., Gonzalez-Meler, M. A., Jastrow, J. D., Norby, R. J., and Schlesinger, W. H. (2003) Impacts of fine root turnover on forest NPP and soil C sequestration potential. *Science*, **302**, 1385–7.

Matete, M. and Hassan, R. (2006) Integrated ecological economics accounting approach to evaluation of interbasin water transfers: an application to the Lesotho Highlands Water Project. *Ecological Economics*, **60**(1), 246–59.

Matson, P. A. and Vitousek, P. M. (2006) Agricultural intensification: will land spared from farming be land spared for nature? *Conservation Biology*, **20**, 709–10.

Matthews, D. P. and Gonzalez, A. (2007) The inflationary effects of environmental fluctuations ensure the persistence of sink metapopulations. *Ecology*, **88**, 2848–56.

Matthiessen, B. and Hillebrand, H. (2005) Dispersal frequency affects local biomass production by controlling local diversity. *Ecology Letters*, **9**, 652–62.

May, R. M. (1972a) *Stability and Complexity in Model Ecosystems*. Princeton University Press, Princeton, NJ.

May, R. M. (1972b) Will a large complex system be stable? *Nature*, **238**, 413–14.

May, R. M. (1974) *Stability and Complexity in Model Ecosystems*, 2nd edn. Princeton University Press, Princeton.

May, R. M. (2006) Network structure and the biology of populations. *Trends in Ecology and Evolution*, **21**, 394–9.

Mazda, Y., Wolanski, E., King, B., Sase, A., Ohtsuka, D., and Magi, M. (1997) Drag force due to vegetation in mangrove swamps: *Mangroves and Salt Marshes*, **1**, 193–9.

McAusland, C. and Costello, C. (2004) Avoiding invasives: trade related policies for controlling unintentional exotic species introductions. *Journal of Environmental Economics and Management*, **48**, 954–77.

McCann, K. S. (2000) The diversity–stability debate. *Nature*, **405**, 228–33.

McCann, K., Hastings, A., and Huxel, G. R. (1998) Weak trophic interactions and the balance of nature. *Nature*, **395**, 794–7.

McCann, K. S., Rasmussen, J. B., and Umbanhowar, J. (2005) The dynamics of spatially coupled food webs. *Ecology Letters*, **8**, 513–23.

McCrea, A. R., Trueman, I. C., and Fullen, M. A. (2004) Factors relating to soil fertility and species diversity in both semi-natural and created meadows in the West Midlands of England. *European Journal of Soil Science*, **55**, 335–48.

McCullagh, P. and Nelder, J. A. (1989) *Generalized Linear Models*, 2nd edn. Chapman & Hall, London.

McGill, B. J., Enquist, B. J., Weiher, E., and Westoby, M. (2006) Rebuilding community ecology from functional traits. *Trends in Ecology & Evolution*, **21**, 178–85.

McGrady-Steed, J. and Morin, P. (2000) Biodiversity, density compensation, and the dynamics of populations and functional groups. *Ecology*, **81**, 361–73.

McGrady-Steed, J., Harris, P. M., and Morin, P. J. (1997) Biodiversity regulates ecosystem predictability. *Nature*, **390**, 162–5.

McGregor, S. E. (1976) Insect pollination of cultivated crop-plants. *U.S.D.A. Agriculture Handbook No. 496*, 93–8.

McIntyre, S. and Lavorel, S. (1994) How environmental and disturbance factors influence species composition in temperate Australian grasslands. *Journal of Vegetation Science*, **5**, 373–84.

McIntyre, P. B., Jones, L. E., Flecker, A. S., and Vanni, M. J. (2007) Fish extinctions alter nutrient recycling in tropical freshwaters. *Proceedings of the National Academy of Sciences of the USA*, **104**, 4461–6.

McKinney, M. L. (1997) Extinction vulnerability and selectivity: combining ecological and paleontological views. *Annual Review of Ecology and Systematics*, **28**, 495–516.

McKinney, M. L. (2004) Measuring floristic homogenization by non-native plants in North America. *Global Ecology and Biogeography*, **13**, 47–53.

McLaughlin, A. and Mineau, P. (1995) The impact of agricultural practices on biodiversity. *Agriculture Ecosystems & Environment*, **55**, 201–12.

McNaughton, S. J. (1977) Diversity and stability of ecological communities: a comment on the role of empiricism in ecology. *American Naturalist*, **111**, 515–25.

McNaughton, S. J. (1993) Biodiversity and function of grazing ecosystems. In E.-D. Shulze and H. A. Mooney (eds.) *Biodiversity and Ecosystem Function*, pp. 362–83. Springer-Verlag, Berlin.

McNeely, J. A. (1994) Lessons from the past: forests and biodiversity. *Biodiversity and Conservation*, **3**, 3–20.

McNeely, J. A. and Scherr, S. J. (2003) *Ecoagriculture: Strategies to Feed the World and Save Wild Biodiversity*. Island Press, Washington, DC.

MEA (2005) *Millenium Ecosystem Assessment. Ecosystems and Human Well-Being: Biodiversity Synthesis*. World Resources Institute, Washington DC.

Meléndez-Ramirez, V., Magaña-Rueda S., Parra-Tabla, V., Azala, R., and Navarro, J. (2002) Diversity of native bee visitors of cucurbit crops (Cucurbitaceae) in Yucatán, México. *Journal of Insect Conservation*, **6**, 135–47.

Mellinger, M. V. and McNaughton, S. J. (1975) Structure and function of successional vascular plant communities in central New York. *Ecological Monographs*, **45**, 161–82.

Memmott, J. (1999) The structure of plant–pollinator food webs. *Ecology Letters*, **2**, 276–80.

Memmott, J. and Waser, N. M. (2002) Integration of alien plants into native flower–pollinator visitation web. *Proceedings of the Royal Society, Series B*, **269**, 2395–9.

Memmott, J., Waser, N. M., and Price, M. V. (2004) Tolerance of pollination networks to species extinctions. *Proceedings of the Royal Society, Series B*, **271**, 2605–11.

Memmott, J., Craze, P. G., Waser, N. M., and Price, M. V. (2007) Global warming and the disruption of plant–pollinator interactions. *Ecology Letters*, **10**, 710–17.

Menalled, F. D., Gross, K. L., and Hammond, M. (2001) Weed aboveground and seedbank community responses to agricultural management systems. *Ecological Applications*, **11**, 1586–601.

Mendelsohn, R. and Balick, M. J. (1995) The value of undiscovered pharmaceuticals in tropical forests. *Economic Botany*, **49**, 223–38.

Menke, C. A. and Muir, P. S. (2004) Patterns and influences of exotic species invasion into the grassland habitat of the threatened plant *Silene spaldingii*. *Natural Areas Journal*, **24**, 119–28.

Merila, J., Kruuk, L. E. B., and Sheldon, B. C. (2001) Cryptic evolution in a wild bird population. *Nature*, **412**, 76–9.

Merkl, N., Schultze-Kraft, R., and Infante, C. (2005) Assessment of tropical grasses and legumes for phytoremediation of petroleum-contaminated soils. *Water, Air and Soil Pollution*, **165**, 195–209.

Merrifield, J. (1996) A market approach to conserving biodiversity. *Ecological Economics*, **16**, 217–26.

Metrick, A. and Weitzman, M. L. (1996) Conflicts and choices in biodiversity preservation. *Journal of Economic Perspectives*, **12**, 21–34.

Micheli, F. and Halpern, B. S. (2005) Low functional redundancy in coastal marine assemblages. *Ecology Letters*, **8**, 391–400.

Michener, R. and Tighe, C. (1992) A Poisson regression model of highway fatalities, *American Economic Review*, **82**(2), 452–6.

Mikkelson, G. M. (1993) How do food webs fall apart? – a study of changes in trophic structure during relaxation on habitat fragments. *Oikos*, **67**, 539–47.

Mikola, J. and Setälä, H. (1998) Relating species diversity to ecosystem functioning: mechanistic backgrounds and experimental approach with a decomposer food web. *Oikos*, **83**.

Milcu, A., Partsch, S., Langel, R., and Scheu, S. (2006) The response of decomposers (earthworms, springtails and microorganisms) to variations in species and functional group diversity of plants. *Oikos*, **112**, 513–24.

Millennium Ecosystem Assessment (2003) *Ecosystems and Human Well-Being: a Framework for Assessment*. Island Press, Washington, DC.

Millennium Ecosystem Assessment (2005a). *Ecosystems and Human Well-Being*. Island Press, Washington, DC.

Millennium Ecosystem Assessment (2005b). *Ecosystems and Human Well-Being: Biodiversity Synthesis Report*. Island Press, Washington, DC.

Millennium Ecosystem Assessment (2005c). *Ecosystems and Human Well-Being: Current State and Trends: Findings of the Condition and Trends Working Group (Millennium Ecosystem Assessment Series)*, Island Press, Washington, DC.

Miller, T. E. and Werner, P. A. (1987) Competitive effects and responses between plant species in a first-year old-field community. *Ecology*, **68**, 1201–10.

Mills, D. E. (1980) Transferable development rights markets. *Journal of Urban Economics*, **7**, 63–74.

Mills, L. S., Soule, M. E., and Doak, D. F. (1993) The keystone-species concept in ecology and conservation. *BioScience*, **43**, 219–24.

Minns, A., Finn, J., Hector, A., et al. (2001) The functioning of European grassland ecosystems: potential benefits of biodiversity to agriculture. *Outlook on Agriculture*, **30**, 179–85.

Mitchell, C. A., Reich, P. B., Tilman, D., and Groth, J. V. (2003) Effects of elevated CO_2, nitrogen deposition, and decreased species diversity on foliar fungal plant disease. *Global Change Biology*, **9**, 438–51.

Mitchell, C. A., Tilman, D., and Groth, J. V. (2002) Effects of grassland plant species diversity, abundance, and composition on foliar fungal disease. *Ecology*, **83**, 1713–26.

Mitchell, C. E. and Power, A. G. (2003) Release of invasive plants from fungal and viral pathogens. *Nature*, **421**, 625–7.

Mitchell, J. E., Ffolliott, P. F., and Patton-Mallory, M. (2005) Back to the future: Forest Service rangeland research and management. *Rangelands*, **27**, 19–28.

Mitsch, W. J. (1993) Ecological engineering – a cooperative role with the planetary life-support-system. *Environmental Science & Technology*, **27**, 438–45.

Moeltner, K., Boyle, K. J., and Paterson, R. W. (2007) Meta-analysis and benefit transfer for resource valuation, addressing classical challenges with Bayesian modeling. *Journal of Environmental Economics and Management*, **53**, 250–69.

Moen, J., Aune, K., Edenius, L., and Angerbjorn, A. (2004) Potential effects of climate change on treeline position in the Swedish mountains. *Ecology and Society*, **9**.

Moles, A. T., Ackerly, D. D., Webb, C. O., Tweddle, J. C., Dickie, J. B., and Westoby, M. (2005) A brief history of seed size. *Science*, **307**, 576–80.

Moller, D. A. (2004) Facilitative interactions among plants via shared pollinators. *Ecology*, **85**, 3289–301.

Molyneux, D., Ostfeld, R. S., Bernstein, A., and Chivian, E. (2008) Ecosystem disturbance, biodiversity loss, and human infectious disease. In E. Chivian and A. Bernstein (eds.) *Sustaining Life: How Human Health Depends on Biodiversity*. Oxford University Press, Oxford.

Montagnini, F., Cusack, D., Petit, B., and Kanninen, M. (2005) Environmental services of native tree plantations and agroforestry systems in Central America. *Journal of Sustainable Forestry*, **21**, 51–67.

Montoya, J. M. and Sole, R. V. (2003) Topological properties of food webs: from real data to community assembly models. *Oikos*, **102**, 614–22.

Montoya, J. M., Pimm, S. L., and Solé, R. V. (2006) Ecological networks and their fragility. *Nature*, **442**, 259–64.

Mooney H. A. (2002) The debate on the role of biodiversity in ecosystem functioning. In M. Loreau, S. Naeem, and P. Inchausti (eds.) *Biodiversity and Ecosystem Functioning. Synthesis and Perspectives*, pp. 12–17. Oxford University Press, Oxford.

Mooney, H. A., Cooper, A., and Reid, W. (2005) Confronting the human dilemma. *Science*, **434**, 561–2.

Moragues, E. and Traveset, A. (2005) Effect of *Carpobrotus* spp. on the pollination success of native plant species of the Balearic Islands. *Biological Conservation*, **122**, 611–19.

Morales, C. L. and Aizen, M. A. (2006) Invasive mutualisms and the structure of plant–pollinator interactions in the temperate forests of north-west Patagonia, Argentina. *Journal of Ecology*, **94**, 171–80.

Moretti, M., Duelli, P., and Obrist, M. K. (2006) Biodiversity and resilience of arthropod communities after fire disturbance in temperate forests. *Oecologia*, **149**, 312–27.

Morin, P. J. and McGrady-Steed, J. (2004) Biodiversity and ecosystem functioning in aquatic microbial systems: a new analysis of temporal variation and species richness–predictability relations. *Oikos*, **104**, 458–66.

Morin, X. and Chuine, I. (2006) Niche breadth, competitive strength and range size of tree species: a trade-off based framework to understand species distribution. *Ecology Letters*, **9**, 185–95.

Morris, S. E. (1995) Geomorphic aspects of stream-channel restoration. *Physical Geography*, **16**.

Morris, W. (2003) Which mutualists are most essential? Buffering of plant reproduction against the extinction of pollinators. In P. Kareiva and S. A. Levin (eds.) *The Importance of Species: Perspectives on Expendability and Triage*, pp. 260–80. Princeton University Press, Princeton.

Morton, D. C., DeFries, R. S., Shimabukuro, Y. E., *et al.* (2006) Cropland expansion changes deforestation dynamics in the southern Brazilian Amazon. *Proceedings of the National Academy of Sciences of the USA*, **103**, 14637–41.

Mouchet, M., Guilhaumon, F., Villeger, S., Mason, N. W. H., Tomasini, J.-A., and Mouillot, D. (2008) Towards a consensus for calculating dendrogram-based functional diversity indices. *Oikos*, **117**, 794–800.

Mouillot, D., Mason, N. W. H., Dumay, O. and Wilson, J. B. (2005a). Functional regularity: a neglected aspect of functional diversity. *Oecologia*, **142**, 353–9.

Mouillot, D., Laune, J., Tomasini, J. A., *et al.* (2005b). Assessment of coastal lagoon quality with taxonomic diversity indices of fish, zoobenthos and macrophyte communities. *Hydrobiologia*, **550**, 121–30.

Mouillot, D., Stubbs, W., Faure, M., *et al.* (2005c). Niche overlap estimates based on quantitative functional traits: a new family of non-parametric indices. *Oecologia*, **145**, 345–53.

Mouillot, D., Spatharis, S., Reizopoulou, S., *et al.* (2006) Alternatives to taxonomic-based approaches to assess changes in transitional water communities. *Aquatic Conservation–Marine and Freshwater Ecosystems*, **16**, 469–82.

Mouillot, D., Dumay, O., and Tomasini, J. A. (2007) Limiting similarity, niche filtering and functional diversity in coastal lagoon fish communities. *Estuarine Coastal and Shelf Science*, **71**, 443–56.

Mouquet, N. and Loreau, M. (2002) Coexistence in metacommunities: the regional similarity hypothesis. *American Naturalist*, **159**, 420–6.

Mouquet, N. and Loreau, M. (2003) Community patterns in source–sink metacommunities. *American Naturalist*, **162**, 544–57.

Mouquet, N., Moore, J. L., and Loreau, M. (2002) Plant species richness and community productivity: why the mechanism that promotes coexistence matters. *Ecology Letters*, **5**, 56–65.

Mouquet N., Hoopes M. F., and Amarasekare, P. (2005) The world is patchy and heterogeneous! Trade-off and source sink dynamics in competitive metacommunities. In M. Holyoak, M. A. Leibold, and R. Holt (eds.) *Metacommunities: Spatial Dynamics and Ecological Communities*. Chicago University Press, Chicago.

Mrozek, J. R. and Taylor, L. O. (2002) What determines the value of life? A meta-analysis. *Journal of Policy Analysis and Management*, **21**, 253–70.

Mueller, S. C. (1999) Tons of value in a pound of seed. *Proceedings of the 29th California alfalfa Symposium*, pp. 76–81. Available online at http://alfalfa.ucdavis.edu/-files/pdf/2001NAAICSymposiumAbstracts.pdf

Mulder, C. P. H., Koricheva, J., Huss-Danell, K., Högberg, P., and Joshi, J. (1999) Insects affect relationships between plant species richness and ecosystem processes. *Ecology Letters*, **2**, 237–46.

Mulder, C. P. H., Uliassi, D. D., and Doak, D. F. (2001) Physical stress and diversity–productivity relationships: the role of positive interactions. *Proceedings of the National Academy of Sciences of the USA*, **98**, 6704–8.

Muller, T., Magid, J., Jensen, L. S., and Nielsen, N. E. (2003) Decomposition of plant residues of different quality in soil – DAISY model calibration and simulation based on experimental data. *Ecological Modelling*, **166**, 3–18.

Muotka, T. and Laasonen, P. (2002) Ecosystem recovery in restored headwater streams: the role of enhanced leaf retention. *Journal of Applied Ecology*, **39**, 145–56.

Murray J. D. (2002) *Mathematical Biology*. Springer-Verlag, New York.

Muth, M. K. and Thurman, W. N. (1995) Why support the price of honey. *Choices*, **10**, 19–22.

Muyzer, G. and Smalla, K. (1998) Application of denaturing gradient gel electrophoresis (DGGE) and temperature gradient gel electrophoresis (TGGE) in microbial ecology. *Antonie Van Leeuwenhoek International Journal of General and Molecular Microbiology*, **73**, 127–41.

Mwangi, P. N., Schmitz, M., Scherber, C., *et al.* (2007) Niche pre-emption increases with species richness in experimental plant communities. *Journal of Ecology*, **95**, 65–78.

Myers, N., Mittermeier, R. A., Mittermeier, C. G., De Fonesca, G. A., and Kent, J. (2000) Biodiversity hotspots for conservation priorities. *Nature*, **403**, 853–8.

Myers, R. A. and Worm, B. (2005) Extinction, survival or recovery of large predatory fishes. *Philosophical Transactions of the Royal Society B: Biological Sciences*, **360**, 13–20.

Naelsund, B. and Norberg, J. (2006) Ecosystem consequences of the regional species pool. *Oikos*, **115**, 504–12.

Naeem, S. (1998) Species redundancy and ecosystem reliability. *Conservation Biology*, **12**, 39–45.

Naeem, S. (2000) Reply to Wardle *et al*. *Bulletin of the Ecological Society of America*, **81**, 241–6.

Naeem, S. (2001a) Experimental validity and ecological scale as tools for evaluating research programs. In R. H. Gardner, W. M. Kemp, V. S. Kennedy, and J. E. Petersen (eds.) *Scaling Relationships in Experimental Ecology*. Columbia University Press, New York.

Naeem, S. (2001b) How changes in biodiversity may affect the provision of ecosystem services. In V. C. Hollowell (ed.) *Managing Human Dominated Ecosystems*. Missouri Botanical Garden Press, St Louis.

Naeem, S. (2002a) Disentangling the impacts of diversity on ecosystem functioning in combinatorial experiments. *Ecology*, **83**, 2925–35.

Naeem, S. (2002b) Ecosystem consequences of biodiversity loss: the evolution of a paradigm. *Ecology*, **83**, 1537–52.

Naeem, S. (2003) Models of ecosystem reliability and their implications for species expendability. In P. Kareiva and S. A. Levin (eds.) *The Importance of Species: Perspectives on Expendability and Triage*. Princeton University Press, Princeton.

Naeem, S. (2006a) Biodiversity and ecosystem functioning in restored ecosystems: Extracting principals for a synthetic perspective. In D. A. Falk, M. A. Palmer, and J. B. Zedler (eds.) *Foundations of Restoration Ecology: the Science and Practice of Ecological Restoration*. Island Press, New York.

Naeem, S. (2006b) Expanding scales in biodiversity-based research: challenges and solutions for marine systems. *Marine Ecology Progress Series*, **311**, 273–83.

Naeem, S. (2008) Advancing realism in biodiversity research. *Trends in Ecology & Evolution*, **23**, 414–16.

Naeem, S. and Li, S. (1997) Biodiversity enhances ecosystem reliability. *Nature*, **390**, 507–9.

Naeem, S. and Li, S. (1998) Consumer species richness and autotrophic biomass. *Ecology*, **79**, 2603–15.

Naeem, S. and Wright, J. P. (2003) Disentangling biodiversity effects on ecosystem functioning: Deriving solutions to a seemingly insurmountable problem. *Ecology Letters*, **6**, 567–79.

Naeem, S., Thompson, L. J., Lawler, S. P., Lawton, J. H., and Woodfin, R. M. (1994) Declining biodiversity can alter the performance of ecosystems. *Nature*, **368**, 734–7.

Naeem, S., Thompson, L. J., Lawler, S. P., Lawton, J. H., and Woodfin, R. M. (1995) Biodiversity and ecosystem functioning: empirical evidence from experimental microcosms. *Philosophical Transactions of the Royal Society of London B*, **347**, 249–62.

Naeem, S., Haakenson, K., Thompson, L. J., Lawton, J. H., and Crawley, M. J. (1996) Biodiversity and plant productivity in a model assemblage of plant species. *Oikos*, **76**, 259–64.

Naeem, S., Byers, D., Tjossem, S. F., Bristow, C., and Li, S. (1999a) Plant neighborhood diversity and production. *Ecoscience*, **6**, 355–65.

Naeem, S., Chapin, F. S., *et al.* (1999b) Biodiversity and ecosystem functioning: maintaining natural life support processes. *Ecological Society of America, Issues in Ecology Series* No. 4. 14pp.

Naeem, S., Hahn, D. R., and Schuurman, G. (2000) Producer–decomposer co-dependency influences biodiversity effects. *Nature*, **403**, 762–4.

Naeem, S., Loreau, M., and Inchausti, P. (2002) Biodiversity and ecosystem functioning: the emergence of a synthetic ecological framework. In M. Loreau, S. Naeem, and P. Inchausti (eds.) *Biodiversity and Ecosystem Functioning: Synthesis and Perspectives*. Oxford University Press, Oxford.

Naeem, S., Colwell, R., Díaz, S., *et al.* (2007) Predicting the ecosystem consequences of biodiversity loss: the biomerge framework. In J. G. Canadell, D. E. Pataki, and L. F. Pitelka (eds.) *Terrestrial Ecosystems in a Changing World*. Springer-Verlag, New York.

Nagel, J. M. and Griffin, K. L. (2001) Construction cost and invasive potential: Comparing Lythrum salicaria (Lythraceae) with co-occurring native species along pond banks. *American Journal of Botany*, **88**, 2252–8.

National Research Council (2000) *Our Common Journey: a Transition Toward Sustainability*. National Academy Press, Washington, DC.

Navas, M. L. and Moreau-Richard, J. (2005) Can traits predict the competitive response of herbaceous Mediterranean species? *Acta Oecologica – International Journal Of Ecology*, **27**, 107–14.

Nee, S. and Stone, G. (2003) The end of the beginning for neutral theory. *Trends in Ecology & Evolution*, **18**, 433–4.

Neutel, A. M., Heesterbeek, J. A. P., and de Ruiter, P. C. (2002) Stability in real food webs: weak links in long loops. *Science*, **296**, 1120–3.

Newman, M. E. J. (2003) The structure and function of complex networks. *SIAM Review*, **45**, 167–256.

Newsome, A. E. and Noble, I. R. (1986) Ecological and physiological characters of invading species. In R. H. Groves and J. J. Burdon (eds.) *Ecology of Biological Invasions*, pp. 1–20. Cambridge University Press, Cambridge.

Ni, J. (2003) Plant functional types and climate along a precipitation gradient in temperate grasslands, northeast China and south-east Mongolia. *Journal of Arid Environments*, **53**, 501–16.

Niesten, E., Frumhoff, P., Manion, M., and Hardner, J. (2002) Designing a carbon market that protects forests in developing countries. *Philosophical Transactions of the Royal Society of London, Series A: Mathematical, Physical and Engineering Sciences*, **360**, 1875–88.

Niles, J. O., Brown, S., Pretty, J., Ball, A. S., and Fay, J. (2002) Potential carbon mitigation and income in developing countries from changes in use and management of agricultural and forest lands. *Philosophical Transactions of the Royal Society of London Series A*, **360**, 1621–39.

Nilsson, M. C. and Wardle, D. A. (2005) Understory vegetation as a forest ecosystem driver: evidence from the northern Swedish boreal forest. *Frontiers in Ecology and the Environment*, **3**, 421–8.

Norberg, J. (1999) Linking nature's services to ecosystems: some general ecological concepts. *Ecological Economics* **29**, 183–202.

Norberg, J. (2000) Resource–niche complementarity and autotrophic compensation determines ecosystem-level responses to increased cladoceran species richness. *Oecologia*, **122**, 264–72.

Norberg, J. (2004) Biodiversity and ecosystem functioning: a complex adaptive systems approach. *Limnology and Oceanography*, **49**, 1269–77.

Norberg, J., Swaney, D. P., Dushoff, J., Lin, J., Casagrandi, R., and Levin, S. A. (2001) Phenotypic diversity and ecosystem functioning in changing environments: a theoretical framework. *Proceedings of the National Academy of Sciences of the USA*, **98**, 11376–81.

Nordhaus, W. D. and Kokkelenberg, E. C. (eds.) (1999) *Nature's Numbers*. National Academy Press, Washington, DC.

Nosil, P., Crespi, B. J., and Sandoval, C. P. (2002) Host–plant adaptation drives the parallel evolution of reproductive isolation. *Nature*, **417**, 440–3.

Noss, R. F. (2001) Beyond Kyoto: forest management in a time of rapid climate change. *Conservation Biology*, **15**, 578–90.

NRC (2001) *Compensating for Wetland Losses Under the Clean Water Act*. National Academy Press, Washington, DC.

NRC, National Research Council of the National Academies (2007) *Status of Pollinators in North America*. National Academy of Science, Washington, DC.

Nunes, P. A. L. D. and van den Bergh, J. C. J. M. (2001) Economic valuation of biodiversity: sense or nonsense? *Ecological Economics*, **39**(2), 203–22.

Nupp, T. E. and Swihart, R. K. (1996) Effect of forest patch area on population attributes of white-footed mice (*Peromyscus leucopus*) in fragmented landscapes. *Canadian Journal of Zoology*, **74**, 467–72.

O'Connor, N. E. and Crowe, T. P. (2005) Biodiversity loss and ecosystem functioning: distinguishing between number and identity of species. *Ecology*, **86**, 1783–96.

O'Connor, T. G., Haines, L. M., and Snyman, H. A. (2001) Influence of precipitation and species composition on phytomass of a semi-arid African grassland. *The Journal of Ecology*, **89**, 850–60.

Odum, E. P. (1953) *Fundamentals of Ecology*. Saunders, Philadelphia.

Odum, E. P. (1969) The strategy of ecosystem development. *Science*, **164**, 262–70.

OECD (2003) *Agriculture and Biodiversity: Developing Indicators for Policy Analysis (Summary and Recommendations)*. OECD, Paris.

OECD (2004) *Handbook of Market Creation for Biodiversity: Issues in Implementation*. OECD, Paris.

Oki, T. and Kanae, S. (2006) Global hydrological cycles and world water resources. *Science*, **313**, 1068–72.

Olden, J. D., Hogan, Z. S., and Vander Zanden, M. J. (2007) Small fish, big fish, red fish, blue fish: size-biased extinction risk of the world's freshwater and marine fishes. *Global Ecology and Biogeography*, **16**, 694–701.

Olden, J. D., Poff, N. L., and Bestgen, K. R. (2008) Trait synergisms and the rarity, extirpation, and extinction risk of desert fishes. *Ecology*, **89**, 847–56.

Olmstead, A. L. and Wooten, D. B. (1987) Bee pollination and productivity growth: the case of alfalfa. *American Journal of Agricultural Economy*, **69**, 56–63.

Olschewski, R., Tscharntke, T., Benítez, P., Schwarze, C. S., and Klein, A. M. (2006) Economic evaluation of pollination services comparing coffee landscapes in Ecuador and Indonesia. *Ecology and Society*, **11**, 7.

Olson, M. K. (2004) Are novel drugs more risky for patients than less novel drugs? *Journal of Health Economics* **23**, 1135–58.

Omer, A., Pascual, U., and Russell, N. P. (2007) Biodiversity conservation and productivity in intensive agricultural systems. *Journal of Agricultural Economics*, **58**, 308–29.

Ortega, Y. K., Pearson, D. E., and McKelvey, K. S. (2004) Effects of biological control agents and exotic plant invasion on deer mouse populations. *Ecological Applications*, **14**, 241–53.

Orwin, K. H., Wardle, D. A., and Greenfield, L. G. (2006) Ecological consequences of carbon substrate identity and diversity in a laboratory study. *Ecology*, **87**, 580–93.

Osenberg, C. W., Sarnelle, O., Cooper, S. D., and Holt, R. D. (1999) Resolving ecological questions through meta-analysis: goals, metrics, and models. *Ecology*, **80**, 1105–17.

Ostfeld, R. S. and Holt, R. D. (2004) Are predators good for your health? Evaluating evidence for top-down regulation of zoonotic disease reservoirs. *Frontiers in Ecology and the Environment*, **2**, 13–20.

Ostfeld, R. S. and Logiudice, K. (2003) Community disassembly, biodiversity loss, and the erosion of an ecosystem service. *Ecology*, **84**, 1421–7.

Ostfeld, R. S. and Keesing, F. (2000a) Biodiversity and disease risk: the case of Lyme disease. *Conservation Biology*, **14**, 722–8.

Ostfeld, R. S. and Keesing, F. (2000b) The role of biodiversity in the ecology of vector-borne zoonotic diseases. *Canadian Journal of Zoology*, **78**, 2061–78.

Ostfeld, R. S. and Holt, R. D. (2004) Are predators good for your health? Evaluating evidence for top-down regulation of zoonotic disease reservoirs. *Frontiers in Ecology and the Environment*, **2**, 13–20.

Ostfeld, R. S., Keesing, F., and LoGiudice, K. (2006) Community ecology meets epidemiology: the case of Lyme disease. In S. Collinge and C. Ray (eds.) *Disease Ecology: Community Structure and Pathogen Dynamics*, pp. 28–40. Oxford University Press, Oxford.

Ostling, A. (2005) Ecology – Neutral theory tested by birds. *Nature*, **436**, 635–6.

Otway, S. J., Hector, A., and Lawton, J. H. (2005) Resource dilution effects on specialist insect herbivores in a grassland biodiversity experiment. *Journal of Animal Ecology*, **74**, 234–40.

Owens, I. P. F. and Bennett, P. M. (2000) Ecological basis of extinction risk in birds: habitat loss versus human persecution and introduced predators. *Proceedings of the National Academy of Science USA*, **97**, 12144–8.

Owensby, C. E., Ham, J. M., Knapp, A. K., and Auen, L. M. (1999) Biomass production and species composition change in a tallgrass prairie ecosystem after long-term exposure to elevated atmospheric CO_2. *Global Change Biology*, **5**, 497–506.

Pacala, S. and Tilman, D. (1994) Limiting similarity in mechanistic and spatial models of plant competition in heterogeneous environments. *The American Naturalist*, **143**, 222–57.

Pacala, S. W. and Deutschman, D. H. (1995) Details that matter: the spatial distribution of individual trees maintains forest ecosystem function. *Oikos*, **74**, 357–65.

Pacala, S. W., Canham, C. D., Saponara, J., Silander, J. A., Kobe, R. K., and Ribbens, E. (1996) Forest models defined by field measurements: estimation, error analysis and dynamics. *Ecological Monographs*, **66**, 1–43.

Pace, M. L., Cole, J. J., Carpenter, S. R., and Kitchell, J. F. (1999) Trophic cascades revealed in diverse ecosystems. *Trends in Ecology & Evolution*, **14**, 483–8.

Pachepsky, E., Bown, J. L., Eberst, A., *et al.* (2007) Consequences of intraspecific variation for the structure and function of ecological communities Part 2: Linking diversity and function. *Ecological Modelling*, **207**, 277–85.

Pagiola, S. (2002) Paying for water services in Central America: learning from Costa Rica. In S. Pagiola, J. Bishop, and N. Landell-Mills (eds.) *Selling Forest Environmental Services: Market-Based Mechanisms for Conservation and Development*, pp. 37–62. Earthscan, London.

Pagiola, S., von Ritter, K., and Bishop, J. (2004) *How Much is an Ecosystem Worth? Assessing the Economic Value of Conservation*. The World Bank, Washington, DC.

Paine, R. T. (1966) Food web complexity and species diversity. *American Naturalist*, **100**, 65–7.

Paine, R. T. (2002) Trophic control of production in a rocky intertidal community. *Science*, **296**, 736–9.

Palm, C. P., Vosti, S. A., Sanchez, P. A., and Ericksen, P. J. (2005) *Slash-and-Burn Agriculture: the Search for Alternatives*. Columbia University Press, New York.

Palmer, M. A., Ambrose, R. F., and Poff, N. L. (1997) Ecological theory and community restoration ecology. *Restoration Ecology*, **5**, 291–300.

Palmer, M., Bernhardt, E., Chornesky, E., *et al.* (2004) Ecology: ecology for a crowded planet. *Science*, **304**, 1251–2.

Panayotou, T. (1994) Economic instruments for environmental management and sustainable development. Prepared for the United Nations Environment Programme's Consultative Expert Group Meeting on them Use and Application of Economic Policy Instruments for Environmental Management and Sustainable Development, Nairobi, 23–24 February 1995. *Environmental Economics Series Paper* No. 16

Parmesan, C. and Yohe, G. (2003) A globally coherent fingerprint of climate change impacts across natural systems. *Nature*, **421**, 37–42.

Pascual, U. and Perrings, C. P. (2007) Developing incentives and economic mechanisms for *in situ* biodiversity conservation in agricultural landscapes. *Agriculture, Ecosystems, and Environment*, **121**, 256–68.

Pauly, D., Christensen, V., Dalsgaard, J., Froese, R., and Torres, F. (1998) Fishing down marine food webs. *Science*, **279**, 860–3.

Pausas, J. G., Bradstock, R. A., Keith, D. A., Keeley, J. E., and GTCE (2004) Plant functional traits in relation to fire in crown-fire ecosystems. *Ecology*, **85**, 1085–100.

Pautasso, M., Holdenrieder, O., and Stenlid, J. (2005) Susceptibility to fungal pathogens of forests differing in tree diversity. In M. Scherer-Lorenzen, C. Körner, and E.-D. Schulze (eds.) *The Functional Significance of Forest Diversity*. Springer-Verlag, Berlin.

Pavoine, S. and Doledec, S. (2005) The apportionment of quadratic entropy: a useful alternative for partitioning diversity in ecological data. *Environmental and Ecological Statistics*, **12**, 125–38.

Peacock, L., Hunter, T., Turner, H., and Brain, P. (2001) Does host genotype diversity affect the distribution of insect and disease damage in willow cropping systems? *Journal of Applied Ecology*, **38**, 1070–81.

Pearce, D. W. and Puroshothamon, S. (1995) The economic value of plant-based pharmaceuticals. In T. Swanson (ed.) *Intellectual Property Rights and Biodiversity Conservation*, pp. 127–38. Cambridge University Press, Cambridge.

Pearce, D. W. (1993) *Economic Values and the Natural World*. Earthscan, London.

Pearce, D. W., Moran, D., and Krug, W. (1999) *The Global Value of Biological Diversity: A Report to UNEP*. Centre for Social and Economic Research on the Global Environment, University College London.

Pellant, M., Abbey, B., and Karl, S. (2004) Restoring the Great Basin Desert, USA: integrating science, management, and people. *Environmental Monitoring and Assessment*, **99**, 169–79.

Pérez-Harguindeguy, N., Diaz, S., Cornelissen, J. H. C., Vendramini, F., Cabido, M., and Castellanos, A. (2000) Chemistry and toughness predict leaf litter decomposition rates over a wide spectrum of functional types and taxa in central Argentina. *Plant and Soil*, **218**, 21–30.

Pérez Harguindeguy, N., Blundo, C., Gurvich, D., Díaz, S., and Cuevas, E. (2008) More than the sum of its parts? Assessing litter heterogeneity effects on the decomposition of litter mixtures through leaf chemistry. *Plant and Soil*, **303**, 151–9.

Perfecto, I., Vandermeer, J. H., Bautista, G. L., *et al.* (2004) Greater predation in shaded coffee farms: the role of resident neotropical birds. *Ecology*, **85**, 2677–81.

Perner, J. and Malt, S. (2003) Assessment of changing agricultural land use: response of vegetation, ground-dwelling spiders and beetles to the conversion of arable land into grassland. *Agriculture Ecosystems & Environment*, **98**, 169–81.

Perrings, C. (1995) Biodiversity conservation and insurance. In T. M. Swanson (ed.) *The Economics and Ecology of Biodiversity Loss*. Cambridge University Press, Cambridge.

Perrings, C. (2001) The economics of biodiversity loss and agricultural development in low income countries. In D. R. Lee and C. B. Barrett (eds.) *Tradeoffs or Synergies? Agricultural Intensification, Economic Development and the Environment*, pp. 57–72. CAB International, Wallingford.

Perrings, C. and Gadgil, M. (2003) Conserving biodiversity: reconciling local and global public benefits. In

I. Kaul, P. Conceicao, K. le Goulven, and R. L. Mendoza (eds.) *Providing Global Public Goods: Managing Globalization*, pp. 532–55. Oxford University Press, Oxford.

Perrings, C. and Vincent, J. (eds.) (2003) *Natural Resource Accounting and Economic Development*. Edward Elgar, Cheltenham.

Perrings, C., Mäler, K.-G., Folke, C., Holling, C. S., and Jansson, B.-O. (eds.) (1995) *Biodiversity Loss: Economic and Ecological Issues*, Cambridge University Press, Cambridge.

Perrings C., Williamson, M., and Dalmazzone, S. (eds.) (2000) *The Economics of Biological Invasions*. Edward Elgar, Cheltenham.

Perrings, C., Williamson, M., Barbier, E. B., *et al.* (2002) Biological invasion risks and the public good: an economic perspective. *Conservation Ecology*, **6**, 1. Available online at http://www.consecol.org/vol6/iss1/art1.

Perrings, C., Dehnen-Schmutz, K., Touza, J., and Williamson, M. (2005) How to manage biological invasions under globalization. *Trends in Ecology and Evolution*, **20**(5), 212–15.

Petchey, O. L. (2004) On the statistical significance of functional diversity. *Functional Ecology*, **18**, 297–303.

Petchey, O. L. and Gaston, K. J. (2002a) Functional diversity (fd), species richness, and community composition. *Ecology Letters*, **5**, 402–11.

Petchey, O. L. and Gaston, K. J. (2002b) Extinction and the loss of functional diversity. *Proceedings of the Royal Society of London Series B: Biological Sciences*, **269**, 1721–7.

Petchey, O. L. and Gaston, K. J. (2006) Functional diversity: back to basics and looking forward. *Ecology Letters*, **9**, 741–58.

Petchey, O. L. and Gaston, K. J. (2007) Dendrograms and measuring functional diversity. *Oikos*, **116**, 1422–6.

Petchey, O. L., McPhearson, P. T., Casey, T. M., and Morin, P. J. (1999) Environmental warming alters food-web structure and ecosystem function. *Nature*, **402**, 69–72.

Petchey, O. L., Morin, P. J., Hulot, F. D., Loreau, M., McGrady-Steed, J., and Naeem, S. (2002a) Contributions of aquatic model systems to our understanding of biodiversity and ecosystem functioning. In M. Loreau, S. Naeem, and P. Inchausti (eds.) *Biodiversity and Ecosystem Functioning: Syntheses and Perspectives*, pp. 127–39. Oxford University Press, Oxford.

Petchey, O. L., Casey, T., Jiang, L., McPhearson, P. T., and Price, J. (2002b) Species richness, environmental fluctuations, and temporal change in total community biomass. *Oikos*, **99**, 231–40.

Petchey, O. L., Downing, A. L., Mittelbach, G. G., *et al.* (2004a) Species loss and the structure and functioning of multitrophic aquatic systems. *Oikos*, **104**, 467–78.

Petchey, O. L., Hector, A., and Gaston, K. J. (2004b) How do different measures of functional diversity perform? *Ecology*, **85**, 847–57.

Petchey, O. L., Evans, K. L., Fishburn, I. S., and Gaston, K. J. (2007) Low functional diversity and no redundancy in British avian assemblages. *Journal of Animal Ecology*, **76**, 977–85.

Petchey, O. L., Beckerman, A. P., Riede, J. O., and Warren, P. H. (2008a) Size, foraging, and food web structure. *Proceedings of the National Academy of Sciences of the USA*, **105**, 4191–6.

Petchey, O. L., Eklof, A., Borrvall, C., and Ebenman, B. (2008b) Trophically unique species are vulnerable to cascading extinction. *American Naturalist*, **171**, 568–79.

Petermann, J., Fergus, A. J., Turnbull, L. A., and Schmid, B. (2008) Janzen-Connell effects are both widespread and strong enough to maintain functional diversity in grasslands. *Ecology*, **89**(9), 2399–406.

Peters, C. M., Balick, M. J., Kahn, F., and Anderson, A. B. (1989) Oligarchic forests of economic plants in ammonia: Utilization and conservation of an important tropical resource. *Conservation Biology*, **3**(4), 341–9.

Peterson, G., Allen, C. R., and Holling, C. S. (1998) Ecological resilience, biodiversity, and scale. *Ecosystems*, **1**, 6–18.

Pfisterer, A. B. and Schmid, B. (2002) Diversity-dependent production can decrease the stability of ecosystem functioning. *Nature*, **416**, 84–6.

Pfisterer, A. B., Diemer, M., and Schmid, B. (2003) Dietary shift and lowered biomass gain of a generalist herbivore in species-poor experimental plant communities. *Oecologia*, **135**, 234–41.

Philpott, S. M. and Armbrecht, I. (2006) Biodiversity in tropical agroforests and the ecological role of ants and ant diversity in predatory function. *Ecological Entomology*, **31**, 369–77.

Philpott, S. M., Uno, S., and Maldonado, J. (2006) The importance of ants and high-shade management to coffee pollination and fruit weight in Chiapas, Mexico. *Biodiversity and Conservation*, **15**, 487–501.

Pías, B. and Guitián, P. (2006) Breeding system and pollen limitation in the masting tree *Sorbus aucuparia* L. (Rosaceae) in the NW Iberian Peninsula. *Acta Oecologica*, **29**, 97–103.

Pimentel, D. (1961) Species diversity and insect population outbreaks. *Annals of the Entomological Society of America*, **54**, 76–86.

Pimentel, D., Wilson, C., McCullum, C., *et al.* (1997) Economic and environmental benefits of biodiversity. *BioScience*, **47**, 747–57.

Pimentel, D., Berger, B., Filiberto, D., *et al.* (2004) Water resources: agricultural and environmental issues. *Bioscience*, **54**, 909–18.

Pimm, S. L. (1980) Food web design and the effect of species deletion. *Oikos*, **35**, 139–49.

Pimm, S. L. (1982) *Food Webs*. Chapman & Hall, London, UK.

Pimm, S. L. (1984) The complexity and stability of ecosystems. *Nature*, **307**, 321–6.

Pimm, S. L. and Lawton, J. H. (1977) Number of trophic levels in ecological communities. *Nature*, **268**, 329–31.

Pimm, S. L. and Lawton, J. H. (1978) Feeding on more than one trophic level. *Nature*, **275**, 542–4.

Pimm, S. L., Russell, G. J., Gittleman, J. L., and Brooks, T. M. (1995) The future of biodiversity. *Science*, **269**, 347–50.

Pinheiro, J. C. and Bates, D. M. (2000) *Mixed Effects Models in S And S-Plus*. Springer-Verlag, Berlin.

Pinkus-Rendon, M. A., Parra-Tabla, V., and Melendez-Ramirez, V. (2005) Floral resource use and interactions between *Apis mellifera* and native bees in cucurbit crops in Yucatan, Mexico. *Canadian Entomologist*, **137**, 441–9.

Piotto, D., Viquez, E., Montagnini, F., and Kanninen, M. (2004) Pure and mixed forest plantations with native species of the dry tropics of Costa Rica: a comparison of growth and productivity. *Forest Ecology and Management*, **190**, 359–72.

Podani, J. and Schmera, D. (2006) On dendrogram-based measures of functional diversity. *Oikos*, **115**, 179–85.

Poff, N. L., Olden, J. D., Vieira, N. K. M., Finn, D. S., Simmons, M. P., and Kondratieff, B. C. (2006) Functional trait niches of North American lotic insects: traits-based ecological applications in light of phylogenetic relationships. *Journal of the North American Benthological Society*, **25**, 730–55.

Polasky, S. and Doremus, H. (1998) When the truth hurts: endangered species policy on private land with imperfect information. *Journal of Environmental Economics and Management*, **35**, 22–47.

Polasky, S. and Solow, A. R. (1995) On the value of a collection of species. *Journal of Environmental Economics and Management*, **29**, 298–303.

Polasky, S., Solow, A. R., and Broadus, J. M. (1993) Searching for uncertain benefits and the conservation of biological diversity. *Environmental and Resource Economics*, **3**, 171–81.

Polasky, S., Nelson, E., Lonsdorf, E., Fackler, P., and Starfield, A. (2003) Conserving species in a working landscape: land use with biological and economic objectives. *Ecological Applications*, **15**(4), 1387–401.

Polasky, S., Costello, C., and McAusland, C. (2004) On trade, land-use and biodiversity. *Journal of Environmental Economics and Management*, **48**, 911–25.

Polis, G. A. (1991) Complex trophic interactions in deserts – an empirical critique of food-web theory. *American Naturalist*, **138**, 123–55.

Polis, G. A. and Holt, R. D. (1992) Intraguild predation – the dynamics of complex trophic interactions. *Trends in Ecology & Evolution*, **7**, 151–4.

Polis, G. A. and Strong, D. R. (1996) Food web complexity and community dynamics. *American Naturalist*, **147**, 813–46.

Polis, G. A., Anderson, W. B., and Holt, R. D. (1997) Toward an integration of landscape and food web ecology: the dynamics of spatially subsidized food webs. *Annual Review of Ecology and Systematics*, **28**, 289–316.

Polley, H. W., Mayeux, H. S., Johnson, H. B., and Tischler, C. R. (1997) Viewpoint: atmospheric CO_2, soil water, and shrub/grass ratios on rangelands. *Journal of Range Management*, **50**, 278–84.

Polley, H. W., Johnson, H. B., and Derner, J. D. (2003) Increasing CO_2 from subambient to superambient concentrations alters species composition and increases above-ground biomass in a C-3/C-4 grassland. *New Phytologist*, **160**, 319–27.

Popper, D. E. and Popper, F. J. (1987) The Great Plains: from dust to dust. *Planning*, **53**, 12–18.

Popper, F. J. and Popper, D. E. (2006) The onset of the Buffalo Commons. *Journal of the West*, **45**, 29–34.

Potthoff, M., Jackson, L. E., Steenwerth, K. L., Ramirez, I., Stromberg, M. R., and Rolston, D. E. (2005) Soil biological and chemical properties in restored perennial grassland in California. *Restoration Ecology*, **13**, 61–73.

Potts, S. G., Petanidou, T., Roberts, S., O'Toole, C., Hulbert, A., and Willmer, P. (2006) Plant–pollinator biodiversity and pollination services in a complex Mediterranean landscape. *Biological Conservation*, **129**, 519–29.

Potvin, C. and Gotelli, N. J. (2008) Biodiversity enhances individual performance but does not affect survivorship in tropical trees. *Ecology Letters*, **11**, 217–23.

Potvin, C. and Vasseur, L. (1997) Long-term CO_2 enrichment of a pasture community: species richness, dominance, and succession. *Ecology*, **78**, 666–77.

Poulin, R. (2004) Macroecological patterns of species richness in parasite assemblages. *Basic and Applied Ecology*, **5**, 423–34.

Poulin, J., Sakai, A. K., Weller, S. G., and Nguyen, T. (2007) Phenotypic plasticity, precipitation, and invasiveness in the fire-promoting grass *Pennisetum* setaceum (Poaceae). *American Journal of Botany*, **94**, 533–41.

Power, M. E., Tilman, D., Estes, J. A., Menge, B. A., Bond, W. J., Mills, L. S., Daily, G., Castilla, J. C., Lubchenco, J., and Paine, R. T. (1996) Challenges in the quest for keystones. *BioScience*, **46**, 609–20.

Powers, J. S., Haggar, J. P., and Fisher, R. F. (1997) The effect of overstory composition on understory woody regeneration and species richness in 7-year old plantations in Costa Rica. *Forest Ecology and Management*, **99**, 43–54.

Prance, G. T. (2002) Species survival and carbon retention in commercially exploited tropical rainforest. *Philosophical Transactions of the Royal Society of London Series A: Mathematical Physical and Engineering Sciences*, **360**, 1777–85.

Prasad, R. P. and Snyder, W. E. (2006) Polyphagy complicates conservation biological control that targets generalist predators. *Journal of Applied Ecology*, **43**, 343–52.

Pregitzer, K. S. and Euskirchen, E. S. (2004) Carbon cycling and storage in world forests: biome patterns related to forest age. *Global Change Biology*, **10**, 2052–77.

Prentice, I. C., Farquhar, G. D., Fasham, M. J. R., *et al.* (2001) The carbon cycle and atmospheric carbon dioxide. In J. T. Houghton, Y. Ding, D. J. Griggs, M. Noguer, P. J. van der Linden, X. Dai, K. Maskell, C. A. Johnson (eds.) *Climate Change 2001: The Scientific Basis. Contribution of Working Group I to the Third Assessment Report of the Intergovernmental Panel on Climate Change*. Cambridge University Press, Cambridge.

Pretty, J. (1995) *Regenerating Agriculture*. Earthscan, London.

Pretty, J. and Ball, A. (2001) *Agricultural Influences on Emissions and Sequestration of Carbon and Emerging Trading Options*. University of Essex, Colchester.

Pretzsch, H. (2005) Diversity and productivity in forests: evidence from long-term experimental plots. In M. Scherer-Lorenzen, C. Körner, E.-D. Schulze (eds.) *The Functional Significance of Forest Diversity*. Springer-Verlag, Berlin.

Price, G. R. (1970) Selection and covariance. *Nature*, **227**, 520–1.

Price, G. R. (1995) The nature of selection. *Journal of Theoretical Biology*, **175**, 389–96.

Priess, J. A., Mimler, M., Klein, A. M., Schwarze, S., Tscharntke, T., and Steffan-Dewenter, I. (2007) Linking deforestation scenarios to pollination services and economic returns in coffee agroforestry systems. *Ecological Applications*, **17**, 407–17.

Prieur-Richard, A. H., Lavorel, S., Linhart, Y. B., and Dos Santos, A. (2002) Plant diversity, herbivory and resistance of a plant community to invasion in Mediterranean annual communities. *Oecologia*, **130**, 96–104.

Prosser, J. I., Bohannan, B. J. M., Curtis, T. P., *et al.* (2007) The role of ecological theory in microbial ecology. *Nature Reviews Microbiology*, **5**, 384–92.

Pulford, I. D. and Watson, C. (2003) Phytoremediation of heavy metal-contaminated land by trees – a review. *Environment International*, **29**, 529–40.

Pulkkinen, K. (2007) Microparasite transmission to *Daphnia magna* decreases in the presence of conspecifics. *Oecologia*, **154**, 45–53.

Pullin, A., Knight, T., Stone, D., *et al.* (2004) Do conservation managers use scientific evidence to support their decision making? *Biological Conservation*, **119**, 245–52.

Purvis, A., Agapow, P.-M., Gittleman, J. L., and Mace, G. M. (2000a). Nonrandom extinction and the loss of evolutionary history. *Science*, **288**, 328–30.

Purvis, A., Jones, K. E., and Mace, G. M. (2000c) Extinction. *Bioessays*, **22**, 1123–33.

Pywell, R. F., Bullock, J. M., Hopkins, A., *et al.* (2002) Restoration of species-rich grassland on arable land: assessing the limiting processes using a multi-site experiment. *Journal of Applied Ecology*, **39**, 294–309.

Pywell, R. F., Bullock, J. M., Roy, D. B., Warman, L. I. Z., Walker, K. J., and Rothery, P. (2003) Plant traits as predictors of performance in ecological restoration. *Journal of Applied Ecology*, **40**, 65–77.

Pywell, R. F., Bullock, J. M., Tallowin, J. B. R., Walker, K. J., Warman, E. A., and Masters, G. J. (2007) Enhancing diversity of species-poor grasslands: An experimental assessment of multiple constraints. *Journal of Applied Ecology*, **44**, 81–94.

Qaim, M. and Zilberman, D. (2003) Yield effects of genetically modified crops in developing countries. *Science*, **299**, 900–2.

Quaas, M. F. and Baumgärtner, S. (2008) Natural vs. financial insurance in the management of public-good ecosystems. *Ecological Economics*, **65**, 397–406.

Quince, C., Curtis, T. P., and Sloan, W. T. (2008) The rational exploration of microbial diversity. *The ISME Journal*, **2**(10), 997–1006.

Quintana, X. D., Brucet, S., Boix, D., *et al.* (2008) A nonparametric method for the measurement of size diversity with emphasis on data standardization. *Limnology and Oceanography – Methods*, **6**, 75–86.

R Development Core Team (2006) *R: a Language and Environment for Statistical Computing*. R Foundation for Statistical Computing, Vienna, Austria.

R Development Core Team (2008) *R: a Language and Environment for Statistical Computing*. R foundation for statistical computing. Vienna, Austria.

Raffaelli, D. (2004) How extinction patterns affect ecosystems. *Science*, **306**, 1141–2.

Raffaelli, D. G. (2006) Biodiversity and ecosystem functioning: issues of scale and trophic complexity. *Marine Ecology Progress Series*, **311**, 285–94.

Raffaelli, D., Van der Putten, W. H., Persson, L., *et al.* (2002) Multi-trophic dynamics and ecosystem processes. In M. Loreau, S. Naeem, and P. Inchausti (eds.) *Biodiversity and Ecosystem Functioning: Syntheses and Perspectives*. Oxford University Press, Oxford.

Raffaelli, D., Emmerson, M., Solan, M., *et al.* (2003) Biodiversity and ecosystem functioning in shallow coastal waters: an experimental approach. *Journal of Sea Research*, **49**, 133–41.

Raffaelli, D., Solan, M., and Webb, T. J. (2005a). Do marine ecologists do it differently? *Marine Ecology Progress Series*, **304**, 283–9.

Raffaelli, D., Cardinale, B. J., Downing, A. L., *et al.* (2005b) Reinventing the wheel in ecology research? – Response. *Science*, **307**, 1875–6.

Rand, T. A., Tylianakis, J. M., and Tscharntke, T. (2006) Spillover edge effects: the dispersal of agriculturally subsidized insect natural enemies into adjacent natural habitats. *Ecology Letters*, **9**, 603–14.

Randall, A. (2002) Valuing the outputs of multifunctional agriculture. *European Review of Agricultural Economics*, **29**(3), 289–307.

Rantalainen, M. L., Fritze, H., Haimi, J., Pennanen, T., and Setälä, H. (2005) Species richness and food web structure of soil decomposer community as affected by size of habitat and habitat corridors. *Global Change Biology*, **11**, 1614–27.

Rappé, M. S. and Giavannoni, S. J. (2002) The uncultured microbial majority. *Annual Review of Microbiology*, **57**, 369–94.

Rasko, D. A., Altherr, M. R., Han, C. S., and Ravel, J. (2005) Genomics of the *Bacillus cereus* group of organisms. *FEMS Microbiology Reviews*, **29**, 303–29.

Rausser, G. C. and Small, A. (2000) Valuing research leads: bioprospecting and the conservation of genetic resources, *Journal of Political Economy*, **108**(1), 173–206.

Raven, P. H. (2002) Science, sustainability, and the human prospect. *Science*, **297**, 954–8.

Raviraja, N. S., Sridhar, K. R., and Bärlocher, F. (1998) Breakdown of *Ficus* and *Eucalyptus* leaves in an organically polluted river in India: fungal diversity and ecological functions. *Freshwater Biology*, **39**, 537–45.

Rayner, M. J., Hauber, M. E., Imber, M. J., Stamp, R. K., and Clout, M. N. (2007) Spatial heterogeneity of mesopredator release within an oceanic island system. *Proceedings of the National Academy of Sciences of the USA*, **104**, 20862–5.

Redondo-Brenes, A. and Montagnini, F. (2006) Growth, productivity, aboveground biomass, and carbon sequestration of pure and mixed native tree plantations in the Caribbean lowlands of Costa Rica. *Forest Ecology and Management*, **232**, 168–78.

Reich, P. B., Ellsworth, D. S., Walters, M. B., *et al.* (1999) Generality of leaf trait relationships: a test across six biomes. *Ecology*, **80**, 1955–69.

Reich, P. B., Knops, J., Tilman, D., *et al.* (2001a) Plant diversity influences ecosystem responses to elevated CO_2 and nitrogen enrichment. *Nature*, **410**, 809–12.

Reich, P. B., Tilman, D., Craine, J., Ellsworth, D., Tjoelker, M. G., Knops, J., Wedin, D., Naeem, S., Bahauddin, D., Goth, J., Bengtson, W., and Lee, T. D. (2001b) Do species and functional groups differ in acquisition and use of C, N and water under varying atmospheric CO_2 and N availability regimes? A field test with 16 grassland species. *New Phytologist*, **150**, 435–48.

Reich, P. B., Tilman, D., Naeem, S., *et al.* (2004) Species and functional group diversity independently influence biomass accumulation and its response to CO_2 and N. *Proceedings of the National Academy of Sciences of the USA*, **101**, 10101–6.

Reich, P. B., Tjoelker, M. G., Machado, J. L., and Oleksyn, J. (2006) Universal scaling of respiratory metabolism, size and nitrogen in plants. *Nature*, **439**, 457–61.

Reid, W. V. C. (1989) Sustainable development – lessons from success. *Environment*, **31**, 7–9.

Rejmanek, M. and Richardson, D. M. (1996) What attributes make some plant species more invasive? *Ecology*, **77**, 1655–61.

Rejmankova, E., Rejmanek, M., Djohan, T., and Goldman, C. R. (1999) Resistance and resilience of subalpine wetlands with respect to prolonged drought. *Folia Geobotanica*, **34**, 175–88.

Resh, S. C., Binkley, D., and Parotta, J. A. (2002) Greater soil carbon sequestration under nitrogen-fixing trees compared with *Eucalyptus* species. *Ecosystems*, **5**, 217–31.

Reusch, T. B. H., Ehlers, A., Hammerli, A., and Worm, B. (2005) Ecosystem recovery after climatic extremes enhanced by genotypic diversity. *Proceedings of the National Academy of Sciences of the USA*, **102**, 2826–31.

Reynolds, J. D., Dulvy, N. K., Goodwin, N. B., and Hutchings, J. A. (2005) Biology of extinction risk in marine fishes. *Proceedings of the Royal Society B: Biological Sciences*, **272**, 2337–44.

Rhoades, C. C., Ekert, G. E., and Coleman, D. C. (2000) Soil carbon differences among forest, agriculture and secondary vegetation in lower montane Ecuador. *Ecological Applications*, **10**, 497–505.

Richardson, C. J. and Hussain, N. A. (2006) Restoring the Garden of Eden: an ecological assessment of the marshes of Iraq. *Bioscience*, **56**, 477–89.

Richardson, D. M. and Pysek, P. (2006) Plant invasions: merging the concepts of species invasiveness and community invasibility. *Progress in Physical Geography*, **30**, 409–31.

Richardson, D. M. and Rejmanek, M. (2004) Conifers as invasive aliens: a global survey and predictive framework. diversity and distributions, **10**, 321–31.

Richardson, S. J., Press, M. C., Parsons, A. N., and Hartley, S. E. (2002) How do nutrients and warming impact on plant communities and their insect herbivores? A 9-year study from a sub-Arctic heath. *Journal of Ecology*, **90**, 544–56.

Richardson, S. J., Peltzer, D. A., Allen, R. B., McGlone, M. S., and Parfitt, R. L. (2004) Rapid development of phosphorus limitation in temperate rainforest along the Franz Josef soil chronosequence. *Oecologia*, **139**, 267–76.

Richerson, P., Armstrong, R., and Goldman, C. R. (1970) Contemporaneous disequilibrium, a new hypothesis to explain the paradox of the plankton. *Proceedings of the National Academy of Sciences of the USA*, **67**, 1710–14.

Ricketts, T. H. (2004) Tropical forest fragments enhance pollinator activity in nearby coffee crops. *Conservation Biology*, **18**, 1262–71.

Ricketts, T. H., Daily, G. C., Ehrlich, P. R., and Michener, C. D. (2004) Economic value of tropical forest to coffee production. *Proceedings of the National Academy of Sciences of the USA*, **101**, 12579–82.

Ricketts, T. H., Regetz, J., Steffan-Dewenter, I. et al. (2008) Landscape effects on crop pollination services: are there general patterns? *Ecology Letters*, **11**, 499–515.

Ricklefs, R. E. and Miller, G. (1999) *Ecology*. W. H. Freeman, New York.

Ricotta, C. (2004) A parametric diversity measure combining the relative abundances and taxonomic distinctiveness of species. *Diversity and Distributions*, **10**, 143–46.

Ricotta, C. (2005a) A note on functional diversity measures. *Basic and Applied Ecology*, **6**, 479–86.

Ricotta, C. (2005b) Through the jungle of biological diversity. *Acta Biotheoretica*, **53**, 29–38.

Ricotta, C. (2007) A semantic taxonomy for diversity measures. *Acta Biotheoretica*, **55**, 23–33.

Robertson, G. P. and Swinton, S. M. (2005) Reconciling agricultural productivity and environmental integrity: a grand challenge for agriculture. *Frontiers in Ecology and the Environment*, **3**, 38–46.

Robinson, G. R., Holt, R. D., Gaines, M. S., et al. (1992) Diverse and contrasting effects of habitat fragmentation. *Science*, **257**, 524–6.

Robinson, W. S., Nowogrodzki, R., and Morse, R. A. (1989) The value of honey bees as pollinators of U.S. crops. *American Bee Journal*, **129**, 411–23.

Rodrıguez, L. C., Pascual, U., and Niemeyer, H. M. (2006) Local identification and valuation of ecosystem goods and services from Opuntia scrublands of Ayacucho, Peru. *Ecological Economics*, **57**, 30–44.

Romanuk, T. N., Vogt, R. J., and Kolasa, J. (2006) Nutrient enrichment weakens the stabilizing effect of species richness. *Oikos*, **114**, 291–302.

Rooney, N., McCann, K., Gellner, G., and Moore, J. C. (2006) Structural asymmetry and the stability of diverse food webs. *Nature*, **442**, 265–9.

Root, R. B. (1967) The niche exploitation pattern of the blue–gray gnatcatcher. *Ecological Monographs*, **37**, 317–50.

Root, R. B. (1973) Organization of plant–arthropod association in simple and diverse habitats: the fauna of collards (i. *Brassica* oleracea). *Ecological monographs*, **43**, 95–124.

Root, T. L., Price, J. T., Hall, K. R., Schneider, S. H., Rosenzweig, C., and Pounds, J. A. (2003) Fingerprints of global warming on wild animals and plants. *Nature*, **421**, 57–60.

Roscher, C., Schumacher, J., Baade, J., et al. (2004) The role of biodiversity for element cycling and trophic interactions: and experimental approach in a grassland community. *Basic and Applied Ecology*, **5**, 107–21.

Roscher, C., Temperton, V. M., Scherer-Lorenzen, M., et al. (2005) Overyielding in experimental grassland communities – irrespective of species pool or spatial scale. *Ecology Letters*, **8**, 419–29.

Rose, N. L. (1990) Profitability and product quality: economic determinants of airline safety performance. *Journal of Political Economy*, **98**(5), 944–64.

Rosenberg, M. J., Adams, D. C., and Gurevitch, J. (2000) *Metawin 2.0 User's Manual: Statistical Software for Meta-Analysis*. Sinauer Associates, Sunderland, MA.

Rosenberger, R. S. and Loomis, J. B. (2000) Using meta-analysis for benefit transfer: in-sample convergent validity tests of an outdoor recreation database. *Water Resources Research*, **36**, 1097–107.

Rosenheim, J. A. (2007) Intraguild predation: new theoretical and empirical perspectives. *Ecology*, **88**, 2679–80.

Rosenheim, J. A., Kaya, H. K., Ehler, L. E., Marois, J. J., and Jaffee, B. A. (1995) Intraguild predation among biological control agents – theory and evidence. *Biological Control*, **5**, 303–35.

Rosenzweig, M. L. (1971) Paradox of enrichment: destabilization of exploitation ecosystems in ecological time. *Science*, **171**, 385–7.

Rosenzweig, M. L. (1987) Restoration ecology: a tool to study population interactions. In W. R. Jordan, M. E. Gilpin, and J. D. Aber (eds.) *Restoration Ecology: a Synthetic Appraoch to Ecological Research*. Cambridge University Press, New York.

Rosgen, D. L. (1994) A classification of natural rivers. *Catena*, **22**, 169–99.

Roubik, D. W. (2002) The value of bees to the coffee harvest. *Nature*, **417**, 708.

Rouget, M., Cowling, R. M., Vlok, J., et al. (2006) Getting the biodiversity intactness index right: the importance of habitat degradation data. *Global Change Biology*, **12**, 2032–6.

Rowe, E. C., Van Noordwijk, M., Suprayogo, D., and Cadisch, G. (2005) Nitrogen use efficiency of monoculture and hedgerow intercropping in the humid tropics. *Plant and Soil*, **268**, 61–74.

Rowlands, I. H. (1996) South Africa and global climate change. *Journal of Modern African Studies* **34**(1), 163–78.

Roy, M., Holt, R. D., and Barfield. M. (2005) Temporal autocorrelation can enhance the persistence and abundance of metapopulations comprised of coupled sinks. *American Naturalist*, **166**, 246–61.

Royal Society (2008) Biodiversity–climate interactions: adaptation, mitigation and human ivelihoods. *Policy document 30/07*. The Royal Society, London.

Royer, D. L. and Wilf, P. (2006) Why do toothed leaves correlate with cold climates? Gas exchange at leaf margins provides new insights into a classic paleo-temperature proxy. *International Journal of Plant Sciences*, **167**, 11–18.

Royer, D. L., Wilf, P., Janesko, D. A., Kowalski, E. A., and Dilcher, D. L. (2005) Correlations of climate and plant ecology to leaf size and shape: potential proxies for the fossil record. *American Journal of Botany*, **92**, 1141–51.

Rozzi, R. (2004) Ethical implications of yahgan and mapuche indigenous narratives about the birds of the austral temperate forests of South America. *Ornitologia Neotropical*, **15**, 435–44.

Rudolf, V. H. and Antonovics, J. (2005) Species coexistence and pathogens with frequency-dependent transmission. *American Naturalist*, **166**, 112–18.

Ruel, J. J. and Ayres. M. P. (1999) Jensen's inequality predicts effects of environmental variation. *Trends in Ecology and Evolution*, **14**, 361–6.

Ruesink, J. L., Lenihan, H. S., Trimble, A. C., et al. (2005) Introduction of non-native oysters: ecosystem effects and restoration implications. *Annual Review of Ecology Evolution and Systematics*, **36**, 643–89.

Ruesink, J. L., Feist, B. E., Harvey, C. J., et al. (2006) Changes in productivity associated with four introduced species: ecosystem transformation of a 'pristine' estuary. *Marine Ecology Progress Series*, **311**, 203–15.

Ruiz, G. M., Fofonoff, P., Hines, A. H., and Grosholz, E. D. (1999) Non-indigenous species as stressors in estuarine and marine communities: assessing invasion impacts and interactions. *Limnology and Oceanography*, **44**, 950–72.

Rundlof, M. and Smith, H. G. (2006) The effect of organic farming on butterfly diversity depends on landscape context. *Journal of Applied Ecology*, **43**, 1121–7.

Runge, C. F. (2001) *A Global Environment Organization (GEO) and the World Trading System: Prospects and Problems*. Working Paper WP01-1, Center for International Food and Agricultural Policy, University of Minnesota, St Paul, MN.

Russell, A. E., Cambardella, J. A., Ewel, J. J., and Parkin, T. B. (2004) Species, rotation and life-form diversity effects on soil carbon in experimental tropical ecosystems. *Ecological Applications*, **14**, 47–60.

Sabine, C. L., Heimann, M., Artaxo, P., et al. (2004) Current status and past trends of the global carbon cycle. In C. B. Field and M. R. Raupach (eds.) *Global Carbon Cycle: Integrating Humans, Climate, and the Natural World*. Island Press, Washington, DC.

Sachs, J. D. (2004) Sustainable development. *Science*, **304**, 649.

Saha, S. and Howe, H. F. (2003) Species composition and fire in a dry deciduous forest. *Ecology*, **84**, 3118–23.

Sala, O. E., Chapin, F. S., Armesto, J. J., et al. (2000) Biodiversity: global biodiversity scenarios for the year 2100. *Science*, **287**, 1770–4.

Sall, S. N., Masse, D., Ndour, N. Y. B., and Chotte, J. L. (2006) Does cropping modify the decomposition function and the diversity of the soil microbial community of tropical fallow soil? *Applied Soil Ecology*, **31**, 211–19.

Sampath, P. G. (2005) *Regulating Bioprospecting: Institutions for Drug Research, Access and Benefit-Sharing*. United Nations University Press, New York, NY.

Sanderson, M. A., Soder, K. J., Muller, L. D. and Klement, K. D. (2005) Forage mixture productivity and botanical composition in pastures grazed by dairy cattle. *Agronomy Journal*, **97**, 1465–71.

Sandhu, H. P., Wrattan, S. D., and Cullen, R. (2008) *Evaluating Ecosystem Services on Farmland: a Novel, Experimental, 'Bottom-Up' Approach*. Comité Interne pour l'Agriculture Biologique, CIAB-INRA.

Sandler, T. (2001) *Economic Concepts in the New Century*. Cambridge University Press, New York.

Sanz, M., Schulze, E.-D., and Valentini, R. (2004) International policy framework on climate change: sinks in recent international agreements. In C. Field and M. Raupach (eds.) *The Global Carbon Cycle: Integrating Humans, Climate, and the Natural World*. Island Press, Washington DC.

SAS Institute (1995) *JMP Statistics and Graphics Guide*, Ver. 3.2. SAS Institute, Cary, NC.

Saunders, L., Hanbury-Tenison, R., and Swingland, I. (2002) Social capital from carbon property: creating equity for indigenous people. *Philosophical Transactions of the Royal Society of London Series A*, **360**, 1763–75.

Sax, D. F. and Gaines, S. D. (2003) Species diversity: from global decreases to local increases. *Trends in Ecology & Evolution*, **18**, 561–6.

Scarborough, C. L., Ferrari, J., and Godfray, H. C. J. (2005) Aphid protected from pathogen by enodsymbiont. *Science*, **310**, 1781.

Schamp, B. S., Chau, J., and Aarssen, L. W. (2008) Dispersion of traits related to competitive ability in an old-field plant community. *Journal of Ecology*, **96**, 204–12.

Scheffer, M. and Carpenter, S. R. (2003) Catastrophic regime shifts in ecosystems: linking theory to observation. *Trends in Ecology & Evolution*, **18**, 648–56.

Scherer-Lorenzen, M., Koerner, C., and Schulze, E.-D. (2005a) *Forest Diversity and Function – Temperate and Boreal Systems*. Springer-Verlag, Berlin.

Scherer-Lorenzen, M., Potvin, C., Koricheva, J., et al. (2005b). The design of experimental tree plantations for functional biodiversity research. In M. Scherer-Lorenzen, C. Körner, and E.-D. Schulze (eds.) *The Functional Significance of Forest Diversity*. Springer-Verlag, Berlin.

Scherer-Lorenzen, M., Schulze, E.-D., Don, A., Schumacher, J., and Weller, E. (2007a) Exploring the functional significance of forest diversity: a new long-term experiment with temperate tree species (biotree). *Perspectives in Plant Ecology Evolution and Systematics*, **9**, 53–70.

Scherer-Lorenzen, M., Bonilla, J. L., and Potvin, C. (2007b) Tree species richness affects litter production and decomposition rates in a tropical biodiversity experiment. *Oikos*, **116**, 2108–24.

Scheu, S. (2001) Plants and generalist predators as links between the below-ground and above-ground system. *Basic and Applied Ecology*, **2**, 3–13.

Schimel, D. (2007) Carbon cycle conundrums. *Proceedings of the National Academy of Sciences of the USA*, **104**, 18353–4.

Schimel, D. S., House, J. I., Hibbard, K. A., et al. (2001) Recent patterns and mechanisms of carbon exchange by terrestial ecosystems. *Nature*, **414**, 169–72.

Schläpfer, F. and Schmid, B. (1999) Ecosystem effects of biodiversity: a classification of hypotheses and exploration of empirical results. *Ecological Applications*, **9**, 893–912.

Schläpfer, F., Schmid, B., and Seidl, I. (1999) Expert estimates about effects of biodiversity on ecosystem processes and services. *Oikos*, **84**, 346–52.

Schläpfer, F., Tucker, M., and Seidl, I. (2002) Returns from hay cultivation in fertilized low diversity and non-fertilized high diversity grassland. *Environmental and Resource Economics*, **21**, 89–100.

Schläpfer, F., Pfisterer, A. B., and Schmid, B. (2005) Non-random species extinction and plant production: implications for ecosystem functioning. *Journal of Applied Ecology*, **42**, 13–24.

Schmera, D., Erős, T., and Podani, J. (2009) A measure for assessing functional diversity in ecological communities. *Aquatic Ecology*, **43**(1), 157–67.

Schmid, B. (2002) The species richness–productivity controversy. *Trends in Ecology and Evolution*, **17**, 113–14.

Schmid, B., Hector, A., Huston, M. A., et al. (2002a) The design and analysis of biodiversity experiments. In M. Loreau, S. Naeem, and P. Inchausti (eds.) *Biodiversity and Ecosystem Functioning. Synthesis and Perspectives*, pp. 61–75. Oxford University Press, Oxford.

Schmid, B., Joshi, J., and Schläpfer, F. (2002b). Empirical evidence for biodiversity–ecosystem functioning relationships. In A. P. Kinzig, S. W. Pacala, and D. Tilman (eds.) *Functional Consequences of Biodiversity: Empirical Progress and Theoretical Extensions*, pp. 120–50. Princeton University Press, Princeton.

Schmid, B., Hector, A., Saha, P., and Loreau, M. (2008) Biodiversity effects and transgressive overyielding. *Journal of Plant Ecology*, **1**, 95–102.

Schmid, B., Pfisterer, A. B., Balvanera, P., (2009) Effects of biodiversity on ecosystem, community and population variables reported 1974-2004. *Ecology*, **90**, 853.

Schmida, A. and Wilson. M. V. (1985) Biological determinants of species diversity. *Journal of Biogeography*, **12**, 1–20.

Schmidt, K. A. and Ostfeld, R. S. (2001) Biodiversity and the dilution effect in disease ecology. *Ecology*, **82**, 609–19.

Schmitz, O. J. (2008) Effects of predator hunting mode on grassland ecosystem function. *Science*, **319**, 952–4.

Schmitz, O. J., Hamback, P. A., and Beckerman, A. P. (2000) Trophic cascades in terrestrial systems: a review of the effects of carnivore removals on plants. *American Naturalist*, **155**, 141–53.

Schmitz, O. J., Krivan, V., and Ovadia, O. (2004) Trophic cascades: the primacy of trait-mediated indirect interactions. *Ecology Letters*, **7**, 153–63.

Schmitzberger, I., Wrbka, T., Steurer, B., Aschenbrenner, G., Peterseil, J., and Zechmeister, H. G. (2005) How farming styles influence biodiversity maintenance in Austrian agricultural landscapes. *Agriculture Ecosystems & Environment*, **108**, 274–90.

Scholes, R. J. and Biggs, R. (2005) A biodiversity intactness index. *Nature*, **434**, 45–9.

Scholes, R. J. and van der Merwe, M. L. (1995) South African green house inventory. *CSIR Report FOR-DEA 918*, Pretoria, CSIR.

Schroth, G., D'Angelo, S. A., Teixeira, W. G., Haag, D., and Lieberei, R. (2002) Conversion of secondary forest into agroforestry and monoculture plantations in Amazonia: Consequences for biomass, litter and soil carbon stocks after 7 years. *Forest Ecology and Management*, **163**, 131–50.

Schulze, E.-D. (2005) Biological control of the terrestrial carbon sink. *Biogesociences Discussions*, **2**, 1283–329.

Schulze, E.-D. and Mooney, H. A. (eds.) (1993) *Biodiversity and Ecosystem Function*. Springer-Verlag, New York.

Schulze, E.-D., Valentini, R., and Sanz, M.-J. (2002) The long way from Kyoto to Marrakesh: implications of the Kyoto Protocol negotiations for global ecology. *Global Change Biology*, **8**, 505–18.

Schulze, E., Mollicone, D., Achard, F., et al. (2003) Climate change – making deforestation pay under the Kyoto Protocol? *Science*, **299**, 1669.

Schussman, H., Geiger, E., Mau-Crimmins, T., and Ward, J. (2006) Spread and current potential distribution of an alien grass, *Eragrostis lehmanniana nees*, in the southwestern USA: comparing historical data and ecological niche models. *Diversity and Distributions*, **12**, 582–92.

Schwartz, M. W., Brigham, C. A., Hoeksema, J. D., Lyons, K. G., Mills, M. H., and Van Mantgem, P. J. (2000) Linking biodiversity to ecosystem function: implications for conservation ecology. *Oecologia*, **122**, 297–305.

Schweiger, O., Maelfait, J. P., Van Wingerden, W. *et al.* (2005) Quantifying the impact of environmental factors on arthropod communities in agricultural landscapes across organizational levels and spatial scales. *Journal of Applied Ecology*, **42**, 1129–39.

Schweitzer, J. A., Bailey, J. K., Rehill, B. J., *et al.* (2004) Genetically based trait in a dominant tree affects ecosystem processes. *Ecology Letters*, **7**, 127–34.

Schweitzer, J. A., Bailey, J. K., Hart, S. C., and Whitman, T. G. (2005a) Nonadditive effects of mixing cottonwood genotypes on litter decomposition and nutrient dyamics. *Ecology*, **86**, 2834–40.

Schweitzer, J. A., Bailey, J. K., Hart, S. C., Wimp, G. M., Chapman, S. K., and Whitham, T. G. (2005a) The interaction of plant genotype and herbivory decelerate leaf litter decomposition and alter nutrient dynamics. *Oikos*, **110**, 133–45.

Scott, J. C. (1998) *Seeing Like a State*. Yale University Press, New Haven.

Seabloom, E. W., Harpole, W. S., Reichman, O. J., and Tilman, D. (2003) Invasion, competitive dominance, and resource use by exotic and native California grassland species. *Proceedings of the National Academy of Sciences of the USA*, **100**, 13384–9.

Segura, C., Feriche, M., Pleguezuelos, J. M., and Santos, X. (2007) Specialist and generalist species in habitat use: implications for conservation assessment in snakes. *Journal of Natural History*, **41**, 2765–74.

Sekericioglu, C. H., Ehrlich, P. R., Daily, G. C., Aygen, D., Goehring, D., and Sandi, R. F. (2002) Disappearance of insectivorous birds from tropical forest fragments. *Proceedings of the National Academy of Sciences of the USA*, **99**, 263–7.

Sekericioglu, C. H., Schneider, S. H., Fay, J. P., and Loarie, S. R. (2008) Climate change, elevational range shifts, and bird extinctions. *Conservation Biology*, **22**, 140–50.

SER (2004) *The SER Primer on Ecological Restoration*, version 2. Society for Ecological Restoration Science and Policy Working Group.

Setälä, H. and McLean, M. A. (2004) Decomposition rate of organic substrates in relation to the species diversity of soil saprophytic fungi. *Oecologia*, **139**, 98–107.

Sheehan, C., Kirwan, L., Connolly, J. and Bolger, T. (2006) The effects of earthworm functional group diversity on nitrogen dynamics in soils. *Soil Biology and Biochemistry*, **38**, 2629–36.

Shmida, A. and Wilson, M. V. (1985) Biological determinants of species diversity. *Journal of Biogeography*, **12**, 1–20.

Shrestha, R. K. and Loomis, J. B. (2001) Testing a meta-analysis model for benefit transfer in international outdoor recreation. *Ecological Economics*, **39**, 67–83.

Shuler, R. E., Roulston, T. H., and Farris, G. E. (2005) Farming practices influence wild pollinator populations on squash and pumpkin. *Journal of Economic Entomology*, **98**, 790–5.

Shultz, S., Bradbury, R. B., Evans, K. L., Gregory, R. D., and Blackburn, T. M. (2005) Brain size and resource specialization predict long-term population trends in British birds. *Proceedings of the Royal Society B: Biological Sciences*, **272**, 2305–11.

Shvidenko, A., CooBarber, C., Persson, R., *et al.* (2005) Forest and woodland systems. In R. Hassan, R. Scholes, and N. Ash (eds.) *Ecosystems and Human Well-Being. Current State and Trends – Findings of the Condition and Trends Working Group of the Millennium Ecosystem Assessment*. Island Press, Washington, DC.

Siemann, E. and Rogers, W. E. (2001) Genetic differences in growth of an invasive tree species. *Ecology Letters*, **4**, 514–18.

Simberloff, D. and Dayan, T. (1991) The guild concept and the structure of ecological communities. *Annual Review of Ecology and Systematics*, **22**, 115–43.

Simberloff, D. and Stiling, P. (1996) How risky is biological control? *Ecology*, **77**, 1965–74.

Simon, K. S., Townsend, C. R., Biggs, B. J. F., Bowden, W. B., and Frew, R. D. (2004) Habitat-specific nitrogen dynamics in New Zealand streams containing native or invasive fish. *Ecosystems*, **7**, 777–92.

Simpson, R. D., Sedjo, R. A., and Reid, J. W. (1996) Valuing biodiversity for use in pharmaceutical research. *Journal of Political Economy*, **104**(1), 163–85.

Skelton, L. E. and Barrett, G. W. (2005) A comparison of conventional and alternative agroecosystems using alfalfa (*Medicago sativa*) and winter wheat (*Triticum aestivum*). *Renewable Agriculture and Food Systems*, **20**, 38–47.

Smale, M., Hartell, J., Heisey, P. W., and Senauer, B. (1998) The contribution of genetic resources and diversity to wheat production in the Punjab of Pakistan. *American Journal of Agricultural Economics*, **80**, 482–93.

Smedes, G. W. and Hurd, L. E. (1981) An empirical-test of community stability – resistance of a fouling community to a biological patch-forming disturbance. *Ecology*, **62**, 1561–72.

Smith, M. D. and Knapp, A. K. (2003) Dominant species maintain ecosystem function with non-random species loss. *Ecology Letters*, **6**, 509–17.

Smith, R. S., Shiel, R. S., Bardgett, R. D., *et al.* (2003) Soil microbial community, fertility, vegetation and diversity as targets in the restoration management of a meadow grassland. *Journal of Applied Ecology*, **40**, 51–64.

Smith, T. M. and Smith, R. L. (2005) *Elements of Ecology*, 6th edn. Benjamin Cummings, San Francisco.

Smith, V. K. and Huang, J.-C. (1995) Can markets value air quality? A meta-analysis of hedonic property value model. *Journal of Political Economy*, **103**, 209–27.

Smith, V. K. and Pattanayak, S. K. (2002) Is meta-analysis a Noah's ark for non-market valuation? *Environmental and Resource Economics*, **22**, 271–96.

Smukler, S. M., Jackson, L. E., Murphree, L., Yokota, R., Koike, S. T., and Smith, R. F. (2008) Transition to large-scale organic vegetable production in the Salinas Valley, California. *Agriculture Ecosystems & Environment*, **126**, 168–88.

Snelder, D. J. (2001) Forest patches in *Imperata* grassland and prospects for their preservation under agricultural intensification in Northeast Luzon, The Philippines. *Agroforestry Systems*, **52**, 207–17.

Snyder, R. E. and Chesson, C. P. (2004) How the spatial scales of dispersal, competition, and environmental heterogeneity interact to affect coexistence. *American Naturalist*, **164**, 633–50.

Snyder, W. E. and Ives, A. R. (2003) Interactions between specialist and generalist natural enemies: parasitoids, predators, and pea aphid biocontrol. *Ecology*, **84**, 91–107.

Snyder, W. E., Snyder, G. B., Finke, D. L., and Straub, C. S. (2006) Predator biodiversity strengthens herbivore suppression. *Ecology Letters*, **9**, 789–96.

Sodhi, N. S., Koh, L. P., Peh, K. S. H., Tan, H. T. W., Chazdon, R. L., Corlett, R. T., Lee, T. M., Colwell, R. K., Brook, B. W., Sekercioglu, C. H., and Bradshaw, C. J. A. (2008) Correlates of extinction proneness in tropical angiosperms. *Diversity and Distributions*, **14**, 1–10.

Sogin, M. L., Morrison, H. G., Huber, J. A., Welch, D. M., Huse, S. M., Neal, P. R., Arrieta, J. M. and Herndl, G. J. (2006) Microbial diversity in the deep sea and the underexplored 'rare biosphere'. *Proceedings of the National Academy of Sciences of the USA*, **103**, 12115.

Sohngen, B. and Brown, S. (2006) The influence of conversion of forest types on carbon sequestration and other ecosystem services in the South Central United States. *Ecological Economics*, **57**, 698–708.

Solan, M., Cardinale, B. J., Downing, A. L., Engelhardt, K. A. M., Ruesink, J. L., and Srivastava, D. S. (2004) Extinction and ecosystem function in the marine benthos. *Science*, **306**, 1177–80.

Soldaat, L. L. and Auge, H. (1998) Interactions between an invasive plant, *Mahonia aquifolium*, and a native phytophagous insect, *Rhagoletis meigenii*. In U. Starfinger, K. Edwards, I. Kowarik, and M. Williamson (eds.) *Plant Invasions: Ecological Mechanisms and Human Responses*, pp. 347–60. Backhuys Publishers, Leiden, Netherlands.

Solé, R. V. and Montoya, J. M. (2001) Complexity and fragility in ecological networks. *Proceedings of the Royal Society of London Series B: Biological Sciences*, **268**, 2039–45.

Solow, R. M. (1974) Intergenerational equity and exhaustible resources. *Review of Economic Studies* (Symposium), **41**, 29–46.

Solow, R. M. (1986) On the intergenerational allocation of exhaustible resources. *Scandinavian Journal of Economics*, **88**, 141–9.

Soule, J. D. and Piper, J. K. (1992) Ecological crisis of modern agriculture: damage and depletion. In J. D. Soule and J. K. Piper (eds.) *Farming in Nature's Image*, pp. 11–30. Island Press, Washington, DC.

Spehn, E. M., Hector, A., Joshi, J., *et al.* (2005) Ecosystem effects of biodiversity manipulations in European grasslands. *Ecological Monographs*, **75**, 37–63.

Srinivasan, U. T., Dunne, J. A., Harte, J., and Martinez, N. D. (2007) Response of complex food webs to realistic extinction sequences. *Ecology*, **88**, 671–82.

Srivastava, D. S. (2002) The role of conservation in expanding biodiversity research. *Oikos*, **98**, 351–60.

Srivastava, D. S. and Vellend, M. (2005) Biodiversity–ecosystem function research: Is it relevant to conservation? *Annual Review of Ecology Evolution and Systematics*, **36**, 267–94.

Srivastava, D. S., Kolasa, J., Bengtsson, J., *et al.* (2004) Are natural microcosms useful model systems for ecology? *Trends in Ecology and Evolution*, **19**, 379–84.

Srivastava, D. S., Cardinale, B. J., Downing, A. L., *et al.* (2009) Diversity has consistent top-down, but not bottom-up, effects on decomposition. *Ecology*, **90**, 1073–1083.

Srivastava, D. S., Cardinale, B. J., Downing, A. L., *et al.*, (2009) Diversity has stronger top-down than bottom-up effects on decomposition. *Ecology*, **90**, 1073–1083.

Stacey, D. A., Thomas, M. B., Blanford, S., Pell., J. K., Pugh, C., and Fellowes, M. D. E. (2003) Genotype and temperature influence pea aphid resistance to a fungal entomopathogen. *Physiological Entomology*, **28**, 75–81.

Stachowicz, J. J., Fried, H., Osman, R. W., and Whitlatch, R. B. (2002) Biodiversity, invasion resistance, and marine ecosystem function. Reconciling pattern and process. *Ecology*, **83**, 2575–90.

Stampe, E. D. and Daehler, C. C. (2003) Mycorrhizal species identity affects plant community structure and invasion: a microcosm study. *Oikos*, **100**, 362–72.

Stanley, W. G. and Montagnini, F. (1999) Biomass and nutrient accumulation in pure and mixed plantations of indigenous tree species grown on poor soils in the humid tropics of Costa Rica. *Forest Ecology and Management*, **113**, 91–103.

Stark, S. C., Bunker, D. E., and Carson, W. P. (2006) A null model of exotic plant diversity tested with exotic and native species–area relationships. *Ecology Letters*, **9**, 136–41.

Starzomski, B. M. and Srivastava, D. S. (2007) Landscape geometry determines community response to disturbance. *Oikos*, **116**, 690–9.

Statzner, B. and Moss, B. (2004) Linking ecological function, biodiversity and habitat: a mini-review focusing on older ecological literature. *Basic and Applied Ecology*, **5**, 97–106.

Stauffer, R. C. (ed.) (1975) *Charles Darwin's Natural Selection, Being the Second Part of His Big Species Book Written from 1856 to 1858*. Cambridge University Press, London.

Steenwerth, K. L., Jackson, L. E., Calderon, F. J., Stromberg, M. R., and Scow, K. M. (2003) Soil community composition and land use history in cultivated and grassland ecosystems of coastal California. *Soil Biology & Biochemistry*, **35**, 489–500.

Steiner, C. F. (2001) The effects of prey heterogeneity and consumer identity on the limitation of trophic-level biomass. *Ecology*, **82**, 2495–506.

Steiner, C. F. (2005a) Impacts of density-independent mortality and productivity on the strength and outcome of competition. *Ecology*, **86**, 727–39.

Steiner, C. F. (2005b) Temporal stability of pond zooplankton assemblages. *Freshwater Biology*, **50**, 105–12.

Steiner, C. F., Darcy-Hall, T. L., Dorn, N. J., Garcia, E. A., Mittelbach, G. G., and Wojdak, J. M. (2005a) The influence of consumer diversity and indirect facilitation on trophic level biomass and stability. *Oikos*, **110**, 556–66.

Steiner, C. F., Long, Z. T., Krumins, J. A., and Morin, P. J. (2005b) Temporal stability of aquatic food webs: partitioning the effects of species diversity, species composition and enrichment. *Ecology Letters*, **8**, 819–28.

Steiner, C. F., Long, Z. T., Krumins, J. A., and Morin, P. J. (2006) Population and community resilience in multitrophic communities. *Ecology*, **87**, 996–1007.

Stephan, A., Meyer, A. H., and Schmid, B. (2000) Plant diversity positively affects soil bacterial diversity in experimental grassland ecosystems. *Journal of Ecology*, **88**, 988–98.

Stephens, B. B., Gurney, K. R., Tans, P. P., et al. (2007) Weak northern and strong tropical land carbon uptake from vertical profiles of atmospheric CO_2. *Science*, **316**, 1732–5.

Stephens, D. W. and Krebs, J. R. (1986) *Foraging theory*. Princeton University Press, Princeton, NJ.

Stern, N. (2006) *Stern Review on the Economics of Climate Change*. Cambridge University Press, Cambridge.

Stern, R. A., Eisikowitch, D., and Dag, A. (2001) Sequential introduction of honeybee colonies and doubling their density increases cross-pollination, fruit-set and yield in 'Red Delicious' apple. *Journal of Horticultural Science and Biotechnology*, **76**, 17–23.

Stirling, G. and Wilsey, B. (2001) Empirical relationships between species richness, evenness, and proportional diversity. *The American Naturalist*, **158**, 286–99.

Stokstad, E. (2007) The case of the empty hives. *Science*, **316**, 970–2.

Stone, G. N. (1994) Activity patterns of females of the solitary bee *Anthophora-plumipes* in relation to temperature, nectar supplies and body-size. *Ecological Entomology*, **19**, 177–89.

Stone, G. N., Gilbert, F., Willmer, P., Potts, S. G., Semida, F., and Zalat, S. (1999) Windows of opportunity and the temporal structuring of foraging activity in a desert solitary bee. *Ecological Entomology*, **24**, 208–21.

Stoneham, G., Chaudhri, V., Strappazzon, L., and Ha, A. (2007) Auctioning biodiversity conservation contracts. In A. Kontoleon, U. Pascual, and T. Swanson (eds.) *Biodiversity Economics: Principles, Methods and Applications*, pp. 389–416. Cambridge University Press, Cambridge.

Strauss, J. (1996) Implications of the TRIPS agreement in the field of patent law. In K. Beier and G. Schricker (eds.) *From GATT to TRIPS – The Agreement on Trade Related Aspects of Intellectual Property Rights*. IIC Studies 18, Max Planck Institute for Foreign and international Patent, Copyright and Competition Law, Munich.

Strong, D. R. (1992) Are trophic cascades all wet? Differentiation and donor-control in speciose ecosystems. *Ecology*, **73**, 747–54.

Stroup, R. (1995) The Endangered Species Act: making endangered species the enemy. *Political Economy Research Center Policy Series PS-3*.

Suding, K. N., Collins, S. L., Gough, L., et al. (2005) Functional- and abundance-based mechanisms explain diversity loss due to N fertilization. *Proceedings of the National Academy of Sciences of the USA*, **102**, 4387–92.

Suding, K. N., Goldberg, D. E., and Hartman, K. M. (2003) Relationships among species traits: separating levels of response and identifying linkages to abundance. *Ecology*, **84**, 1–16.

Suding, K. N., Lavorel, S., Chapin, F. S., Cornelissen, J. H. C., Diaz, S., Garnier, E., Goldberg, D., Hooper, D. U., Jackson, S. T., and Navas, M. L. (2008) Scaling environmental change through the community-level: a trait-based response-and-effect framework for plants. *Global Change Biology*, **14**, 1125–40.

Sumner, D. A. and Boriss, H. (2006) Bee-economics and the leap in pollination fees. *Giannini Foundation of Agricultural Economics Update*, **9**, 9–11.

Sutherland, G. D., Harestad, A. S., Price, K., and Lertzman, K. P. (2000) Scaling of natal dispersal distances in terrestrial birds and mammals. *Conservation Ecology*, **4**(1), 16.

Sutherland, W., Pullin, J., Dolman, P., *et al.* (2004) The need for evidence-based conservation. *Trends in Ecology & Evolution*, **19**, 305–8.

Sutherst, R. W. (1993) Arthropods as disease vectors in a changing environment. In J. V. Lake, G. R. Bock, and K. Ackrill (eds.) *Environmental Change and Human Health*, pp. 124–39. Wiley, New York.

Suttle, C. A. (2007) Marine viruses – major players in the global ecosystem. *Nature Reviews Microbiology*, **5**, 801–12.

Sutton-Grier, A. E., Wright, J. P., McGill, B. Richardson, C., (in review). Environmental conditions influence plant functional diversity effects on potential denitrification. *Ecology*.

Swallow, S. K. (1990) Depletion of the environmental basis for renewable resources: the economics of interdependent renewable and nonrenewable resources. *Journal of Environmental Economics and Management*, **19**, 281–96.

Swan, C. M. and Palmer, M. A. (2005) Leaf litter diversity leads to non-additivity in stream detritivore colonization dynamics. *Oceanological and Hydrobiological Studies*, **34**, 19–38.

Swan, C. M. and Palmer, M. A. (2006) Composition of speciose leaf litter alters stream detritivore growth, feeding activity and leaf breakdown. *Oecologia*, **147**, 469–78.

Swanson, T. M. (ed.) (1995) *The Economics and Ecology of Biodiversity Loss.* Cambridge University Press, Cambridge.

Swart, J. A. A. (2003) Will direct payments help biodiversity? *Science*, **299**, 1981.

Swift, M. J., Izac, A.-M. N., and van Noordwijk, M. (2004) Biodiversity and ecosystem services in agricultural landscapes – are we asking the right questions? *Agriculture Ecosystems and Environment*, **104**, 113–34.

Symstad, A. J. and Tilman, D. (2001) Diversity loss, recruitment limitation, and ecosystem functioning: lessons learned from a removal experiment. *Oikos*, **92**, 424–435.

Symstad, A. J., Siemann, E., and Haarstad, J. (2000) An experimental test of the effect of plant functional group diversity on arthropod diversity. *Oikos*, **89**, 243–53.

Symstad, A. J., Wienk, C. L., and Thorstenson, A. (2006) Field-based evaluation of two herbaceous plant community sampling methods for long-term monitoring in northern great plains national parks. *Open-file report 2006–1282*. U.S. Geological Survey, Helena MT.

Taki, H. and Kevan, P. G. (2007) Does habitat loss affect the communities of plants and insects equally in plant–pollinator interactions? Preliminary findings. *Biodiversity and Conservation*, **16**, 3147–61.

Ter Steege, H. and Hammond, D. S. (2001) Character convergence, diversity, and disturbance in tropical rain forest in Guyana. *Ecology*, **82**, 3197–212.

Teyssonneyre, F., Picon-Cochard, C., Falcimagne, R., and Soussana, J. F. (2002) Effects of elevated CO_2 and cutting frequency on plant community structure in a temperate grassland. *Global Change Biology*, **8**, 1034–46.

Thébault, E. and Loreau, M. (2003) Food-web constraints on biodiversity–ecosystem functioning relationships. *Proceedings of the National Academy of Sciences of the USA*, **100**, 14949–54.

Thébault, E. and Loreau, M. (2005) Trophic interactions and the relationship between species diversity and ecosystem stability. *American Naturalist*, **166**, 95–114.

Thébault, E. and Loreau, M. (2006) The relationship between biodiversity and ecosystem functioning in food webs. *Ecological Research*, **21**, 17–25.

Thébault, E., Huber, V., and Loreau, M. (2007) Cascading extinctions and ecosystem functioning: contrasting effects of diversity depending on food web structure. *Oikos*, **116**, 163–73.

Thies, J. E. and Devare, M. H. (2007) An ecological assessment of transgenic crops. *Journal of Development Studies*, **43**, 97–129.

Thies, C. and Tscharntke, T. (1999) Landscape structure and biological control in agroecosystems. *Science*, **285**, 893–95.

Thies, C., Roschewitz, I., and Tscharntke, T. (2005) The landscape context of cereal aphid–parasitoid interactions. *Proceedings of the Royal Society B: Biological Sciences*, **272**, 203–10.

Thomas, M. B. and Reid, A. M. (2007) Are exotic natural enemies an effective way of controlling invasive plants? *Trends in Ecology & Evolution*, **22**, 447–53.

Thompson, C., Beringer, J., Chapin, F. S., and McGuire, A. D. (2004) Structural complexity and land-surface energy exchange along a gradient from Arctic tundra to boreal forest. *Journal of Vegetation Science*, **15**, 397–406.

Thompson, J. N. (2006) Mutualistic webs of species. *Science*, **312**, 372–3.

Thompson, K., Askew, A. P., Grime, J. P., Dunnett, N. P., and Willis, A. J. (2005) Biodiversity, ecosystem function and plant traits in mature and immature plant communities. *Functional Ecology*, **19**, 355–8.

Thompson, R. M., Hemberg, M., Starzomski, B. M., and Shurin, J. B. (2007) Trophic levels and trophic tangles: the prevalence of omnivory in real food webs. *Ecology*, **88**, 612–17.

Thuiller, W., Lavorel, S., Midgley, G., Lavergne, S., and Rebelo, T. (2004) Relating plant traits and species distributions along bioclimatic gradients for 88 Leucadendron taxa. *Ecology*, **85**, 1688–99.

Thuiller, W., Lavorel, S., Sykes, M. T., and Araujo, M. B. (2006a) Using niche-based modelling to assess the impact of climate change on tree functional diversity in Europe. *Diversity and Distributions*, **12**, 49–60.

Thuiller, W., Richardson, D. M., Rouget, M., Proches, S., and Wilson, J. R. U. (2006b) Interactions between environment, species traits, and human uses describe patterns of plant invasions. *Ecology*, **87**, 1755–69.

Tilman, D. (1982) *Resource Competition and Community Structure*, Princeton University Press, Princeton.

Tilman, D. (1988) *Plant Strategies and the Dynamics and Structure of Plant Communities*. Princeton University Press, Princeton, NJ.

Tilman, D. (1994) Competition and biodiversity and spatially structured habitats. *Ecology*, **75**, 2–16.

Tilman, D. (1996) Biodiversity: population versus ecosystem stability. *Ecology*, **77**, 350–63.

Tilman, D. (1997) Distinguishing between the effects of species diversity and species composition. *Oikos*, **80**, 185.

Tilman, D. (1999a) Diversity and production in European grasslands. *Science*, **286**, 1099–100.

Tilman, D. (1999b) The ecological consequences of changes in biodiversity: a search for general principles. *Ecology*, **80**, 1455–74.

Tilman, D. (2000) What *Issues in Ecology* is, and isn't. *Bulletin of the Ecological Society of America*, **81**, 240.

Tilman, D. (2001) Functional diversity. In S. A. Levin (ed.) *Encyclopaedia of Biodiversity*, pp. 109–20. Academic Press, San Diego.

Tilman, D. and Downing, J. A. (1994) Biodiversity and stability in grasslands. *Nature*, **367**, 363–5.

Tilman, D. and Kareiva, P. (eds.) (1997) *Spatial Ecology*. Princeton University Press, Princeton.

Tilman, D. and Wedin, D. (1991) Plant traits and resource reduction for five grasses growing on a nitrogen gradient. *Ecology*, **72**(2), 685–700.

Tilman, D., May, R. M., Lehman, C. L., and Nowak, M. A. (1994) Habitat destruction and the extinction debt. *Nature*, **371**, 65–6.

Tilman, D., Wedin, D., and Knops, J. (1996) Productivity and sustainability influenced by biodiversity in grassland ecosystems. *Nature*, **379**, 718–20.

Tilman, D., Naeem, S., Knops, J., et al. (1997a) Biodiversity and ecosystem properties. *Science*, **278**, 1865–9.

Tilman, D., Knops, J., Wedin, D., Reich, P., Ritchie, M., and Sieman, E. (1997b) The influence of functional diversity and composition on ecosystem processes. *Science*, **277**, 1300–2.

Tilman, D., Lehman, C. L., and Thomson, K. T. (1997c) Plant diversity and ecosystem productivity: theoretical considerations. *Proceedings of the National Academy of Sciences of the USA*, **94**, 1857–61.

Tilman, D., Lehman, C. L., and Bristow, C. E. (1998) Diversity–stability relationships: statistical inevitability or ecological consequence? *American Naturalist*, **151**, 277–82.

Tilman, D., Reich, P. B., Knops, J. Wedin, D., Mielke, T., and Lehman, C. (2001) Diversity and productivity in a long-term grassland experiment. *Science*, **294**, 843–5.

Tilman, D., Polasky, S., and Lehman, C. (2005) Diversity, productivity and temporal stability in the economies of humans and nature. *Journal of Environmental Economics and Management*, **49**, 405–26.

Tilman, D., Hill, J., and Lehman, C. (2006a) Carbon-negative biofuels from low-input high-diversity grassland biomass. *Science*, **314**, 1598–600.

Tilman, D., Reich, P. B., and Knops, J. M. H. (2006b). biodiversity and ecosystem stability in a decade-long grassland experiment. *Nature*, **441**, 629–32.

Titus, J. H. and Tsuyuzaki, S. (2003) Influence of a non-native invasive tree on primary succession at Mt. Koma, Hokkaido, Japan. *Plant Ecology*, **169**, 307–15.

Tiunov, A. V. and Scheu, S. (2005) Facilitative interactions rather than resource partitioning drive diversity–functioning relationships in laboratory fungal communities. *Ecology Letters*, **8**, 618–25.

Tomich, T. P., van Noordwijk, M., Budidarsono, S., et al. (2001) Agricultural intensification, deforestation, and the environment: assessing tradeoffs in Sumatra, Indonesia. In D. R. Lee and C. B. Barrett (eds.) *Tradeoffs or Synergies? Agricultural Intensification, Economic Development and the Environment*, pp. 221–44. CAB-International, Wallingford.

Torchin, M. E., Lafferty, K. D., Dobson, A. P., McKenzie, V. J., and Kuris, A. M. (2003) Introduced species and their missing parasites. *Nature*, **421**, 628–30.

Traill, L. W., Bradshaw, C. J. A., and Brook, B. W. (2007) Minimum viable population size: a meta-analysis of 30 years of published estimates. *Biological Conservation*, **139**, 159–66.

Traveset, A. and Richardson, D. M. (2006) Biological invasions as disruptors of plant reproductive mutualisms. *Trends in Ecology & Evolution*, **21**, 208–16.

Travis, J. M. J. (2003) Climate change and habitat destruction: a deadly anthropogenic cocktail. *Proceedings of the Royal Society B: Biological Sciences*. **270**, 467–73.

Treberg, M. A. and Husband, B. C. (1999) Relationship between the abundance of *Lythrum salicaria* (Purple Loosestrife) and plant species richness along the Bar River, Canada. *Wetlands*, **19**, 118–25.

Trenbath, B. R. (1974) Biomass productivity of mixtures. In N. Brady (ed.) *Advances in Agronomy*. Academic Press, New York & London.

Treseder, K. and Vitousek, P. (2001) Potential ecosystem-level effects of genetic variation among populations of *Metrosideros polymorpha* from a soil fertility gradient in Hawaii. *Oecologia*, **126**, 266–75.

Treton, C., Chauvet, E. and Charcosset, J. Y. (2004) Competitive interaction between two aquatic hyphomycete species and increase in leaf litter breakdown. *Microbial Ecology*, **48**, 439–46.

Troumbis, A. Y. and Memtsas, D. (2000) Observational evidence that diversity may increase productivity in Mediterranean shrublands. *Oecologia*, **125**, 101–8.

Trumbore, S. (2000) Age of soil organic matter and soil respiration: radiocarbon constraints on belowground C dynamics. *Ecological Applications*, **10**, 399–411.

Truscott, A. M., Soulsby, C., Palmer, S. C. F., Newell, L., and Hulme, P. E. (2006) The dispersal characteristics of the invasive plant *Mimulus guttatus* and the ecological significance of increased occurrence of high-flow events. *Journal of Ecology*, **94**, 1080–91.

Tscharntke, T., Klein, A. M., Kruess, A., Steffan-Dewenter, I., and Thies, C. (2005) Landscape perspectives on agricultural intensification and biodiversity – ecosystem service management. *Ecology Letters*, **8**, 857–74.

Tschirhart, J. (2000) General equilibrium of an ecosystem. *Journal of Theoretical Biology*, **203**, 13–32.

Turner, R. K. (1999) The place of economic values in environmental valuation. In I. Batemen and K. Willis (eds.) *Valuing Environmental Preferences*, pp. 17–41. Oxford University Press, Oxford.

Turner, R. K., Paavola, J., Cooper, P., Farber, S., Jessamy, V., and Georgiou S. (2003) Valuing nature: Lessons learned and future research directions. *Ecological Economics*, **46**, 493–510.

Turpie, J., Heydenrych, B., and Hassan, R. (2001) Accounting for fynbos: A preliminary assessment of the status and economic value of fynbos vegetation in the Western Cape. In R. M. Hassan (ed.), *Accounting for Stock and Flow Values of Wooded Land Resources, Methods and Results from South Africa*. Centre for Environmental Economics and Policy in Africa (CEEPA), University of Pretoria.

Tylianakis, J. M. (2008) Understanding the web of life: the birds, the bees, and sex with aliens. *PloS Biology*, **6**, e47, 224–8.

Tylianakis, J. M., Tscharntke, T., and Klein, A. (2006) Diversity, ecosystem function, and the stability of parasitoid–host interactions across a tropical habitat gradient. *Ecology*, **87**, 3047–57.

Tylianakis, J. M., Tscharntke, T., and Lewis, O. T. (2007) Habitat modification alters the structure of tropical host–parasitoid food webs. *Nature*, **445**, 202–5.

Tylianakis, J. M., Rand, T. A., Kahmen, A., *et al.* (2008) Resource heterogeneity moderates the biodiversity–function relationship in real world ecosystems. *PLoS Biology*, **6**, e122, 947–56.

UNEP (United Nations Environment Program) (2005) *After the Tsunami, Rapid Environmental Assessment Report*. UNEP, Nairobi, 22 February; http://www.unep.org/tsunami/reports.

United Nations Environmental Programme (1999) *Global Environmental Outlook*. Earthscan, London.

United Nations Environmental Program (2007) *Global Environmental Outlook 4*. UNEP, New York.

United States Department of Agriculture National Agricultural Statistics Service, USDA-NASS (2008) *Alfalfa Seed 2007*. Available online at http://www.nass.usda.gov/Statistics_by_State/Montana/Publications/Press_Releases_Crops/alfaseed.htm.

USDA (2006) *Conservation Reserve Program General Sign-Up 33 Environmental Benefits Index*. Fact sheet. USDA Farm Service Agency.

USDA Forest Service (1997) *Final Environmental Impact Statement to Accompany the 1997 Revised Land and Resource Management Plan, Arapaho and Roosevelt National Forests and Pawnee National Grassland*. US Department of Agriculture, Forest Service, Rocky Mountain Region.

USDA Forest Service (2001a) *Final Environmental Impact Statement for the Northern Great Plains Management Plans Revision*. US Department of Agriculture, Forest Service, Rocky Mountain Region.

USDA Forest Service (2001b) *Final Environmental Impact Statement to Accompany the Sierra Nevada Forest Plan Amendment*. US Department of Agriculture Forest Service, Pacific Southwest Region.

USDA-NRCS (1997) *National Range and Pasture Handbook*. United States Department of Agriculture, Natural Resources Conservation Service, Grazing Lands Institute, Washington, DC.

USDI National Park Service (2006a) *Final General Management Plan and Comprehensive River Management Plan/*

Environmental Impact Statement, Sequoia and Kings Canyon National Parks, Middle and South Forks of the Kings River and North Fork of the Kern River. US Department of the Interior, National Park Service.

USDI National Park Service (2006b) *Final General Management Plan Environmental Impact Statement, Badlands National Park North Unit*. US Department of the Interior, National Park Service.

USDI National Park Service (2007) *Final General Management Plan/Wilderness Study/Environmental Impact Statement, Great Sand Dunes National Park and Preserve*. US Department of the Interior, National Park Service.

Valentini, R., Matteucci, G., Dolman, A., et al. (2000) Respiration as the main determinant of carbon balance in European forests. *Nature*, **404**, 861–5.

Valett, H. M., Crenshaw, C. L., and Wagner, P. F. (2002) Stream nutrient uptake, forest succession, and biogeochemical theory. *Ecology*, **83**, 2888–901.

Valiela, I., Bowen, J., and York, J. (2001) Mangrove forests, one of the world's threatened major tropical environments. *BioScience*, **51**, 807–15.

Valone, T. J. and Hoffman, C. D. (2003a) A mechanistic examination of diversity–stability relationships in annual plant communities. *Oikos*, **103**, 519–27.

Valone, T. J. and Hoffman, C. D. (2003b) Population stability is higher in more diverse annual plant communities. *Ecology Letters*, **6**, 90–5.

Van Andel, J. and Aronson, J. (eds.) (2006) *Restoration Ecology: the New Frontier*. Blackwell, Malden, MA.

Vandenberg, P. T., Poe, G. L., and Powell, J. R. (2001) Assessing the accuracy of benefits transfers: evidence from a multi-site contingent valuation study of groundwater quality. In J. C. Bergstrom, K. J. Boyle, and G. L. Poe (eds.) *The Economic Value of Water Quality*. Edward Elgar, Cheltenham.

van der Gast, C. J., Whiteley, A. S., Lilley, A. K., Knowles, C. J., and Thompson, I. P. (2003) Bacterial community structure and function in a metal-working fluid. *Environmental Microbiology*, **5**, 453–61.

Van der Heijden, M. G. A., Boller, T., Wiemken, A., and Sanders, I. R. (1998a) Different arbuscular mycorrhizal fungal species are potential determinants of plant community structure. *Ecology*, **79**, 2082–91.

van der Heijden, M. G. A., Klironomos, J. N., Margot, U., Moutoglis, P., Streitwolf-Engel, R., Boller, T., Wiemken, A., and Sanders, I. R. (1998b) Mycorrhizal fungal diversity determines plant biodiversity, ecosystem variability and productivity. *Nature*, **396**, 69–72.

van der Heijden, M. G. A., Bakker, R., Verwaal, J., Scheublin, T. R., Rutten, M., van Logtestijn, R., and Staehelin, C. (2006) Symbiotic bacteria as a determinant of plant community structure and plant productivity in dune grassland. *FEMS Microbiology Ecology*, **56**, 178–87.

van der Heijden, M. G. A., Bardgett, R. D., and van Straalen, N. M. (2008) The unseen majority: soil microbes as drivers of plant diversity and productivity in terrestrial ecosystems. *Ecology Letters*, **11**, 296–310.

Vandermeer, J. (1989) *The Ecology of Intercropping*. Cambridge University Press, Cambridge.

Van Deynze, A. E., Sundstrom, F. J., and Bradford, K. J. (2005) Pollen-mediated gene flow in California cotton depends on pollinator activity. *Crop Science*, **45**, 1565–70.

Van Diggelen, R. (2006) Landscape: spatial interactions. In J. Van Andel and J. Aronson (eds.) *Restoration Ecology: the New Frontier*. Blackwell Publishing, Malden, MA.

Van Kooten, J. and Bulte, E. H. (2000) *The Economics of Nature: Managing Biological Assets*. Blackwell, Oxford.

van Kooten, G. C., Eagle, A. J., Manley, J., and Smolak, T. (2004) How costly are carbon offsets? A meta-analysis of carbon forest sinks. *Environmental Science & Policy*, **7**, 239–51.

van Noordwijk, M., Kuncoro, S., Martin, E., Joshi, L., Saipothong, P., Areskoug, V., and O'Connor, T. (2005) Donkeys, carrots, sticks and roads to a market for environmental services: rapid agrobiodiversity appraisal for the PES – ICDP continuum. 2005. *Paper presented at the DIVERSTIAS First Open Science Conference*, Oaxaca, November.

Van Peer, L., Nijs, I., Reheul, D., and De Cauwer, B. (2004) Species richness and susceptibility to heat and drought extremes in synthesized grassland ecosystems: compositional vs physiological effects. *Functional Ecology*, **18**, 769–78.

van Ruijven, J., De Deyn, G. B., and Berendse, F. (2003) Diversity reduces invasibility in experimental plant communities: the role of plant species. *Ecology Letters*, **6**, 910–18.

Vasseur, D. A. and Gaedke, U. (2007) Spectral analysis unmasks synchronous and compensatory dynamics in plankton communities. *Ecology* **88**, 2058–71.

Vasseur, D. A., Gaedke, U., and McCann, K. S. (2005) A seasonal alternation of coherent and compensatory dynamics occurs in phytoplankton. *Oikos*, **110**, 507–14.

Vaught, D. (2007) *After the Gold Rush: Tarnished dreams in the Sacramento Valley*. The Johns Hopkins Univeristy Press, Baltimore, MD.

Vázquez, D. P. and Aizen, M. A. (2004) Asymmetric specialization: a pervasive feature of plant–pollinator interactions. *Ecology*, **85**, 1251–7.

Vázquez, D. P., Morris, W. F., and Jordano, P. (2005) Interaction frequency as a surrogate for the total effect of animal mutualists on plants. *Ecology Letters*, **8**, 1088–94.

Vehvilaäinen, H., Koricheva, J., Ruohomaki, K., Johansson, T., and Valkonen, S. (2006) Effects of tree stand species composition on insect herbivory of silver birch in boreal forests. *Basic and Applied Ecology*, **7**, 1–11.

Vellend, M., Verheyen, K., and Jacquemyn, H. *et al.* (2006) Extinction debt of forest plants persists for more than a century following habitat fragmentation. *Ecology*, **87**, 542–8.

Veltman, C. J., Nee, S., and Crawley, M. J. (1996) Correlates of introduction success in exotic New Zealand birds. *American Naturalist*, **147**, 542–57.

Venail, P. A., MacLean, R. C., Bouvier, T., Brockhurst, M. A., Hochberg, M. E., and Mouquet, N. (2008) Functional diversity and productivity peak at intermediate levels of dispersal in evolving metacommunities. *Nature* **452**, 210–15.

Venter, J. C., Remington, K., Heidelberg, J. F., Halpern, A. L., Rusch, D., Eisen, J. A., Wu, D., Paulsen, I., Nelson, K. E., and Nelson, W. (2004) Environmental genome shotgun sequencing of the Sargasso Sea. *Science*, **304**, 66–74.

Vermeij, G. J. (2004) Ecological avalanches and the two kinds of extinction. *Evolutionary Ecology Research*, **6**, 315–37.

Vesterdal, L., Ritter, E., and Gundersen, P. (2002) Change in soil organic carbon following afforestation of former arable land. *Forest Ecology and Management*, **169**, 137–47.

Viketoft, M., Palmborg, C., Sohlenius, B., Huss-Danell, K., and Bengtsson, J. (2005) Plant species effects on soil nematode communities in experimental grasslands. *Applied Soil Ecology*, **30**, 90–103.

Vilà, M., Vayreda, J., Gracia, C., and Ibanez, J. J. (2003) Does tree diversity increase wood production in pine forests? *Oecologia*, **135**, 299–303.

Vilà, M., Vayreda, J., Gracia, C., and Ibanez, J. (2004) Biodiversity correlates with regional patterns of forest litter pools. *Oecologia*, **139**, 641–6.

Vilà, M., Inchausti, P., Vayreda, J., *et al.* (2005) Confounding factors in the observed productivity–diversity relationship in forests. In M. Scherer-Lorenzen, C. Körner, and E.-D. Schulze (eds.) *The Functional Significance of Forest Diversity*. Springer-Verlag, Berlin.

Vilà, M., Vayreda, J., Comas, L., Ibanez, J. J., Mata, T., and Obon, B. (2007) Species richness and wood production: a positive association in Mediterranean forests. *Ecology Letters*, **10**, 241–50.

Vile, D., Shipley, B., and Garnier, E. (2006) Ecosystem productivity can be predicted from potential relative growth rate and species abundance. *Ecology Letters*, **9**, 1061–7.

Villéger, S., Mason, N. W. H., and Mouillot, D. (2008) New multidimensional functional diversity indices for a multifaceted framework in functional ecology. *Ecology*, **89**, 2290–301.

Vinebrooke, R. D., Schindler, D. W., Findlay, D. L., Turner, M. A., Paterson, M., and Milis, K. H. (2003) Trophic dependence of ecosystem resistance and species compensation in experimentally acidified lake 302S (Canada). *Ecosystems*, **6**, 101–13.

Vinebrooke, R. D., Cottingham, K. L., Norberg, J., Scheffer, M., Dodson, S. I., Maberly, S. C., and Sommer, U. (2004) Impacts of multiple stressors on biodiversity and ecosystem functioning: the role of species co-tolerance. *Oikos*, **104**, 451–7.

Violle, C., Navas, M. L., Vile, D., *et al.* (2007) Let the concept of trait be functional! *Oikos*, **116**, 882–92.

Vitousek, P. M. (1990) Biological invasions and ecosystem processes – towards an integration of population biology and ecosystem studies. *Oikos*, **57**, 7–13.

Vitousek, P. M. and Hooper, D. U. (1993) Biological diversity and terrestrial ecosystem biogeochemistry. In E.-D. Schulze and H. A. Mooney (eds.) *Biodiversity and Ecosystem Function*. Springer-Verlag, New York.

Vitousek, P. and Walker, L. (1989) Biological invasion by *Myrica faya* in Hawaii – plant demography, nirogen nixation, ecosystem effects. *Ecological Monographs*, **59**, 247–65.

Vitousek, P. M., Turner, D. R., Parton, W. J., *et al.* (1994) Litter decomposition on the Mauna Loa environmental matrix, Hawaii: patterns, mechanisms and models. *Ecology*, **75**, 418–29.

Vitousek, P. M., Mooney, H. A., Lubchenco, J., and Melillo, J. M. (1997) Human domination of Earth's ecosystems. *Science*, **277**, 494–9.

Vitousek, P. M., Cassman, K., Cleveland, C., *et al.* (2002) Towards an ecological understanding of biological nitrogen fixation. *Biogeochemistry*, **57**, 1–45.

Vittor, A. Y., Gilman, R. H., Tielsch, J., Glass, G., Shields, T., Lozano, W. S., Pinedo-Cancino, V., and Patz, J. A. (2006) The effect of deforestation on the human-biting rate of *Anopheles darlingi*, the primary vector of falciparum malaria in the Peruvian Amazon. *American Journal of Tropical Medicine and Hygiene*, **74**, 3–11.

Vivrette, N. J. and Muller, C. H. (1977) Mechanism of invasion and dominance of coastal grassland by *Mesembryanthemum-crystallinum*. *Ecological Monographs*, **47**, 301–18.

Vogelsang, K. M., Reynolds, H. L., and Bever, J. D. (2006) Mycorrhizal fungal identity and richness determine the diversity and productivity of a tallgrass prairie system. *New Phytologist*, **172**, 554–62.

Vogt, R. J., Romanuk, T. N. and Kolasa, J. (2006) Species richness–variability relationships in multi-trophic aquatic microcosms. *Oikos*, **113**, 55–66.

Vojtech, E., Loreau, M., Yachi, S., Spehn, E. M., and Hector, A. (2008) Light partitioning in experimental grass communities. *Oikos*, **117**, 1351–61.

Volkov, I., Banavar, J. R., He, F. L., Hubbell, S. P., and Maritan, A. (2005) Density dependence explains tree species abundance and diversity in tropical forests. *Nature*, **438**, 658–61.

von Canstein, H., Kelly, S., Li, Y., and Wagner-Döbler, I. (2002) Species diversity improves the efficiency of mercury-reducing biofilms under hanging environmental conditions. *Applied and Environmental Microbiology*, **68**, 2829–37.

Vörösmarty, C. J., Green, P., Salisbury, J., and Lammers, R. B. (2000) Global water resources: vulnerability from climate change and population growth. *Science*, **289**, 284–8.

Wagner, M., and Loy, A. (2002) Bacterial community composition and function in sewage treatment systems. *Current Opinion in Biotechnology*, **13**, 218–27.

Walker, B. (1995) Conserving biological diversity through ecosystem resilience. *Conservation Biology*, **9**, 747–52.

Walker, B. H. (1992) Biodiversity and ecological redundancy. *Conservation Biology*, **6**, 18–23.

Walker, B. H. and Langridge, J. L. (2002) Measuring functional diversity in plant communities with mixed life forms: a problem of hard and soft attributes. *Ecosystems*, **5**, 529–38.

Walker, B., Kinzig, A., and Langridge, J. (1999) Plant attribute diversity, resilience, and ecosystem function: the nature and significance of dominant and minor species. *Ecosystems*, **2**, 95–113.

Walker, J., Thompson, C. H., Reddell, P., and Rapport, D. J. (2001) The importance of landscape age in influencing landscape health. *Ecosystem Health*, **7**, 7–14.

Walker, K. J., Stevens, P. A., Stevens, D. P., Mountford, J. O., Manchester, S. J., and Pywell, R. F. (2004) The restoration and re-creation of species-rich lowland grassland on land formerly managed for intensive agriculture in the UK. *Biological Conservation*, **119**, 1–18.

Walker, S. C., Poos, M. S., and Jackson, D. A. (2008) Functional rarefaction: estimating functional diversity from field data. *Oikos*, **117**, 286–96.

Wallace, J. B., Webster, J. R., and Meyer, J. L. (1995) Influence of log additions on physical and biotic characteristics of a mountain stream. *Canadian Journal of Fisheries and Aquatic Sciences*, **52**, 2120–37.

Walsh, P. D., Abernethy, K. A., Bermejo, M., Beyers, R., De Wachter, P., Akou, M. E., Huljbregis, B., Mambounga, D. I., Toham, A. K., Kilbourn, A. M., Lahm, S. A., Latour, S., Maisels, F., Mbina, C., Mihindou, Y., Obiang, S. N., Effa, E. N., Starkey, M. P., Telfer, P., Thibault, M, Tutin, C. E. G., White, L. T. J., and Wilkie, D. S. (2003) Catastrophic ape decline in western equatorial Africa. *Nature*, **422**, 611–14.

Wardle, D. A. (1998) A more reliable design for biodiversity study? *Nature*, **394**, 30.

Wardle, D. A. (1999) Is 'sampling effect' a problem for experiments investigating biodiversity–ecosystem function relationships? *Oikos*, **87**, 403–7.

Wardle, D. A. (2002) *Communities and Ecosystems: Linking the Aboveground and Belowground Components*. Princeton University Press, Princeton, NJ.

Wardle, D. A. and Grime, J. P. (2003) Biodiversity and stability of grassland ecosystem functioning. *Oikos*, **100**, 622–3.

Wardle, D. A. and Zackrisson, O. (2005) Effects of species and functional group loss on island ecosystem properties. *Nature*, **435**, 806–10.

Wardle, D. A., Bonner, K. I., and Nicholson, K. S. (1997a) Biodiversity and plant litter: experimental evidence which does not support the view that enhanced species richness improves ecosystem function. *Oikos*, **79**, 247–58.

Wardle, D. A., Zackrisson, O., Hörnberg, G., and Gallet, C. (1997b) The influence of island area on ecosystem properties. *Science*, **277**, 1296–9.

Wardle, D. A., Barker, G. M., Bonner, K. I., and Nicholson, K. S. (1998) Can comparative approaches based on plant ecophysiological traits predict the nature of biotic interactions and individual plant species effects in ecosystems? *Journal of Ecology*, **86**, 405–20.

Wardle, D. A., Nicholson, K. S., Bonner, K. I., and Yeates, G. W. (1999) Effects of agricultural intensification on soil-associated arthropod population dynamics, community structure, diversity and temporal variability over a seven-year period. *Soil Biology & Biochemistry*, **31**, 1691–706.

Wardle, D. A., Bonner, K. I., and Barker, G. M. (2000a). Stability of ecosystem properties in response to aboveground functional group richness and composition. *Oikos*, **89**, 11–23.

Wardle, D. A., Huston, M. A., Grime, J. P., *et al.* (2000b) Biodiversity and ecosystem function: an issue in ecology. *Bulletin of the Ecological Society of America*, **81**, 235–9.

Wardle, D. A., Bonner, K. I., and Barker, G. M. (2002) Linkages between plant litter decomposition, litter quality, and vegetation responses to herbivores. *Functional Ecology*, **16**, 585–95.

Wardle, D. A., Hornberg, G., Zackrisson, O., Kalela-Brundin, M., and Coomes, D. A. (2003a) Long-term effects of wildfire on ecosystem properties across an island area gradient. *Science*, **300**, 972–5.

Wardle, D. A., Yeates, G. W., Barker, G. M., Bellingham, P. J., Bonner, K. I., and Williamson, W. M. (2003b) Island biology and ecosystem functioning in epiphytic soil communities. *Science*, **301**, 1717–20.

Wardle, D. A., Bardgett, R. D., Klironomos, J. N., Setala, H., van der Putten, W. H., and Wall, D. H. (2004a). Ecological linkages between aboveground and belowground biota. *Science*, **304**, 1629–33.

Wardle, D. A., Walker, L. R., and Bardgett, R. D. (2004b) Ecosystem properties and forest decline in contrasting long-term chronosequences. *Science*, **305**, 509–13.

Warren, D. and Kraft, C. (2006) Invertebrate community and stream substrate responses to woody debris removal from an ice storm-impacted stream system, ny USA. *Hydrobiologia*, **568**, 477–88.

Watling, J. I. and Donnelly, M. A. (2007) Multivariate correlates of extinction proneness in a naturally fragmented landscape. *Diversity and Distributions*, **13**, 372–8.

Watson, R. T., Noble, I. R., Bolin, B., Ravindranath, N. H., Verado, D. J., and Dokken, D. J. (2000) *Land Use, Land Use Change, and Forestry*. WMO/UNEP. Intergovernmental Panel on Climate Change.

Wätzold, F., Drechsler, M., Armstrong, C. W., et al. (2006) Ecological-economic modeling for biodiversity management: potential, pitfalls, and prospects. *Conservation Biology*, **20**(4), 1034–41.

Webb, S. L., Dwyer, M., Kaunzinger, C. K., and Wyckoff, P. H. (2000) The myth of the resilient forest: case study of the invasive Norway Maple (*Acer platanoides*). *Rhodora*, **102**, 332–54.

Weber, S. (1999) Designing seed mixes for prairie restorations: revisiting the formula. *Ecological Restoration*, **17**, 196–201.

Weibull, A. C., Östman, O., and Granqvist, A. (2003) Species richness in agroecosystems: the effect of landscape, habitat and farm management. *Biodiversity Conservation*, **12**, 1335–55.

Weigelt, A., Schumacher, J., Roscher, C., and Schmid, B. (2008) Does biodiversity increase spatial stability in plant community biomass? *Ecology Letters*, **11**, 338–47.

Weiher, E. and Keddy, P.-A. (1995a) The assembly of experimental wetland plant communities. *Oikos*, **73**, 323–35.

Weiher, E. and Keddy, P.-A. (1995b). Assembly rules, null models, and trait dispersion – new questions front old patterns. *Oikos*, **74**, 159–64.

Weiher, E. and Keddy, P.-A. (eds.) (1999) *Ecological Assembly Rules*. Cambridge University Press, Cambridge.

Weis, J. J., Cardinale, B. J., Forshay, K. J., and Ives, A. R. (2007) Effects of species diversity on community biomass production change over the course of succession. *Ecology*, **88**, 929–39.

Weisbrod, B. A. (1964) Collective-consumption services of individualized-consumption goods. *Quarterly Journal of Economics*, **LXXVIII**, 471–7.

Weithoff, G. (2003) The concepts of 'plant functional types' and 'functional diversity' in lake phytoplankton – a new understanding of phytoplankton ecology? *Freshwater Biology*, **48**, 1669–75.

Weitzman, M. (2000) Economic profitability versus ecological entropy. *Quarterly Journal of Economics* **115**(1), 237–63.

Wells, M. and Brandon, K. (1992) *People and Parks: Linking Protected Area Management with Local Communities*. World Bank, Washington, DC.

Wenum, J., Buys J., and Wossink, A. (1999) Nature quality indicators in agriculture. In F. Brouwer and B. Crabtree (eds.) *Environmental Indicators and Agricultural Policy*. CABI Publishers, Wallingford.

Wertz, S., Dégrange, V., Prosser, J. I., Poly, P., Commeaux, C., Freitag, T., Guillaumaud, N., and Le Roux, X. (2006) Maintenance of soil functioning following erosion of microbial diversity. *Environmental Microbiology*, **8**, 2162–9.

West, G. B., Brown, J. H., and Enquist, B. J. (1997) A general model for the origin of allometric scaling laws in biology. *Science*, **276**, 122–6.

Westoby, M. (1998) A leaf–height–seed (LHS) plant ecology strategy scheme. *Plant and Soil*, **199**, 213–27.

Westoby, M., Falster, D. S., Moles, A. T., Vesk, P. A., and Wright, I. J. (2002) Plant ecological strategies: some leading dimensions of variation between species. *Annual Review of Ecology and Systematics*, **33**, 125–59.

Westphal, C., Steffan-Dewenter, I., and Tscharntke, T. (2003) Mass-flowering crops enhance pollinator densities at a landscape scale. *Ecology Letters*, **6**, 961–5.

White, T. D., Suwa, G., and Asfaw, B. (1994) *Australopithecus ramidus*, a new species of early hominid from Aramis, Ethiopia. *Nature*, **371**, 306–12.

Whitman, W. B., Coleman, D. C., and Wiebe, W. J. (1998) Prokaryotes: the unseen majority. *Proceedings of the National Academy of Sciences of the USA*, **95**, 6578–83.

Whitham, T. G., Young, W. P., Martinsen, G. D., et al. (2003) Community and ecosystem genetics: a consequence of the extended phenotype. *Ecology*, **84**, 559–73.

Whittaker, R. H. (1975) *Communities and Ecosystems*, 2nd edn. MacMillan, New York.

Widawsky, D., and Rozelle, S. (1998) Varietal diversity and yield variability in Chinese rice production. In M. Smale (ed.) *Farmers, Gene Banks, and Crop Breeding*, pp. 159–72. Kluwer, Boston.

Widdicombe, S., Austen, M. C., Kendall, M. A., et al. (2000) Bioturbation as a mechanism for setting and maintaining levels of diversity in subtidal macrobenthic communities. *Hydrobiologia*, **440**, 369–77.

Wiegmann, S. M. and Waller, D. M. (2006) Fifty years of change in northern upland forest understories: identity and traits of 'winner' and 'loser' plant species. *Biological Conservation*, **129**, 109–23.

Wilby, A. and Shachak, M. (2004) Shrubs, granivores and annual plant community stability in an arid ecosystem. *Oikos*, **106**, 209–16.

Wilby, A. and Thomas, M. B. (2002) Natural enemy diversity and pest control: patterns of pest emergence with agricultural intensification. *Ecology Letters*, **5**, 353–60.

Wilby, A. and Thomas, M. B. (2007) Diversity and pest management in agroecosystems – some perspectives from ecology. In D. I. Jarvis, C. Padoch, and H. D. Cooper (eds.) *Managing Biodiversity in Agricultural Ecosystems*, pp. 269–91. Columbia University Press, New York.

Wilby, A., Villareal, S. C., Lan, L. P., Heong, K. L., and Thomas, M. B. (2005) Functional benefits of predator species diversity depend on prey identity. *Ecological Entomology*, **30**, 497–501.

Wilby, A., Lan, L. P., Heong, K. L., et al. (2006) Arthropod diversity and community structure in relation to land use in the Mekong Delta, Vietnam. *Ecosystems*, **9**, 538–49.

Wilkinson, B. H. (2005) Humans as geologic agents: a deep-time perspective. *Geology*, **33**, 161–4.

Williams, S. L. (2001) Reduced genetic diversity in eelgrass transplantations affects both population growth and individual fitness. *Ecological Applications*, **11**, 1472–88.

Williams, J. L. and Crone, E. E. (2006) The impact of invasive grasses on the population growth of anemone patens, a long-lived native forb. *Ecology*, **87**, 3200–8.

Williams, N. M. and Kremen, C. (2007) Resource distributions among habitats determine solitary bee offspring production in a mosaic landscape. *Ecological Applications*, **17**, 910–21.

Williams, N. M., Minckley, R. L., and Silveira, F. A. (2001) Variation in native bee faunas and its implications for detecting community changes. *Conservation Ecology*, **5**, 7.

Williams, N. S. G., Morgan, J. W., McDonnell, M. J., and McCarthy, M. A. (2005) Plant traits and local extinctions in natural grasslands along an urban–rural gradient. *Journal of Ecology*, **93**, 1203–13.

Williamson, M. and Fitter, A. (1996) The varying success of invaders. *Ecology*, **77**, 1661–6.

Wilsey, B. J. and Polley, H. W. (2002) Reductions in grassland species evenness increase dicot seedling invasion and spittle bug infestation. *Ecology Letters*, **5**, 676–84.

Wilsey, B. J. and Polley, H. W. (2004) Realistically low species evenness does not alter grassland species-richness–productivity relationships. *Ecology*, **85**, 2693–700.

Wilson, E. O. (1992) *The Diversity of Life*. Princeton University Press, Princeton.

Wilson, K. A., Underwood, E. C., Morrison, S. A., Klausmeyer, K. R., Murdoch, W. W., Reyers, B., Wardell-Johnson, G., Marquet, P. A., Rundel, P. W., McBride, M. F., Pressey, R. L., Bode, M., Hoekstra, J. M., Andelman, S., Looker, M., Rondinini, C., Kareiva, P., Shaw, M. R., and Possingham, H. P. (2007) Conserving biodiversity efficiently: what to do, where and when. *PLoS Biology*, **5**, e223. doi:10. 1371/journal.pbio.0050223(Abstract)

Winfree, R. and Kremen, C. (2008) Are ecosystem services stabilized by differences among species? A test using crop pollination. *Proceedings of the Royal Society, Series B*, published online first.

Winfree, R., Williams, N., Gaines, H., Asher, J., and Kremen, C. (2008) Wild bee pollinators provide the majority of crop visitation across land-use gradients in New Jersey and Pennsylvania, USA. *Journal of Applied Ecology*, **45**, 793–802.

Winkelmann R. (2003) *Econometric Analysis of Count Data*, 4th edn. Springer-Verlag, Berlin.

Winkler, R. (2006a) Valuation of ecosystem goods and services Part 1: An integrated dynamic approach. *Ecological Economics*, **59**(1), 82–93.

Winkler, R. (2006b) Valuation of ecosystem goods and services Part 2: Implications of unpredictable novel change. *Ecological Economics*, **59**(1), 94–105.

Wirth, C. (2005) Fire regime and tree diversity in boreal forests: Implications for the carbon cycle. In M. Scherer-Lorenzen, C. Körner, and E.-D. Schulze (eds.) *The Functional Significance of Forest Diversity*, Springer-Verlag, Berlin.

Wohl, D. L., Arora, S., and Gladstone, J. R. (2004) Functional redundancy supports biodiversity and ecosystem function in a closed and constant environment. *Ecology*, **85**, 1534–40.

Wojdak, J. M. (2005) Relative strength of top-down, bottom-up, and consumer species richness effects on pond ecosystems. *Ecological Monographs*, **75**, 489–504.

Wolszczan, A. and Frail, D. A. (1992) A planetary system around the millisecond pulsar PSR1257+12. *Nature*, **355**, 145–7.

Wood, D. and Lenné, J. M. (2005) 'Received Wisdom' in agricultural land use policy: 10 years on from Rio. *Land Use Policy*, **22**, 75–93.

Wood, S., Sebastian, K., and Scherr, S. J. (2000) *Analysis of Global Ecosystems: Agroecosystems*. International Food Policy Research Institute and World Resources Institute, Washington DC.

Woodcock, S., van der Gast, C. J., Bell, T., Lunn, M., Curtis, T. P., Head, I. M., and Sloan, W. T. (2007) Neutral assembly of bacterial communities. *FEMS Microbiology Ecology*, **62**, 171–80.

Woodhead, T. M. (1906) Ecology of woodland plants in the neighbourhood of Huddersfield. *Linnean Journal of Botany*, **37**, 333–407.

Woodward, R. T. and Yong-Suhk Wui (2001) The economic value of wetland services: a meta-analysis. *Ecological Economics*, **37**(2), 257–70.

Wootton, J. T. (2005) Field parameterization and experimental test of the neutral theory of biodiversity. *Nature*, **433**, 309–12.

Wootton, J. T. and Emmerson, M. (2005) Measurement of interaction strength in nature. *Annual Review of Ecology Evolution and Systematics*, **36**, 419–44.

World Commission on Environment and Development (1987) *Our Common Future*, Oxford University Press, Oxford.

Worm, B., Barbier, E. B., Beaumont, N., *et al.* (2006) Impacts of biodiversity loss on ocean ecosystem services. *Science*, **314**, 787–90.

Worm, B., Sandow, M., Oschlies, A., Lotze, H. K., and Myers, R. A. (2005) Global patterns of predator diversity in the open oceans. *Science*, **309**, 1365–9.

Wright, D. H., Gonzalez, A., and Coleman, D. C. (2007) Changes in nestedness in experimental communities of soil fauna undergoing extinction. *Pedobiologia*, **50**, 497–503.

Wright, I. J., Reich, P. B., Westoby, M., *et al.* (2004) The worldwide leaf economics spectrum. *Nature*, **428**, 821–7.

Wright, J. P. and Flecker, A. S. (2004) Deforesting the riverscape: the effects of wood on fish diversity in a Venezuelan piedmont stream. *Biological Conservation*, **120**, 439–47.

Wurst, S., Allema, B., Duyts, H., and van der Putten, W. H. (2008) Earthworms counterbalance the negative effect of microorganisms on plant diversity and enhance the tolerance of grasses to nematodes. *Oikos*, **117**(5), 718–18.

Yachi, S. and Loreau, M. (1999) Biodiversity and ecosystem functioning in a fluctuating environment: the insurance hypothesis. *Proceedings of the National Academy of Science*, **96**, 1463–8.

Yodzis, P. (1984) How rare is omnivory. *Ecology*, **65**, 321–3.

Young, T. P. (2000) Restoration ecology and conservation biology. *Biological Conservation*, **92**, 73–83.

Young, T. P., Petersen, D. A., and Clary, J. J. (2005) The ecology of restoration: historical links, emerging issues and unexplored realms. *Ecology Letters*, **8**, 662–73.

Zak, D. R., Holmes, W. E., White, D. C., Peacock, A. D., and Tilman, D. (2003) Plant diversity, soil microbial communities, and ecosystem function: are there any links? *Ecology*, **84**, 2042–50.

Zavaleta, E. S. and Hulvey, K. B. (2004) Realistic species losses disproportionately reduce grassland resistance to biological invaders. *Science*, **306**, 1175–7.

Zavaleta, E. S., Shaw, M. R., Chiariello, N. R., Thomas, B. D., Cleland, E. E., Field, C. B., and Mooney, H. A. (2003) Grassland responses to three years of elevated temperature, CO_2, precipitation, and N deposition. *Ecological Monographs*, **73**, 585–604.

Zedler, J. B. (1993) Canopy architecture of natural and planted cordgrass marshes – selecting habitat evaluation criteria. *Ecological Applications*, **3**, 123–38.

Zhang, Q. G. and Zhang, D. Y. (2006b) Species richness destabilizes ecosystem functioning in experimental aquatic microcosms. *Oikos*, **112**, 218–26.

Zhang, Q. G. and Zhang, D. Y. (2006a) Resource availability and biodiversity effects on the productivity, temporal variability and resistance of experimental algal communities. *Oikos*, **114**, 385–96.

Zhang, Q. G. and Zhang, D. Y. (2007) Consequences of individual species loss in biodiversity experiments: an essentiality index. *Acta Oecologica – International Journal of Ecology*, **32**, 236–42.

Zhu, Y., Chen, H., Fan, J., *et al.* (2000) Genetic diversity and disease control in rice. *Nature*, **406**, 718–22.

Index

Note: page numbers in *italics* refer to Figures and Tables.

abundance, influencing factors 5
Acer platanoides invasion, impact on native species 224
adaptability 232
  relationship to establishment success 221
adaptive foraging 90
adaptive systems 265
additive experimental designs 213
additive partitioning methods 96–7, 100–2, *103*
adjusted net savings indicator 259, *261*, 262
aggregate abundance, flower visitors 196, 198
aggregate biodiversity index 263, 272
Agreement on Trade Related Aspects of Intellectual Property Rights (TRIPS Agreement) 235–6
agricultural intensification 73, 179–80, 193, 270
  effect on pollinator communities 195, 196, 197, 206
  meta-analysis of field studies 180–3
  mitigation options 183–4
  see also managed ecosystems
agriculture
  land use 178–9
  microbial functions 124
  overyielding 100
agrobiodiversity, insurance value 232–4, 256
agro-ecological zoning schemes 241
agro-ecosystems 10
  econometric model 269–73
agroforestry 183
  carbon sequestration 164–5
Alaska, GEEM model 266–9
alerting behaviour valuation method 252
alfalfa seed production, economic value of bees 207–8
algae
  dispersal effects 145

  temporal stability 80, *81*
alien species integration, effect on PFv webs 204
Allen, B.P. and Loomis, J.B. 254
Allison, G. *86*, *88*, 89
Andow, D.A. 111
animal traits, associated responses to environmental changes 62, *63–4*, 65
aphids
  biological control 119
  genotypic diversity, relationship to infection risk 214
application of BEF research 32, 34, 41–2, 44–5
aquatic bacteria, microcosm studies 127–8
aquatic fungi 131
Argentine ants (*Linepithema humile*) 221–2
Armbrecht, I. *et al.* 181
Armbrecht, I. and Perfecto, I. *181*
Armington, P. 268
arthropod diversity, relationship to land use gradient *186*
aspen, genotypic richness 176
assassin bugs, value in biological control 119
assembly history, microbial communities 126
Auction Contracts for Conservation (ACCs) 241–2
averting behaviour models *252*, 253

bacteria
  aquatic 127–8
  Bacillus thuringiensis 124–5
  E. coli, presence on spinach 190
  nitrogen-fixing 128–9
  see also microbes
balanced diet hypothesis 109, 112
Bali Road Map 150
Balmford, A. *et al.* 11, 251
Balvanera, P. *et al.* 14, 17, 110, 187

  meta-data set 15
Barbarika, A. 171
Barbier, E.B. 257, 258
Bärlocher, F. and Corkum, M. 128
Baumgärtner, S. 232
Baur, B. *et al.* 181, 191
bean, PFv web *202*
bee communities
  importance of species richness 198
  temporal turnover 196
bees
  decline, consequences 205–6
  parasitoid diversity 83
  pollination services 120, 196
    economic value 207–8
    functional facilitation 201
    PFv webs *202*, 203
    realistic extinction scenarios 73, *74*
  behavioural niche differentiation, flower-visiting communities 201
Bell, G. 4
Bell, T. *et al.* 98, 102, 127, 128, 132
Bellwood, D.R. *et al.* 92
Benedetti-Cecchi, L. 138
benefits from landscape, econometric modelling 276
benefit transfer valuation techniques 255
Benton, T.G. *et al.* 34
bequest use value of ecosystems 251
biocontrol 10, 26
BIODEPTH experiments 8, 54
  drought resistance 87
biodiversity
  definitions 150, 263
  divergence from natural biodiversity (DNB) 264–5
  influencing factors 5
  importance relative to community species composition 210–11
  meta-analyses 14
    data sets 15
    hypotheses 15–17
    meta-data

357

biodiversity (cont.)
    distribution of studies 19–21, 20
    hypotheses to explain variation in biodiversity effects 21–4
    methods of analysis 17–18
    multivariate mixed-model analyses 19
biodiversity effects 15
    differences between ecosystem, community and population responses 16, 22, 23, 27
    differences between ecosystem types 15–16, 21–2, 27
    differences between residents and invaders 17, 22, 23, 24, 28
    differences between standing stocks and rates 16, 22–4, 27
    future studies 28–9
    multitrophic studies 16–17, 22, 23, 24, 27–8
    shapes 17, 24–6, 28
Biodiversity effects on Ecosystem Functioning (BEF) 105–6, 120, 219–20
    BEF-6 36–8
        information flow 39
    BEF approach to restoration 169–70
    BEF database 33–4, 34–6, 35
    citation rates 36, 37
    biodiversity gradients 292, 293
    impacts of colonizing species 220–6
    trait-based research 283
biodiversity index 263, 272
biodiversity markets 234–8
bioeconomic risk analysis, colonization 226–8
BioFlor 286
biological control 119, 120, 187–8, 215
    economic value 191–2
    scaling effects 192–3
    value of landscape mosaics 189
biomass, relationship to microbial community diversity 126
biomass compensation 66–7, 76
biomass production, roles 10
BioMERGE (Biotic Mechanisms of Ecosystem Regulation in the Global Environment) xii, 9, 281
BioPop 286
bioprospecting 235–6, 246
bioremediation, trait-based research 283
biosecurity 227
biosphere, human domination 297

biotic feedback 3–4
biotic functions, impacts of simplification 10–11
bioturbation, effect of benthic marine invertebrates 73, 74
birds, as indicators of biodiversity 190–1
bison, restoration to Great Plains 190
bivariate plots 4
body size
    animals 71–2
        relationship to extinction risk 64, 65, 77
    plants, relationship to ecosystem function 70
Bolker, B.M. et al. 165
Booth, R.E. and Grime, J.P. 155
Botta-Dukát, Z. 51, 52, 53
bottom-up biodiversity effects 16, 17, 23, 24, 27, 107, 108, 109
    empirical evidence 112
Boyd, J. and Simpson, R.D. 237
Bracken, M.E. et al. 73, 74
Bradshaw, A.D. 167
Brandle, J.R. et al. 165
Brazil, TDR programs 241
Brock, W.A. et al. 295
Brock, W.A. and Xepapadeas, A. 274–7
Brundtland Report (World Commission on Environment and Development 1987) 295–6
Buenos Aires accords 149
Bullock, J.M. et al. 181
bumble bees, declining species 205
Bunker, D.E. et al. 73, 74, 76, 159, 281
BushTender 242
Butler, S.J. et al. 191
Bwindi national park, Uganda 242

calcareous grassland, PFv web 202
Caldeira, M.C. et al. 80, 81, 84, 86, 87
Callaway, J.C. et al. 171
Camill, P. et al. 171
CAMPFIRE (Communal Areas Management Program for Indigenous Resources), Zimbabwe 236
candidate variables, biodiversity experiments 29
carbon cycling 249
    seasonality 172
carbon dioxide increase, plant responses 63, 65
carbon loss, experimental studies 161–2

carbon offset arrangements 246–7
carbon sequestration 149–51, 165–6
    biodiversity effects 151
    empirical evidence 156–7
    experimental studies 158–62
    mass ratio hypothesis 152–5
    neotropical trees 73, 74
    neutral hypothesis 152
    niche complementarity hypothesis 155–6
    suites of plant attributes 152, 153
carbon sequestration initiatives 163–5
carbon stocks, forest and woodlands, South Africa 260, 261
carbon uptake (capture) 151
Cardinale, B.J. et al. 14, 99, 100–1, 110, 113, 119, 171, 212
    meta-data set 15
Carlander, K.D. 4
Carney, K.M. et al. 181
Carolina Vegetation Survey 177
cascading (secondary) extinction 67–8, 69, 90–1
case study approach 11
Caspersen, J.P. and Pacala, S.W. 158
cellulose utilization, bacteria 127
Center for Tropical Forest Science Trait database 286
central construct 11–12
Chagnon, M. et al. 200
Chesapeake Bay, exotic species 222
Chesson, P. 113–14, 144
China, demand for natural resources 295
choice experiments 253
Chopra, K. and Kumar, P. 237–8
citation rates 33–4, 36, 37
    BEF-6 36–8
    functional diversity measures 52–4
    in management plans 42, 43
Clarke, P.J. et al. 222
Clean Development Mechanisms (CMDs) 149, 164, 166
Clean Water Act, USA 177
climate
    effect of microbes 124
    effect of plant traits 153–4
climate change
    animal responses 64
    effect on pollinator services 206
    plant responses 63, 65
    prediction of responses 286
climate change mitigation 149–50, 151, 165–6
    see also carbon sequestration
clone libraries 122, 123

Cobb–Douglas production function 268
coefficient of variation (CV) 80
  meta-communities *142*, 143
coexistence, mechanisms of 113–15
coffee production
  financial incentives 193
  pollination services, spatial variation *198*, 200
coinfections 214
cold resistance *86*
colonization 217, 220
  biodiversity effects 17, *22*, *23*, 24, 28
  establishment success 218–19, 221–2
  impact on BEF relationship 224–6
  impact on native ecosystem 222–4, 227–8
  likelihood of 220–1
  patch dynamics 117–18
  prediction of impact 228–9
  risk assessment 226–8
  terminology 218
communication of BEF concepts 31, 43–5
community approach to restoration 168
community assembly/disassembly, natural versus random, effect on disease dynamics 212–14
community-based conservation 236
community disassembly 135
community ecology, trait-based research *283*
community fingerprints *123*
community-level functions 16
community-level responses, biodiversity effects 16, *22*, *23*, 27
community organization, link to traits 66
comparative microbial studies 132
  soil communities 130
comparative trait-based research *283*
compensation, as response to extinctions 66–7, 76
compensation for conservation measures 243–4
compensatory dynamics, meta-communities 143, *144*
competition, Lotka–Volterra models 114–15
competitor diversity, role in establishment success 218
complementarity effects 95, 96, 99–100, 101, 188, 219, 281, 296
  role in establishment success 222
Complex Agroecosystems *181*

biodiversity 183
Computable General Equilibrium (CGE) models 265
confirmatory experiments 8
conjoint analysis 253
connectance, food webs 90–1
connected species, extinction risk 68, *69*
connections 105
conservation 230
  biodiversity markets 234–8
  community-based 236
  economic instruments 238–42, 244
  Endangered Species Act, United States 243–4
  government and NGO involvement 236–7
  of primary forests 163, 166
  value 251
conservation biology, trait-based research *283*
conservation easements 237
conservation investment 271
Conservation Reserve Program (CRP) 171, 177, 184
consumer diversity, relationship to resource fluxes *108*, *109*
consumptive use 264
contemporaneous disequilibrium 140–1
context dependence effect (CDE) 49, 97
contingent valuation 253
Convention on Biological Diversity 40, 209
Convention on International Trade in Endangered Species (CITES) 245
convex hull volume (CHV) 51, 55
  applications *52*, 53
  relationship to species richness *56*, *57*, *58*
Cook, D.C. *et al.* 226
coral reefs
  bioerosion 54
  stability 92
Cornwell, W.K. *et al.* 51, *52*
correlation coefficients, conversion to Z_r values 18
Costa Rica, payments for ecosystem services 239
Costello, C. and McAusland, C. 244
cottonwood, genotypic richness 176
Countryside Surveys, UK 272
coupled ecological–economic systems 263, 277–8
  agro-ecosystems 269–73
  generalized model 273–7

'naturalness' of ecosystems 264–9
covariances 84–5
Craft, C. *et al.* 168
Crocker, T.D. and Tschirhart, J. 231, 267
crop diseases 190
crop diversity 270
cropland increases 178
crop pollination services *see* pollination
crop production function 271
cross-ecosystem biodiversity exchanges 189–90
Crutsinger, G.M. *et al.* 155, 176
cultural services 248
  'naturalness' of ecosystems 264–9
  valuation 253

Dang, C.K. *et al.* 81, 83, 128
Daphnia magna, genotypic diversity, relationship to infection risk 214
Darwin, C. 4, 94–5
Dasgupta, P. 259
databases 281–2
  trait data *286*
  see also TraitNet
data integration, TraitNet 286–8
Davis, M.A. and Thompson, K. 218
De Bello, F. *et al.* 53
Debras, J.F. *et al.* 181
DeClerck, F. *et al.* 81, 82, *86*, *88*, 89
decomposition
  carbon loss 154
  effect of fungal diversity 83, 128
    aquatic fungi 131
  litter-mixing studies 156
  relationship to plant traits 71, *72*
  by soil microbes 129–30
deforestation
  effects on infectious disease risk 214
  mangroves 258
Degens, B.P. 130
dendrogram-based measures of functional diversity 52
Dennehy, J.J. *et al.* 214
detrital food webs, biodiversity effects 17
detritivore diversity effects, comparison with detrital diversity effects 112
detritus processing, stream insects 73, *74*
Dhôte, J.-F. *161*
diffusion 218
Dillard, W. 113
dilution experiments, soil microbes 130
dilution hosts 212, 213

dilution hypothesis 109, 111
Dimitrakolpoulos, P.G. and
    Schmid, B. 101
direct compensation payments (DCP)
    240
direct use value of ecosystems 251
direct utility function 270
discrete functional groupings 54
diseases *see* infectious disease risk;
    infectious diseases
dispersal, disruption by colonizing
    species 223
dispersal abilities, likelihood of
    colonization 220
dispersal effects 117, 135, 138–9,
    143–4, 146
  empirical tests 145–6
  meta-community dynamics 141
  meta-community productivity
    141–3, 144–5
  spatial insurance hypothesis 139–40
distance-based measures, functional
    diversity 51–2
disturbance regimes, impact of
    colonizing species 223
disturbances
  effect on carbon sequestration 153
  role in establishment
    success 219
divergence from natural biodiversity
    (DNB) 264–5
DIVERSITAS xii
diversity–disease curves 210
diversity–interaction models 98–9,
    102–3
Dobson, A.P. *et al.* 137
domesticated species 9
dominance effect (DE) *see* sampling
    (selection) effect
'dose–response' analysis 252–3, 256
drinking water provision, New York
    City 252
drought resilience *88*
drought resistance 85, *86*, 87, 172
drugs, bioprospecting 235–6, 246
dual zone TDR programs 241
Duarte, C.M. *et al.* 128
Duffy, J.E. 107, 109
Dukes, J.S. 73, 76
dung beetles, realistic extinction
    scenarios 73, *74*
Dyer, A.R. and Rice, K.J. 223

Earth Summit, (Rio de Janeiro 1992)
    296
Ebola virus infection risk, effect of
    deforestation 214

ecoagriculture 184
eco-engineered approach, sterilized
    Earth experiment 6
ecoinformatics frameworks,
    integration of TraitNet 288
E. coli, presence on spinach 190
ecological engineering 168–9
Ecological Flora of the British Isles
    286
Ecological Flora of California *286*
Ecological Metadata
    Language (EML) 288
ecological resilience 91
economic decision modelling 12
economic instruments 238–42, 244
economic literature 12
economics 246–7
  biodiversity as insurance 232–4
  biodiversity markets 234–8
  ecosystem externalities 231–2
  international dimension 244–6
  sustainability indicators 259
  value of ecosystems 249–51
  see also coupled ecological–
    economic systems
economic value 174–5, 230–1
  bees as crop pollinators 207–8
  microbial functions *125*
  natural pest control 191–2
ecosystem approach to restoration
    168–9
ecosystem ecology, trait-based
    research *283*
ecosystem engineers 105, 155
ecosystem externalities 231, 244–5,
    246–7, 263, 277
  international 244–5
ecosystem function 6–7, 170–1
  measures 83
  relationship to effect traits 70–2
  species effects 173–4
  trajectories 4, 17, 24–6
ecosystem-level functions 16
ecosystem-level responses,
    biodiversity effects *22*, *23*, 27
ecosystem services 10, 11, 248–9
  in managed ecosystems 191–2
  payment services 238–40
  relationship to species richness 67,
    68, 76
  restoration of multiple services
    173–4, 176
  socio-economic impacts 118–20
  stability 92–3
    in restoration 172–3
  valuation 11, 12, 174–5, 230–1,
    249–51, 261–2

  bees as crop pollinators 207–8
  forest and woodlands, South
    Africa 160–1
  Keoladeo National Park 254
  mangrove storm buffering 257–8
  methods *252*
  microbial functions *125*
  natural pest control 191–2
  provisioning and cultural
    services 251–6
  regulating services 256–7
ecosystem types, variability of
    biodiversity effects 15–16,
    21–2, 27
ecotourism 235, 248
Ecotron experiments 7–8, 59
edge effects 134, 189–90
effective dimensionality of trait space 55
effect traits 66
  relationship to ecosystem function
    70–2
efficacy of research 290, 291
Eichner, T. and Tschirhart, J. 264, 266
Ellis, G.M. and Fisher, A.C. 268
Elmqvist, T. *et al.* 144
emerging infections *215*
endangered species, CITES 245
Endangered Species Act, United
    States 243–4
endemism, trait-based research *283*
enemy release, role in establishment
    success 218–19, 221, 285
energy flow, GEEM model 266–7
energy prices 267
Engelhardt, K.A.M. and Ritchie, M.E.
    225
Enquist, B.J. and Niklas, K.J. *158*
Environmental Kuznets Curve 245
environmental microbiology 121
Environmental Stewardship Scheme,
    UK 177
Epps, K.Y. *et al.* 53
equalizing forces 114
equilibrium species diversity, general
    model 273–7
establishment success 218–19, 221–2,
    225
Eviner, V.T. *et al.* 172
evolutionary potential 296–7
exclusion principle 273–4, 275–6
existence value of
    ecosystems 251
expected damage function
    (EDF) 257, 258
exploitation, animal responses *64*, 65
explorative approach, sterilized Earth
    experiment 6, 7

externalities 231, 246–7, 263, 277
  international 244–5
extinction
  biomass compensation 66–7
  cascading 67–8, 69, 90–1
  consequences 60–2
  effect on PFv webs 203–4
  realistic scenarios 72–5, 115–17
  reversed causation 14
extinction debts 136, 213
extinction order prediction 66
extinction research, future
    directions 77
extinction of resources 115
extinction risk
  relationship to response traits 62–5
  relationship to trophic level 107

facilitation
  bacterial species 127
  by colonizing species 222
Farm Business Survey, UK 272
FD 52
  applications 53
  relationship to species richness 56,
    57, 58
FD$_{LD}$ 52
  relationship to species richness 56,
    57, 58
FD$_{var}$ 51
  applications 52, 53
  relationship to species richness 56,
    57, 58
fecundity, relationship to
    invasiveness 220
Fédoroff, E. et al. 181
feedbacks 3–4, 12
Ferrari, J. et al. 214
Ferraro, P.J. and Simpson, R.D. 237
'Field of Dreams' approach to
    restoration 169
field experiments 32
Final Environmental Impact
    Statements (FEISs), application
    of BEF research 34, 42, 43
fire ecology, trait-based
    research 283
fire frequency, impact of colonizing
    species 223
FishBase 286
fisheries 230
  GEEM model 268–9
  production function models 255
  'roving bandit' phenomenon 295
  stability 93
fitness net energy flow, GEEM
    models 266–8

flower visitors
  aggregate abundance 196
  species richness 196–8
food supply, consequences of
    pollinator decline 205–6
food web diversity, socio-economic
    impacts 118–20
food webs 105
  complexity 106
  diversity effects across trophic
    levels 109–10
    empirical evidence 112–13
  diversity effects within trophic
    levels 107, 108, 109
    empirical evidence 110–12
  effects of extinctions 116–17
  protistan 126–7
  stabilizing properties 89–91
  trait-based research 283
  see also plant–flower visitor (PFv)
    interaction webs
forestry, timber extraction study,
    Uttar Pradesh 237–8
forests
  carbon sequestration 163
  temporal stability 81, 82
Forest Service, USA, application
    of BEF concepts 41, 42
fossil fuel emissions 165
fossil resources 296
fragment area distribution 138
fragmentation of habitat 134–5, 146
  animal responses 64, 65
  functioning debt 136
  plant responses 63
  source–sink meta-community
    model 140–1
  spatial variance of biodiversity
    137–8, 139
  species loss estimation 136–7
  see also habitat loss; meta-
    community models
Framework Convention on Climate
    Change (FCCC) 246–7
framework incentives 238, 239
France, K.E. and Duffy, J.E. 145
Frivold, L. and Frank, J. 158
Fukami, T. and Morin, P.J. 126
fully combinatorial studies 292
fumigation experiments, soil microbes
    130
functional attribute diversity (FAD) 51
functional diversity 8, 49, 58–9, 211, 281
  applications 52–4
  choice of measure 58
  effects of species richness 54–8
  measures 51–2

non-grouping measures 50–1
relationship to plant productivity 273
trait-based research 283
functional facilitation, pollinator
    communities 199, 201
functional group richness 50, 53
  applications 53–4
functional groups 49–50, 70
functional redundancy 79, 83, 173,
    188, 296
  aquatic bacteria 127
  soil microbes 129–30
functional reorganization 61
functional response diversity 79
functional traits 5
  invasive species 226
  plants, effects on climate 154
functioning debt, habitat
    fragmentation 136–7, 146
functioning ecosystems 294–5
funding of BEF research 34, 38–40
funding of restoration 177
fungi
  aquatic 131
  diversity, effect on decomposition 83
  microcosm studies 128

Gabriel, D. et al. 181
Gallai, N. et al. 206
Gamfeldt, L. et al. 173
gene pool 230
General Agreement on Tariffs and
    Trade (GATT) 245–6
General Equilibrium Ecosystem
    Model (GEEM) 265–9
general equilibrium models 264, 265
Genghini, M. et al. 181
genotypic richness 155, 176
  relationship to disease risk
    209, 214
Giller, P.S. et al. 15
Gillison, A.N. et al. 181
Global Biodiversity Information
    Facility 288
Global Environmental Organization
    (GEO) 246
globalization 294, 295
global warming
  effect on pollinator services 206
  plant responses 63, 65
  prediction of responses 286
Glopnet 286
Glor, R.E. et al. 181
Gonzalez, A. and Chaneton, E. 136, 145
Gonzalez, A. and Descamps-Julien, B.
    81
Gordon, C. et al. 181, 191

Gould, A.M.A. and Gorchov, D.L. 223
governments, policies on
    conservation 237
gradient analysis, trait-based research 283
grassland plants
    realistic extinction scenarios 73, 74
    variations in fungal disease 211
grasslands
    effect of nitrogen pollution 231
    experimental variation of biodiversity 231
    restoration 168, 169, 176–7
        diversity of plantings 171
        multiple ecosystem services 173, 174
    temporal stability 80, 81, 82
grazing, plant responses 66, 71
Great Plains, restoration of bison 190
Greenleaf, S.A. and Kremen, C. 201
'green payments' 240
Griffiths, B.S. et al. 85, 86, 88, 130
Griffiths, G.J.K. et al. 192
Grime, J.P. 152
gross domestic product (GDP) 259
Gross, K. and Cardinale, B.J. 116
growth equation 275
growth form, plants, relationship to extinction risk 63, 65
growth rate, relationship to establishment success 221
guild analysis, trait-based research 283

Habitat Conservation Plans 244
habitat loss 134–5
    animal responses 64, 65
    community disassembly 135
    plant responses 63
    see also fragmentation of habitat
Haggar, J.P. and Ewel, J.J. 158
Hairston–Smith–Slobodkin hypothesis 17
Hannon, B. 267
Hartwick, J.M. 259
harvesting, extinction risk 62
Harvey, C.A. et al. 181, 191
Hättenschwiler, S. et al. 161
Hawaii Plant Trait Database 286
hazard × likelihood product, disease risk 215
heat resilience 88, 173
heat resistance 86, 172–3
Hector, A. and Bagchi, R. 28, 173
Hector, A. et al. 257
    citations 38
hedonic price valuation method 252, 253

Heemsbergen, D.A. et al. 51
herbivory, effect on impact of plant diversity 109, 112, 113
Herendeen, R. 267
heritability, trait-based research 283
Hillebrand, H. and Cardinale, B.J. 111
Hiremath, A.J. and Ewel, J.J. 158
Hoehn, P. et al. 199–200
holistic approach to research 31
Holling, C.S. 91, 232
Holt, R.D. and Loreau, M. 109
Holway, D. 221
Holzschuh, A. et al. 181
homogenization 217
homologous characters 5
honey bee (Apis mellifera)
    decline, consequences 205–6
    pollination services 196
        economic value 207–8
        functional facilitation 201
    Varroa mite infestation 204
    see also bees
Hooper, D.U. et al. 298
Hooper, D.U. and Vitousek, P.M. 8
    citations 38
horizontal diversity 109
horizontal gene transfer 125
Hortus Gramineus Woburnensis experiment 290
Hubbell, S.P. 114, 152
Hughes, A.R. and Stachowicz, J.J. 87
Hughes, T.P. 92
Hulme, P.E. and Bremner, E.T. 223
human capital 296
Human Development Index (HDI) 259
human manipulations 9–11
human-mediated introductions 221
Hunter, T. et al. 161
Huston, M.A., citations 38
Hutchinson, G.E. 130
Hutton, S.A. and Giller, P.S. 181

Impatiens glandulifera, removal study 223
incentives for biodiversity management 193–4, 209, 243–4, 246, 247
incentives for conservation 238–42, 244
included (nested) niches 101
index of biodiversity 263, 272
indicators of biodiversity 190–1
indirect use value of ecosystems 251
infectious disease risk 214–16
    coinfections 214
    deforestation effects 214

diversity–disease curves 210
diversity versus species compositions 210–11
    monocultures versus polycultures 212
    natural versus random community assemblies 212–13
    relationship to genotypic diversity 214
    species diversity versus functional diversity 211
infectious diseases 209–10
information flow 31, 39, 43–5, 298
    in restoration ecology 175
in silico studies 292
insurance hypothesis 79, 87, 91, 144, 146, 155, 188
    economic aspects 232–4
    spatial insurance hypothesis 118, 139–40, 144, 145
integrated conservation–development projects (ICDPs) 236
integration, effect on PFv webs 204
intellectual property rights 235–6
    TraitNet 287
intensification 73, 179–80, 193, 270
    effect on pollinator communities 195, 196, 197, 206
    meta-analysis of field studies 180–3
    mitigation options 183–4
    see also managed ecosystems
intentional introductions 221
interactions 5, 6
    ecosystem services 250
    effect on extinction risk 76
    effects on stability 90
intermediate disturbance hypothesis 183
intermediate inputs, ecosystem services as 250
international trade 244–6
inter-specific competition 16
intra-specific variation 59
introduced species 221, 229, 244
invaders 60, 169, 218, 224
    biodiversity effects 17, 22, 23, 24, 28
    effects of removal 223, 228
    establishment success 218–19
    as an externality of trade 244
    impact on native ecosystem 222–4, 226, 227–8
    mechanisms of invasion 285–6
    prediction of impact 228–9
    risk assessment 226–8
    trait-based research 284
    see also colonization
invasiveness, related traits 220–1

inverse sampling effect 219, 224–5
investment rule 259
Ives, A.R. and Cardinale, B.J. 66, 116–17
ivory trade ban 245

Jactel, H. et al. 161
Jensen's inequality 138, *139*
Jiang, L. et al. 53
joint implementation projects 246–7
Jones, H.E. et al. 159
Jones, K.E. et al. 215
Jonsson, M. et al. 74
Joshi, J. et al. 85, *86*, 155
Joyce, C. *181*

Kahmen, A. et al. *86*, 87
Keller, R.P. et al. 227
Kemp, D.R. et al. 92
Keoladeo National Park, ecosystem valuation 254
keystone species 105, 155, 264–5
Kindscher, K. and Tieszen, L.L. 171
Kirwan, L. et al. 98–9, 102, *186*
Klein, A.M. et al. 120
Knowler, D. 268
Kolasa, J. and Li, B. 82–3
Kolmogorov model 274, 275
Kondoh, M. 90
Koo, B. and Wright, B.D. 236
Korhonen, K.P.C. et al. 161
Koricheva, J. et al. 162
Körner, C. 154
Kremen, C. et al. 93, *181*
Kyoto Protocol 149, 166

laboratory studies 32
  limitations 30
ladybeetles, value in biological control 119
land acquisition, conservation promotion 237
land conversion, carbon loss 163
landscape complexity, relationship to natural pest control 187
landscape gradients 184, 185
  relationship to arthropod diversity *186*
landscape mosaics, value 189–90
land sparing 183–4
land use change 178–*9*, 230–1
land use and land cover change (LUCC) 269
Larsen, T.H. et al. 67, *68*, 73, 74, 199
leaf characteristics, relationship to ecosystem function *71*
leafcutter bees, pollination services, economic value 207–8

LEDA Traitbase *286*
Leibold, M.A. et al. 117
leishmaniasis risk, effect of deforestation 214
lettuce mosaic virus 190
Leung, B. et al. 226
life expectancy at birth 259
life-history, relationship to extinction risk
  animals, *64*
  plants *63*
life-history trade-offs 115
  dimensionality 285
lignin degradation 129
litter mixing studies 156
LoGiudice, K. et al. 211
Long, Z.T. et al. 126
Lonicera mackii, removal studies 223
Loranger, G. et al. 181, *186*
Loreau-Hector additive partitioning method 96
Loreau, M. 96
Loreau, M. and Behera, N. 87
Loreau, M. et al. 118, 140
  citations 38
Losey, J.E. and Vaughan, M. 208
lost species
  consequences 60–2
  effect on functional diversity 49
  see also extinctions
Lotka–Volterra models 114–15, 274
lupine, PFv web *202*
Lyme disease 73, 74, 120
  diversity–disease curve 210
  effect of dilution hosts 212
  impact of white-footed mouse 213
  variations in blacklegged tick infection 211

McAusland, C. and Costello, C. 244
McCann, K.S. 90
McGrady-Steed, J. et al. 126–7
McIntyre, P.B. et al. 73, 74, 281
macroecology, trait-based research *283*
Madritch, M.D. and Hunter, M.D. 161
maize (*Zea mays*), introduction to Old World 221
malaria 120
  effect of deforestation 214
managed ecosystems 178–9, 194
  BEF relationships 184–5
    cross-ecosystem exchanges of biodiversity 189–90
    indicators for functions of biodiversity 190–1
    insurance value of biodiversity 188
    pest control 187–8

  relevance of controlled experiments 185
  research approaches 185–6
  ecosystem functions 191–2
    scaling effects *192–3*
  incentives 193–4
management decisions, application of BEF research 34, 41–2, *43*, 44–5
management intensification 179–80
  meta-analysis of field studies 180–3
  mitigation options 183–4
mangrove storm buffering, valuation 257–8
manipulative experiments, microbial communities 130, 131, 132
manmade capital 296
marine benthic invertebrates, extinction scenarios 73, *74*, 75
marine protected areas, economic value 174
market failures 246
markets in biodiversity 234–8
Marrakech Accord 149
Mason, N.W.H. et al. 51, *52*, 53
mass ratio hypothesis 152–5, 156, 163–4
  experimental studies 157
Matthiessen, B. and Hillebrand, H. 145
maximal function, number of species required 26, 28
meadows, restoration 168, 169, 176
mean dissimilarity (MD) 51
  applications 52–3
  relationship to species richness 56, 57, 58
mechanisms of coexistence 113–15
mechanistic statistical models 94
media coverage of BEF concepts 40–1
Meléndez-Ramirez, V. et al. 197
Memmott, J. et al. 203–4
Mendelsohn, R. and Balick, M.J. 236
Merrifield, J. 240–1
Mesopotamian marshes, restoration of ecosystem services 173
metabolically inactive microbe cells 132
metabolic theory of ecology, trait-based research *284*
metabolome, microbial 124
meta-community models 117–18, 143–6
  dynamics 141, *142*
  productivity 141–3
  source–sink model 140–1
  stability 143
meta-genomics *123*
meta-regression models 255

364 INDEX

Metrick, A. and Weitzman, M.L. 243
Mgahinga and Bwindi Impenetrable Forest Conservation Trust Fund 242
Michaelis–Menten curves, limitations 26
microarrays *123*
microbes 8
  biodiversity estimation 122–4, *123*, 132
  diversity patterns, similarity to those of larger organisms 133
  species identification 124
microbial communities 121
  restoration 175–6
microbial functions 124–5
  in the wild 129, 131–2
    aquatic fungi 131
    microbe-plant interactions 130–1
    soil microbes 129–30
microcosm studies 121, *124*, 131–3, 291
  algae, dispersal effects 145
  aquatic bacteria 127–8
  fungi 128
  model systems 125–6
  mutualistic microbe–plant interactions 128–9
  protistan food webs 126–7
  spatial averaging 145
  temporal stability *81*, 82–3
Millennium Ecosystem Assessment 11, 59, 178, 230, 234, 248–9, 250, 261–2, 293–5
mixed forestry systems, carbon sequestration 164
model systems, microbial communities 125–6
monocultures, carbon sequestration 164, 165
Montoya, J.M. and Solè, R.V. 91
Morin, P.J. McGradySteed, J. *81*
moss communities, dispersal effects 145
Mouillot, D. *et al.* 53
Mulder, C.P. *et al.* 85, *86*, 87, 113
multifunctionality 10, 28, 178, 188
  in restorations 173–4
multifunctional redundancy 173
multiple stable states 91–2
multiple traits, FD$_{var}$ values 51
multi-resource model 276
multitaxon biodiversity studies 191
multi-trophic biodiversity effects 16–17, *22, 23, 24*, 27–8
multi-trophic communities, temporal stability *81*, 82–3
mutualistic microbe-plant interactions 128–9

mycorrhizal fungi 128–9
Myrica faya, colonization in Hawaii 224

Naeem, S. *et al.* 7, 59, 168
  citations 38
National Center for Ecological Analysis and Synthesis (NCEAS) 286
  Knowledge Network for Biocomplexity project 288
National Marine Fisheries Service (NMFS) policies 269
National Park Service, USA, application of BEF concepts 41–2
national product measures (GNP, NNP) 259
natural capital 296
  depletion 9
'naturalness' of ecosystems 264–9
natural selection, trait-based research 284
Nature Conservancy 237
NatureServe databases *286*
negative covariance effect 84
negative selection effects 96, 100–1
nested niches 101, *102*
net biodiversity effect 96
net carbon sequestration 150–1
net primary productivity (NPP), relationship to carbon sequestration 154
net product measures (NDP, NNP) 259
neutral theory of biodiversity 114, 152
  experimental studies 157
New York City, drinking water provision 252
niche complementarity (niche differentiation) 114, 155–6, 164, 188, 212, 219
  experimental studies 157
  pollinator communities 198–9, 199–201
  Pseudomonas fluorescens 127
  role in establishment success 222
niches, nesting 101, *102*
niche theory 17
nitrate pollution 231
nitrifier communities, comparative studies 130
nitrogen cycling
  impact of colonizing species 223
  seasonality 172
nitrogen deposition, plant responses 63
nitrogen-fixing bacteria 128–9

nitrogen-fixing trees, carbon sequestration 154
nitrogen oxide pollution 231
nitrogen uptake, seaweeds 73, *74*
nodes 105
non-governmental organizations 41
non-use value 250–1
Noss, R.F. 165
novel, invasive colonizers 218
  see also invaders
novel, noninvasive colonizers 218
numerical compensation 66–7, 76
nutrient cycling 248–9
  freshwater tropical fishes 73, *74*
  seasonality 172
  shapes of biodiversity effects 26, 28
nutrition, consequences of pollinator decline 206

observational studies, microbes *124*
Observation Ontology (OBOE) 288
oilseed rape, PFv web *202*, 203
Olschewski, R. *et al.* 186
Omer, A. *et al.* 270, 272–3
omnivory 105
optimization technique 263–4, 271–2
  private optimization management problem (POMP) 274, 276
  social optimization management problem (SOMP) 274, 276, 277
option value of ecosystems 251
option value of gene combinations 297
Opuntia scrubland 256
organic farms, biodiversity 183
Ostfeld, R.S. *et al.* 212
Ostfeld, R.S. and LoGiudice, K. 73, *74*, 76, 213
Our Common Future (World Commission on Environment and Development 1987) 295–6
overyielding 95–6, 100–2, *186*

Pacala, S. and Tilman, D. 274
Pacala, S.W. and Deutschman, D.H. 165
palaeobiology, trait-based research *284*
PAM (Partitioning Along Metroids) analysis, agricultural landscape studies 180, *181*
Panayotou, T. 240
paradox of the plankton 130
parasite diversity 54
parasitoid diversity 83

parasitoid wasps, value in biological control 119
Pareto criterion 242
parrotfish, bioerosion 54
partitions 98
Pascual, U. and Perrings, C.P. 241
pasture increases 187–8
pasture lands, stability 92
patch dynamics 117–18, 144
patents 235
  application to microbes 125
pathogenicity, microbes 124–5
Pautasso, M. et al. 161
payments (rewards) for ecosystem services (P(R)ES) 238–40
Peacock, L. et al. 161
Pérez Harguindeguy, N. et al. 162
periwinkle, as source of drugs 235, 236
Perner, J. and Malt, S. 181
Perrings, C. et al. 232, 269
pest control see biological control
pesticides use 270
Petchey, O.L. et al. 80, 81, 84
Petchey, O.L. and Gaston, K.J. 50, 52, 53, 74
  species richness, effect on functional diversity 55
Peters, C.M. et al. 251
Pfisterer, A.B. and Schmid, B. 42, 86, 87, 88
pharmaceutical firms, bioprospecting 235–6, 246
phases of BEF research 31, 32
Philpott, S.M. et al. 181
phylogeny 5, 6
phytoplankton, functional diversity 54
Pimm, S.L. 232
planning documents, application of BEF research 34, 41–2, 43
plant community diversity, role in establishment success 218
plant ecological strategies, trait-based research 284
plant–flower visitor (PFv) interaction webs 201
  higher trophic levels 204
  quantitative analysis 204–5
  species extinction and integration 203–4
  structure and characteristics 201–3, 202
plant productivity, relationship to functional diversity 273
plants
  functional traits, effects on climate 154
  interactions with microbes 130–1
  mutualistic interactions 128–9
PLANTS database 286
plant traits
  associated responses to environmental changes 62–3, 65
  relationship to ecosystem function 70–1
Podani, J. and Schmera, D. 52, 53
Polasky, S. et al. 257
policy frameworks 238
pollination 10, 195–6, 206
  economic value of bees 207–8
  flower visitor abundance 196
  flower visitor species richness 196–8
  mechanisms of effects 198–201
  future research areas 207
  plant–flower visitor interaction webs 201
    higher trophic levels 204
    quantitative analysis 204–5
    species extinction and integration 203–4
    structure and characteristics 201–3, 202
  realistic extinction scenarios, bees 73, 74
  socio-economic impacts 120
  stability 93
  valuation 253
  value of landscape mosaics 189–90
pollination characteristics, relationship to extinction risk 63
pollinator decline, consequences for global food supply 205–6
polymerase chain reaction (PCR) 122–3
poor populations, impact of market-like mechanisms 242, 244
population ecology, trait-based research 284
population growth 178
population-level functions 16
population-level responses, biodiversity effects 16, 22, 23, 27
Potvin, C. and Gotelli, N.J. 160
poverty, implications for managed ecosystems 193
practicability of research 290, 291
precipitation changes, plant responses 63, 65
predation, effects on microbial communities 126
predator diversity 113
  socio-economic aspects 119
predator loss, implications 120
predator–prey interactions 187
  GEEM models 265, 266–8
Pretzsch, H. et al. 160
Price equation partition 96, 97–8, 102, 103
primary forests, carbon sequestration 163, 166
primary production, shapes of biodiversity effects 26
principle component axis 1(PCA1) plants, ecosystem function 71, 72
privately optimal management problem (POMP) 274, 276
production function valuation methods 252, 253–5
production–simplification trade-off 10
productivity
  meta-communities 141–3, 144–5
  relationship to biodiversity 4
  relationship to plant traits 71
property rights regimes 240
protective value of ecosystems 253–4
protistan food webs 126–7
provisioning services 248
  valuation 251–3
Pseudomonas species
  bacteriophage infection risk 214
  P. fluorescens 127
public awareness 40–1, 43–4
public goods 230–1, 249
Pulkkinen, K. 214
Pullin, A. et al. 42
pumpkin, pollinator services, spatial stability 199–200
purposive decisions 263

Q (Rao's quadratic entropy) 51–2
  applications 52, 53
  relationship to species richness 56, 57, 58
Quaas, M.S. 232
quantitative PFv web analysis 204–5

race to the bottom, international trade 245
random assembly experimental design 60–1, 213
random-loss scenarios 73–5
random partitions design and analysis 98, 103
rangelands, stability 92
Rao's quadratic entropy see Q

rates, biodiversity effects 16, 22–4, 27
Raviraja, N.S. et al. 128
realistic extinction scenarios 72–5
realistic research 291–3, 298
reassociation kinetics 123
reclamation 169
recreation, GEEM model 268, 269
red brome (*Bromus madritensis* ssp. *rubens*), establishment success 222
Redondo-Brenes, A. and Montagnini, F. 160
Reducing Emissions from Deforestation and Degradation (REDD) initiatives 149–50
regulating services 249
   valuation 256–7
rehabilitation 168
Reich, P.B. et al. 173, 174
relative yields (RYs) 96, 103
Relative Yield Total (RYT) 96
removal studies
   colonizing species 223
   microbes 124
replacement cost valuation methods 252, 258
research
   application 34, 41–2
   BEF database 33–4, 34–6, 35
     citation rates 36, 37, 42
     most cited papers (BEF-6) 36–8
   current challenges 298
   efficacy and realism 291–3
   evolution 7–9
   funding 34, 38–40
   history of 4, 30–1
   information flow 31, 39, 43–5, 175, 298
   limitations 95
   in managed ecosystems 185–6
   real-world relevance 176
   scale and duration of experiments 116, 117
   simplified communities 106
research approaches 31–3, 32
reserve species 276–7
reservoir hosts 213
resilience 78, 188, 232
   ecological 91
   empirical findings 87, 88–9
   and insurance hypothesis 79
   link to resistance 89
   reserve species 276–7
   role of mangroves 258

resistance 78, 188, 232
   empirical findings 83, 85, 86–7
   insurance hypothesis 79
   link to resilience 89
resource-based models 273–4
resource concentration hypothesis 109
resource depletion, biodiversity effects 22–4
resource diversity, relationship to resource fluxes 108, 109
resource dynamics equation 275
resource extinction 115
resource generalization 107
resource supplies, impact of colonizing species 223
resource use, flower visitors 199
resource capture and utilization, relationship to plant traits 71
response diversity 144
   pollinator communities 197
response traits 66
   relationship to extinction risk 62–5
restoration 11
   funding 177
restoration ecology 167–8
   BEF approach 169–70
   classical BEF implications 170–2
   community approach 168
   ecological approach 168–9
   economics of BEF 174–5
   multiple ecosystem services restoration 173–4
   recommendations 175–7
   stability of services 172–3
restorative approach, sterilized Earth experiment 6
Reusch, T.B.H. et al. 155
Rewarding Upland Poor for Environmental Services (RUPES) 240
rewards (or payments) for ecosystem services (P(R)ES) 238–40
rice blast infection rate, relationship to genotypic diversity 214
Ricketts, T.H. et al. 206, 253
Ricotta, C. 53
river restoration 170
Rodríguez, L.C. et al. 256
Romanuk, T.N. et al. 81, 82, 84
Rundlof, M. and Smith, H.G. 181
Russell, A.E. 159
ryegrass rust fungal infection, diversity–disease curve 210

Sall, S.N. et al. 181
saltation 218

sampling, microbial communities 122–3
sampling (selection) effect 95, 96, 187, 219
   in colonization 221–2, 224–5
   pollinator communities 198, 199
Sanderson, M.A. et al. 92
Sandhu, H.P. et al. 191
Sanitary and Phytosanitary (SPS) Agreement 246
scale of experiments 116, 117
scaling effects, managed ecosystems 192–3
Schamp, B.S. et al. 53
Scherer-Lorenzen, M. et al. 162, 165
Schläpfer, F., et al. 17, 24, 26, 73, 74
Schmitz, O.J. et al. 113
Schroth, G. et al. 159
Science Environment for Ecological Knowledge (SEEK) 286, 288
   Taxon project 287
sea grass
   genotypic richness 155
   meta-communities, dispersal effects 145
seasonality, nutrient cycling 172
seaweeds, nitrogen uptake 73, 74
secondary extinctions 67–8, 69, 90–1
seed characteristics, relationship to extinction risk 63
Seed Information Database, Kew Botanic Gardens 286
selection effect 95, 96, 100–1, 118
self-citation 33, 36
semi-mechanistic statistical models 94
service flows 274
set-aside land 183–4
shapes of biodiversity effects 4, 17, 24–6, 25, 28
Sierra Nevada forest plan 42
simplifications 10
Simpson, R.D. Sedjo, R.A., and Reid, J.W. 235
simulation studies, extinction scenarios 73–6
single TDR programs 241
'slow life history', relationship to extinction risk 63, 64, 65
Snelder, D.J. et al. 181
socially optimal management problem (SOMP) 274, 276, 277
social welfare function 265
societal will 290, 291
socio-economic aspects
   disease risk 215
   food web diversity 118–20

soil functioning, effect of plant species richness 131
soil microbes, field experiments and observations 129–30
soil organic carbon (SOC) 154, 164
Solan, M. *et al.* 66, 73, *74*, 281
Solidago altissima
    clonal diversity 176
    genotypic richness 155
Solow, R.M. 259
Soule, J.D. and Piper, J.K. 270
source-sink meta-community model 140–1
South Africa, forest and woodland asset values 260–1
Spartina anglica, facilitation of native species 222
Spartina marshes, construction 168
spatial autocorrelation 144
spatial averaging 138, 140, *143*, 145
spatial insurance hypothesis 118, 139–40, 144, 145
spatial stability, pollinator services 197, 199–201
spatial storage effect 144
spatial variance
    fragmented landscapes 137–8, *139*
    pollinator services *198*
specialization, animals, relationship to extinction risk *64*
species, composition, relative importance 210–11
species–area relationship 136–7
species complementarity 95, 99–100, 101
species composition effect (SCE) 97
species identification, microbes 124
species richness
    effects on functional diversity 54–8
    relationship to disease risk 209
species richness effect (SRE) 97
species selections in research 292–3
species-sorting models 117–18
spillover, emerging diseases 215
spinach, presence of *E. coli* 190
Srinivasan, U.T. *et al.* 73, 74, 76, 112
Srivastava, D.S. *et al.* 27
stability 78, 93, 188
    of ecosystem services 92–3
    in restoration 172–3
    empirical findings 89
        resilience 87, *88*–9
        resistance 83, 85, *86*–7
        temporal stability 79–83
    food web properties 89–91
    measures of ecosystem function 83

meta-communities *142*, 143
multiple stable states 91–2
relationship to biodiversity 10
theoretical links to biodiversity 79
see also temporal stability
stabilizing forces 114
Stanley, W.G. and Montagnini, F. *158*
stated preference valuation methods *252*, 253
statistical averaging 79
statistical modelling 94, *103*
    diversity–interactions approach 98–9
    meta-analysis of results 99–103
    random partitions design 98
Steenwerth, K.L. *et al.* 181
Steiner, C.F. *et al.* 80, *81*, 82, *84*, 85, 88, 126
sterilized Earth thought experiment 5–7
stochastic production frontier (SPF) approach 272–3
stocks, biodiversity effects 22–4, 27
Stoneham, G. *et al.* 242
strawberry, spatial stability of pollinator services 200
stream insects, detritus processing 73, 74
Strong, D.R. 110
strong interactors 105
structural reorganization 61, 65–6
    biomass compensation 66–7
    cascading extinction 67–8, *69*
Sub-Saharan Africa, wealth 259
succession
    carbon sequestration 152–3, 164
    trait-based research 284
successional colonizers 218, 224
suites of traits 70, 71, 152, *153*, 173–4
Sumatra, biodiversity 183, 190
sunflower pollination, functional facilitation 201
super colony formation, Argentine ants 221
super-competition, invasive species 224
support services 248–9, 293, *294*, 295
sustainability 178, 249
sustainability indicators 259
sustainable development 295–7
Symstad, A.J. *et al.* 171
system-driven studies 293
tariffs, effect on accidental introductions 244

Taxonomic Concept Schema (TCS) 287
Taxonomic Object Service (TOS) 287

taxonomic standardization 287
taxonomy of plants, relationship to extinction risk *63*
temperature, effect on decomposition rate 154
temporal stability 78, 79–80, 172, 188
    mechanisms of diversity's effects 84–5
    in multi-trophic communities *81*, 82–3
    pollinator services 196–7
    within a single trophic level 80, *81*, 82
temporal variability (CV) 80
    BEF effects 219
    in impact of colonizing species 226
    protistan food webs 126
Thébault, E. and Loreau, M. 109
theory-driven studies 293
tick-borne encephalitis 120
Tilman, D., resource-based model 274, 275
Tilman, D. and Downing, J.A. 85, *86*, *88*, 89
Tilman, D. *et al.* 80, *81*, *84*, 231, 257
    citations *38*
timber extraction study, Uttar Pradesh 237–8
top-down biodiversity effects 16–17, 23, 24, 27, 107, *108*, 109, 137
    empirical evidence 111–12
tourism 235
    GEEM model 268
tradeoffs 115, 155, 174
    in carbon sequestration 163–4
    life histories 285
trade restrictions 245
traditional approach 3
TraitBank 288
trait-based extinction scenarios 72–5, 76, 292–3
trait-based research 281, *283*–4
    see also TraitNet
trait data, databases *286*
trait-dependent complementarity effect (TDCE) 96, 101–2
trait diversity 211
    colonizing species 224, 226, 228
trait-independent complementarity effect (TICE) 96
TraitNet 282, 285, 288–9
    areas of research 285–6
    architecture *285*
    data integration 286–8
    integration into ecoinformatics frameworks 288
    trait observation schema 287

trait protocol standardization 288
traits 54, 59, 61
　inclusion in measurement of functional diversity 58
　relationship to ecosystem function 70–2
　relationship to extinction risk 62–5
　relationship to secondary extinction risk 69
　relationship to structural reorganization 65–70
trait space
　dimension reduction 70
　effective dimensionality 55
trait syndromes 70, 71
trajectories of ecosystem function 4, 17, 24–6, 25, 28
transferable development rights (TDR) 240–1
transgressive overyielding 95, 103, 186
　meta-analysis 99–100
travel cost valuation method 252, 253
trees
　carbon loss, experimental studies 161–2
　carbon sequestration 73, 74, 154–5, 163–5
　　experimental studies 158–60
　　primary forests 163
　　genotypic richness 155–6
Treton, C. et al. 128
tripartite additive partitioning 96–7, 101
TRIPS Agreement 235–6
trophic cascades 61, 108, 110
　experimental data 113
trophic complexity 8
trophic groupings 53–4
trophic interactions, effects on stability 90
trophic levels
　biodiversity effects 16–17, 22, 23, 24, 27–8, 107–10
　　empirical evidence 110–13
　relationship to cascading extinction 68, 69, 70
　relationship to extinction risk 63, 64, 65, 115–16
tropical freshwater fishes, extinction scenarios 74, 75, 76
TRY database 286
Tschirhart, J. 267
two-way additive partitioning 96
Tylianakis, J.M. et al. 81, 83, 204

Unified Neutral Theory of Biodiversity and Biogeography 114, 152
　trait-based research 284
USDA PLANTS database 286
use value 250
utility index 277
Uttar Pradesh, timber extraction study 237–8

Valone, T.J. and Hoffman, C.D. 81, 82, 84
valuation of ecosystems 11, 12, 174–5, 230–1, 249–51, 261–2
　bees as crop pollinators 207–8
　forest and woodland, South Africa 260–1
　Keoladeo National Park 254
　mangrove storm buffering 257–8
　methods 252
　microbial functions 125
　natural pest control 191–2
　provisioning and cultural services 251–6
　regulating services 256–7
value production function 271
Van Kooten, J. and Bulte, E.H. 245
Van Peer, L. et al. 86, 87
variance in edibility hypothesis 107, 111
Varroa destructor, bioeconomic risk analysis 226–7
Vehviläinen, H. et al. 162
vertical diversity 109
vicarious use value of ecosystems 251
Vilà, M. et al. 159, 160
vincristine 235, 236
Vitousek, P.M. and Hooper, D.U. 4
Vittor, A.Y. et al. 214
Vogt, R.J. et al. 81, 82, 84, 85

Walker, B. et al. 50, 51, 52, 53, 276–7
Wardle, D.A. et al. 85, 86, 158, 161, 181
Wardle, D.A. and Zackrisson, O. 158, 161
wasps, parasitoid diversity 83
water cycling, shapes of biodiversity effects 26, 28
Wätzold, F. et al. 263
wealth 259
Weibull, A.C. et al. 187
Wenum, J. et al. 272
Wertz, S. et al. 124, 130

West Nile virus 120
　diversity–disease curve 210
wetlands
　changes, expected damage function 257
　ecosystem services 173
　edge effects 190
white-footed mouse, impact on Lyme disease risk 213
Wilby, A. et al. 181
Wilby, A. and Thomas, M.B. 188, 189
Wilderness Act (1964), United States 264
wildlife-friendly farming 183
Williams, S.L. 176
Wirth, C. 162
Wohl, D.L. et al. 127
wood density measurement 288
Wootton, J.T. 157
working groups, TraitNet 285
World Bank, adjusted net savings indicator 259, 261, 262
World Environmental Organization (WEO) 246
World Summit on Sustainable Development (Johannesburg 2002) 296
World Tourism Organization 235
World Trade Agreement 125
World Trade Organization (WTO) 246
Worm, B. et al. 93, 174, 230

yellow fever risk, effect of deforestation 214

Zavaleta, E.S. and Hulvey, K.B. 73, 74
zebra mussels (Dreissena polymorpha), bioeconomic risk analysis 226
Zedler, J.B. 168
Zhang, Q.G. and Zhang, D.Y. 80, 81, 82, 84, 86, 87
Zhu, Y. et al. 214
zoonotic diseases 120, 215
　reservoir hosts 213
　see also Lyme disease
zooplankton, temporal stability 80, 81
Zostera marina
　clonal diversity 176
　genotypic richness 155